Lecture Notes in Physics

Springer
Berlin
Heidelberg
New York
Barcelona
Hong Kong
London
Milan
Paris
Singapore
Tokyo

Physics and Astronomy ONLINE LIBRARY

http://www.springer.de/phys/

Editorial Policy

The series *Lecture Notes in Physics* (LNP), founded in 1969, reports new developments in physics research and teaching -- quickly, informally but with a high quality. Manuscripts to be considered for publication are topical volumes consisting of a limited number of contributions, carefully edited and closely related to each other. Each contribution should contain at least partly original and previously unpublished material, be written in a clear, pedagogical style and aimed at a broader readership, especially graduate students and nonspecialist researchers wishing to familiarize themselves with the topic concerned. For this reason, traditional proceedings cannot be considered for this series though volumes to appear in this series are often based on material presented at conferences, workshops and schools (in exceptional cases the original papers and/or those not included in the printed book may be added on an accompanying CD ROM, together with the abstracts of posters and other material suitable for publication, e.g. large tables, colour pictures, program codes, etc.).

Acceptance

A project can only be accepted tentatively for publication, by both the editorial board and the publisher, following thorough examination of the material submitted. The book proposal sent to the publisher should consist at least of a preliminary table of contents outlining the structure of the book together with abstracts of all contributions to be included.
Final acceptance is issued by the series editor in charge, in consultation with the publisher, only after receiving the complete manuscript. Final acceptance, possibly requiring minor corrections, usually follows the tentative acceptance unless the final manuscript differs significantly from expectations (project outline). In particular, the series editors are entitled to reject individual contributions if they do not meet the high quality standards of this series. The final manuscript must be camera-ready, and should include both an informative introduction and a sufficiently detailed subject index.

Contractual Aspects

Publication in LNP is free of charge. There is no formal contract, no royalties are paid, and no bulk orders are required, although special discounts are offered in this case. The volume editors receive jointly 30 free copies for their personal use and are entitled, as are the contributing authors, to purchase Springer books at a reduced rate. The publisher secures the copyright for each volume. As a rule, no reprints of individual contributions can be supplied.

Manuscript Submission

The manuscript in its final and approved version must be submitted in camera-ready form. The corresponding electronic source files are also required for the production process, in particular the online version. Technical assistance in compiling the final manuscript can be provided by the publisher's production editor(s), especially with regard to the publisher's own Latex macro package which has been specially designed for this series.

Online Version/ LNP Homepage

LNP homepage (list of available titles, aims and scope, editorial contacts etc.):
http://www.springer.de/phys/books/lnpp/

LNP online (abstracts, full-texts, subscriptions etc.):
http://link.springer.de/series/lnpp/

M. Orszag J. C. Retamal (Eds.)

Modern Challenges in Quantum Optics

Selected Papers of the First International Meeting in Quantum Optics
Held in Santiago, Chile, 13-16 August 2000

Springer

Editors

Miguel Orszag
Dept. de Física
Pontificia Universidad Católica de Chile
Casilla 306 Santiago 22, Chile

Juan Carlos Retamal
Dept. de Física
Universidad de Santiago de Chile
Avda. Ecuador, 3493, correo 2, Santiago, Chile

Cover picture: see contribution by E. Massoni, M. Orszag in this volume

Library of Congress Cataloging-in-Publication Data

International Meeting in Quantum Optics (1 st : 2000 : Santiago, Chile)
Modern challenges in quantum optics : selected papers of the First International Meeting in Quantum Optics, held in Santiago, Chile, 13-16 August 2000 / M. Orszag, J.C. Retamal (eds.).
p. cm. -- (Lecture notes in physics, ISSN 0075-8450 ;
575) Includes bibliographical references.
ISBN 3540419578 (alk. paper)
1. Quantum optics--Congresses. I. Orszag, Miguel, 1944- II. Retamal, J. C. (Juan Carlos), 1962- III. Title. IV. Series.

QC446.15 .1625 2000
535--dc21

ISSN 0075-8450
ISBN 3-540-41957-8 Springer-Verlag Berlin Heidelberg New York

Typesetting: Camera-ready by the authors/editors
Camera-data conversion by Steingraeber Satztechnik GmbH Heidelberg
Cover design: *design & production*, Heidelberg

Printed on acid-free paper
SPIN: 10834150 57/3141/du - 5 4 3 2 1 0

Foreword

Quantum optics, in a sense, is precisely as old as the quantum theory, that is to say 100 years old this year. The very first quanta, after all, and for a long time the only ones, were the indivisible units of radiation energy discovered by Planck. Those oddly constrained light waves were indeed quanta, and our coming to grips with their baffling particle-like behavior consumed another 27 years. That understanding was not, in fact, fully achieved until well after all the particles of matter were understood to have the same dual nature. The quantum mechanical theory of the atom, based on that duality, brought many dramatic successes, but it was only in 1927, after many of these, that Dirac was able to extend the theory to include light itself – and thus lay the true groundwork for quantum optics.

The decades that followed provided many verifications of Dirac's fundamental theory. Most of them involved the particle-like behavior of high-energy quanta in processes that never involved more than one or two quanta at a time. The calculations could be carried out by the standard method of perturbation theory, which turned out to work quite well.

Although light quanta are normally present in vast abundance, our interest in multi-quanta phenomena was slow in developing. The first hint of the need for a more comprehensive theory came from the photon correlation measurements of Hanbury Brown and Twiss in 1956. But it was not long before the brilliant and monochromatic beams of the laser began to change the optical landscape completely, revealing new worlds of experimental wonders. It was only in that context that we began to develop a range of mathematical tools sufficiently versatile to deal quantum mechanically with light of all intensities. And only then did we begin seriously to apply the name "quantum optics" to such studies.

The ensuing years have seen the introduction of a vast range of new experimental techniques. These have been guided by ever more sophisticated theory, exploring in the most intimate detail the ways in which light interacts with matter. Isolated atoms, suspended in vacuum, can be made to perform many of the stunts that were hardly even imaginable as "thought experiments" a few years ago.

And now it has become possible, partly by quantum optical means, to reduce the temperature of atomic gases to so low a level that the wave-like properties of the atoms dominate their particle-like properties. We are witnessing in this work the development of a remarkable new field of wave optics that we call for the moment “atom optics”. As the offspring of quantum optics, it possesses many resemblances to its parent field. It can be approached and understood by using many of the same mathematical tools.

It is the most recent advances in the related fields of quantum and atom optics that concern us here, and their description in one form or another has occupied most of this conference. But there is another development that is worth pointing out at this meeting. The gospel of quantum mechanics, which was in its earliest days localized to three or four countries of Europe, has spread gradually over the globe in the last century, eliciting contributions from the United States and many countries of the Northern Hemisphere. Within the last several years we have seen significant research in quantum optics appear from the southern countries as well, and we can now point to contributions from groups in Argentina, Brazil, Chile, Colombia, Peru and Uruguay, not to mention Australia and New Zealand. Those groups are now advancing the gospel as well as spreading it.

Cambridge, March 2001 *Roy J. Glauber*

Preface

QUANTUM OPTICS I was the 1st International Meeting in Quantum Optics held in Santiago, Chile, in 23-26 August, 2000 at the Pontificia Universidad Catolica de Chile.

Just under 60 participants gathered together in Santiago, to discuss problems in Non-Linear Optics, Quantum Information, Trapped Ions, Matter Waves and Fundamentals of Quantum Mechanics.

There was a strong participation by Latin-American Scientists and students, in particular from Brazil, Chile, Argentina, Peru, Uruguay and Colombia, as well as by Scientists from USA, Europe and Asia.

New experimental and theoretical results were presented at the Conference.

Prof. Walther talked about quantum phenomena of single trapped atoms and the experimental demonstration of pure Fock states in the micromaser. Prof. Eberly talked about quantum entanglement(not included here) and determination of the localization of a single photon in a space time volume. Professor Meystre referred to matter-wave amplification, Prof. Glauber talked about the Polarium Model(not included here) and coherent trapping. Prof. Kenkre talked about oscillations between coupled Bose-Einstein condensates and also did some beautiful sketches of several participants (included in this volume), Prof. Becker addressed the topic of above- threshold ionization and harmonic generation, Prof. Marangos gave us an interesting talk on Electromagnetically Induced Transparency, Prof. Swain referred to the quantum interference effects in fluorescence and Prof. Kaige talked about QND measurements in trapped ions.

From the Southern Hemisphere there were talks about dwell time and tunneling, quite relevant for several recent experiments on " superluminal effects" (Prof. M.Nussezweig), quantum error correction (Aldo Delgado et. al.), manipulation of motional states (Prof. Zagury et. al.),ion-trapped lasers and squeezing(Prof. Orszag et. al.), decoherence and state protection (Prof. Davidovich), radiation pressure and decoherence(Professor Maia Neto et al.),and many more exciting results.

The Conference took place during the Chilean Winter, in a relaxing atmosphere with a beautiful background of the snow covered Andes Cordillera.

It was a good opportunity to establish new collaborations and contacts.

The editors would like to thank the participants for providing their papers promptly after the Conference and also the following organizations who provided financial or other forms of support: Pontificia Universidad Catolica de Chile(PUC), Universidad de Santiagode Chile (USACH), Conicyt(Fondecyt), Consortium of the Americas for Interdisciplinary Science(UNM), Centro Latinoamericano de Fisica(CLAF), The Abdus Salam International Centre for Theoretical Physics(Trieste), Fundacion Andes, United Airlines.

Finally, Ms. Carolina Leon was a great help in the local organization and also we would like to thank Ms. Marta Montoya for her support in all the local events as well as the general organization of this Conference.

Santiago,
February 2001

Miguel Orszag
Juan Carlos Retamal

Table of Contents

Part I Trapped Ions and Cavity QED

List of Contributors

M. Anscombe
Laser Optics and Spectroscopy
Group,Blackett Laboratory
Imperial College of Science,
Technology and Medicine,
London, SW7 2BW UNited Kingdom

V. Ahufinger
Departamento de Física,
Universitat Autònoma de Barcelona,
E-08193, Bellaterra, Spain

Wilhelm Becker
Max-Born-Institut,
Max-Born-Str. 2a,12489
Berlin, Germany

Jean Bellissard
Université Paul Sabatier,
Institut de Recherche sur les
Systèmes Atomiques ou
Moléculaires Complexes,
F-31062 Toulouse cedex 4, France,

A.R.R. Carvalho
Instituto de Física, Universidade
Federal do Rio de Janeiro,
Caixa Postal 68528, 21945-970
Rio de Janeiro, RJ, Brazil

R. Corbalán
Departamento de Física,
Universitat Autònoma de Barcelona,
E-08193, Bellaterra, Spain

D.A.R. Dalvit
T-6, Theoretical Division,
MS B288, Los Alamos National
Laboratory
Los Alamos, NM 87545, USA

L. Davidovich
Instituto de Física, Universidade
Federal do Rio de Janeiro,
Caixa Postal 68528,21945-970
Rio de Janeiro, RJ, Brazil

Aldo Delgado
Abteilung für Quantenphysick,
Universität Ulm, D-89069 Ulm,
Germany

Sergio M. Dutra
Huygens Laboratory,
Leiden University
P.O. Box 9504, 2300 RA Leiden,
The Netherlands

Marcelo França Santos
Instituto de Física,Universidade
Federal do Rio de Janeiro,

Caixa Postal 68528, 21945-970
Rio de Janeiro, RJ, Brazil

Z. Ficek
Department of Physics,
The University of Queensland,
Brisbane, QLD 4072, Australia

Victo S. Filho
Instituto de Física Teórica
Universidade Estadual Paulista,
Rua Pamplona, 145, 01405-900,
São Paulo, SP, Brazil.

Tobias Frederico
Departamento de Física,
Instituto Tecnológico da Aeronáutica,
Centro Técnico Aeroespacial,
12228-900, São José dos Campos,
SP, Brazil

Arnaldo Gammal
Departamento de Física,
Instituto Tecnológico da Aeronáutica,
Centro Técnico Aeroespacial,
12228-900, São José dos Campos,
SP, Brazil

K. M. Gheri
Institut für Theoretische Physik,
Universität Innsbruck,
Technikerstrasse 25/2, A-6020
Innsbruck, Austria

Angela Guzmán
Depto. de Física,
Universidad Nacional de Colombia,
Bogotá, Colombia

Wang Kaige
CCAST (World Lab.),
P. O. Box, 8730,Beijing 100080;
and Department of Physics,
Beijing Normal University,
Beijing 100875, China

J. I. Kim
Instituto de Física,
Universidade de São Paulo,
Caixa Postal 66318, CEP 05315-970,
São Paulo, SP, Brazil

Richard Kopold
Max-Born-Institut,
Max-Born-Str. 2a, 12489
Berlin, Germany

I. Kuçukkara
Laser Optics and Spectroscopy
Group, Blackett Laboratory
Imperial College of Science,
Technology and Medicine,
London, SW7 2BW United Kingdom

M. L. Ladrón de Guevara
Departamento de Física,
Facultad de Ciencias,
Universidad de Chile,
Casilla 653, Santiago, Chile

P.A. Maia Neto
Instituto de Física, UFRJ,
Caixa Postal 68528 21945-970
Rio de Janeiro, Brazil

J.P. Marangos,
Laser Optics and Spectroscopy
Group, Blackett Laboratory
Imperial College of Science,

Technology and Medicine,
London, SW7 2BW

R.L. de Matos Filho
Instituto de Física, Universidade
Federal do Rio de Janeiro,
Caixa Postal 68528, 21945-970
Rio de Janeiro, RJ, Brazil

Han Pu
Optical Sciences Center,
University of Arizona,
Tucson, AZ 85721, USA

S. Maniscalco
INFM, MURST and
Dipartimento di Scienze
Fisiche ed Astronomiche
dell'Università di Palermo,
Palermo 90123, Italy

Eduardo Massoni
Facultad de Física, Pontificia
Universidad Católica de Chile,
Casilla 306, Santiago, Chile

Dejan B. Milošević
Faculty of Science,
University of Sarajevo,
Zmaja od Bosne 35, 71000 Sarajevo,
Bosnia-Hercegovina

P. Milman
Instituto de Física, Universidade
Federal do Rio de Janeiro,
Caixa Postal 68528, 21945-970
Rio de Janeiro, RJ, Brazil

Pierre Meystre
Optical Sciences Center,
University of Arizona,
Tucson, AZ 85721, USA

A. Messina
INFM, MURST and
Dipartimento di Scienze
Fisiche ed Astronomiche
dell'Università di Palermo,
Palermo 90123, Italy

Michael G. Moore
ITAMP, Harvard-Smithonian
Institute for Astrophysics,
Cambridge, MA, USA

J. Mompart
Departament de Física,
Universitat Autònoma de Barcelona,
E-08193, Bellaterra, Spain

A. Napoli
INFM, MURST and
Dipartimento di Scienze
Fisiche ed Astronomiche
dell'Università di Palermo,
Palermo 90123, Italy

Gerard Nienhuis
Huygens Laboratory,
Leiden University
P.O. Box 9504, 2300 RA Leiden,
The Netherlands

H.M. Nussenzweig
Instituto de Fisica da UFRJ
C.P. 68528, Rio de Janeiro,
R.J., Brazil

P. Nussenzveig
Instituto de Física,
Universidade de São Paulo,
Caixa Postal 66318, CEP 05315-970,
São Paulo, SP, Brazil

Miguel Orszag
Facultad de Física, Pontificia
Universidad Católica de Chile,
Casilla 306, Santiago, Chile

J. C. Retamal
Departamento de Física
Universidad de Santiago de Chile,
Avda. Ecuador 3493, casilla 3493
correo 2 Santiago, Chile
jretamal@lauca.usach.cl

Luis Roa
Departamento de Física,
Universidad de Concepción,
Casilla 160-C, Concepción

Carlos Saavedra
Departamento de Física,
Universidad de Concepción,
Casilla 160-C, Concepción

R.B.B. Santos
Instituto de Física,
Universidade de São Paulo,
Caixa Postal 66318, CEP 05315-970,
São Paulo, SP, Brazil

F. Silva
Departament d'Òptica,
Universitat de València,
E-46100 Burjassot, Spain

E. Solano
Sección Física,
Departamento de Ciencias, Pontificia
Universidad Católica del Perú,
Apartado 1761, Lima, Peru

Dominique Spehner
Facultad de Física, Pontificia
Universidad Católica de Chile,
Casilla 306, Santiago 22, Chile
dspehner@alfven.fis.puc.cl

S. Swain
School of Mathematics and Physics,
The Queen's University of Belfast,
Belfast BT7 1NN, The United
Kingdom

Lauro Tomio
Instituto de Física Teórica -
Universidade Estadual Paulista,
Rua Pamplona, 145, 01405-900,
São Paulo, SP, Brazil

Wilhelm von Waldenfels
Institut für Angewandte Mathematik
der Universität Heidelberg,
Im Neuenheimer Feld 294 69120
Heidelberg Germany

Herbert Walther
Sektion Physik der Universität
München and Max-Planck-Institut
für Quantenoptik
85748 Garching,
Fed. Rep. of Germany

Nicim Zagury
Instituto de Física, Universidade
Federal do Rio de Janeiro,
Caixa Postal 68528, 21945-970
Rio de Janeiro, RJ, Brazil

Jesús Zapata
Depto. de Física,
Universidad Nacional de Colombia,
Bogotá, Colombia

F. De Zela
Sección Física,
Departamento de Ciencias, Pontificia
Universidad Católica del Perú,
Apartado 1761, Lima, Peru

Part I

Trapped Ions and Cavity QED

Generation of Fock States in the One-Atom Maser

H. Walther

Sektion Physik der Universität München and
Max-Planck-Institut für Quantenoptik
85748 Garching, Fed. Rep. of Germany

Abstract. In this paper we give a survey of our experiments performed with the one-atom maser or micromaser on the generation of Fock states. Three methods can be used for this purpose: the trapping states leading to Fock states in a continuous wave operation; state reduction of a pulsed pumping beam and finally using a pulsed pumping beam to produce Fock states on demand. In the latter case trapping states stabilize the photon number.

1 Introduction

The quantum treatment of the radiation field uses the number of photons in a particular mode to characterize the quantum states. In the ideal case the modes are defined by the boundary conditions of a cavity giving a discrete set of eigenfrequencies. The ground state of the quantum field is represented by the vacuum state consisting of field fluctuations with no residual energy. The states with fixed photon number are usually called Fock or number states. They are used as a basis in which any general radiation field state can be expressed. Fock states thus represent the most basic quantum states and differ maximally from what one would call a classical field. Although Fock states of vibrational motion are routinely observed for single ions in traps [1], Fock states of the radiation field are very fragile and very difficult to produce and maintain. They are perfectly number-squeezed, extreme sub-Poissonian states in which intensity fluctuations vanish completely. In order to generate these states it is necessary that the mode considered has minimal losses and the thermal field, always present at finite temperatures, has to be eliminated to a large extent since it causes photon number fluctuations.

The one-atom maser or micromaser [42] is the ideal system to realize Fock states. In the micromaser highly excited Rydberg atoms interact with a single mode of a superconducting cavity which can have a quality factor as high as 4×10^{10}, leading to a photon lifetime in the cavity of 0.3s. The steady-state field generated in the cavity has already been the object of detailed studies of the sub-Poissonian statistical distribution of the field [3], the quantum dynamics of the atom-field photon exchange represented in the collapse and revivals of the Rabi nutation [4], atomic interference [5], bistability and quantum jumps of the

M. Orszag, J.C. Retamal (Eds.): LNP 575, pp. 3–13, 2001.

field [6], atom-field and atom-atom entanglement [7]. The cavity is operated at a temperature of 0.2 K leading to a thermal field of about 5×10^{-2} photons per mode.

There have been several experiments published in which the strong coupling between atoms and a single cavity mode is exploited (see e.g. Ref. [8]). The setup described here is the only one where maser action can be observed and the maser field investigated. In our setup the threshold for maser action is as small as 1.5 atoms/s. This is a consequence of the high value of the quality factor of the cavity which is three orders of magnitude larger than that of other experiments with Rydberg atoms and cavities [9].

In this paper we present three methods of creating number states in the micromaser. The first is by way of the well known trapping states, which are generated in a c.w. operation of the pumping beam and lead to Fock states with high purity. We also present a second method where the field is prepared by state reduction and the purity of the states generated is investigated by a probing atom. It turns out that the two methods of preparation of Fock states are in fact equivalent and lead to a similar result for the purity of the Fock states. The third method pumps the cavity with a pulsed beam using the trapping condition to stabilize the photon number in the cavity. This method produces Fock states on demand.

2 The One-Atom Maser and the Generation of Fock-States Using Trapping States

The one-atom maser or micromaser is the experimental realisation of the Jaynes-Cummings model [1], as it allows to study the interaction of a single atom with a single mode of a high Q cavity. The setup used for the experiments is shown in Fig. 1 and has been described in detail previously [61]. Briefly, in this experiment, a $^3He - ^4He$ dilution refrigerator houses the microwave cavity which is a closed superconducting niobium cavity. A rubidium oven provides two collimated atomic beams: a central one passing directly into the cryostat and a second one directed to an additional excitation region. The second beam was used as a frequency reference. A frequency doubled dye laser ($\lambda = 294$ nm) was used to excite rubidium (^{85}Rb) atoms to the 63 $P_{3/2}$ Rydberg state from the 5 $S_{1/2}$ (F = 3) ground state.

Velocity selection is provided by angling the excitation laser towards the main atomic beam at about 11° to the normal. The dye laser was locked, using an external computer control, to the 5 $S_{1/2}$ (F = 3)-63 $P_{3/2}$ transition of the reference atomic beam excited under normal incidence. The reference transition was detuned by Stark shifting the resonance frequency using a stabilized power supply. This enabled the laser to be tuned while remaining locked to an atomic transition. The maser frequency corresponds to the transition between 63 $P_{3/2}$ and 61 $D_{5/2}$. The Rydberg atoms are detected by field ionization in two detectors

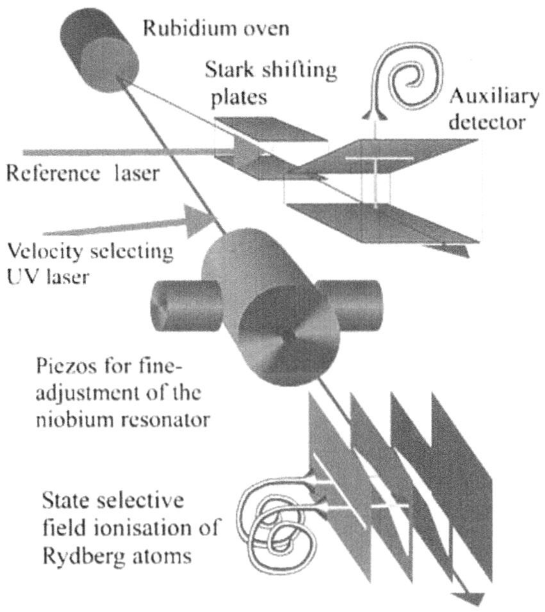

Fig. 1. The micromaser setup. For details see Ref. [9]

set at different voltages so that the upper and lower states of the maser transition can be investigated separately.

The trapping states are a steady-state feature of the maser field peaked in a single photon number, they occur in the micromaser as a direct consequence of field quantisation. At low cavity temperatures the number of blackbody photons in the cavity mode is reduced and trapping states begin to appear [61,62]. They occur when the atom field coupling constant given by the Rabi frequency Ω, and the interaction time, $t_{\rm int}$, are chosen such that in a cavity field with n photons each atom undergoes an integer number, k, of Rabi cycles. This is summarised by the condition,

$$\Omega t_{\rm int}\sqrt{n+1} = k\pi. \tag{1}$$

When Eq.1 is fulfilled the cavity photon number is left unchanged after the interaction of an atom and hence the photon number is "trapped". This will occur regardless of the atomic pump rate $N_{\rm ex}$, where $N_{\rm ex}$ is the rate of pumping atoms in the excited state per decay time of the cavity. The trapping state is therefore characterised by the photon number n and the number of integer multiples of full Rabi cycles k.

The build-up of the cavity field can be seen in Fig. 2, where the emerging atom inversion I = P_g - P_e is plotted against interaction time and pump rate; $P_{g(e)}$ is the probability of finding a ground (excited) state atom. At low atomic pump rates (low $N_{\rm ex}$) the maser field cannot build up and the maser exhibits

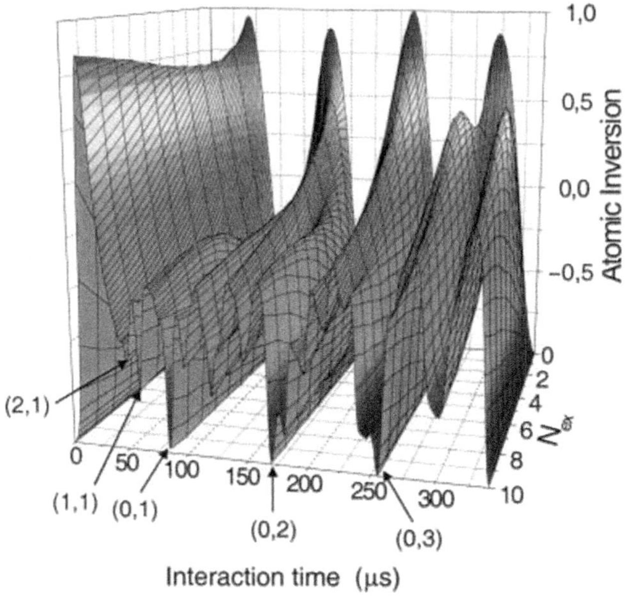

Fig. 2. A theoretical plot, in which the trapping states can be seen as valleys in the N_{ex} direction. As the pump rate is increased, the formation of the trapped states from the vacuum can be seen. The positions of trapping states are indicated by arrows with the respective designation (n,k).

Rabi oscillations due to the interaction with the vacuum field. At the positions of the trapping states, the field increases until it reaches the trapping state condition. This manifests itself as a reduced emission probability and hence as a dip in the atomic inversion. Once in a trapping state the maser will remain there regardless of the pump rate. The trapping states show up therefore as valleys in the N_{ex} direction. Figure 3 shows the photon number distribution as the pump rate is increased for the special condition of the five photon trapping state. The photon distribution develops from a thermal distribution towards higher photon numbers until the pump rate is high enough for the atomic emission to be stabilized by the trapping state condition. As the pump rate is further increased, and in the limit of a low thermal photon number, the field continues to build up to a single trapped photon number and the steady-state distribution approaches a Fock state.

Owing to blackbody radiation at finite temperatures, there is always a small probability of having a thermal photon enter the mode. The presence of a thermal photon in the cavity disturbs the trapping state condition and an atom can emit a photon. This causes the field to change around the trapping condition.

Note that under readily achievable experimental conditions, it is possible for the steady-state field in the cavity to approach a Fock state with a high

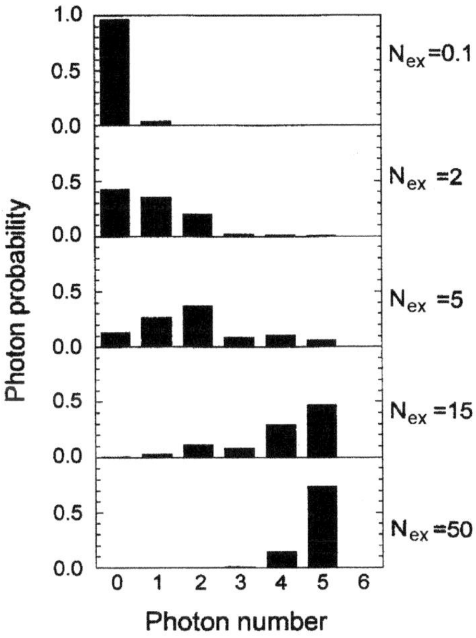

Fig. 3. A numerical simulation of the photon number distribution as the atomic pump rate (N_{ex}) is increased until the cavity field is in a Fock state with a high probability. Shown is the example of the n = 5 Fock state. A further increase of N_{ex} beyond 50 does not change the distribution.

fidelity. Under the present experimental conditions the main deviation from a pure Fock state results from dissipation of the field in the cavity. If a photon disappears it takes a little while until the next incoming excited atom can be used to replace the lost photon. Therefore smaller photon numbers show up besides the considered Fock state. Figure 4 shows micromaser simulations, for achievable experimental conditions, in which Fock states with high purity are created from $n = 0$ to $n = 5$. The experimental realisation requires a pump rate of $N_{\text{ex}} = 50$, a temperature of about 100mK [6], a high selectivity of atomic velocity and very low mechanical noise of the system [61,13]. The details on the generation of Fock states using the trapping condition are described in Ref. [61].

3 Dynamical Preparation of Number States in a Cavity

When the atoms leave the cavity in a micromaser experiment they are in an entangled state with the field. A method of state reduction was suggested by Krause et al. [14] to observe the build up of the cavity field to a known Fock state. State reduction uses the entanglement produced by the interaction of an atom with a cavity field to project the field onto a well defined number state. If

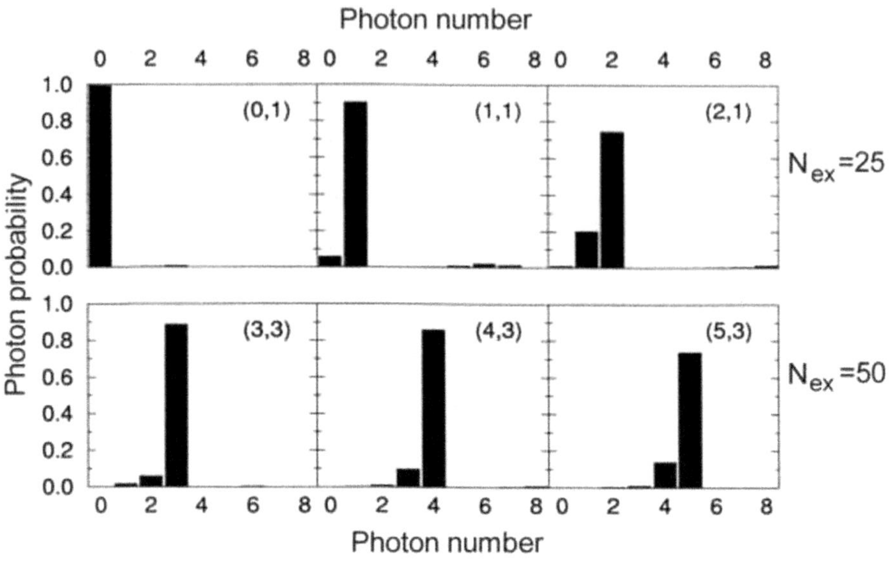

Fig. 4. In simulations of the maser operation Fock states from n=0 to n=5 can be readily generated for achievable experimental conditions ($n_{\text{th}} = 10^{-4}$).

the field is in an initial state $|n\rangle$ then an interaction of an atom with the cavity leaves the cavity field in a superposition of the states $|n\rangle$ and $|n+1\rangle$ and the atom in a superposition of the internal atomic states $|e\rangle$ and $|g\rangle$.

$$\Psi = \cos(\phi)|e\rangle|n\rangle - i\sin(\phi)|g\rangle|n+1\rangle \tag{2}$$

where ϕ is an arbitrary phase. The state selective field ionisation measurement of the internal atomic state, reduces the field to one of the states $|n\rangle$ or $|n+1\rangle$. State reduction is independent of the interaction time, hence a ground state atom always projects the field onto the $|n+1\rangle$ state independent of the time spent in the cavity. This results in an *a priori* probability of the maser field being in a specific but unknown number state [14]. If the initial state of the cavity is the vacuum, $|0\rangle$, then a number state created is equal to the number of ground state atoms that were collected within a suitably small fraction of the cavity decay time.

In a system governed by the Jaynes-Cummings Hamiltonian, spontaneous emission is reversible and an atom in the presence of a resonant quantum field undergoes Rabi oscillations. That is the relative populations of the excited and ground states of the atom oscillate at a frequency $\Omega\sqrt{n+1}$, where Ω is the atom field coupling constant. Experimentally we measure the atomic inversion. In the presence of dissipation a fixed photon number n in a particular mode is not observed and the field always evolves into a mixture of such states. Therefore the inversion is generally given by,

$$I(n, t_{\text{int}}) = -c \sum_{\text{n}} P_{\text{n}} \cos(2\Omega\sqrt{n+1}t_{\text{int}}) \tag{3}$$

where P_{n} is the probability of finding n photons in the mode, t_{int} is the interaction time of the atoms with the cavity field. The factor c considers the reduction of the signal amplitude as a result of dark counts.

The experimental verification of the presence of Fock states in the cavity corresponds to a pump-probe experiment in which a pump atom prepares a quantum state in the cavity and the Rabi phase of the emerging probe atom measures the quantum state. The signature that the quantum state of interest has been prepared is simply the detection of a defined number of ground state atoms. To verify that the correct quantum state has been projected onto the cavity a probe atom is sent into the cavity with a variable, but well defined, interaction time. As the formation of the quantum state is independent of interaction time we need not to change the relative velocity of the pump and probe atoms, thus reducing the complexity of the experiment. In this sense we are performing a reconstruction of a quantum state in the cavity using a similar method to that described by Bardoff et al. [15]. This experiment reveals the maximum amount of information that can be found relating to the cavity photon number. We have recently used this method to demonstrate the existence of Fock states up to n=2 in the cavity [16].

When the interaction time corresponding to the trapping state condition is met in this experiment, the formation of the cavity field is identical to that which occurs in the steady-state, hence the probe atom should perform an integer number of Rabi cycles. In fact this was observed experimentally [16], which indicates that the pulsed experiment is actually the formation stage of the steady-state experiment. One would therefore expect that the measured photon number distribution, in the dynamical measurement, would be the same as that predicted for the trapping states. State reduction is simply a method of observation that determines the appropriate moment for a measurement. In this sense the observation of a lower emission probability in the steady state is also a field-state measurement as the steady-state inversion measurement occurs for practically the same conditions as for the dynamical measurement described here [16].

4 Preparation of Fock States on Demand

In the following we describe another variant of a dynamical Fock-state preparation with the micromaser [17]. To demonstrate the principle of the source described here; Fig. 5 shows the simulation of a sequence of five arbitrary atom pulses using a Monte Carlo calculation in which the micromaser is operated in the (1, 1) trapping state. In each pulse there is a single emission event, producing a single lower state atom and leaving a single photon in the cavity. The atom-cavity system is then in the trapping condition, as a consequence the emission

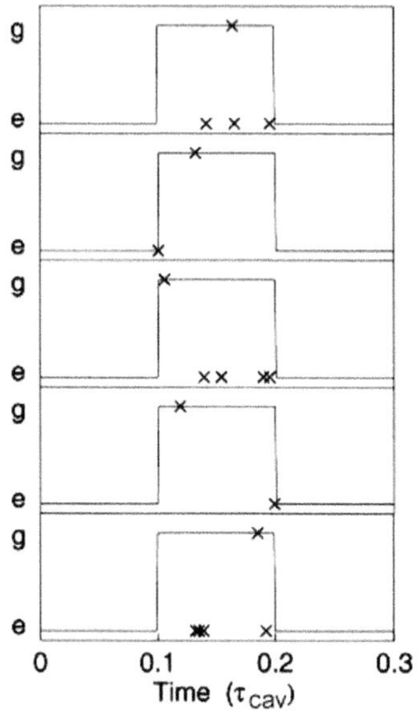

Fig. 5. A computer simulation of five sequential excitation pulses for Rydberg atoms. The interaction time is tuned to the (1, 1) trapping state. The cavity is initially in the vacuum state. Rydberg atoms are marked by crosses; those in the upper or lower maser level are marked on the lower and upper row respectively. For high pump rates, every pulse contains at least one lower-state atom with a deviation occurring once in every 50 pulses. The parameters used are: $\tau_{cav} = 100$ ms, $n_{th} = 0.03$, $\tau_{pulse}=10$ ms, $N_{ex} = 40$ or $N_a=4$ atoms per pulse.

probability is reduced to zero and the photon number is stabilized. In steady state operation, the influence of thermal photons and variations in interaction time or cavity tuning complicates this picture, resulting in deviations from Fock states [18]. Pulsed excitation however reduces the influence of such effects and the generated Fock states show a high purity.

Figure 6(a-c) presents three curves obtained from the computer simulation, that illustrate the behaviour of the maser under pulsed excitation as a function of interaction time for more ideal (but achievable) experimental parameters. The simulations show the probability of finding; no lower-state atom per pulse ($P^{(0)}$); finding exactly one lower-state atom per pulse ($P^{(1)}$); and the conditional probability of finding a second lower-state atom in a pulse already containing one ($P^{(>1;1)}$). The latter plot of the conditional probability, $P^{(>1;1)}$, is relatively insensitive to the absolute values of the atomic detection efficiency and therefore has advantages when comparisons with experimental data are performed [17].

It follows from Fig. 6 that with an interaction time corresponding to the (1, 1) trapping state, both one photon in the cavity and a single atom in the lower-state are produced with a 98% probability. In order to maintain an experimentally

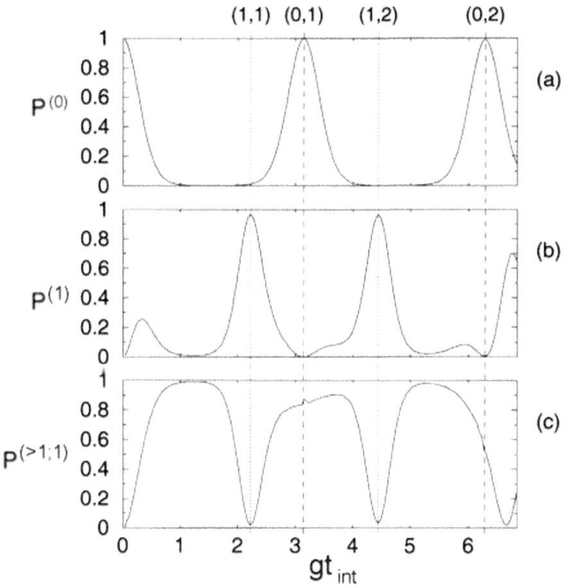

Fig. 6. The figure presents the probability of finding; (a) no lower-state atoms per pulse $P^{(0)}$, (b) exactly one lower-state atom per pulse $P^{(1)}$, and (c) of finding a second lower-state atom, if one has already been detected $P^{(>1;1)}$. Parameters for these simulations were $\tau_{\text{pulse}}=0.02\tau_{\text{cav}}$, $N_{\text{a}}=7$ atoms and $n_{\text{th}} = 10^{-4}$. The maximum value of $P^{(1)}$ is 98% for the (1, 1) trapping state.

verifiable quantity, the simulations presented here relate to the production of lower-state atoms rather than to the Fock state left in the cavity. However, pulse lengths are rather short ($0.01\tau_{\text{cav}} \leq \tau_{\text{pulse}} \leq 0.1\tau_{\text{cav}}$) so there is little dissipation and the probability of finding a one photon state in the cavity following the pulse is very close to the probability of finding an atom in the lower-state. Atomic beam densities must also be chosen with care to avoid short pulses with high atom density that could violate the one-atom-at-a-time condition.

Note that at no time in this process is a detector event required to project the field, the field evolves to the target photon number state, when a suitable interaction time has been chosen so that the trapping condition is fulfilled.

It should be noted that for thermal photon numbers as high as $n_{\text{th}} = 0.1$ or for t_{int} fluctuations of up to 10% (both beyond the current experimental parameters), simulations show that Fock states are still prepared with an 80-90% fidelity. This is considerably better than for steady state trapping states, where highly stable conditions with low thermal photon numbers are required [62,61,18].

The present setup of the micromaser was specifically designed for steady state operation. Nevertheless the current apparatus does permit a comparison between theory and experiment in a relatively small parameter range.

A method was used for the comparison with the experiment which is described briefly in the following. During the interaction, strong coupling between pumping atoms and the cavity field creates entanglements between internal atomic levels and the cavity field. Subsequent pumping atoms will therefore also become entangled both with the field and a previous pumping atom. The correlations between subsequent atomic levels are determined by the dynamics of the atom-cavity interaction. The connection between population correlations and the micromaser dynamics, has been studied in detail in previous papers [7,19]. It is important to note that even in the presence of lost counts, the correlations between subsequently detected atoms are maintained. Thus rather than a measurement of the single-atom-per-pulse probability $P^{(1)}$ which is heavily dependent on exact knowledge of detection efficiency, it is more useful to measure atom pair correlations given by $P^{(>1;1)}$, owing to the insensitivity of the parameter to detector efficiencies. Experimentally the parameter is obtained via,

$$P^{(>1;1)} = \frac{N_{\mathrm{gg}}}{N_{\mathrm{gg}} + N_{\mathrm{eg}} + N_{\mathrm{ge}}} \tag{4}$$

where for example N_{eg} is the probability of detecting a pair of atoms containing first an upper state atom (e) and then a lower-state atom (g) within a pulse. Eq. 4 provides both a value appropriate to the existent correlation and is directly related to the total probability of finding one atom per pulse. Although $P^{(>1;1)}$ is insensitive to the absolute detector efficiency it does depend on the relative detector efficiencies (which are nearly equal) and the miscount probability (the probability that a given atomic level is detected in the wrong detector). Each has been measured experimentally.

The source presented here has the significant advantage over our previous method of Fock state creation [16] of being unconditional and therefore significantly faster in preparing a target quantum state. Previously state reduction by detection of a predefined number of lower state atoms was used to prepare the state with 95% fidelity. However, this method has the disadvantage that it is affected by non perfect detectors. In the current experiment however the cavity field is correctly prepared in 83.2% of the pulses and is independent of any detector efficiencies. Improving the experimental parameters we can expect to reach conditions for which 98% of the pulses prepare single photon Fock states and a single atom in the lower-state.

5 Conclusion

In this paper we gave a survey of the possibilities for generating Fock states in the micromaser. The generation of Fock states on demand has been experimentally confirmed and will be published elsewhere [17]. The possibility to generate Fock states will allow us to perform the reconstruction of a single photon field or other Fock states in a next step.

References

1. D. Leibfried, D.M. Meekhof, B.E. King, C. Monroe, W. M. Itano, and D. J. Wineland: Phys. Rev. Lett **77**, 4281 (1996)
2. D. Meschede, H. Walther, and G. Müller: Phys. Rev. Lett. **54**, 551 (1985)
3. G. Rempe and H. Walther: Phys. Rev. A **42**, 1650 (1990)
4. G. Rempe, H. Walther, and N. Klein: Phys. Rev. Lett. **58**, 353 (1987)
5. G. Raithel, O. Benson, and H. Walther: Phys. Rev. Lett. **75**, 3446 (1995)
6. O. Benson, G. Raithel, and H. Walther: Phys. Rev. Lett. **72**, 3506 (1994)
7. B.-G. Englert, M. Löffler, O. Benson, B. Varcoe, M. Weidinger, and H. Walther: Fortschr. Phys. **46**, 897 (1998)
8. H. J. Kimble, O. Carnal, N. Georgiades, H. Mabuchi, E. S. Polzik, R. J. Thomson and Q. A. Turchettte: Atomic Physics **14**, ed. by D. J. Wineland, C. E. Wieman, and S. J. Smith (AIP Press 1995)pp. 314–335.
9. G. Nogues, A. Rauschenbeutel, S. Osnaghi, M. Brune, J. M. Raimond, and S. Haroche: Nature **400**, 239 (1999
10. E.T. Jaynes and F.W. Cummings: Proc. IEEE **51**, 89 (1963)
11. M. Weidinger, B.T.H. Varcoe, R. Heerlein, and H. Walther: Phys. Rev. Lett. **82**, 3795 (1999)
12. P. Meystre, G. Rempe, and H. Walther: Optics Lett., **13**, 1078 (1988)
13. G. Raithel, *et al.*: 'The Micromaser: a Proving Ground for Quantum Physics'. In: *Advances in Atomic, Molecular and Optical Physics*, Supplement 2, ed. by P. Berman (Academic Press, New York, 1994) pp. 57–121
14. J. Krause, M. O. Scully, and H. Walther: Phys. Rev. A **36**, 4547 (1987)
15. P. J. Bardoff, E. Mayr, and W.P. Schleich: Phys. Rev. A **51**, 4963 (1995)
16. B. T. H. Varcoe, S. Brattke, M. Weidinger, and H. Walther: Nature **403**, 743 (2000)
17. Brattke, S., Varcoe, B. T. H. and Walther, H.: Publication in preparation.
18. B. T. H. Varcoe, S. Brattke, and H. Walther: Journ. of Optics B: Quantum Semiclass. Opt. **2**, 154 (2000)
19. H.-J. Briegel, B.-G. Englert, N. Sterpi, and H. Walther: Phys. Rev. A **49**, 2962 (1994)

Coherent Manipulation of Two Trapped Ions with Bichromatic Light

E. Solano[1,2], R.L. de Matos Filho[1], and N. Zagury[1]

[1] Instituto de Física, Universidade Federal do Rio de Janeiro, Caixa Postal 68528, 21945-970 Rio de Janeiro, RJ, Brazil
[2] Sección Física, Departamento de Ciencias, Pontificia Universidad Católica del Perú, Apartado 1761, Lima, Peru

Abstract. We study different aspects of the coherent manipulation of two trapped ions with bichromatic light. Using a dispersive bichromatic interaction we propose a procedure to generate the four electronic Bell states. We also make proposals for implementing a reliable protocol for teleporting electronic states and for measuring the Wigner function associated with the collective motional state of both ions. We study a second kind of bichromatic dispersive interaction that effectively promotes resonant transitions inside a previously chosen vibronic subspace. Using this selective interaction, we propose a procedure for generating Fock states and entanglement in the collective motional states of the two ions. Finally, we discuss a resonant bichromatic interaction that may be used for generating efficiently different kinds of Schrödinger's cat states and squeezing in the collective motional state of both ions.

1 Introduction

In the last decade, the study of a single trapped ion cooled down to almost zero temperature was realized with success, from a theoretical and experimental point of view [23]. In this system it is established a dialogue between the internal states and the external movement of the ion by means of the application of laser beams, reproducing under certain conditions an interaction similar to the one appearing in cavity quantum eletrodynamics (CQED) [2], the so called Jaynes-Cummings interaction. The role of the bosonic field of electromagnetic radiation is played by the bosonic field of the ion motion. The system of trapped ions presents a richer dynamics and better conditions for scalability. Furthermore, as long as the ions are trapped, we may profit of their long coherence times having them at our disposal during longer times. In the search of scalability in trapped ions systems, the study of two ions has a relevant place. Reproducing what was already done with a single ion has a natural importance if we think, for instance, in the generation of nonclassical mesoscopic states, as bigger Schrödinger cats and monitoring decoherence. The two ion system presents other features and may offer us non local entanglement for eventual fundamental tests of quantum physics [3,4,5], implementation of a teleportation scheme [6], selective interactions [7,8], different kinds of mesoscopic superpositions [9,10], more elaborated quantum logic gates [4,11], among many others.

M. Orszag, J.C. Retamal (Eds.): LNP 575, pp. 14–28, 2001.

In Section 2, we discuss a general bichromatic dispersive interaction and show how to generate the four Bell states in the electronic states of two trapped ions, as well as how to implement a reliable teleportation scheme. We discuss, also, how this dispersive interaction leads to the determination of the Wigner function of the motional state of two trapped ions. In Section 3, we present a bichromatic selective interaction and make two different proposals for generating Fock states in the vibrational modes. In Section 4, we discuss briefly a bichromatic resonant interaction that produces efficiently coherent states, superposition of mesoscopic states (Schrödinger cat states) of different kinds and squeezing in the collective motion. In Section 5, we present our brief conclusion.

2 Dispersive Interaction

2.1 The model

Collective coordinates. Let us consider two two-level ions cooled down to very low temperatures [12] in a linear Paul trap of secular frequency ν, with the same mass m, the same charge e and positions z_1 and z_2. The classical equilibrium positions of the ions are $z_{10} = -d/2$, $z_{20} = d/2$, and are realized when the attractive force of the trap on the ions is compensated by their coulombian repulsion. In the equilibrium situation the center-of-mass (CM) coordinate, $Z = (z_1 + z_2)/2$, is zero and the relative coordinate, $z = z_1 - z_2$, is equal to $z_o = -d$ $(d > 0)$. The classical Hamiltonian describing this system is

$$H_o = \frac{p_1^2}{2m} + \frac{1}{2}m\nu^2 z_1^2 + \frac{p_2^2}{2m} + \frac{1}{2}m\nu^2 z_2^2 + \frac{e^2}{|z_2 - z_1|}, \tag{1}$$

where p_1 and p_2 are the conjugated momenta associated with the ionic positions, z_1 and z_2. We rewrite this Hamiltonian in the collective coordinates [13], CM and relative as defined above, and consider only small oscillations of the ions around their equilibrium positions. After taking into account the Taylor expansion of the coulombian term around $|z_2 - z_1| = d$ we have

$$H_0 = \frac{P^2}{2M} + \frac{1}{2}M\nu^2 Z^2 + \frac{p^2}{2\mu} + \frac{1}{2}\mu(\sqrt{3}\nu)^2(\delta z)^2 + \frac{1}{2}\mu(\sqrt{3}\nu)^2 d^2, \tag{2}$$

where $M = 2m$, $\mu = m/2$ are the total and reduced mass of the system, P and p are the conjugated momenta associated with their corresponding CM and relative coordinates. We observe that Z (the CM coordinate) and $\delta z = z - z_0$ (the relative coordinate z subtracted by the equilibrium distance z_0) oscillate harmonically around their equilibrium values with frequencies ν and $\nu_r \equiv \sqrt{3}\nu$, respectively. The last term in this Hamiltonian is constant and corresponds to the minimal energy, $H_{\min}$, that occurs when the system is at rest in the equilibrium situation. we introduce the usual annihilation and creation harmonic oscillator operators $\hat{a}$,$\hat{a}^\dagger$ and $\hat{b}$, $\hat{b}^\dagger$, associated with the excitations of the center of mass and relative motion modes, respectively, by

$$\begin{aligned}\hat{Z} &= \sqrt{\frac{\hbar}{2M\nu}}(\hat{a}^\dagger + \hat{a}) \ , \ \hat{P} = i\sqrt{\frac{M\hbar\nu}{2}}(\hat{a}^\dagger - \hat{a}), \\ \hat{\delta z} &= \sqrt{\frac{\hbar}{2\mu\nu_r}}(\hat{b}^\dagger + \hat{b}) \ , \ \hat{p} = i\sqrt{\frac{\mu\hbar\nu_r}{2}}.(\hat{b}^\dagger - \hat{b}),\end{aligned} \tag{3}$$

In this way, the quantized Hamiltonian reads

$$\widehat{H}_o = \hbar\nu\hat{a}^\dagger\hat{a} + \hbar\nu_r\hat{b}^\dagger\hat{b}, \tag{4}$$

where constant terms have been discarded.

Effective Hamiltonian. We apply, homogeneously [1] on both ions, two classical laser fields, $\boldsymbol{E}_I = \boldsymbol{E}_{0I}e^{i(q_I z - \omega_I t + \phi_I)}$ and $\boldsymbol{E}_{II} = \boldsymbol{E}_{0II}e^{i(q_{II} z - \omega_{II} t + \phi_{II})}$, which are quasiresonant with a long living electronic transition among two ionic hyperfine levels $|\downarrow\rangle$ and $|\uparrow\rangle$. The respective interaction Hamiltonian, in the Schrödinger picture, reads

$$\begin{aligned}\widehat{H}_{int} = \sum_{\alpha=I,II} & \hbar\Omega_\alpha\left(\widehat{S}_{+1} + \widehat{S}_{-1}\right)\left[e^{i(q_\alpha\hat{z}_1 - \omega_\alpha t + \phi_\alpha)} + \text{H.c.}\right] \\ & +\hbar\Omega_\alpha\left(\widehat{S}_{+2} + \widehat{S}_{-2}\right)\left[e^{i(q_\alpha\hat{z}_2 - \omega_\alpha t + \phi_\alpha)} + \text{H.c.}\right],\end{aligned} \tag{5}$$

where Ω_α ($\alpha = I, II$) are the Rabi classical frequencies (chosen here as real numbers) corresponding to the one photon transitions, $\widehat{S}_{+j} = |\uparrow_j\rangle\langle\downarrow_j|$ and $\widehat{S}_{-j} = |\downarrow_j\rangle\langle\uparrow_j|$ are the raising and lowering operators associated with the two electronic levels, $|\downarrow\rangle$ and $|\uparrow\rangle$, of frequencies $\omega_\downarrow$ and $\omega_\uparrow = \omega_o + \omega_\downarrow$, respectively. We may rewrite this interaction Hamiltonian in terms of the operators associated with the normal collective modes and in the rotating wave approximation (RWA) as

$$\begin{aligned}\widehat{H}^{rwa}_{int} = \sum_{\alpha=I,II} & \hbar\Omega_\alpha\widehat{S}_{+1}e^{i\left[\eta_\alpha(\hat{a}+\hat{a}^\dagger)+\eta_{r\alpha}(\hat{b}+\hat{b}^\dagger)-\omega_\alpha t+(\phi_\alpha+\frac{\phi_o}{2})\right]} \\ & +\hbar\Omega_\alpha\widehat{S}_{+2}e^{i\left[\eta_\alpha(\hat{a}+\hat{a}^\dagger)-\eta_{r\alpha}(\hat{b}+\hat{b}^\dagger)-\omega_\alpha t+(\phi_\alpha-\frac{\phi_o}{2})\right]} + \text{H.c.},\end{aligned} \tag{6}$$

where $\phi_0 = qz_0 = -qd < 0$ is the phase difference due to the equilibrium spacing between the ions and $\eta_\alpha = q_\alpha\sqrt{\frac{\hbar}{2M\nu}}$ and $\eta_{r\alpha} = \frac{q_\alpha}{2}\sqrt{\frac{\hbar}{2\mu\nu_r}}$ are the Lamb-Dicke parameters of the laser excitations associated to the CM and relative vibrational modes, respectively. Note that $\eta_{r\alpha} = \eta_\alpha/\sqrt[4]{3}$. Without losing generality, we may consider $q \equiv q_I = q_{II}$ and, in consequence, we may define $\eta \equiv \eta_I = \eta_{II}$ and $\eta_r \equiv \eta_{rI} = \eta_{rII}$. Additionally, we may choose $\Omega \equiv \Omega_I = \Omega_{II}$, and $\phi \equiv \phi_I = \phi_{II}$.

[1] We take into account the technical limitations for individually addressing the ions. At the moment, big efforts are being done for overcoming this problem [14,15].

We discuss a situation where the considered effective laser excitations are quasiresonant to one of the vibrational modes, e.g. the k-th upper or lower sideband of the center-of-mass mode. Analogous results are obtained if we consider the excitations of the relative mode. A similar excitation scheme has been used before [4] under more restricted conditions to realize quantum computing with ions in thermal motion. We choose, then, $\omega_I = \omega_0 + k\nu - \delta$ and $\omega_{II} = \omega_0 - k\nu + \delta$, where $\delta \ll \nu$, such that $\omega_I + \omega_{II} = 2\omega_0$. We replace these values in Eq. (6) with all previous considerations, rewrite it in the interaction picture and, after discarding the terms that oscillate rapidly with frequency ν and its harmonics, we have, following a well known procedure[10],

$$\begin{aligned}\widetilde{\widehat{H}}_{int} &= \hbar\Omega[(i\eta)^k(\widehat{S}_{+1}e^{i\phi_o/2} + \widehat{S}_{+2}e^{-i\phi_o/2})e^{i\phi} + \mathrm{H.c.}] \\ &\quad \times[\hat{a}^{\dagger k}\widehat{F}_k(\hat{n},\hat{n}_r)\, e^{i\delta t} + \widehat{F}_k(\hat{n},\hat{n}_r)\, \hat{a}^k e^{-i\delta t}],\end{aligned} \tag{7}$$

The corresponding number operators are $\hat{n} = \hat{a}^\dagger\hat{a}$, $\hat{n}_r = \hat{b}^\dagger\hat{b}$, and

$$\widehat{F}_k(\hat{n}) = \sum f_k(n)|n\rangle\langle n| \quad , \quad \widehat{G}_{k_r}(\hat{n}_r) = \sum g_{k_r}(n_r)|n_r\rangle\langle n_r|, \tag{8}$$

with

$$f_k(n) = e^{-\eta^2/2}\frac{n!}{(n+k)!}L_n^k(\eta^2) \quad , \quad g_{k_r}(n_r) = e^{-\eta_r^2/2}\frac{n_r!}{(n_r+k_r)!}L_{n_r}^{k_r}(\eta_r^2), \tag{9}$$

where $L_m^k(x) = \sum_{\ell=0}^m(-1)^\ell\binom{m+k}{m-\ell}\frac{x^\ell}{\ell!}$ are Laguerre polynomials.

After all these considerations, we may turn to the derivation of an effective time-independent Hamiltonian. For achieving this goal, we consider the case when δ is much larger than the vibronic Rabi frequency, $\eta^k\Omega f_k(n, n_r)$, and the interaction times are such that $\delta t \gg 1$. In such a situation, the Hilbert space of the electronic states effectively decouples into two subspaces: one containing only one ion in the excited state $\{|\downarrow\uparrow\rangle, |\uparrow\downarrow\rangle\}$ and the other containing both ions either in the excited or in the ground state $\{|\downarrow\downarrow\rangle, |\uparrow\uparrow\rangle\}$. Then, the second order effective Hamiltonian reads [5]

$$\begin{aligned}\widehat{H}_{\mathrm{eff}} &= \hbar\Omega_k[\widehat{S}_{+1}\widehat{S}_{+2}e^{2i\phi} + (-1)^k\left(\widehat{S}_{+1}\widehat{S}_{-2}e^{i\phi_o} + \frac{1}{2}\right)] \\ &\quad \times[\widehat{F}_k^2(\hat{n}-k,\hat{n}_r)\frac{\hat{n}!}{(\hat{n}-k)!} - \widehat{F}_k^2(\hat{n},\hat{n}_r)\frac{(\hat{n}+k)!}{\hat{n}!}] + \mathrm{H.c.}\end{aligned} \tag{10}$$

where $\Omega_k = 2|\Omega|^2(i\eta)^{2k}/\delta$. The first term and its Hermitian conjugate, acting only in the subspace determined by the states $\{|\downarrow\downarrow\rangle, |\uparrow\uparrow\rangle\}$, describe two-photon processes leading to the simultaneous excitation or deexcitation of the electronic states of the two ions. The second term and its Hermitian conjugate, acting exclusively in the subspace spanned by the states $\{|\downarrow\uparrow\rangle, |\uparrow\downarrow\rangle\}$, describe processes where one ion undergoes a transition from the ground to the excited electronic state and the other ion makes a transition in the inverse direction, both processes taking place simultaneously. The third term is a self energy term where, remarkably, the energy shifts in the two effective ionic levels are equal and in the same direction, just producing a time-dependent overall phase in the states evolution, as we will see below. The vibrational part of this Hamiltonian, completely factorized on the right, contains only number operators that will not change the population of a two mode Fock state basis $|n, n_r\rangle$. The first (second) term of the vibrational part has its origin in a virtual two step process: in the first one, k vibrational quanta are annihilated (created) and in the second one, the same number of vibrational quanta are created (annihilated), preserving the number of motional quanta.

As a representative example of the dynamical evolution under the interaction described above, we consider the temporal evolution of states of the form $|\downarrow,\downarrow\rangle_{n,n_r} \equiv |\downarrow,\downarrow\rangle \otimes |n, n_r\rangle$, that is,

$$e^{-i(-1)^k \Omega^k_{nn_r} t}[\cos(|\Omega^k_{nn_r}| t)\,|\downarrow,\downarrow\rangle + i(-1)^k e^{2i\phi} \sin(|\Omega^k_{nn_r}| t)\,|\uparrow,\uparrow\rangle]|n, n_r\rangle\,. \quad (11)$$

Here

$$\Omega^k_{nn_r} = \Omega_k[f_k^2(n-k, n_r)\frac{n!}{(n-k)!} - f_k^2(n, n_r)\frac{(n+k)!}{n!}] \quad (12)$$

are effective Rabi frequencies for transitions in the Hilbert subspace spanned by the states $|\downarrow,\downarrow\rangle_{n,n_r}$ and $|\uparrow,\uparrow\rangle_{n,n_r}$. Note that, in this approximation, the interaction does not change the number of vibrational excitations.

2.2 Bell states

We will show now that with the help of the effective Hamiltonian in Eq. (10) and an additional adequate carrier interaction we can generate the four Bell states in the ionic internal degrees of freedom

$$|\Phi^{(\pm)}\rangle = \frac{1}{\sqrt{2}}[|\downarrow,\downarrow\rangle \pm |\uparrow,\uparrow\rangle] \quad , \quad |\Psi^{(\pm)}\rangle = \frac{1}{\sqrt{2}}[|\downarrow,\uparrow\rangle \pm |\uparrow,\downarrow\rangle]\,, \quad (13)$$

avoiding any undesired entanglement with the vibrational modes. For this purpose, we consider the following initial state $|\downarrow,\downarrow\rangle_{n,n_r}$. Then, if we consider the times $t_+ = \pi/(4|\Omega^k_{nn_r}|)$ and $t_- = 3\pi/(4|\Omega^k_{nn_r}|)$, in the temporal evolution of Eq.(11), we generate the first pair of Bell states

$$|\Phi^{(\pm)}\rangle = \frac{1}{\sqrt{2}}[|\downarrow,\downarrow\rangle \pm i(-1)^k e^{2i\phi}\,|\uparrow,\uparrow\rangle]\,. \quad (14)$$

Alternatively, we could maintain the same time of interaction and change the value of ϕ by $\pi/2$. Note that k and ϕ are free parameters that we may control and manipulate adequately.

Once one of the states $|\Phi^{(\pm)}\rangle$ is generated, we may apply on both ions, homogeneously, a laser pulse resonant to the single ion electronic transition $|\uparrow\rangle \leftrightarrow |\downarrow\rangle$ (carrier pulse). Always aiming to a realistic proposal, as in the previous dispersive interaction, we do not claim individual addressing of the ions. This conventional carrier pulse may be implemented with one effective laser $\boldsymbol{E} = \boldsymbol{E_o} e^{i(q_o z - \omega_o t + \varphi)}$. Then, the corresponding effective interaction Hamiltonian is

$$\widehat{H} = \hbar\Omega[(\widehat{S}_{+1}e^{i\varphi_0/2} + \widehat{S}_{+2}e^{-i\varphi_0/2})e^{i\varphi} + \mathrm{H.c.}]\widehat{F}_o(\hat{n}, \hat{n}_r)\,, \tag{15}$$

where $\varphi_o = -q_o d$. Now, we let the pair of Raman lasers interact with the ions, in the initial state $|\Phi^{(\pm)}\rangle_{n,n_r}$, during a time $t_o = \pi/(4|\Omega f_o(n, n_r)|)$. Then, we may choose their relative phase φ equal to 0 (modulus π) or $\pi/2$ (modulus π), accordingly if we start with $|\Phi^{(+)}\rangle$ or $|\Phi^{(-)}\rangle$, to generate the other couple of Bell states

$$|\Psi^{\pm}\rangle = \frac{1}{\sqrt{2}}[|\uparrow, \downarrow\rangle + e^{i\varphi_o^{\pm}}\,|\uparrow, \downarrow\rangle]\;. \tag{16}$$

Here, the condition $\varphi_o^+ = \varphi_o^- + \pi$ is enough for producing the other two orthogonal maximally entangled states.

2.3 Reliable teleportation

Scheme. Our procedure [6], along the lines of the BBCJPW protocol [17], can be sketched as follows (see Fig. 1): Trap A (Alice station) contains ion 1 with an arbitrary electronic state, $\alpha\,|\downarrow_1\rangle + \beta\,|\uparrow_1\rangle$ and trap B (Bob station) contains ions 2 and 3 in a maximally entangled state, e.g. any of the above generated four Bell states. Ion 2 is then transferred adiabatically to Alice station, who performs a suitable operation on the joint Hilbert space of ions 1 and 2 transforming unitarily the Bell base into a disentangled base. Subsequently, Alice measures their individual electronic states, completing the so called Bell analyzer. Then, Alice informs to Bob the result of her measurements, consisting in two bits of classical information. Bob uses them to perform one, out of four, specific unitary transformation on ion 3, whose electronic state is left at the original state of ion 1.

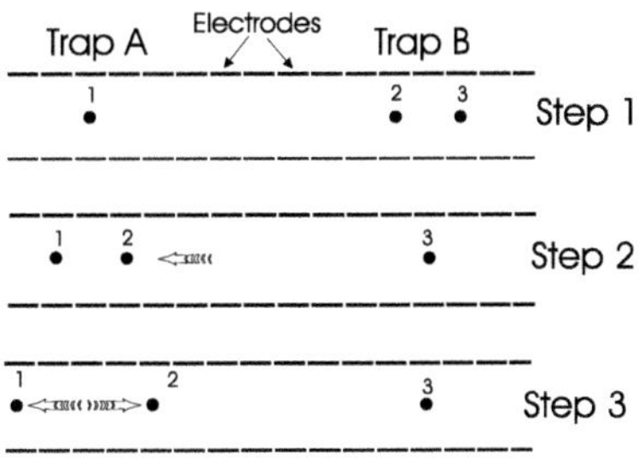

Fig. 1. Trap configuration for teleportation of the electronic state of ion 1 to ion 3.

Quantum channel. The first task in the implementation of our teleportation scheme is the preparation of the quantum channel, i.e., the deterministic generation of a Bell state in trap B. It was shown that such a state can be generated, with present techniques, through the use of two pairs of Raman lasers interacting dispersively with both ions. In a real situation we will deal with motional thermal states. Therefore, it is convenient to work in the first side band ($k = 1$) and in the Lamb-Dicke regime. In this case, the general effective Hamiltonian of Eq. (10) reduces to

$$H = \hbar\,|\Omega| \left[S_{+j} S_{+m} e^{2i\phi} - S_{+j} S_{-m} e^{i\phi_o} - \frac{1}{2} \right] + \text{H.c.} \tag{17}$$

Notice that, in this limit, the effective Hamiltonian H is identical to the one discussed in Ref. [4] and does not depend on the vibrational quantum state of the ions. For the preparation of the Bell state in trap B, we consider that the two ions were previously cooled down to the Lamb–Dicke regime and stay in their electronic ground states. By letting the two Raman laser pairs interact with the ions during a time $\tau = \pi/(4|\Omega|)$, we prepare the Bell electronic state

$$|\Phi_{23}\rangle = \frac{1}{\sqrt{2}} \left(|{\downarrow_2}, {\downarrow_3}\rangle - i e^{2i\phi_B} |{\uparrow_2}, {\uparrow_3}\rangle \right), \tag{18}$$

where ϕ_B is the effective phase of both Raman lasers pairs in trap B. We then transfer adiabatically ion 2 from trap B to trap A, where it interacts with ion 1. A possible mechanism to make this transfer will be discussed below. No entanglement should occur between the electronic and motional states, if this transfer is done adiabatically in a time larger than $1/\omega_o$. We assume that in the transfer process the mean quantum number $\bar{n}$ associated to the vibrational modes are kept small, so that the Lamb–Dicke approximation, $\bar{n}\eta^2 \ll 1$, remains valid. If it is not the case, sympathetic cooling techniques may be used. The total electronic state of ions 1, 2 and 3 is then given by

$$|\Psi_o(1,2,3)\rangle = (\alpha\,|{\downarrow_1}\rangle + \beta\,|{\uparrow_1}\rangle)\,|\Phi_{23}\rangle\,. \tag{19}$$

Bell analyzer. If we follow closely the recipe of the BBCJPW protocol [17], our second task would be to perform a complete measurement, of the von Neumann type, on the subsystem of particles 1 and 2 in the Bell operator basis, that will confidently provide two bits of classical information. We propose here a practical way of obtaining this classical information by suitably entangling the electronic state of particles 1 and 2 and then measuring the individual electronic state of each ion. This substitutes the non-trivial requirement of a direct Bell measurement by that of monitoring the individual ion fluorescence. For achieving this goal, we first apply, on ions 1 and 2, a pulse of duration $\tau = \pi/(4|\Omega|)$, using the same excitation scheme as used before in trap B, transforming the total electronic state $|\Psi_0(1,2,3)\rangle$ into

$$
\begin{aligned}
|\Psi(1,2,3)\rangle = \frac{1}{2}\Big\{ -i \quad e^{2i\phi_A} \quad & |\uparrow_1\uparrow_2\rangle \otimes \Big[\alpha\,|\downarrow_3\rangle + e^{2i(\phi_B-\phi_A)}\beta\,|\uparrow_3\rangle\Big] \\
+ \quad & |\downarrow_1\downarrow_2\rangle \otimes \Big[\alpha\,|\downarrow_3\rangle - e^{2i(\phi_B-\phi_A)}\beta\,|\uparrow_3\rangle\Big] \\
+ \quad & |\uparrow_1\downarrow_2\rangle \otimes \Big[\beta\,|\downarrow_3\rangle + e^{i(2\phi_B+\phi_o)}\alpha\,|\uparrow_3\rangle\Big] \\
+ie^{-i\phi_o} \quad & |\downarrow_1\uparrow_2\rangle \otimes \Big[\beta\,|\downarrow_3\rangle - e^{i(2\phi_B+\phi_o)}\alpha\,|\uparrow_3\rangle\Big] \Big\}, \qquad (20)
\end{aligned}
$$

where ϕ_A is the effective phase of both Raman laser pairs in trap A and ϕ_o is the phase due to the equilibrium separation of the ions 1 and 2 in trap A. If we set, for simplicity, $\phi_A = \phi_B = \pi - \phi_o/2$, it is clear that the determination of the energy state of ions 1 and 2 projects ion 3 on one of the four states $\alpha\,|\downarrow_3\rangle + \beta\,|\uparrow_3\rangle$, $\alpha\,|\downarrow_3\rangle - \beta\,|\uparrow_3\rangle$, $\beta\,|\downarrow_3\rangle + \alpha\,|\uparrow_3\rangle$, and $\beta\,|\downarrow_3\rangle - \alpha\,|\uparrow_3\rangle$. These four states corresponds to the original state or a π rotation of it around the z axis, the x axis or the y axis, respectively. If now Alice measures the electronic energy of each ion in her trap and send the result (2 bits of classical information) to Bob, he will need or not to make an additional operation (the corresponding inverse rotation around the z, x or y axis) on ion 3. This completes the teleportation of the original arbitrary state from particle 1 to particle 3. Our scheme provides reliable teleportation [18,19], where we have theoretically 100% of success in distinguishing the four Bell states in our analyzer. This is also the case for atom-cavity systems, as in the proposal discussed in Ref. [10], and only very recently for photons [21].

It is important to note that, initially, ions 1 and 2 were not correlated and that the applied pulse entangles these particles, communicating ion 1 with the quantum channel already established in ions 2 and 3. This is not in contradiction with the fact that this pulse, formally, transforms unitarily the Bell basis into a completely disentangled one, as we may see in Eq. (20).

2.4 Wigner function of the collective motion

Let us now turn our attention to the determination of a general vibrational quantum state, $\hat{\varrho}_v$, of the ions. Our method [5] is based on the determination of the two-mode Wigner function $W(\alpha_1, \alpha_2)$, characterizing the joint state of two harmonic quantum oscillators at the point (α_1, α_2) of their extended phase space, which can be written as [22]

$$
W(\alpha_1, \alpha_2) = \frac{4}{\pi^2} \sum_{n_1,n_2}^{\infty} (-1)^{n_1+n_2}\, \Pi_{n_1,n_2}(-\alpha_1, -\alpha_2)\,, \qquad (21)
$$

where $\Pi_{n_1,n_2}(-\alpha_1, -\alpha_2)$ is the population corresponding to the two-mode harmonic oscillator state displaced coherently in the extended phase space by $-\alpha_1$ and $-\alpha_2$, respectively. To determine the vibrational Wigner function, $W(\alpha, \alpha_r)$, using the relation in Eq. (21), we first displace coherently the state $\hat{\varrho}_v$ in the extended phase space

$$\hat{\varrho}_v(-\alpha, -\alpha_r) = \hat{D}^\dagger(\alpha)\hat{D}_r^\dagger(\alpha_r)\hat{\varrho}_v\hat{D}_r(\alpha_r)\hat{D}(\alpha). \tag{22}$$

The displacement of the CM and relative vibrational modes can be done, independently of each other, by using the excitation mechanism applied in Ref. [24]. Our problem, then, is to find a way to measure the displaced vibrational population, $\Pi_{n,n_r}(-\alpha, -\alpha_r)$. For this purpose, we use the same dispersive interaction scheme that led to the generation of the $|\Phi^\pm\rangle$ Bell states. Consider that, after the vibrational displacements, the state of the ions is described by $\hat{\rho}(0) = \hat{\varrho}_v(-\alpha, -\alpha_r) \otimes |\downarrow, \downarrow\rangle\langle\downarrow, \downarrow|$. From Eqs. (10) and (11), the probability $P_{\downarrow\downarrow}(\tau)$ of finding both ions in their lower electronic states, after an interaction time τ with the laser fields, is

$$\begin{aligned} P_{\downarrow\downarrow}(\tau) &= \langle\downarrow, \downarrow| \operatorname{Tr}_v \left\{ e^{-\frac{i}{\hbar}\hat{H}_{\text{eff}}\tau}\hat{\rho}(0)e^{\frac{i}{\hbar}\hat{H}_{\text{eff}}\tau} \right\} |\downarrow, \downarrow\rangle \\ &= \sum_{n,n_r} \cos^2\left(|\Omega^k_{nn_r}|\,\tau\right) \Pi_{n,n_r}(-\alpha, -\alpha_r)\,, \end{aligned} \tag{23}$$

where Tr_v means the partial tracing over the vibrational degrees of freedom. From Eq. (12), one can verify that, for $k \neq 0$, the dependence of the vibronic Rabi frequencies $\Omega^k_{nn_r}$ on n and n_r discriminates efficiently the different two-mode Fock states $|n, n_r\rangle$. For this reason, the vibrational populations $\Pi_{n,n_r}(-\alpha, -\alpha_r)$ may be easily extracted from $P_{\downarrow\downarrow}(\tau)$ [23].

We can measure $P_{\downarrow\downarrow}(\tau)$ by monitoring the fluorescence produced in driving the auxiliary cycling transition $|\downarrow\rangle \leftrightarrow |d\rangle$ of the ions with resonant laser light, once the interaction leading to Eq. (23) has been turned off. The presence of fluorescence assures us that the ions are in the joint electronic state $|\downarrow, \downarrow\rangle$, and its absence indicates the occupation of the state $|\uparrow, \uparrow\rangle$. This is only the case because the states $|\downarrow, \uparrow\rangle$ and $|\uparrow, \downarrow\rangle$ are not populated before the cycling transition is driven. That is, the use of the dispersive interaction of Eq. (10) opens the possibility of measuring $P_{\downarrow\downarrow}(\tau)$ in as simple and highly efficient manner, as the one used to measure the corresponding quantity in a single ion [24].

3 Selective Interaction

In this section, we present an effective interaction consisting of two dispersive Raman pulses, as was done in Section 2, homogeneously illuminating two trapped ions. This time, the beam frequencies are initially chosen as $\omega_I = \omega_o + k\nu + k_r\nu_r - \delta$ and $\omega_{II} = \omega_o + \delta$, in such a way that $\omega_I + \omega_{II} = 2\omega_o + k\nu + k_r\nu_r$. This effective interaction may be transformed into a selective one [7], that will act resonantly only in a previously chosen subspace and dispersively in all others. This interaction opens possibilities for manipulating coherently the vibronic states. In what follows we will only consider the particular and illustrative case $k = 1, k_r = 0$.

We may follow the same procedure as the one developped in Section 2, for these particular new beam frequencies, and deduce the following interaction Hamiltonian

$$H_{\text{eff}} = \hbar\Omega_o \hat{G}_o^2 \Big\{ 2(i\eta)\hat{S}_{+1}\hat{S}_{+2}[\hat{a}^\dagger \hat{F}_1, \hat{F}_o] + \eta^2 \hat{S}_{+1}\hat{S}_{-2}[\hat{a}^\dagger \hat{F}_1, \hat{F}_1 \hat{a}]$$
$$+\frac{1}{2}(\hat{S}_{+1}\hat{S}_{-1} + \hat{S}_{+2}\hat{S}_{-2})(\eta^2 \hat{a}^\dagger \hat{F}_1^2 \hat{a} - \hat{F}_o^2)$$
$$-\frac{1}{2}(\hat{S}_{-1}\hat{S}_{+1} + \hat{S}_{-2}\hat{S}_{+2})(\eta^2 \hat{F}_1 \hat{a}\hat{a}^\dagger \hat{F}_1 - \hat{F}_o^2) + \text{H.c.} \Big\} . \quad (24)$$

The first term and its hermitian conjugate give rise to an anti Jaynes-Cummings dynamics, leading to a simultaneous excitation (or de-excitation) of the electronic states of the two ions, accompanied by the creation (or annihilation) of one vibrational quanta in the CM mode. The second term and its hermitian conjugate generate a dynamics where, simultaneously, one ion undergoes a transition from the ground to the excited electronic state and the other ion makes a transition in the inverse direction. This process is not accompanied by any excitation of the vibrational modes. The third and fourth terms, together with their hermitian conjugates, generate motional dependent dynamical energy shifts in the electronic levels. Due to this dependence, this terms turn the anti Jaynes-Cummings dynamics more or less resonant, depending on the particular level of excitation of the vibrational modes. Note that the sensitivity of the energy shifts to the vibrational state of the ions increases with increasing values of the Lamb–Dicke parameters. For not too small values of these parameters, it is possible to make the anti Jaynes-Cummings interaction completely resonant inside a previously chosen subspace $\{|\downarrow,\downarrow,N,N_r\rangle, |\uparrow,\uparrow,N+1,N_r\rangle\}$, whereas remaining largely non resonant inside other subspaces. This may occur, for example, when the initial Raman lasers frequencies are modified and correctly tuned to take into account the motional dependent energy shifts.

As can be easily seen from Eq.24, the energy shifts of levels $|\downarrow,\downarrow,n,n_r\rangle$ and $|\uparrow,\uparrow,n+1,n_r\rangle$ are given by

$$\Delta_{\downarrow\downarrow}^{n,n_r} = 2\hbar\Omega_o g_o^2(n_r)\left[f_o^2(n) - \eta^2(n+1)f_1^2(n)\right]$$
$$\Delta_{\uparrow\uparrow}^{n+1,n_r} = 2\hbar\Omega_o g_o^2(n_r)\left[\eta^2(n+1)f_1^2(n) - f_o^2(n+1)\right] , \quad (25)$$

respectively. By properly adjusting the laser frequencies we may put them in resonance with the Stark shifted levels associated with a previously chosen vibronic subspace $\{|\downarrow,\downarrow,N,N_r\rangle , |\uparrow,\uparrow,N+1,N_r\rangle\}$, while preventing resonant transitions in other subspaces with $n \neq N$ and $n_r \neq N_r$. This can be done by modifying the laser frequencies to

$$\omega_I = \tilde{\omega}_0 + \nu - \delta, \quad \omega_{II} = \tilde{\omega}_0 + \delta, \quad (26)$$

where $2\tilde{\omega}_0 = 2\omega_0 + \Delta_{\uparrow\uparrow}^{N+1,N_r} - \Delta_{\downarrow\downarrow}^{N,N_r}$ is the renormalized splitting of the levels. Notice that, for very small values of the Lamb-Dicke parameters, the motional dependence of the dynamical Stark shift disappears. For this reason it is important to work beyond the Lamb-Dicke regime in order to effectively select a chosen subspace out of the whole vibronic Hilbert space. It is noteworthy to mention that for special values of the Lamb-Dicke parameter η, it may happen that energy shifts $\Delta_{\uparrow\uparrow}^{n+1,n_r}$ and $\Delta_{\downarrow\downarrow}^{n,n_r}$ are equal for certain values of n, irrespectively of the state of the relative vibrational mode. For example, for $\eta \simeq 0.51, 0.42, ..., 0.24$, resonant transitions inside the chosen subspaces $\{|\downarrow,\downarrow,N,N_r\rangle, |\uparrow,\uparrow,N+1,N_r\rangle\}$, naturally appear for $N = 1,2,...,8$, and $N_r = 0$, respectively. Clearly, in this case it is not necessary to correct the laser frequencies.

The Hamiltonian (24) could be used to generate a large set of motional states. For example, any Fock state associated with the CM motion could be obtained from the initial state $|\downarrow,\downarrow,0,0\rangle$ by successively applying π pulses with different frequencies. However, if one is interested in generating a highly excited Fock state, this process could take an unsatisfactory long time. A more efficient, non unitary, way of producing such states, as well as engineering other vibrational states, is to start from a product of the electronic ground state and any motional state $|\psi_{\rm vib}\rangle$. By selecting the laser frequencies, we excite only a chosen vibronic transition $\{k, k_r\}$ with a π pulse. Ideally, we would end up with a superposition of the two states $|\uparrow,\uparrow,N+k,N_r+k_r\rangle$ and $|\downarrow,\downarrow\rangle \otimes (|\psi_{\rm vib}\rangle - |N,N_r\rangle\langle N,N_r|\psi_{\rm vib}\rangle)$. Measurement of the electronic levels projects out either the Fock state $|N+k,N_r+k_r\rangle$ or the original state with a "hole" in the N,N_r component. As a numerical example we have taken the initial state to be the product of the electronic ground state with a coherent state of the CM motion and the vacuum of the relative motion. We start from a coherent state with $\bar{n} = 4.0$ and excite the center of mass transition from $n = 4$ to $n = 5$. A dark event in a fluorescent measurement should leave us with an state that is close to the Fock state with $n = 5$. In Fig. 2 we show the results for the phonon distribution. As expected, it is possible to make a "hole" in the vibrational quanta distribution (Fig. 2a) while creating a quasi Fock state (Fig. 2b) using a π pulse. The state with a"hole" will be associated with the ground electronic state, while the approximate Fock state of Fig. 2b will be associated with the excited electronic state. As shown

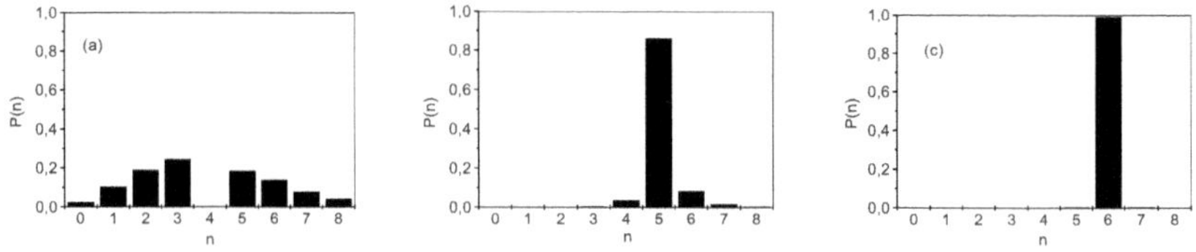

Fig. 2. Vibrational distributions $P(n)$ for the center of mass motion after a π pulse resonant to the transition from $n = 4$ to $n = 5$. The initial state is the product of a coherent state with $\bar{n} = 4.0$ and $|\downarrow,\downarrow\rangle$. (a) Distribution correlated to $|\downarrow,\downarrow\rangle$. (b) Distribution correlated to $|\uparrow,\uparrow\rangle$. (c) Distribution correlated to $|\downarrow,\downarrow\rangle$ after the second pulse. $\eta = 0.3$ and $\delta = 40\eta\Omega$.

in Fig. 2b, small contamination occurs around the target Fock state because transitions to levels other than $n = 5$ are not totally suppressed. For the case studied, the *a priori* probability of finding the approximated Fock state after fluorescence is about 30%. A Fock state with $n = 6$, for example, can now be obtained if we apply subsequently to the ions another π pulse resonant to the transition $|\uparrow, \uparrow, 5, 0\rangle \leftrightarrow |\downarrow, \downarrow, 6, 0\rangle$. If by measuring the electronic states we find $|\downarrow, \downarrow\rangle$ (*a priori* probability of $\approx 87\%$), we are left with a state very close to the Fock state $|6, 0\rangle$. In this case the fidelity is $\approx 99\%$ (See Fig. 2c).

Other applications, like generation of entanglement in the collective motion and entanglement transfer between the internal and external ionic degrees of freedom are discussed in Ref. [7]. The implementation of a quantum phase gate (QPG) in trapped ions by using a similar selective interaction is discussed in Ref. [11]. The implementation of a selective interaction scheme in the domain of CQED is presented in Ref. [8].

4 Resonant Interaction

In this section, we will discuss briefly some applications of a bichromatic resonant interaction [10], where laser beams of frequencies [2] $\omega_I = \omega_o - k\nu$ and $\omega_{II} = \omega_o + k\nu$, illuminate homogeneously two trapped ions.

4.1 Conditional vibrational displacement

We consider the case $k = 1$, which consists in tuning the laser beams to the first red and blue CM vibrational sidebands, respectively. At the same time, we choose the laser beams phases in order to have $\Omega_I = \Omega_{II}^* = \Omega$ and consider the Lamb-Dicke regime by taking $\{\eta\sqrt{\bar{n}}, \eta_r\sqrt{\bar{n}_r}\} \ll 1$. In this way, following similar steps as in Section 2, we may deduce the effective interaction Hamiltonian

$$\widehat{H} = \hbar(\widehat{S}_{+1} + \widehat{S}_{+2} - \widehat{S}_{-1} - \widehat{S}_{-2})(\beta\hat{a} - \beta^*\hat{a}^\dagger) = \hbar\widehat{H}_{\mathrm{e}}(\beta\hat{a} - \beta^*\hat{a}^\dagger) \tag{27}$$

with $\beta = i\eta\Omega$ and $\widehat{H}_{\mathrm{e}}$ is an antihermitian operator and its eigenvalues may be pure imaginary or null. Consider now the ions initially in the state $|\psi\rangle = |\Phi\rangle \otimes |\Psi_{\mathrm{vib}}\rangle$, where $|\Phi\rangle$ is an eigenstate of $\widehat{H}_{\mathrm{e}}$ with eigenvalue $i\lambda$. After an interaction time τ the initial sate $|\psi\rangle$ evolves to $|\psi(\tau)\rangle = |\Phi\rangle \otimes \widehat{D}(\lambda\beta\tau)|\Psi_{\mathrm{vib}}\rangle$, where $\widehat{D}(\alpha) \equiv e^{\alpha\hat{a} - \alpha^*\hat{a}^\dagger}$ is the CM mode displacement operator. We can conclude that this bichromatic resonant interaction generates a displacement operator conditioned to the initial electronic state.

---

[2] We could have chosen, with similar consequences, to excite the strech mode. More general considerations and other relevant results are discussed in Ref. [10].

4.2 Motional Schrödinger's cat states

Notice that any pure electronic state $|\Phi_e\rangle$ may be written as a superposition of the eigenstates of $\widehat{H}_e$ in the form $|\Phi_e\rangle = \sum_j c_j|\Phi_j\rangle$, where $\widehat{H}_e|\Phi_j\rangle = i\lambda_j|\Phi_j\rangle$. An initial state $|\psi\rangle = |\Phi_e\rangle \otimes |\Psi_{\rm vib}\rangle$ will evolve, after an interaction time τ, to the state $|\psi(\tau)\rangle = \sum_j c_j|\Phi_j\rangle \otimes \widehat{D}(\lambda_j\beta\tau)|\Psi_{\rm vib}\rangle$. To see this explicitly, we write the eigenvectors and eigenvalues of $\widehat{H}_e$

$$|\Phi_{1,2}\rangle = \frac{1}{2}\left[|\downarrow\downarrow\rangle \mp i|\downarrow\uparrow\rangle \mp i|\uparrow\downarrow\rangle - |\uparrow\uparrow\rangle\right] \rightarrow \lambda_{1,2} = \pm 2$$

$$|\Phi_{3,4}\rangle = \frac{1}{2}\left[|\downarrow\downarrow\rangle \pm i|\downarrow\uparrow\rangle \mp i|\uparrow\downarrow\rangle + |\uparrow\uparrow\rangle\right] \rightarrow \lambda_{3,4} = 0 \ .$$

We will present two practical examples of how to use this interaction for generating two different kind of Schrödinger's cat states. First, let us consider the following initial state

$$|\psi(0)\rangle = \frac{1}{\sqrt{2}}(|\Phi_1\rangle + |\Phi_2\rangle) \otimes |0,0\rangle = \frac{1}{\sqrt{2}}(|\downarrow\downarrow\rangle - |\uparrow\uparrow\rangle) \otimes |0,0\rangle. \qquad (28)$$

In this case, after an interaction time τ the final state will be $|\psi(\tau)\rangle = \frac{1}{2}\left[|\Phi_1\rangle|\alpha,0\rangle + |\Phi_2\rangle|-\alpha,0\rangle\right]$ with $\alpha = 2i\eta\Omega\tau$. This state can also be written as

$$\begin{aligned}|\psi(\tau)\rangle = \frac{1}{2\sqrt{2}}\Big\{ & |\downarrow\downarrow\rangle\,[|\alpha,0\rangle + |-\alpha,0\rangle] - i|\downarrow\uparrow\rangle\,[|\alpha,0\rangle - |-\alpha,0\rangle] \\ & -i|\uparrow\downarrow\rangle\,[|\alpha,0\rangle - |-\alpha,0\rangle] - |\uparrow\uparrow\rangle\,[|\alpha,0\rangle + |-\alpha,0\rangle]\Big\} \qquad (29)\end{aligned}$$

By testing the electronic state $|\uparrow\uparrow\rangle$ via the fluorescence technique [25], after a dark event, we obtain the Schrödinger's cat state $\mathcal{N}(|\alpha,0\rangle + |-\alpha,0\rangle)$ with probability $P = \frac{1+e^{-2|\alpha|^2}}{4}$, $\frac{1}{4} \leq P \leq \frac{1}{2}$ (P diminishes with $|\alpha|^2$). If now the initial state is

$$|\psi(0)\rangle = \frac{1}{2}(|\Phi_1\rangle + |\Phi_2\rangle + |\Phi_3\rangle + |\Phi_4\rangle) \otimes |0,0\rangle = |\downarrow\downarrow\rangle \otimes |0,0\rangle, \qquad (30)$$

after an interaction time τ, the final state will be

$$|\psi(\tau)\rangle = \frac{1}{4}\Big\{|\Phi_1\rangle|\alpha,0\rangle + |\Phi_2\rangle|-\alpha,0\rangle + |\Phi_3\rangle|0,0\rangle + |\Phi_4\rangle|0,0\rangle\Big\}, \qquad (31)$$

with $\alpha = 2i\eta\Omega\tau$. This state can also be written as

$$\begin{aligned}|\psi(\tau)\rangle = \frac{1}{4}\Big\{ & |\downarrow\downarrow\rangle\,[|\alpha,0\rangle + |-\alpha,0\rangle + 2|0,0\rangle] - i|\downarrow\uparrow\rangle\,[|\alpha,0\rangle - |-\alpha,0\rangle] \\ & -i|\uparrow\downarrow\rangle\,[|\alpha,0\rangle - |-\alpha,0\rangle] - |\uparrow\uparrow\rangle\,[|\alpha,0\rangle + |-\alpha,0\rangle - 2|0,0\rangle]\Big\}\end{aligned} \qquad (32)$$

By testing the electronic state $|\uparrow\uparrow\rangle$ via fluorescence, after a dark event, we obtain obtain a different kind of Schrödinger's cat state $\mathcal{N}(|\alpha,0\rangle + |-\alpha,0\rangle - 2|0,0\rangle)$ with probability $P = \frac{6-8e^{-|\alpha|^2/2}+2e^{-2|\alpha|^2}}{16}$, $0 \leq P \leq \frac{3}{8}$ (P increases with $|\alpha|^2$).

4.3 Motional squeezed states

Consider both ions interacting with two laser beams which are tuned to the second red and blue CM vibrational sidebands, that is $k = 2$, $\omega_I = \omega_o - 2\nu$ and $\omega_{II} = \omega_o + 2\nu$. Now, we choose the laser beams phases so as to have $\Omega_I = -\Omega_{II}^* = \Omega$ and consider the Lamb-Dicke regime by taking $\{\eta\sqrt{\bar{n}}, \eta_r\sqrt{\bar{n}_r}\} \ll 1$. In this way, following similar steps as in Section 2, we may deduce the effective interaction Hamiltonian

$$\widehat{H} = \hbar(\widehat{S}_{+1} + \widehat{S}_{+2} - \widehat{S}_{-1} - \widehat{S}_{-2})(\beta\hat{a}^2 - \beta^*\hat{a}^{\dagger 2}) = \hbar\widehat{H}_{\mathrm{e}}(\beta\hat{a}^2 - \beta^*\hat{a}^{\dagger 2}) \quad (33)$$

with $\beta = -\eta^2\Omega$. An initial state $|\psi\rangle = |\Phi_{\mathrm{e}}\rangle \otimes |\Psi_{\mathrm{vib}}\rangle$ will evolve, after an interaction time τ, to the state $|\psi(\tau)\rangle = \sum_j c_j|\Phi_j\rangle \otimes \widehat{S}(\lambda_j\beta\tau)|\Psi_{\mathrm{vib}}\rangle$, where $\widehat{S}(\xi) \equiv e^{\xi\hat{a}^2 - \xi^*\hat{a}^{\dagger 2}}$ is the squeezing operator. Note that, similar to what happened with the displacement operator in subsection 4.1, the squeezing of the motional degrees of freedom is conditioned to the initial electronic state. Naturally, following the steps of subsection 4.2, we expect to be able to generate coherent superpositions of two different squeezed states.

More applications and a generalization to N trapped ions is discussed in Ref. [10].

5 Conclusions

We presented realistic proposals for producing nonclassical states, their measurement, and schemes of quantum communication in two trapped ions, illuminating them with bichromatic light. We wanted to show that due to its fundamental interest and in the search of scalability in the trapped ions system, the study of two trapped ions has a fundamental place.

6 Acknowledgments

This work was partially supported by the Conselho Nacional de Desenvolvimento Científico e Tecnológico (CNPq), Fundação de Amparo à Pesquisa do Estado do Rio de Janeiro (FAPERJ), Fundação Universitária José Bonifácio (FUJB) and the Programa de Apoio a Núcleos de Excelência (PRONEX).

References

1. D. J. Wineland, C. R. Monroe, W. M. Itano, D. Leibfried, B. E. King, and D. M. Meekhof, NIST J. Research **103** (3), 259 (1998) (Available at http://nvl.nist.gov/pub/nistpubs/jres/jres.htm); D. J. Wineland, C. Monroe, W. M. Itano, D. Kielpinski, B. E. King, C. J. Myatt, Q. A. Turchette, and C. S. Wood, to appear in Proc. 6th Int. Symp. on Foundations of Quantum Mechanics in the Light of New Technology, K. Fujikawa, and Y. A. Ono, eds. (Elsevier Science press).
2. S. Haroche and J. M. Raimond, in *Advances in Atomic an Molecular Physics vol. XX*, edited by D. R. Bates and B. Bederson (Academic, New York, 1985).
3. Q. A. Turchette, C. S. Wood, B. E. King, C. J. Myatt, D. Leibfried, W. M. Itano, C. Monroe, and D. J. Wineland, Phys. Rev. Lett. **81**, 3631 (1998).
4. A. Sørensen and K. Mølmer, Phys. Rev. Lett. **82**, 1971 (1999).
5. E. Solano, R. L. de Matos Filho, and N. Zagury, Phys. Rev. A, **59**, R2539 (1999); **61**, 029903(E) (2000).
6. E. Solano, C.L. Cesar, R. L. de Matos Filho, and N. Zagury, accepted for publication in Eur. Phys. J. D (availabe in quant-ph/9903029).
7. E. Solano, P. Milman, R. L. de Matos Filho, and N. Zagury, Phys. Rev. A **62**, 021401(R) (2000).
8. M. França Santos, R. L. de Matos Filho, and E. Solano. Submitted for publication.
9. K. Mølmer and A. Sørensen, Phys. Rev. Lett. **82**, 1835 (1999).
10. E. Solano, R. L. de Matos Filho, and N. Zagury. Submitted for publication.
11. E. Solano, M. França Santos, and P. Milman. Submitted for publication.
12. B. E. King, C. S. Wood, C. J. Myatt, Q. A. Turchette, D. Leibfried, W. M. Itano, C. Monroe, and D. J. Wineland, Phys. Rev. Lett. **81**, 1525 (1998).
13. D. F. V. James, Appl. Phys. B **66**, 181 (1998).
14. D. Leibfried, Phys. Rev. A **60**, R3335 (1999).
15. H.C. Nägerl, D. Leibfried, H. Rohde, G. Thalhammer, J. Eschner, F. Schmidt-Kaler, and R. Blatt, Phys. Rev. A **60**, 145 (1999).
16. W. Vogel and R.L. de Matos Filho, Phys. Rev. A**52**, 4214 (1995).
17. C. H. Bennett, G. Brassard, C. Crépeau, R. Jozsa, A. Peres, and W. Wootters, Phys. Rev. Lett. **70**, 1895 (1993).
18. L. Vaidman and N. Yoran, Phys. Rev. A **59**, 116 (1999).
19. N. Lütkenhaus, J. Calsamiglia, and K.-A. Suominen, Phys. Rev. A **59**, 3295 (1999).
20. L. Davidovich, N. Zagury, M. Brune, J. M. Raimond, and S. Haroche, Phys. Rev. A **50**, R895 (1994).
21. Yoon-Ho Kim, Sergei P. Kulik, and Yanhua Shih, e-print quant-ph/0010046 (2000).
22. S. Wallentowitz, R. L. de Matos Filho, S.-C. Gou and W. Vogel, Eur. Phys. J. D **6**, 97 (1999).
23. D. Leibfried, D. M. Meekhof, B. E. King, C. Monroe, W. M. Itano, and D. J. Wineland, Phys. Rev. Lett. **77**, 4281 (1996).
24. D. M. Meekhof, C. Monroe, B. E. King, W. M. Itano, and D. J. Wineland, Phys. Rev. Lett. **76**, 1796 (1996).
25. W. Nagourney, J. Sandberg, and H. Dehmelt, Phys. Rev. Lett. **56**, 2797 (1986).

Quantum Nondemolition Measurement and Quantum State Manipulation in Two Dimensional Trapped Ion

W. Kaige[1], S. Maniscalco[2], A. Napoli[2], and A. Messina[2]

[1] CCAST (World Lab.), P.O. Box, 8730, Beijing 100080; and Department of Physics, Beijing Normal University, Beijing 100875, China
[2] INFM, MURST and Dipartimento di Scienze Fisiche ed Astronomichedell' Università di Palermo, Palermo 90123, Italy

Abstract. An extension of QND measurement of the vibrational energy of the trapped ion from one dimensional case to the bidimensional one is presented. Our approach exploits the fixed phase difference existing between the two orthogonal and appropriately configured classical laser beams determining the vibronic coupling. We in fact show that this phase difference may play the role of an adjustable external parameter which allows to optimize the measurement scheme itself in terms of both precision and sensitivity. Our proposal provides a cooling method for the trapped ion from the vibrational thermal state. Due to the coherent superposition of two sub Rabi oscillations, the Rabi frequency degeneration and offset may occur in this model. This provides a possibility to generate various entangled superposition of Fock states, pair coherent state and entangled pair coherent states.

1 Introduction

It is well know that quantum measurement disturbs the system in a rather unpredictable manner. The process of measurement of an observable may introduce noise so that the outcomes of further measurements of the same quantity will become more or less unpredictable (feedback action). It is thus desirable to construct quantum non-demolition (QND) measurement procedures allowing one to avoid the contamination of the measured physical variable. These schemes have to be designed in such a way that the results of successive measurements of a given observable become predictable. In order to reach this goal, the existence of a QND variable such that its free evolution is not contaminated by uncertainties in its conjugate observable, is required. Thus, for instance, a system initially prepared in an eigenstate of the QND variable remains in that eigenstate for all time during its free evolution. Moreover the measurement should be arranged in an indirect way, that is, the "signal" and the "meter" observables must be coupled by an interaction mechanism. The detection of the "meter" will give information about the "signal" without perturbing it. It has been proved that when the signal variable commutes with both the free and the interaction Hamiltonian, the measurement is QND type [1]. In mid of 1970s, QND measurement schemes have been proposed to detect gravitational wave [2]. Since then, QND

M. Orszag, J.C. Retamal (Eds.): LNP 575, pp. 29–42, 2001.

procedures have been presented and experimentally demonstrated in the optical domain [3,4]. More recently, laser cooling techniques have been successfully applied to trapped ions allowing to cool them down near their zero point vibrational energy [5–8]. In this regime the ionic center of mass motion is quantized as harmonic oscillator and can be coupled to internal ionic degrees of freedom by classical laser beams. It has been shown that, by appropriately choosing the driving lasers, it is possible to realize a huge class of interaction mechanisms [9,10]. It is moreover of particular relevance that, under appropriate conditions, the quantum dynamics of such systems may be deduced from Hamiltonian models formally equivalent to that describing the interaction of single two level atom coupled to few modes of a high-Q cavity field. This prominent feature directly leads to the possibility of testing the rich dynamics predicted in the realm of CQED by using trapped ions [11–15]. One of the advantages of exploiting such systems is related to the circumstance that typical dissipative effects limiting the performance of experiments in cavities, can be significantly suppressed for the ionic motion.

Quite recently, a QND scheme for measuring the vibrational energy of a one-dimensionally (1D) trapped ion has been proposed [16,17]. This scheme follows the original QND idea. In this case the "signal" QND observable coincides with the ionic vibrational energy whereas the occupation probability of the ionic internal levels is the "meter". When the driving laser beam is exactly resonant with the electronic transition (i. e. carrier resonance), the vibronic effective interaction is described by a nonlinear Jaynes–Cummings model (JCM) [18]. This system provides an example of a QND interaction since its Hamiltonian commutes with the signal observable, i.e. the vibrational number operator. In the case of a 2D trapped ion an analogous method is applied to measure the Wigner function matrix [19].

In this paper we study QND measurements of vibrational energy and quantum state manipulation via QND schemes in a 2D model of trapped ion proposed in [16,17,19,20]. In principle one might extend directly the 1D method to two or even three dimensions. In fact, since the quantized vibrational motions along the x, y and z axis are decoupled, one could perform the 1D measurement sequentially along the three directions. On the contrary, we use two laser beams along the x and y directions simultaneously and coherently. This means that the two lasers have to keep a fixed phase difference so, in practice, in the experiments one can use a single laser source. We show for the first time that, due to the coherence between the two driving beams, the atomic Rabi oscillation in the bidimensional case is the coherent superposition of two sub-Rabi oscillations along the x and y directions and thus can be varied by changing the phase difference ϕ between the two laser beams. In particular we find that ϕ plays the role of an adjustable external parameter which allows to optimize the measurement scheme, in terms of both precision and sensitivity. Similarly to the one dimensional case, our proposal leads to interesting applications as, for example, cooling and preparation of arbitrary bimodal Fock states. Furthermore we have brought to light two new effects which are peculiar of our 2D model: the Rabi frequency degeneration and the offset of Rabi oscillation. These two effects come strongly into play in that part of a QND measurement scheme known as quantum state manipulation (QSM). Moreover, such effects make it possible to generate, with practical feasibility, various superpositions of bimodal Fock states, pair coherent states and superpositions of pair coherent states.

2 Description of the Model

We consider a two-level ion of mass m confined in a 2D harmonic potential [19,20]. If the ion is irradiated by two classical laser beams propagating along the x and y directions, the Hamiltonian model describing the ionic dynamics can be written as follows

$$\begin{aligned} H &= H_0 + H_I \\ H_0 &= \hbar\nu_x a_x^\dagger a_x + \hbar\nu_y a_y^\dagger a_y + \frac{1}{2}\hbar\omega_0\sigma_z \\ H_I &= d\epsilon^{(-)}(\widehat{x},\widehat{y},t)\sigma_- + \text{h.c.} \ , \end{aligned} \tag{1}$$

where ν_x and ν_y are the vibrational frequencies in x and y directions, respectively. $a_i^\dagger$ and a_i $(i = x, y)$ are the boson operators whereas σ_z and σ_- describe the internal degrees of freedom. The position operators $\widehat{x}$ and $\widehat{y}$ are defined as

$$\widehat{x} = \sqrt{\hbar/(2\nu_x m)}(a_x^\dagger + a_x) \ , \qquad \widehat{y} = \sqrt{\hbar/(2\nu_y m)}(a_y^\dagger + a_y) \ . \tag{2}$$

In (3) ω_0 is the electronic transition frequency, and d is the appropriate dipole matrix element assumed real for simplicity. The driving field is

$$\epsilon^{(-)}(\widehat{x},\widehat{y},t) = E_x \exp[\mathrm{i}(\omega_L t - k\widehat{x})] + E_y \exp[\mathrm{i}(\omega_L t - k\widehat{y} + \phi)] \tag{3}$$

where ϕ is the phase difference between the two laser beams and $\omega_L \simeq \omega_0$ the lasers frequency.

We set the carrier resonance $\omega_L = \omega_0$ and expand the two exponential terms appearing in (5) as

$$\exp(-\mathrm{i}k\widehat{j}) = 1 - \mathrm{i}k\widehat{j} + \frac{1}{2}(-\mathrm{i}k\widehat{j})^2 + \cdots \approx 1 - \frac{1}{2}\eta_j^2 - \eta_j^2 a_j^\dagger a_j \ , \qquad j = x, y \tag{4}$$

where the Lamb–Dicke parameters are defined as

$$\eta_j \equiv k\sqrt{\hbar/(2\nu_j m)} \ . \tag{5}$$

In (6) we have applied two approximations: the Lamb–Dicke limit $\eta_j \ll 1$ and the weak coupling regime (i.e. the vibration frequencies are much larger than the Rabi frequency). For the latter, one may neglect all the non-resonant terms such as $a_j, a_j^\dagger, a_j^2$ and $(a_j^\dagger)^2$ in (6). Therefore in the interaction picture the Hamiltonian of the system can be written as follows

$$H_I = \hbar\Omega(n_x, n_y, \phi)\mathrm{e}^{\mathrm{i}\Theta(n_x,n_y,\phi)}\sigma_- + \text{h.c.} \tag{6}$$

where

$$\begin{aligned}\Omega(n_x, n_y, \phi)\mathrm{e}^{\mathrm{i}\Theta(n_x,n_y,\phi)} &\equiv \Omega_x(\eta_x, n_x) + \Omega_y(\eta_y, n_y)\mathrm{e}^{\mathrm{i}\phi} \qquad (7)\\ &\equiv [\Omega_{Lx}(1 - \frac{1}{2}\eta_x^2 - \eta_x^2 a_x^\dagger a_x) + \Omega_{Ly}\mathrm{e}^{\mathrm{i}\phi}(1 - \frac{1}{2}\eta_y^2 - \eta_y^2 a_y^\dagger a_y)]\,.\end{aligned}$$

In (9) the laser Rabi frequencies are defined as $\Omega_{Li} \equiv dE_i/\hbar, (i = x, y)$. We wish to underline that, as shown in [16,17,19], for the purpose of QND measurement the Lamb–Dicke limit is not necessary. We will however show that when this limit is applied, the simple degeneration relations (linear and quadratic) of the Rabi frequency with respect to the vibrational quanta occur. With the aid of these degeneration relations, one may produce useful entanglement of bimodal vibrational states, as will be shown in Sects. 3 and 5.

The nonlinear JCM described by (8) can be exactly solved. Consider, in fact, an initial state of the form $|n_x, n_y, g\rangle$ or $|n_x, n_y, e\rangle$ where $|n_x, n_y\rangle$ is the vibrational bimodal Fock state and g and e indicate the ground and excited electronic states, respectively. The time evolution operator can be written as follows

$$\begin{aligned}U(t) &= \exp(-\mathrm{i}H_I t/\hbar) \qquad (8)\\ &= \cos[\Omega(n_x, n_y, \phi)t] - \mathrm{i}\sin[\Omega(n_x, n_y, \phi)t](\mathrm{e}^{\mathrm{i}\Theta(n_x,n_y,\phi)}\sigma_- + \mathrm{e}^{-\mathrm{i}\Theta(n_x,n_y,\phi)}\sigma_+)\,.\end{aligned}$$

Therefore the wavefunction evolutions for the initial states $|n_x, n_y, g(e)\rangle$ are respectively

$$\begin{aligned}\Psi_g\,(t) &= U(t)|n_x, n_y, g\rangle \qquad (9a)\\ &= \cos[\Omega(n_x, n_y, \phi)t]|n_x, n_y, g\rangle - \mathrm{i}\sin[\Omega(n_x, n_y, \phi)t]\mathrm{e}^{-\mathrm{i}\Theta(n_x,n_y,\phi)}|n_x, n_y, e\rangle\,,\\ \Psi_e\,(t) &= U(t)|n_x, n_y, e\rangle \qquad (9a)\\ &= \cos[\Omega(n_x, n_y, \phi)t]|n_x, n_y, e\rangle - \mathrm{i}\sin[\Omega(n_x, n_y, \phi)t]\mathrm{e}^{\mathrm{i}\Theta(n_x,n_y,\phi)}|n_x, n_y, g\rangle\,.\end{aligned}$$

3 Properties of the Model

We observe that the operators $a_x^\dagger a_x$ and $a_y^\dagger a_y$ commute with both the free Hamiltonian H_0 and the interaction Hamiltonian H_I. As a consequence, Rabi oscillations of the electronic occupation occur without changing the vibrational state, as shown in (9). This circumstance is of particular relevance in view of the fact that detections of the electronic states of the ion do not disturb its vibrational state. However, the atomic Rabi frequency $\Omega(n_x, n_y, \phi)$, which can be measured in experiment, includes the information about the vibrational quanta n_x and n_y.

According to (9), the Rabi oscillations experienced by the system under scrutiny can be considered as composed by two sub-Rabi oscillations at frequencies
$\Omega_x(\eta_x, n_x)$ and $\Omega_y(\eta_y, n_y)$. It is convenient to recast (9) into the modulus and the phase

$$\Omega(n_x, n_y, \phi) = [\Omega_x(\eta_x, n_x)^2 + \Omega_y(\eta_y, n_y)^2 + 2\Omega_x(\eta_x, n_x)\Omega_y(\eta_y, n_y)\cos\phi]^{1/2}, \tag{10}$$

$$\tan\Theta(n_x, n_y, \phi) = \frac{\Omega_y(\eta_y, n_y)\sin\phi}{\Omega_x(\eta_x, n_x) + \Omega_y(\eta_y, n_y)\cos\phi}. \tag{11}$$

Equation (11) clearly shows that the phase difference ϕ of the two driving beams may enhance or reduce the atomic Rabi frequency. In what follows we consider three special cases correspondent to $\phi = 0, \pi$, and $\pi/2$, respectively.
(i) in phase laser beams, $\phi = 0$

$$\begin{aligned} \Omega(n_x, n_y, 0) &= \Omega_x(\eta_x, n_x) + \Omega_y(\eta_y, n_y) \\ &= \Omega_{Lx}(1 - \frac{1}{2}\eta_x^2 - \eta_x^2 n_x) + \Omega_{Ly}(1 - \frac{1}{2}\eta_y^2 - \eta_y^2 n_y), \\ \Theta(n_x, n_y, 0) &= 0. \end{aligned} \tag{12}$$

(ii) out of phase laser beams, $\phi = \pi$

$$\begin{aligned} \Omega(n_x, n_y, \pi) &= |\Omega_x(\eta_x, n_x) - \Omega_y(\eta_y, n_y)| \\ &= |\Omega_{Lx}(1 - \frac{1}{2}\eta_x^2 - \eta_x^2 n_x) - \Omega_{Ly}(1 - \frac{1}{2}\eta_y^2 - \eta_y^2 n_y)|, \\ \Theta(n_x, n_y, \pi) &= 0. \end{aligned} \tag{13}$$

(iii) $\pi/2$ out of phase laser beams, $\phi = \pi/2$

$$\begin{aligned} \Omega(n_x, n_y, \pi/2) &= [\Omega_x(\eta_x, n_x)^2 + \Omega_y(\eta_y, n_y)^2]^{1/2} \\ &= [\Omega_{Lx}^2(1 - \frac{1}{2}\eta_x^2 - \eta_x^2 n_x)^2 + \Omega_{Ly}^2(1 - \frac{1}{2}\eta_y^2 - \eta_y^2 n_y)^2]^{1/2}, \\ \Theta(n_x, n_y, \pi/2) &= \arctan\left[\frac{\Omega_y(\eta_y, n_y)}{\Omega_x(\eta_x, n_x)}\right] = \arctan\left[\frac{\Omega_{Ly}(1 - \eta_y^2/2 - \eta_y^2 n_y)}{\Omega_{Lx}(1 - \eta_x^2/2 - \eta_x^2 n_x)}\right]. \end{aligned} \tag{14}$$

In the cases of $\phi = 0$ and $\phi = \pi$, one may obtain the maximum and the minimum Rabi frequencies, respectively.

At this point we wish to stress that, in the 2D trapped ion system, two special effects arise which cannot appear in 1D case: the Rabi frequency degeneration and the offset of Rabi oscillation.

The Rabi frequency degeneration happens when different bimodal Fock states $|n_x, n_y\rangle$ share the same Rabi frequency, i. e. $\Omega(n_x, n_y, \phi) = \Omega(m_x, m_y, \phi)$. The occurrence of this degeneration depends on the Lamb–Dicke parameters and on the strength of the driving laser beams. Fixing these parameters we may plot, in the plane (n_x, n_y), a 'degeneration curve' in correspondence to a given Rabi frequency. This curve allows us to find the existence of Fock states sharing the same Rabi frequency. Figures 1(a), (b) and (c) show respectively the 'degeneration' curves obtained for the three cases of $\phi = 0, \pi$, and $\pi/2$. As we mentioned above, these simple linear and quadratic curves are related to the Lamb–Dicke limit.

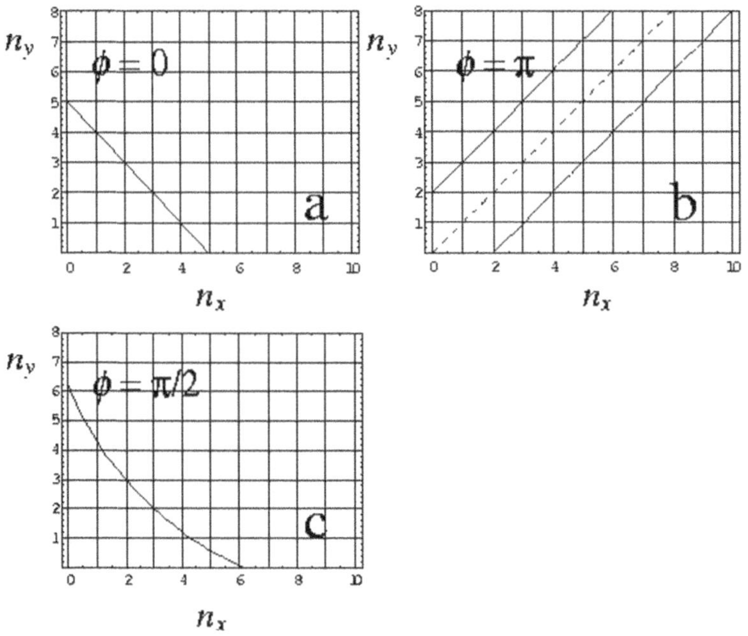

Fig. 1. Schemes of the degeneration curves in the space (n_x, n_y) with $\Omega_{Lx} = \Omega_{Ly}$ and $\eta_x = \eta_y$ for **(a)** $\phi = 0$, **(b)** $\phi = \pi$, and **(c)** $\phi = \pi/2$. Dashed line in **(b)** indicates the offset of Rabi oscillations

In the next section we will see that, in order to perform a QND measurement of the number of vibrational quanta along the x and y directions, one should avoid the Rabi frequency degeneration by choosing the parameters appropriately. On the other hand, we will show in Sect. 5 that the Rabi frequency degeneration provides a possibility for quantum state manipulation via the QND measurement.

We now turn our attention on the second effect mentioned above, namely the offset of Rabi oscillation. This effect occurs only in the case $\phi = \pi$ when, according to (14), the Rabi frequency vanishes. In Fig. 1(b), the offset effect is indicated by the dashed line just in the middle of the two degeneration lines. This line indicates the existence of Fock states in correspondence of which the Rabi frequency vanishes. It is interesting to note that if the ion is initially prepared in one of these states, it will be 'trapped' in the excited or ground level at any time instant t.

4 QND Measurement of Vibrational Quanta

We have already mentioned that the atomic Rabi frequency $\Omega(n_x, n_y, \phi)$ is related to the number of vibrational quanta n_x and n_y, so that one may get information about the c.m. oscillatory motion by measuring the time evolution of the electronic occupation probability. Measurements of the internal ionic state are currently performed in the experiments and they are based on the so called quantum jumps technique. The method essentially consists in coupling the ground electronic level $|g\rangle$ to an auxiliary level $|r\rangle$ by means a third laser beam. The transition $|g\rangle - |r\rangle$ is chosen to be a dipole allowed one. Under these conditions, the presence of fluorescence detects the ion in the electronic ground level whereas its absence in the excited level. In other words, the detection of null fluorescence prepares the ion in the excited state since after the measurement the quantum system collapses in the eigenstate correspondent to the measured outcome.

Suppose now to prepare the ion in the state $|m_x, m_y, g\rangle$, according to (9a), after an interaction time t_1 satisfying the condition

$$\Omega(m_x, m_y, \phi)t_1 = (2l_1 + 1)\pi/2, \qquad l_1 = 0, 1, 2, \cdots, \tag{15}$$

the state of the system will coincide with $|m_x, m_y, e\rangle$. In other words, a detection of the electronic state at $t = t_1$ will give null fluorescence. On the other hand, (9a) shows that successive null fluorescence detections should occur after the interaction time intervals t_k $(k = 2, 3, \cdots)$ defined by

$$\Omega(m_x, m_y, \phi)t_k = l_k\pi, \qquad l_k = 1, 2, 3, \cdots \tag{16}$$

while the ion is again in the excited state after the measurement. If, on the contrary, in any of these detections a fluorescence is seen, one may change the detection time to continue the measurement. Therefore, by measuring the Rabi frequency one may determine the vibrational numbers n_x and n_y. In presence of Rabi frequency degeneration it is necessary to change some parameters (for instance, the laser intensity) and to perform the measurement again. The cross point of the two lines corresponding to the two different sets of fixed parameters, in the plane (n_x, n_y), identifies uniquely the detected Fock state.

In principle, in the experimental measurement the phase ϕ can be arbitrarily chosen. However, it is interesting to observe that for $\phi = \pi$, both the precision and the sensitivity of measurement can be greatly improved. To better understand this point we consider for simplicity the case of two laser beams having equal intensity $(\Omega_{Lx} = \Omega_{Ly} = \Omega_L)$ and compare the two different cases obtained in correspondence to $\phi = 0$ and $\phi = \pi$, respectively. In the case $\phi = 0$, the fluctuation of the Rabi frequency induced by the noise of laser beams can be written as

$$\Delta\Omega(n_x, n_y, 0) = 2\Delta\Omega_L - \Delta\Omega_L(\eta_x^2/2 + \eta_y^2/2 + \eta_x^2 n_x + \eta_y^2 n_y) \tag{17}$$

where $\Delta\Omega_L$ is the fluctuation of the parameter Ω_L due to the laser intensity imprecision. In the Lamb-Dicke limit, it is immediate to realize that the contribution affecting the most the precision of measurement is the first term appearing in (18). Moreover, we define the sensitivity of the measurement of vibrational quanta n_x as

$$S_x(\phi) \equiv \frac{|\Omega(n_x \pm 1, n_y, \phi) - \Omega(n_x, n_y, \phi)|}{\Omega(n_x, n_y, \phi)} \,, \tag{18}$$

it is easy to deduce that $S_x(0) \approx \eta_x^2/2$. On the contrary, in the case $\phi = \pi$, the Rabi frequency fluctuation is given by

$$\Delta\Omega(n_x, n_y, \pi) = \Delta\Omega_L|\eta_x^2/2 - \eta_y^2/2 + \eta_x^2 n_x - \eta_y^2 n_y| \tag{19}$$

and the sensitivity,

$$S_x(\pi) = \eta_x^2/|\eta_x^2/2 - \eta_y^2/2 + \eta_x^2 n_x - \eta_y^2 n_y| \,. \tag{20}$$

Comparing (18) and $S_x(0)$ with (20) and (21), respectively, we have $\Delta\Omega(n_x, n_y, \pi) \ll \Delta\Omega(n_x, n_y, 0)$ and $S_x(\pi) \gg S_x(0)$. We may thus state that the growth of both the precision and the sensitivity of the measurement in correspondence to the choice $\phi = \pi$ is due to the coherent counteracting for the parts of the laser beams which do not contribute to the measurement.

5 Quantum State Manipulation

In the last section, we have assumed that the vibrational state at $t = 0$ is a bimodal Fock state. In most cases, this initial condition is not so clear and needs to be proved in the experiment. In general, QND measurement should prepare the quantum system into the state required for the measurement itself. This preliminary step of a QND scheme is called quantum state preparation (QSP). This consideration puts clearly into evidence that the QND procedure to measure the vibrational numbers n_x and n_y can be usefully exploited for engineering arbitrary bimodal Fock states.

5.1 Generation of a bimodal Fock state

Suppose now that at $t = 0$ the ion is prepared in its electronic ground state whereas the vibrational motion is in a generic superposition of two-mode Fock states (pure or mixed), (9a) and (9a) suggest a simple protocol to generate a prefixed bimodal Fock state $|m_x, m_y\rangle$ provided that the probability of finding at

$t = 0$ the center of mass motion in the state $|m_x, m_y\rangle$ is different from zero. In order to generate this state, we imagine to measure several times the internal electronic state at time instants t_i appropriately chosen.

The first measurement is performed at the time t_1 defined by (16). If null fluorescence is detected, according to (9a), the vibrational state $|m_x, m_y\rangle$ keeps the original probability unchanged whereas the other components present at $t = 0$ decrease their probabilities. In other words, after the renormalization of the state, the probability of the state $|m_x, m_y\rangle$ is increased with respect to the other two-mode Fock states. This process can be obviously iterated by detecting the null fluorescence at time instants t_k $(k = 2, 3, \cdots)$ defined in (17). It is easy to convince oneself that the prefixed state $|m_x, m_y\rangle$ can be generated.

Figure 2 shows the steps which lead to the vacuum state $|0, 0\rangle$ by starting from the two-mode thermal state with an equal average intensity $\langle n_x\rangle = \langle n_y\rangle = 1.0$. In this case we have set $\phi = \pi$, $\eta_x = 0.07$, $\eta_y = 0.09$, and $\Omega_{Ly}/\Omega_{Lx} = 0.99$. After only three successfully interaction-detection cycles a bimodal vacuum state is generated with more than 88% probability, and then after the fifth step, 98%

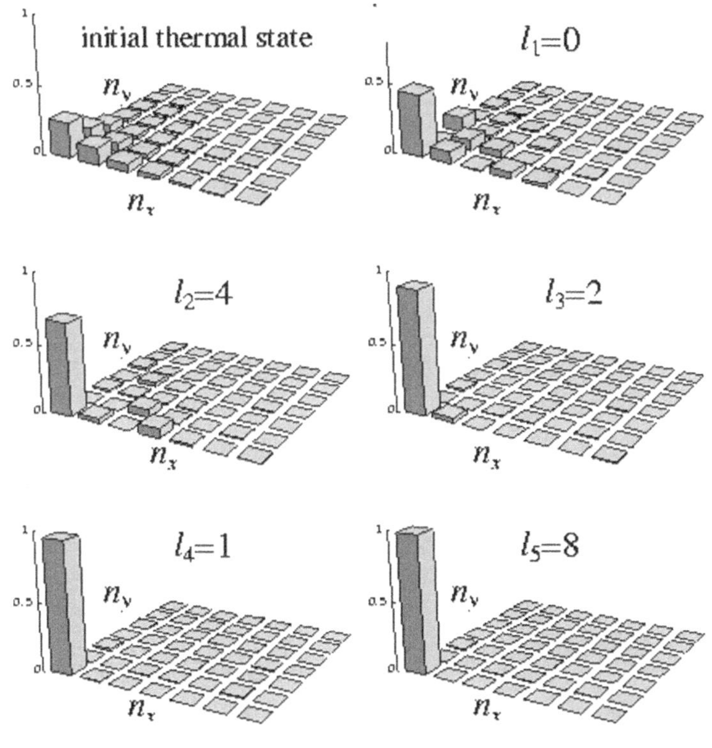

Fig. 2. Probability distributions in the process of generating the vacuum state $|0, 0\rangle$ from the initial thermal state. l_1 corresponds to the interaction time defined by (15), and $l_2 - l_5$, by (16). (These l_i have the same meaning for Figs. 3 and 4)

probability. This example shows that, as in 1D case, our model can be thought of as an useful cooling method.

Due to the periodicity of the Rabi oscillation we can choose different values of the integer number l_k defined by (17) and, consequently, different sequences of the interaction times t_k. Of course an appropriate choice of l_k is requested in order to make the generation process faster and then more effective. Regarding the condition $\phi = \pi$ imposed in Fig. 2, we stress again that this choice makes the detection sensitivity higher and thus it may suppress the unwanted states more efficiently. Finally we remind that for generating bimodal Fock states, the Rabi frequency degeneration should be avoided by changing properly the laser parameters. On the contrary the Rabi frequency degeneration provides the possibility of generating a class of superpositions of bimodal Fock states, as we will show in the next subsection.

5.2 Generation of entangled superposition of Fock states

In order to exploit QND measurement in a 2D trapped ion for engineering entangled Fock state, we should focus our attention on two fundamental aspects: (i) the initial pure state must contain the state we wish to generate; (ii) in correspondence to the Fock states to be superimposed, degeneration of Rabi frequency is required. The experimental procedure is the same as described above for generating a Fock state. After several successful null fluorescence detections, all the unwanted states are greatly suppressed and the remaining states on the degenerate line keep the original entangled relation unchanged. In the cases of $\phi = 0$ and π, indeed, because the condition $\Theta \equiv 0$ is independent of any state, the phase relation between the states to be entangled will not be changed by the measurement. Even for the cases $\phi \neq 0$ and π, Θ can be considered as approximately independent of n_x and n_y in view of the Lamb–Dicke limit (see (11) and (9)). In what follows we show some examples of entangled Fock state generation.
(i) Entangled superposition of all the Fock states with the equal number of total vibrational quanta N

When $\phi = 0$ and $\Omega_{Lx}\eta_x^2 = \Omega_{Ly}\eta_y^2$, the Rabi degeneration condition is simple as $n_x + n_y = N$. The degeneration line has been shown in Fig. 1a. Following the detection steps described above, one may generate entangled states such as $c_1|0,1\rangle + c_2|1,0\rangle$, and $c_1|0,2\rangle + c_2|1,1\rangle + c_3|2,0\rangle, \cdots$ etc. As an example, Fig. 3 shows procedure in which the entangled state $(1/\sqrt{2})[|1,0\rangle + \mathrm{e}^{\mathrm{i}\theta}|0,1\rangle]$ is generated from the initial coherent state $|\alpha, \alpha\mathrm{e}^{\mathrm{i}\theta}\rangle$ with $|\alpha|^2 = 2.5$. In this example we have set $\eta_x = \eta_y = 0.07$, and $\Omega_{Lx} = \Omega_{Ly}$.
(ii)Entangled superposition of Fock states associated with (i)

After (i), an entangled superposition of the Fock states with an equal total vibrational quanta has been generated. Then, we may use the condition $\phi = \pi$ to erase further some Fock states from the generated state. When $\Omega_{Lx} = \Omega_{Ly} \equiv \Omega_L$ and $\eta_x = \eta_y \equiv \eta$, the Rabi frequency for a particular number difference $q = |n_x - n_y|$ becomes

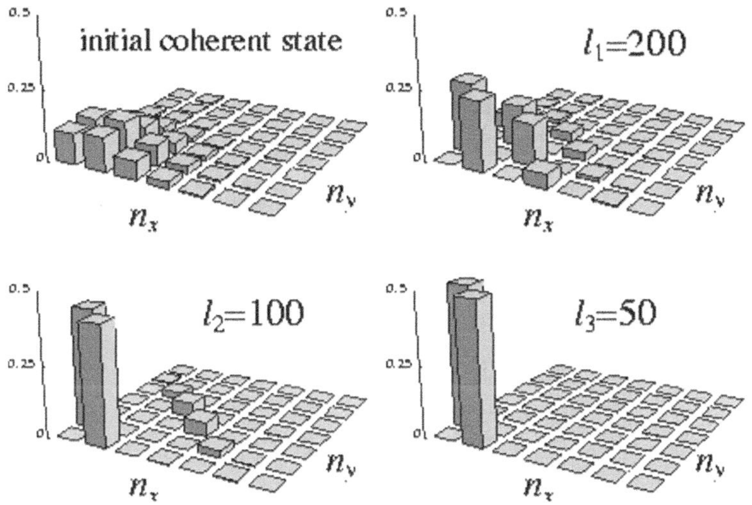

Fig. 3. Probability distributions in the process of generating the entangled superposition of Fock states $(1/\sqrt{2})[|1,0\rangle + e^{i\theta}|0,1\rangle]$ from the initial coherent state $|\alpha, \alpha e^{i\theta}\rangle$. Symmetric parameters $\Omega_{Lx} = \Omega_{Ly}$ and $\eta_x = \eta_y = 0.07$ are taken

$$\Omega(n_x, n_y, \pi) = \Omega_L \eta^2 |n_x - n_y| = \Omega_L \eta^2 q \equiv \Omega_q \, . \tag{21}$$

In this case, for example, after the generation of the entangled state $c_1|4,0\rangle + c_2|3,1\rangle + c_3|2,2\rangle + c_4|1,3\rangle + c_5|4,0\rangle]$, we may suppress further the states $|1,3\rangle$ and $|3,1\rangle$ in only once successful detection by applying the interaction time $t = \pi/\Omega_{q=4}$. Indeed, the entangled state $c_1|4,0\rangle + c_3|2,2\rangle + c_5|4,0\rangle$ is remained. We emphasize that the presence of the state $|2,2\rangle$ in this entangled state is a consequence of the offset of Rabi frequency.

(iii) Entangled state having the form $c_1|m,n\rangle + c_2|n,m\rangle$

In order to generate an entangled state having the form $c_1|m,n\rangle + c_2|n,m\rangle$, we put $\phi = \pi/2$, and the symmetric parameters $\Omega_{Lx} = \Omega_{Ly} \equiv \Omega_L$ and $\eta_x = \eta_y \equiv \eta$. The Rabi degeneration condition is now

$$(1 - \eta^2/2 - \eta^2 n_x)^2 + (1 - \eta^2/2 - \eta^2 n_y)^2 = \text{const.} \, . \tag{22}$$

As shown in Fig. 1(c), the degenerate curve is a circle instead of an ellipse, and the states $|m,n\rangle$ and $|n,m\rangle$ simultaneously satisfy condition (23). Following this method, we may generate any bimodal entangled state $1/\sqrt{2}[|m,n\rangle + e^{i(m-n)\delta}|n,m\rangle]$ from the initial coherent state $|\alpha, \alpha e^{i\delta}\rangle$.

5.3 Generation of a pair coherent state

The pair coherent state (PCS) [21-23] is defined as an eigenstate of the pair annihilation operator $a_x a_y$, such that

$$a_x a_y |\varsigma, q\rangle_{\text{pcs}} = \varsigma |\varsigma, q\rangle_{\text{pcs}}, \qquad (a_x^\dagger a_x - a_y^\dagger a_y)|\varsigma, q\rangle_{\text{pcs}} = q|\varsigma, q\rangle_{\text{pcs}} \, . \tag{23}$$

In the Fock state representation it is expanded as

$$|\varsigma, q\rangle_{\rm pcs} = N_q(|\varsigma|) \sum_{n=0}^{\infty} \frac{\varsigma^n}{\sqrt{n!(n+q)!}} |n+q, n\rangle \tag{24a}$$

$$N_q(|\varsigma|) = [|\varsigma|^{-q} J_q(2|\varsigma|)]^{-1/2} \tag{24b}$$

where J_q is the modified Bessel function of first kind of order q. For our purpose, we deduce that a two-mode coherent state can be written as a superposition of the pair coherent states over all q

$$\begin{aligned} |\alpha, \beta\rangle &= \mathrm{e}^{-\frac{1}{2}(|\alpha|^2+|\beta|^2)} \sum_{m,n=0}^{\infty} \frac{\alpha^m \beta^n}{\sqrt{m!n!}} |m, n\rangle \\ &= \mathrm{e}^{-\frac{1}{2}(|\alpha|^2+|\beta|^2)} \left\{ \sum_{n=0}^{\infty} \frac{(\alpha\beta)^n}{n!} |n, n\rangle \right. \\ &\quad \left. + \sum_{n=0,q=1}^{\infty} \left(\frac{\alpha^{n+q}\beta^n}{\sqrt{n!(n+q)!}} |n+q, n\rangle + \frac{\alpha^n \beta^{n+q}}{\sqrt{n!(n+q)!}} |n, n+q\rangle \right) \right\} \\ &= \mathrm{e}^{-\frac{1}{2}(|\alpha|^2+|\beta|^2)} \left\{ [N_0(|\alpha\beta|)]^{-1} |\alpha\beta, 0\rangle_{\rm pcs} \right. \\ &\quad \left. + \sum_{q=1}^{\infty} [N_q(|\alpha\beta|)]^{-1} \left[\alpha^q |\alpha\beta, q\rangle_{\rm pcs} + \beta^q |\alpha\beta, -q\rangle_{\rm pcs} \right] \right\} . \end{aligned} \tag{25}$$

In order to generate a pair coherent state, we may use the degeneration condition for $\phi = \pi$ to keep only the Fock states $|n+q, n\rangle$ with a particular q in the initial coherent state. In this case, we have to set $\Omega_{Lx}\eta_x^2 = \Omega_{Ly}\eta_y^2 \equiv \chi$ but $\Omega_{Lx} \neq \Omega_{Ly}$. According to (14), the Rabi frequency is given by

$$\Omega(n_x, n_y, \pi) = |\Omega_{Lx} - \Omega_{Ly} - \chi(n_x - n_y)| = |\Omega_{Lx} - \Omega_{Ly} - \chi q| \tag{26}$$

so that all the states $|n+q, n\rangle$ $(n = 0, 1, 2, \cdots)$ share this Rabi frequency. To get the pair coherent state $|\varsigma, q\rangle_{\rm pcs}$, one should initially prepare a coherent state $|\alpha, \beta\rangle$ with $\varsigma = \alpha\beta$, and select the detection times according to the Rabi frequency (27). In view of the considerations made in the previous sections, the successful null fluorescence detections guarantee the generation of the state $|\varsigma, q\rangle_{\rm pcs}$. Figure 4 shows the generation of a pair coherent state $|\alpha\beta, q\rangle_{\rm pcs}$ from an initial coherent state $|\alpha, \beta\rangle$. Analogously, by putting the symmetric parameters $\Omega_{Lx} = \Omega_{Ly}$ and $\eta_x = \eta_y$, and starting from the initial coherent state $|\alpha, \beta\rangle$ with $\varsigma = \alpha\beta$, it is possible to generate two classes of entangled superposition of pair coherent states:
(a) $\sum_{k=1}^{\infty}(c_k^{(+)}|\varsigma, (2k-1)q\rangle_{\rm pcs} + c_k^{(-)}|\varsigma, -(2k-1)q\rangle_{\rm pcs})$;
(b) $c_0\ |\varsigma, 0\rangle_{\rm pcs} + \sum_{k=1}^{\infty}(c_k^{(+)}|\varsigma, kq\rangle_{\rm pcs} + c_k^{(-)}|\varsigma, -kq\rangle_{\rm pcs}$.
The possibility of obtaining class (a) or (b) is related to the initial electronics state.

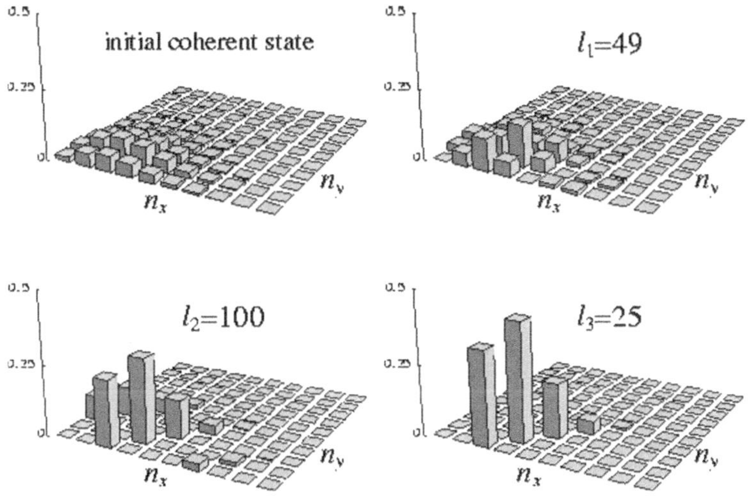

Fig. 4. Probability distributions in the process of generating the pair coherent state $|\alpha\beta, q = 2\rangle_{\rm pcs}$ from the initial coherent state $|\alpha, \beta >$ with $|\alpha|^2 = 2.5$ and $|\beta|^2 = 1.5$. Phase difference of two driving beams is $\phi = \pi$, and the other parameters are $\Omega_{Lx}\eta_x^2 = \Omega_{Ly}\eta_y^2 \equiv \chi$, and $\chi/(\Omega_{Lx} - \Omega_{Ly}) = 0.0049$

6 Conclusion

In this paper we have presented an extension of QND measurement of the vibrational energy of the trapped ion from one dimensional case to the bidimensional one. Our approach, based on the possibility of driving the confined ion by two appropriately configured laser beams, puts into evidence that such a generalization is not trivial. Our results show indeed that the many choices of the parameters relevant in the measurement schemes can be performed, and some of them may improve precision and sensitivity of measurement. The key aspect of our proposal is its application in the manipulation of the quantum state of the trapped ion. We wish to emphasize that the experimental scheme here presented is rather simple and easier to carry out, albeit its success might require a large number of trials. Our procedure provides a simple way to realize a variety of bimodal Fock states, entangled superpositions of Fock states and, for the first time, pair coherent states.

Acknowledgment. The authors acknowledge financial support from Palermo University. One of us, Wang Kaige, acknowledges financial support from the National Natural Science Foundation of China, Nos. 19774013 and 10074008.

References

1. C. M. Caves, K. S. Thorne, R. W. P. Drever, V. D. Sandberg, and M. Zimmerman, Rev. Mod. Phys. **52,** 341 (1980)

2. V. B. Braginsky, Y. I. Vorontsov, and F. Y. Khalili, Zh. Eksp. Teor. Fiz. **73**, 1340 (1977) [Sov. Phys. JETP **46,** 705 (1977)]; W. G. Unruh, Phys. Rev. D **18,** 1764 (1978)
3. M. Brune, S. Haroche, V. Lefevre, J. M. Raimond, and N. Zagury, Phys. Rev. Lett. **65**, 976 (1990); M. Brune, S. Haroche, J. M. Raimond, L. Davidovich, and N. Zagury, Phys. Rev. A **45**, 5193 (1992)
4. See review article, P. Grangier, J. A. Levenson, J-P Poizat: Nature **396**, 537-542 (1998)
5. W. M. Itano and D. J. Wineland, Phys. Rev. A **25**, 35 (1982)
6. D. J. Wineland, W. M. Itano, J. C. Bergquist and R. G. Hulet, Phys. Rev. A **36**, 2220 (1987)
7. F. Diedrich, J. C. Bergquist, W. M. Itano and D. J. Wineland, Phys. Rev. Lett. **62**, 403 (1989)
8. C. Monroe, D. M. Meekhof, B. E. King, J. R. Jefferts, W. M. Itano and D. J. Wineland, and P. Gould, Phys. Rev. Lett. **75**, 4011 (1995)
9. C. A. Blockley, D. F. Walls and H. Risken, Europhys. lett. **17**, 509 (1992)
10. J. I. Cirac, R. Blatt, A. S. Parkins and P. Zoller, Phys. Rev. A **49**, 1202 (1994)
11. D. M. Meekhof, C. Monroe, B. E. King, W. M. Itano and D. J. Wineland, Phys. Rev. Lett. **76**, 1796 (1996)
12. C. Monroe, D. M. Meekhof, B. E. King and D. J. Wineland, Science **212**, 1131 (1996)
13. Ch. Roos, Th. Zeiger, H. Rohde, H. C. Nàgerl, J. Eschner, D. Leibfried, F. Schmidt-Kaler, R. Blatt, Phys. Rev. Lett. **83**, 4713 (1999)
14. F. Schmidt-Kale, Ch. Roos, H. C. Nàgerl et al., Experimental issues in coherent quantum state manipulation of trapped atomic ions, Los Alamos e-print archive, *quant-ph/*0003096
15. D. J. Wineland, C. Monroe, W. M. Itano, D. Leibfried, B. E. King and D. M. Meekhof, J. Res. Natl. Inst. Stand. Technol. **103**, 259 (1998)
16. R. L. de Matos Filho and W. Vogel, Phys. Rev. Lett. **76,** 4520-4523 (1996)
17. L. Davidovich, M. Orszag, and N. Zagury, Phys. Rev. A **54**, 5118-5125 (1996)
18. W. Vogel, and R. L. de Matos Filho, Phys. Rev. A **52**, 4214 (1995)
19. S. Wallentowitz, R. L. de Matos Filho, S. -C. Gou, and W. Vogel, Eur. Phys. J. D **6**, 397-405 (1999)
20. C. C. Gerry, Phys. Rev. A **55,** 2478-2481 (1997)
21. G. S. Agarwal, J. Opt. Soc. Am. B **5**, 1940-1947 (1988)
22. C. C. Gerry, and R. Grobe, Phys. Rev. A **51,** 1698-1701 (1995)
23. S. -C. Gou, J. Steinbach, and P. L. Knight, Phys. Rev. A **54,** R1014-R1017 (1996)

Phonon-Photon Translation with a Trapped Ion in a Cavity

E. Massoni and M. Orszag

Facultad de Física, Pontificia Universidad Católica de Chile, Casilla 306, Santiago, Chile

Abstract. We present a way of using a trapped ion (Paul Trap), placed in a cavity, to translate phonon to photon statistics when pumped incoherently, and trasnfer squeezing from phonons to photons, when the electronic pump is coherent and one has a particular phonon pump.

1 Introduction

Recently, single quantum systems have been investigated based on the advances in atomic cooling and confinment techniques. [1]

An interesting example is the case of single trapped ions. Much work has been devoted to the coupling of the internal degrees of freedom and it's center-of-mass motion. These couplings can be achieved by direct [2] or non-resonant Raman transitions. [3]. In the latter case, the coupling is via two optical fields. [4]

Other examples of single quantum systems are micromasers [5], microlasers [6] and more recently Ion Trap Lasers. [3]

Also, some recent work reported the creation of thermal, Fock, coherent and squeezed states of motion of a harmonically bound Be^+ ion. [8]

In the present review [10], we discuss the general transfer of information from the phonon field of an ion to the photon field inside an electromagnetic cavity.

In Section 2 we discuss the Phonon-Photon Translator(PPT) Model. This particular Translator can be achieved using two-level ions pumped incoherently, and what we get is the phonon statistics of the ion translated into photon statistics in the cavity.Due to the incoherent pumping, only the diagonal elements of the density matrix are transferred to the cavity and the phase information is lost.The next three sections (Sections 3,4,5) are devoted to The Information Transfer,Numerical Simulation and Discussion of the Translator.

In the Section 6 we introduce a model of an Ion-Trap Laser, with three-level ions with their upper electronic levels are coherenty pumped with two resonant lasers and with an external phonon pumping. A correlation is established between the vibrational states of the center-of-mass motion of the ion and it's internal states, which has the effect, when the system is coherently pumped, of translating the quantum properties of the vibrational state of the ion to the photon state inside the cavity.

M. Orszag, J.C. Retamal (Eds.): LNP 575, pp. 43–62, 2001.

In particular, if one has a quadratic phonon pump that generates squeezed vibrations in the trapped ion, this produces squeezing in the cavity field.

Thus, in this system, a flow of quantum information takes place from the ion to the cavity field, which maybe of interest, for example, to establish a communication between two nearby ions, via an optical cavity.Finally, Section 7 we do an analytical calculation, within the Semiclassical Approximation and in Section 8 we compare these approximate results with the exact numerical ones and discuss these results.

2 The Model for a Phonon-Photon Translator

We consider a single trapped ion in a harmonic potential and located inside an optical cavity that consists of two mirrors, as shown in Fig.[1]. The Hamiltonian of the system is given by

$$\begin{aligned} H &= \hbar\omega_a\sigma_z + \hbar\omega_v b^+ b + \hbar\omega_c a^+ a \\ &\quad + \hbar g\cos(kx+\phi)(a^+ + a)(\sigma^+ + \sigma^-). \end{aligned}$$

where b^+, b, a^+, a are the creation and anihilation operators associated with the center of mass of the ion and one electromagnetic mode of the cavity respectively. $\sigma_z, \sigma^+, \sigma^-$ are the usual atomic operators of the two level ion defined in the basis $\{|2\rangle, |1\rangle\}$ as $\sigma_z = (|2\rangle\langle 2| - |1\rangle\langle 1|)/2$, $\sigma^+ = |2\rangle\langle 1|$ and $\sigma^- = |1\rangle\langle 2|$. The position operator is given by $x = x_0(b^+ + b)$, g is the atom-field coupling constant, k is the electromagnetic wavevector and ϕ determines the position of the trap inside the cavity.

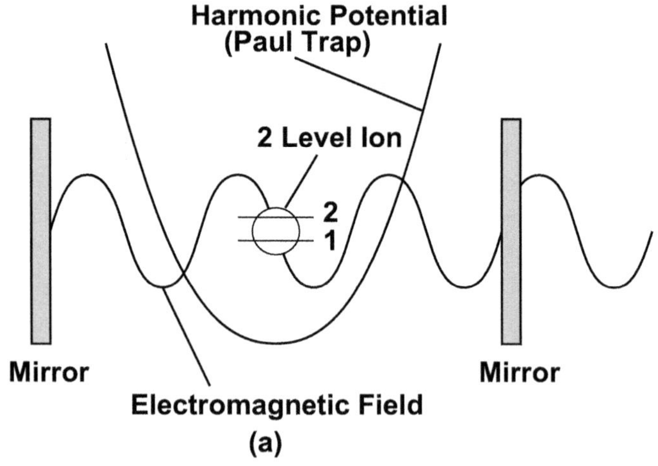

Fig. 1. Schematic Diagram of a Phonon-Photon Translator, considered in this work.

We consider the case $\omega_c = \omega_a + \omega_v$ that selects two possible processes: (i) a photon is created, a phonon is absorbed and the ion decays from state $|2\rangle$ to $|1\rangle$ and (ii) a photon is absorbed, a phonon is created and the atom is excited from state $|1\rangle$ to $|2\rangle$. These two processes are shown in Fig.[2].

In the interaction picture, the Hamiltonian reads as follows:

$$H_I = \hbar\Omega(a\sigma^+ b^+ + a^+\sigma^- b) \tag{1}$$

where $\Omega \equiv gkx_0\sin(\phi)$. We also introduce an incoherent pump from $|1\rangle$ to $|2\rangle$, with intensity Γ. This pump can be modeled through a process that corresponds to an inverse spontaneous decay [11]. The purpose of the pump is to favour process (i) over (ii), in such a way that the conversion from phonons to photons is achieved. We baptize the present system as the 'Phonon-Photon Translator' (PPT).

Now, the evolution of the density operator ρ is given by Liouville's equation:

$$\frac{d\rho}{dt} = \frac{1}{i\hbar}[H_I, \rho] + \mathcal{L}_\Gamma \rho, \tag{2}$$

where $\mathcal{L}_\Gamma \rho \equiv \frac{\Gamma}{2}(2\sigma^+\rho\sigma^- - \sigma^-\sigma^+\rho - \rho\sigma^-\sigma^+)$ is the pumping term. From equations (25) and (2) we get the explicit formulas for the time evolution for the various density matrix elements expressed in the $|j\rangle|nc\rangle|nv\rangle$ (atomic, photon and phonon number) basis:

$$\begin{aligned}\dot{\rho}^{m_c,m_v}_{11n_c,n_v} = \Omega\{&\sqrt{mc(mv+1)}\;\rho^{m_c-1,m_v+1}_{21n_c,n_v}\\ &+\sqrt{nc(nv+1)}\;\rho^{m_c,m_v}_{12n_c-1,n_v+1}\}\end{aligned}$$

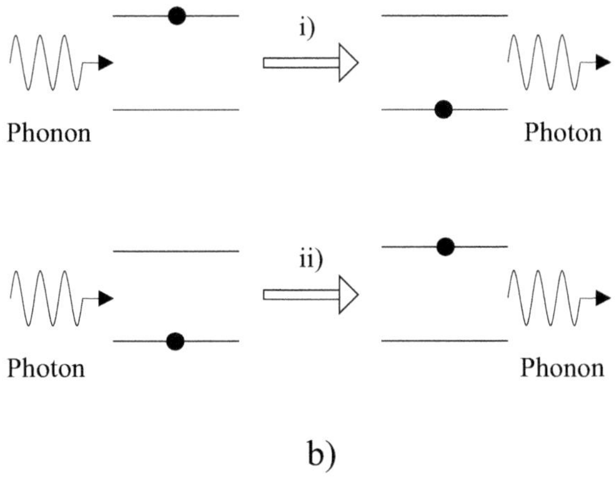

Fig. 2. (i)Absorption of a phonon followed by the creation of a photon and decay of the ion from level 2 to 1.(ii) absorption of a photon followed by the creation of a phonon and the excitation of the ion.

$$-\Gamma\, \rho^{m_c,m_v}_{11n_c,n_v}, \tag{3}$$

$$\begin{aligned}\dot{\rho}^{m_c-1,m_v+1}_{22n_c,n_v} &= -\Omega\{\sqrt{mc(mv+1)}\,\rho^{m_c,m_v}_{12n_c-1,n_v+1} \\ &+\sqrt{nc(nv+1)}\,\rho^{m_c-1,m_v+1}_{21n_c,n_v}\}\end{aligned}$$

$$+\Gamma\, \rho^{m_c-1,m_v+1}_{11n_c-1,n_v+1}, \tag{4}$$

$$\begin{aligned}\dot{\rho}^{m_c,m_v}_{12n_c-1,n_v+1} &= \Omega\{\sqrt{mc(mv+1)}\,\rho^{m_c-1,m_v+1}_{22n_c-1,n_v+1} \\ &-\sqrt{nc(nv+1)}\,\rho^{m_c,m_v}_{11n_c,n_v}\}\end{aligned}$$

$$-\frac{\Gamma}{2}\,\rho^{m_c,m_v}_{12n_c-1,n_v+1}, \tag{5}$$

$$\begin{aligned}\dot{\rho}^{m_c-1,m_v+1}_{21n_c,n_v} &= \Omega\{\sqrt{nc(nv+1)}\,\rho^{m_c-1,m_v+1}_{22n_c-1,n_v+1} \\ &-\sqrt{mc(mv+1)}\,\rho^{m_c,m_v}_{11n_c,n_v}\}\end{aligned}$$

$$-\frac{\Gamma}{2}\,\rho^{m_c-1,m_v+1}_{21n_c,n_v} \tag{6}$$

with $\rho^{m_c,m_v}_{ijn_c,n_v} \equiv \langle mc|\langle mv|\langle i|\rho|j\rangle|nv\rangle|nc\rangle$. In the above system, we have neglected both the atomic and cavity loss terms, assuming that the corresponding lifetimes are much longer than the translation time. We will further discuss this aproximation in section 5. In the next section we study the problem of how to transfer the information of the vibrational state of the ion to the cavity mode.

3 Information Transfer

The PPT works as follows: we initially prepare the cavity in the vacuum state and the ion in a vibrational state $|\psi_v\rangle$. As mentioned in the introduction, the preparation of vibrational Fock, coherent or squeezed states has been achieved experimentally with the interaction of the ion with laser pulses in a Raman Scheme [12]. Also, we choose the initial state of the ion to be $|1\rangle$ for computational convenience. Once the system has been prepared in the initial state, we turn on the incoherent pump that will convert phonons into photons via the process (i) until the phonons are depleted. At this point, it is clear that a phonon-photon conversion takes place. However, the question is: what is the relation between the initial state of the trap and the final state of the cavity?.

In this section, we will show that the initial phonon distribution is identical to the final photon distribution.

Taking the Laplace Transform L of the system (6) for $nc = mc$ y $nv = mv$ we get:

$$
\begin{aligned}
s\, p_{11}^{n_c,n_v}(s) - \rho_{11n_c,n_v}^{n_c,n_v}(0) &= G_{nc,nv}\{p_{12}(s) + p_{21}(s)\} \\
&\quad -\Gamma p_{11}^{n_c,n_v}(s), &(7)\\
s\, p_{22}^{n_c-1,n_v+1}(s) - \rho_{22n_c-1,n_v+1}^{n_c-1,n_v+1}(0) &= -G_{nc,nv}\{p_{12}(s) + p_{21}(s)\} \\
&\quad -\Gamma p_{11}^{n_c-1,n_v+1}(s), &(8)\\
s\, p_{12}(s) - \rho_{12n_c-1,n_v+1}^{n_c,n_v}(0) &= G_{nc,nv}\left\{p_{22}^{n_c-1,n_v+1}(s) - p_{11}^{n_c,n_v}(s)\right\} \\
&\quad -\frac{\Gamma}{2}p_{12}(s), &(9)\\
s\, p_{21}(s) - \rho_{21n_c,n_v}^{n_c-1,n_v+1}(0) &= G_{nc,nv}\left\{p_{22}^{n_c-1,n_v+1}(s) - p_{11}^{n_c,n_v}\right\} \\
&\quad -\frac{\Gamma}{2}p_{21}^{n_c,n_v}(s), &(10)
\end{aligned}
$$

where we have used the following definitions $p_{11}^{n_c,n_v}(s) \equiv L\left(\rho_{11n_c,n_v}^{n_c,n_v}(t)\right)$, $p_{22}^{n_c-1,n_v+1}(s) \equiv L\left(\rho_{22n_c-1,n_v+1}^{n_c-1,n_v+1}(t)\right)$, $p_{12}(s) \equiv L\left(\rho_{12n_c-1,n_v+1}^{n_c,n_v}(t)\right)$, $p_{21}(s) \equiv L\left(\rho_{21n_c,n_v}^{n_c-1,n_v+1}(t)\right)$ and $G_{nc,nv} \equiv \Omega\sqrt{nc(nv+1)}$. With the initial conditions described before, the only non-zero initial elements are $\rho_{110,n_v}^{0,n_v}(0)$. After some algebra we find:

$$
\begin{aligned}
p_{11}^{n_c,n_v}(s) &= \frac{2G_{nc,nv}^2 \Gamma p_{11}^{n_c-1,n_v+1}(s)}{(s+\Gamma/2)(s^2+\Gamma s+4G_{nc,nv}^2)}, &(11)\\
p_{22}^{n_c-1,n_v+1}(s) &= \frac{2G_{nc,nv}^2 p_{11}^{n_c,n_v}(s) + \Gamma(s+\Gamma/2)p_{11}^{n_c-1,n_v+1}(s)}{(s(s+\Gamma/2)+2G_{nc,nv}^2)}, &(12)\\
p_{12}(s) &= \frac{G_{nc,nv}(p_{22}^{n_c-1,n_v+1}(s) - p_{11}^{n_c,n_v}(s))}{(s+\Gamma/2)} = p_{21}(s), &(13)\\
p_{11}^{0,n_v}(s) &= \frac{\rho_{110,n_v}^{0,n_v}}{(s+\Gamma)}, &(14)\\
p_{11}^{n_c,0} &= \frac{\Gamma p_{11}^{n_c,0}}{s}. &(15)
\end{aligned}
$$

From equations (14),(11) and (15) we can calculate the final photon distribution $\rho_{22N_c,0}^{N,0}(t)$. With the initial probability distibution $P_N = \rho_{110,N}^{0,N}(0)$ of having N phonons in the trap and the recursion relation (11), we obtain $p_{11}^{N,0}(s)$. Then from (15), the inverse transform of $p_{22}^{N,0}(s)$ gives us the the final probability distribution $\rho_{22}^{N,0}(t)$ of having N photons in the cavity. Explicitly we have

$$p_{11}^{N,0}(s) = \frac{P_N(2\Omega^2\Gamma)^N(N!)^2}{(s+\Gamma/2)^N(s+\Gamma)} \prod_{i=1}^{N} \frac{1}{(s-\lambda_i^+)(s-\lambda_i^-)}, \tag{16}$$

$$\rho_{22,N,0}^{N,0}(t) = \Gamma \int_0^t dt' \rho_{11,N,0}^{N,0}(t'), \tag{17}$$

with $\lambda_i^{\pm} = \frac{-\Gamma}{2} \pm \frac{\sqrt{\Gamma^2-4\eta_i^2}}{2}$ and $\eta_i^2 = 4\Omega^2 i(N+1-i)$. Before doing the inverse transform of (16), we notice that if N is even, there are $N/2$ factors of multiplicity 2 of the type $(s-\lambda_i^+)^2(s-\lambda_i^-)^2$, while if N is odd, there will be $(N-1)/2$ factors of the same kind plus an extra factor of multiplicity 1 of the type $(s-\lambda_i^+)(s-\lambda_i^-)$. Since both calculations are quite similar we will concentrate in the even N case. We define:

$$F(s) \equiv \frac{1}{(s+\Gamma/2)^N(s+\Gamma)} \prod_{i=1}^{N/2} \frac{1}{(s-\lambda_i^+)^2(s-\lambda_i^-)^2}. \tag{18}$$

For the sake of simplicity we take the limit $\Gamma \gg 4\eta_i^2$. The exact calculation will generate more terms with the same qualitative result in the transfer from a phonon distribution in the trap to a photon distribution in the cavity. In this limit, the factors like $\frac{1}{(s-\lambda_i^-)^2}$ transform to fast decaying exponentials that contribute little to (17), so we can write

$$F(s) = \sum_{i=1}^{N/2} \frac{A_i}{(s-\lambda_i^+)^2} + \frac{B_i}{(s-\lambda_i^+)}, \tag{19}$$

where

$$\begin{aligned} A_i &= F(s)(s-\lambda_i^+)^2/_{s=\lambda_i^+} \\ &= \frac{-2^N}{\lambda_i^-(\Gamma^2-4\eta_i^2)^{N/2+1}} \prod_{i\neq j=1}^{N/2} \frac{1}{(\eta_j^2-\eta_i^2)^2}, \\ B_i &= \frac{d}{ds}\left(F(s)(s-\lambda_i^+)^2\right)/_{s=\lambda_i^+} \\ &= A_i \left\{ \frac{-2(N+1)}{\sqrt{\Gamma^2-4\eta_i^2}} + \frac{1}{\lambda_i^-} - 2\sqrt{\Gamma^2-4\eta_i^2} \sum_{i\neq j=1}^{N/2} \frac{1}{\eta_j^2-\eta_i^2} \right\}. \end{aligned}$$

Now taking the inverse transform of (19) and using (16), we get

$$\rho_{11,N,0}^{N,0}(t) = P_N(2\Omega^2\Gamma)^N(N!)^2 \sum_{i=1}^{N/2} exp(\lambda_i^+ t)\,(A_i t + B_i)\,. \tag{20}$$

We still need to comment about the diagonal matrix elements $\rho^{i,j}_{22,i,j}(t)$. Evidently $\rho^{N,0}_{22,N,0}(t)$ will have a non-zero value when the transfer is finished, as seen from (17). However, intermediate elements like $\rho^{N-j,j}_{22,N-j,j}(t)$ $(j = 1..N)$ will decay to zero. This result can be seen from (12), rewritten as

$$p^{n_c-1,n_v+1}_{22}(s) = p^{n_c,n_v}_{11}(s)h(s) + p^{n_c-1,n_v+1}_{11}(s)g(s), \tag{21}$$

where $h(s)$ and $g(s)$ are easily read out. If we take the inverse Laplace transform to the last equation and using the convolution theorem, equation (21) takes the following form:

$$\rho^{n_c-1,n_v+1}_{22n_c-1,n_v+1}(t) = \int_0^t du\ \{\rho^{n_c,n_v}_{11n_c,n_v}(u)H(t-u) \tag{22}$$

$$+ \rho^{n_c-1,n_v+1}_{11n_c-1,n_v+1}(u)G(t-u)\}. \tag{23}$$

However, the inverse Laplace tranforms $H(t)$ and $G(t)$ of h and g respectively are proportional to $exp(-\Gamma t/4)$, so these terms will vanish for long times.

Now, we go back to the question posed at the begining of this section. First, we notice that all the matrix elements $\rho^{n_c,n_v}_{11n_c,n_v}(t)$ decay to zero for obvious reasons. In the continuous phonon-photon conversion, we end up with no phonons in the trap, thus inhibiting the process i) which allows the $|2\rangle \rightarrow |1\rangle$ transition. If we continue to incoherently pump the system, there is no way to keep the system in state $|1\rangle$. On the other hand, if for example we start with a vibrational Fock state, that is $P_N = 1$ for a given N, then according to the recursion relation (11), the diagonal matrix elements participating in the dynamics are $\rho^{j,N-j}_{11j,N-j}(t)$, $\rho^{j,N-j}_{22j,N-j}(t)$ $(j = 0..N)$. All the matrix elements of this set that we call the 'transfer channel of N' should add up to one. Since all the elements of this channel decay to zero except for $\rho^{N,0}_{22,N,0}(t)$, we conclude that$\rho^{N,0}_{22,N,0}(\infty) = 1$ (a Fock state in the cavity). In the general case of an arbitrary distribution P_N, each N has its own independent transfer channel and there is no probability flux between them. Since P_N is just a factor in (11) and using the previously discussed fact that each Fock state is totally transmitted from the trap to the cavity, we arrive to the conclusion that $\rho^{N,0}_{22,N,0}(\infty) = P_N$.

4 Numerical Simulation

We have solved numerically the set of equations (6). Our analytical solutions of the matrix elements show no difference with the numerical solutions with a high enough pumping rate Γ. As discussed in section 3, we only take this approximation for the sake of simplicity, but exact analytical formulas can be obtained following a similar procedure for all values of Γ. For these reasons, all the following figures are plotted from the numerical calculations. All parameters are scaled by Ω to make them dimensionless. Here, we will compare the effects of cavity and atomic losses in the behaviour of the PPT versus the ideal case where no losses are introduced.

A. Ideal PPT

In Fig.[3] we have plotted the time evolution of some diagonal matrix elements corresponding to a transmission channel of an initial vibrational Fock state $N = 4$.

We observe that there is not a direct transmission of the probability from $\rho_{11,0,4}^{0,4}(t)$ to $\rho_{22,4,0}^{4,0}(t)$ but through intermediate elements. As shown in this figure, these intermediate elements appear and decay in an ordered sequence: $\rho_{22,1,3}^{1,3}(t)$, $\rho_{22,2,2}^{2,2}(t)$ and $\rho_{22,3,1}^{3,1}(t)$. The same occurs to the other $\rho_{11,i,4-i}^{i,4-i}(t)$, $(i = 1..4)$ and the off-diagonal elements, not plotted here. Physically speaking, this just shows that the conversion of phonons to photons is done gradually. For the specific case of a general phonon state, the initial coherences decay to zero in a time scale similar to the diagonal elements. To analize when this conversion is finished we define a characteristic time τ_N as the time it takes for $\rho_{22,4,0}^{4,0}$ to reach 90% of its final value. We found an optimum pumping rate Γ that produces the fastest phonon-photon conversion. This is shown in Fig.[4], where τ_N is plotted versus Γ for various initial vibrational Fock states with $N = 2, 4, 16$.

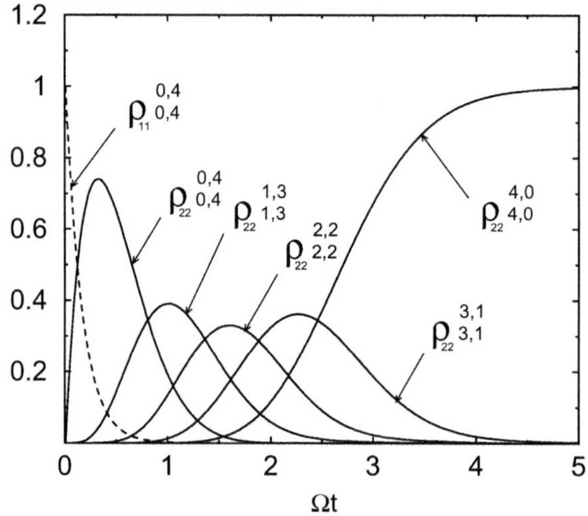

Fig. 3. Time evolution of some density matrix elements corresponding to an initial vibrational Fock state N=4 ,$\Gamma = 6$,without cavity losses. The only surviving element for long times is $\rho_{22,40}^{4,0}$that corresponds to the phonon vacuum state , a photon Fock state with N=4 photons and the ion in the 2 state.

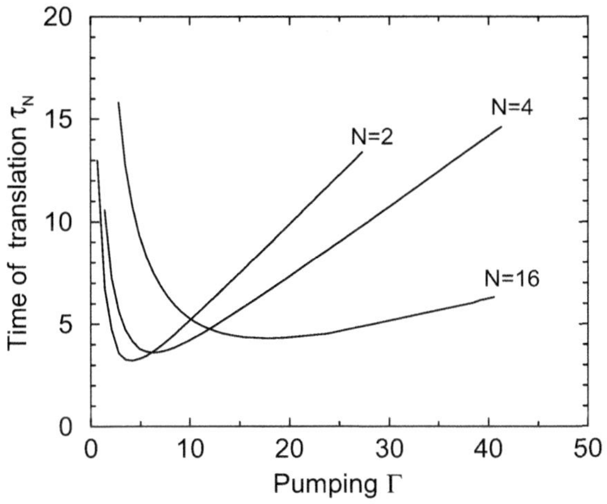

Fig. 4. Translation time versus pumping rate Γ for different initial vibrational Fock states N=2,4,16. Each curve has a minimum which gives the optimal pumping rate for a given N.

This dependence shows, for a fixed Γ, that the larger N portion of a general distribution (e.g. a coherent state) is transmitted first and the lower N values take a longer time.

B. Non ideal PPT

Here, we introduce cavity and atomic losses by adding to (2) the standard Liouville operators $\mathcal{L}_{atom}\rho$ and $\mathcal{L}_{field}\rho$. They are defined as:

$$\mathcal{L}_{atom}\rho \equiv \frac{\gamma_a}{2}(2\sigma^- \rho\sigma^+ - \sigma^+\sigma^-\rho - \rho\sigma^+\sigma^-),$$
$$\mathcal{L}_{field}\rho \equiv \frac{\gamma_c}{2}(2a\rho a^+ - a^+ a\rho - \rho a^+ a),$$

where γ_a is the spontaneous emission rate from level $|2\rangle$ to $|1\rangle$, and γ_c is the cavity damping rate. These losses spoil the final photon distribution in two different ways. The cavity losses will eventually kill the field in the cavity. On the other hand, spontaneous emission increases the time the ion is in state $|1\rangle$, favouring the conversion of photons to phonons. Fig.[5] shows the final photon and phonon distribution with an initial vibrational Fock state $N = 4$, $\gamma_a = 2$, $\gamma_c = 0$ and $\Gamma = 10$.

We see that spontaneous emission spreads the final distributions, so the trap ends up with some phonons and the field is no longer a Fock state with $N = 4$. To minimize the effect of the spontaneous emission we could increase the pumping rate Γ. However this also increases the transmission time τ_N. For example, varying the pumping rate from $\Gamma = 10$ to 25 with $\gamma_a = 0.2$ raises τ_4 by about a factor 2, as seen in Fig.[6].

This dependence limits the value of Γ to keep τ_N quite below the cavity lifetime.

Next, we discuss the experimental realization of the PPT and the possible applications.

5 Discussion

Some realistic values of the parameters are: $g = 10MHz$, $\gamma_c = 1MHz$ to $10MHz$ [15], $\gamma_a = 1MHz$ to $100MHz$, and the cavity wavelenght $\lambda = 866nm$ [14]. We define also the parameter $\eta \equiv kx_0$ such that $\Omega = g\eta$ ($\phi = 0$). If we scale the above frecuencies by Ω we get: $\gamma_a = 0.1/\eta$ to $10/\eta$ and $\gamma_c = 0.1/\eta$ to $1/\eta$. A reasonable value for $\eta = 0.5$ can be obtained for trap parameters as in [12]. These parameters allow us to set $\gamma_a = 0.2$ to achieve the transmission of a vibrational Fock state $N = 4$ to a final cavity Fock state $N = 4$ with probability $= 0.95$ (i.e $\rho^{4,0}_{22,4,0} = 0.95$) for $\Gamma = 10$ as in Fig 6). However, the transmission time $\tau_4 = 5\Omega^{-1}$ is of the same order as the cavity lifetime $t_c = \gamma_c^{-1} = 5\Omega^{-1}$. This shows that we require an optical cavity with higher quality Q-parameter by at least a factor 10 to observe a good translation.

As an application, we may think about a one atom laser with some novel features. In order to make it work, we would need two kinds of pumping. The atomic pumping to favour phonon to photon conversion, and a phonon pumping to maintain the production of photons. As we have shown in this work, the vibrational state of the ion influences the photon distribution in the cavity. So

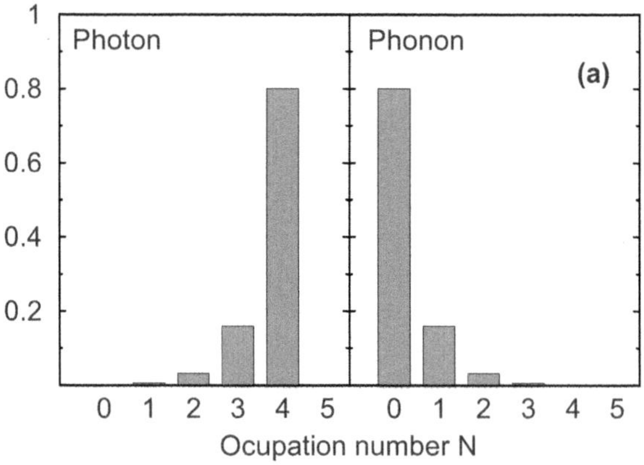

Fig. 5. Final phonon and photon distribution with $\Gamma = 10, \gamma_a = 2$ and $\gamma_c = 0$. The initial vibrational state was a Fock state with N=4.

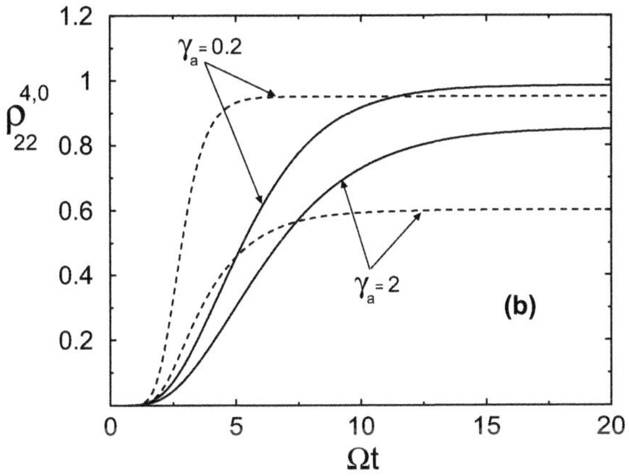

Fig. 6. The time evolution of $\rho^{4,0}_{22,40}$ matrix elements with different spontaneous emission rates $\gamma_a = 0.2$ and $\gamma_a = 2$ .We observe that as we increase the pumping rate from $\Gamma = 10$(dashed line) to $\Gamma = 25$(solid line), τ_N grows.

we could manipulate the field state via the phonon pumping. The phonons can be pumped via Raman coupling, a spatially uniform classical driving field or a parametric driving of the ion [12]. This variety of pumps may lead to new and interesting physics.

In summary, we have demonstrated the possibility of transferring a phonon distribution of an ion in a Paul trap to a photon distribution in a high-Q cavity. This scheme can be used to produce photon Fock states in an optical cavity.

6 The Model for an Ion Trap Laser Producing Transfer of Squeezing

The present model consists of a three-level trapped ion in a Paul trap, represented by a harmonic potential at frecuency ω_v placed inside an optical cavity at frecuency ω_c as shown in Fig.[7].

The Hamiltonian of the above system is given by:

$$H = \hbar\omega_2|2\rangle\langle 2| + \hbar\omega_1|1\rangle\langle 1| + \hbar\omega_0|0\rangle\langle 0| + \hbar\omega_v b^\dagger b + \hbar\omega_c a^\dagger a + \hbar g\cos(kx+\phi)(a^\dagger + a)(\sigma^\dagger_{21} + \sigma^-_{21}). \quad (24)$$

where ω_2, ω_1, and ω_0 represent the ion energy levels, $b^\dagger$, b, $a^\dagger$, a are the creation and anihilation operators associated with the motion of the ion and one electromagnetic mode of the cavity respectively. We also denote the raising and lowering fermionic operators as $\sigma^\dagger_{ij} = |i\rangle\langle j|$ and $\sigma^-_{ij} = |j\rangle\langle i|$ obeying the standard anticommutation relations. The position operator is given by $x = x_0(b^\dagger + b)$,

g is the atom-field coupling constant, k is the electromagnetic wavevector and ϕ determines the position of the trap inside the cavity. We consider the case $\omega_c = \omega_2 - \omega_1 + \omega_v$ that selects two possible processes: (i) a photon is created, a phonon is absorbed and the ion decays from state $|2\rangle$ to $|1\rangle$, and (ii) a photon is absorbed, a phonon is created and the atom is excited from state $|1\rangle$ to $|2\rangle$. These two processes are shown in Fig.[8].

In the interaction picture, we only keep the time-independent terms arising from the cosine expansion. Up to first order in the Lamb-Dicke parameter $\eta = kx_0$, the Hamiltonian reads as follows:

$$H_I = \hbar\Omega(a\sigma_{21}^{\dagger}b^{\dagger} + a^{\dagger}\sigma_{21}^{-}b) \tag{25}$$

where $\Omega \equiv g\eta$ and $\phi = \pi/2$. We estimate the error of this approximation by comparing the above Hamiltonian with the next resonant term:

$$H_I^{(3)} = \hbar\frac{g\eta^3}{6}(a\sigma_{21}^{\dagger}b^{\dagger 2}b + a^{\dagger}\sigma_{21}^{-}b^{\dagger}b^2), \tag{26}$$

so our approximation is valid when (Lamb-Dicke approximation):

$$\frac{\eta^2 N}{6} \ll 1, \tag{27}$$

for all relevant phonon excitation numbers N [16]. Process (i) creates photons in the cavity as long as there are phonons to absorb and level 2 is populated. As a consequence, an interesting feature of this model is the self-cooling. At variance with previous Ion-Trap Laser models, in this one, the heating problem is minimized.

To transfer squeezing, we introduce two lasers E_1 and E_2 which are resonant with levels $|2\rangle \leftrightarrow |0\rangle$ and $|1\rangle \leftrightarrow |0\rangle$ transitions as shown in Fig.[**??**]. Although

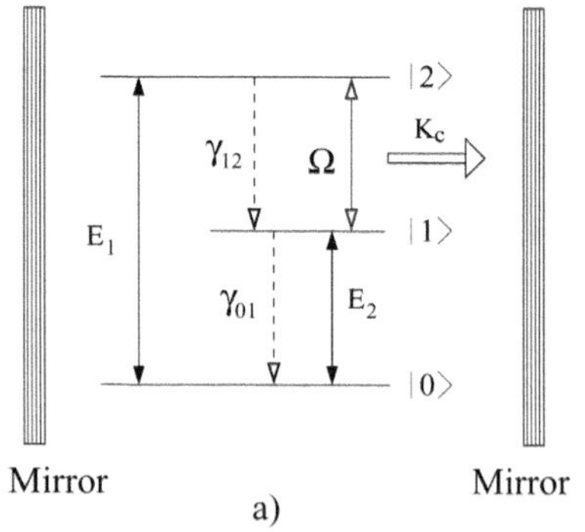

Fig. 7. Schematic diagram showing the three internal levels of an ion. $E_1$and E_2 are the coherent pumps, γ_{12} and $\gamma_{01}$are the spontaneous emission rates and κ_c is the cavity decay rate.

E_1 may couple states with the same parity, which in principle is a forbidden transition for dipole coupling, one can achieve an effective coherent pump from a Raman type configuration resonant with the "carrier transition", where a fourth level was present, and adiabatically eliminated.

These coherent pumps are described by the Hamiltonian in the interaction picture

$$H_E = i\hbar E_1(\sigma_{20}^- - \sigma_{20}^\dagger) + i\hbar E_2(\sigma_{10}^- - \sigma_{10}^\dagger), \tag{28}$$

where E_i are proportional to the field strenghts. Basically, the role of each laser is quite different. E_1 populates level $|2\rangle$ to keep the creation of photons while E_2 is necessary to transfer coherences from the vibrational state to the cavity field. On the other side, phonons are generated driving the ion with a spatially linear classical electric field at frecuency $2\omega_v$. This field puts the ion in a vibrational squeezed state [17]. In the interaction picture, rotating at the trap frequency ω_v, this phonon pump is described by

$$H_q = i\hbar q(b^{\dagger 2} - b^2), \tag{29}$$

where q is proportional to the driving field strength.

We also include in this model, cavity decay and spontaneous emission from the ionic levels following standard techniques. The cavity mode is damped by the the mirrors at the rate of κ_c photons per second. Only spontaneous emission from level 2 to 1 and from 1 to 0 at rates γ_{12} and γ_{01} respectively, have been considered in this calculation. The dynamics of the full system is described by the Lioville's equation [11]:

$$\frac{d\rho}{dt} = \frac{1}{i\hbar}[H_I + H_E + H_q, \rho] + \mathcal{L}_{loss}\rho, \tag{30}$$

where ρ is the density operator for the whole system and $\mathcal{L}_{loss}\rho$ describes the Lindblad terms of the various decay channels.

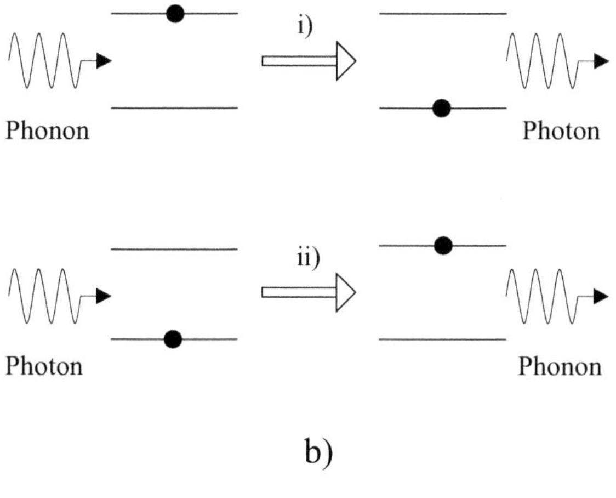

Fig. 8. The two processes allowed by the Ion Trap Laser Hamiltonian.

7 Semiclassical Aproximation

In this section, we find some analytical results concerning to the squeezing transfer and stability of the Ion-Trap Laser. Using (30), one can take moments of the various physical quantities of interest, and get the following system of 12 first order differential equations:

$$\frac{d\langle\sigma_{21}^{+}\rangle}{dt} = \Omega\langle a^{+}b\sigma_{21}^{z}\rangle - E_1\langle\sigma_{10}^{-}\rangle - E_2\langle\sigma_{20}^{+}\rangle - \frac{\gamma_{12}}{2}\langle\sigma_{21}^{+}\rangle - \frac{\gamma_{01}}{2}\langle\sigma_{21}^{+}\rangle \quad (31)$$

$$\frac{d\langle\sigma_{10}^{+}\rangle}{dt} = \Omega\langle ab^{+}\sigma_{20}^{+}\rangle + E_1\langle\sigma_{21}^{-}\rangle + E_2\langle\sigma_{10}^{z}\rangle - \frac{\gamma_{01}}{2}\langle\sigma_{01}^{+}\rangle \quad (32)$$

$$\frac{d\langle\sigma_{20}^{+}\rangle}{dt} = -\Omega\langle a^{+}b\sigma_{10}^{+}\rangle + E_1\langle\sigma_{20}^{z}\rangle + E_2\langle\sigma_{21}^{+}\rangle - \frac{\gamma_{12}}{2}\langle\sigma_{20}^{+}\rangle \quad (33)$$

$$\frac{d\langle\sigma_1\rangle}{dt} = \Omega\langle a^{+}b\sigma_{21}^{-}\rangle - E_2(\langle\sigma_{10}^{+}\rangle + \langle\sigma_{10}^{-}\rangle) + \gamma_{12}\langle\sigma_2\rangle - \gamma_{01}\langle\sigma_1\rangle \quad (34)$$

$$\frac{d\langle\sigma_2\rangle}{dt} = -2\Omega\langle a^{+}b\sigma_{21}^{-}\rangle - E_1(\langle\sigma_{20}^{+}\rangle + \langle\sigma_{20}^{-}\rangle) - \gamma_{12}\langle\sigma_2\rangle \quad (35)$$

$$\frac{d\langle a^{+}a\rangle}{dt} = 2\Omega\langle a^{+}b\sigma_{21}^{-}\rangle - k_c\langle a^{+}a\rangle$$

$$\frac{d\langle b^{+}b\rangle}{dt} = -2\Omega\langle a^{+}b\sigma_{21}^{-}\rangle + 4q\langle b^2\rangle \quad (36)$$

$$\frac{d\langle b^2\rangle}{dt} = -2\Omega\langle ab\sigma_{21}^{+}\rangle + 4q\langle b^{+}b\rangle + 2q$$

$$\frac{d\langle a^2\rangle}{dt} = 2\Omega\langle ab\sigma_{21}^{-}\rangle - k_c\langle a^2\rangle \quad (37)$$

$$\frac{d\langle a^{+}b\rangle}{dt} = \Omega[\langle b^{+}b\sigma_{21}^{+}\rangle - \langle a^{+}a\sigma_{21}^{+}\rangle] + 2q\langle ab\rangle - \frac{k_c}{2}\langle a^{+}b\rangle \quad (38)$$

$$\frac{d\langle a^{+}b\rangle}{dt} = \Omega[\langle b^2\sigma_{21}^{-}\rangle - \langle a^2\sigma_{21}^{+}\rangle] + 2q\langle a^{+}b\rangle - \frac{k_c}{2}\langle ab\rangle \quad (39)$$

$$\frac{d\langle\sigma_0\rangle}{dt} = -\frac{d\langle\sigma_2\rangle}{dt} - \frac{d\langle\sigma_1\rangle}{dt} \quad (40)$$

where $\sigma_i \equiv\mid i\rangle\langle i\mid$, $i = 0, 1, 2$. The above system is not closed, since products of the type $\langle a^{\dagger}b\sigma_{21}^{-}\rangle$ appear. For this reason, we perform a factorization of the type: $\langle a^{\dagger}b\sigma_{21}^{-}\rangle = \langle a^{\dagger}b\rangle\langle\sigma_{21}^{-}\rangle$, assuming that the atomic characteristic time is much shorter than the phonon and photon times (Adiabatic approximation). For $q, \kappa_c \ll E_1, E_2, \gamma_{01}$, and $\gamma_{12} = 0$, we get the following Semiclassical results for the steady state mean values:

$$\langle b^+b\rangle = \frac{s^2p^2+sp+s^2+1}{2(1-s)(1+s)(p-1)(p+1)} \tag{41}$$

$$\langle a^+a\rangle = \frac{sp(1+sp)}{2(1-s)(1+s)(p-1)(p+1)} \tag{42}$$

$$\langle b^2\rangle = \frac{p(1+sp)}{2(1-s)(1+s)(p-1)(p+1)} \tag{43}$$

$$\langle a^2\rangle = \frac{sp(s+p)}{2(1-s)(1+s)(p-1)(p+1)} \tag{44}$$

$$\langle \sigma^{+}_{21}\rangle = \frac{2E_2E_1}{\gamma_{10}^2+4E_1^2+4E_2^2} \tag{45}$$

with $s=\frac{4q}{k_c}$ and $p=\frac{\Omega^2\langle\sigma^+_{21}\rangle^2}{qk_c}$. The variances for the quadratures Y_c and Y_v

$$(\Delta Y_v)^2 \equiv \langle\left(\frac{b-b^\dagger}{2i}\right)^2\rangle = \frac{p}{4(1+s)(p+1)}$$

$$(\Delta Y_c)^2 \equiv \langle\left(\frac{a-a^\dagger}{2i}\right)^2\rangle = \frac{p+s+1}{4(1+s)(p+1)} \tag{46}$$

$$\tag{47}$$

For the case $p \gg 1$ we obtain:

$$(\Delta Y_v)^2 = (\Delta Y_c)^2 \approx \frac{1}{4(1+s)} \tag{48}$$

Equation (48) is our main result and clearly shows that squeezing in the cavity is obtained from the phonon pump q.We notice from (46) that $(\Delta Y_c)^2$ is a decreasing function of p. On the other hand, $p \propto \langle\sigma^+_{21}\rangle^2$ which has an inverse dependence on γ_{01}^2. Thus a large γ_{01} tends to destroy the squeezing effects in the cavity field.

We can also add that for $\kappa_c < 4q$, there is no steady state. This can be easily understood in the sense that we are continuously pumping phonons into the system, thus creating photons in the cavity with very little output, so the cavity would tend to store more and more photons. A steady state is achieved as an equilibrium between gain and loss inside the cavity. There are two divergences in (41-44): The one that originates when $p \to 1$ is not relevant since in our limit $p \gg 1$. However, when $s \to 1$ the system is not longer stationary. To verify this statement, we study the evolution of the system as it aproaches to its steady state. We note that the equation for $\langle\sigma^+_{21}\rangle$ (31) contains a term like $-\frac{\gamma_{01}}{2}\langle\sigma^+_{21}\rangle$. If $\gamma_{01}^{-1} \gg kc^{-1}, \Omega^{-1}$ then $\langle\sigma^+_{21}\rangle$ can be considered a constant. Using the above assumption, which is also verified in the numerical calculations,we define the vector $\boldsymbol{x}$ as

$$\boldsymbol{x} = \begin{bmatrix} \langle a^+a \rangle \\ \langle b^+b \rangle \\ \langle a^2 \rangle \\ \langle b^2 \rangle \\ \langle a^+b \rangle \\ \langle ab \rangle \end{bmatrix}$$

and the matrix M

$$M = \begin{bmatrix} -k_c & 0 & 0 & 0 & 2\,\Omega\langle\sigma_{21}^+\rangle & 0 \\ 0 & 0 & 0 & 4\,q & -2\,\Omega\langle\sigma_{21}^+\rangle & 0 \\ 0 & 0 & -k_c & 0 & 0 & 2\,\Omega\langle\sigma_{21}^+\rangle \\ 0 & 4\,q & 0 & 0 & 0 & -2\,\Omega\langle\sigma_{21}^+\rangle \\ -\Omega\langle\sigma_{21}^+\rangle & \Omega\langle\sigma_{21}^+\rangle & 0 & 0 & -\frac{1}{2}\,k_c & 2\,q \\ 0 & 0 & -\Omega\langle\sigma_{21}^+\rangle & \Omega\langle\sigma_{21}^+\rangle & 2\,q & -\frac{1}{2}\,k_c \end{bmatrix}$$

so the equation of motion of $\boldsymbol{x}$ can be written in a compact form as:

$$\frac{d\boldsymbol{x}}{dt} = M\boldsymbol{x} + \boldsymbol{c} \tag{49}$$

where $\boldsymbol{c}$ is a constant vector arising from the constant term $2q$ in the equation for $\langle b^2 \rangle$.

A stationary state is only achieved if the real part of the eigenvalues of M are all negative.These eigenvalues are given by the following expressions:

$$\begin{aligned} \lambda_{1,2} &= \frac{k_c}{2}(-1 \pm s), \\ \lambda_{3,4} &= \frac{k_c}{2}(-1 \pm s) \pm \frac{k_c}{2}\sqrt{(1 \pm s)^2 - 4ps}, \\ \lambda_{5,6} &= \frac{k_c}{2}(-1 \pm s) \mp \frac{k_c}{2}\sqrt{(1 \pm s)^2 - 4ps}. \end{aligned} \tag{50}$$

In the limit we are working, $s \simeq 1, p \gg 1$,so the square roots in the above expressions are imaginary and the condition of stability for the steady state is satisfied if $s < 1$.

8 Numerical Results

In the stable region, we solved numerically (30), by integrating it in time in small steps Δt, until steady state is reached. By comparing these results with the ones obtained from the Semiclassical theory, we find that both results are quite consistent.

In Fig.[9], we have plotted $\langle a^\dagger a\rangle, \langle b^\dagger b\rangle, \langle a^2\rangle$ and $\langle b^2\rangle$ versus κ_c for the values: $\Omega = 1$, $q = 0.01$, $\gamma_{12} = 0, \gamma_{10} = 20$ and $E_1 = E_2 = 25$, obtained from the numerical solution of Liouville's equation (30).

These curves show that as κ_c decreases to $4q$ the the photon and phonon number tend to increase leaving the steady state region previously discussed. This behaviour does not allow us to numerically integrate near the point $\kappa_c = 4q$.

Fig.[10] shows $(\Delta Y_c)^2$ and $(\Delta Y_v)^2$ versus κ_c, for the same parameters as in Fig.[9].

The dashed lines are from the exact numerical calculation and the solid line corresponds to the Semiclassical theory. We observe a considerable amount of phase squeezing in a large range of the κ_c parameter tending, Semiclassically, to $\frac{1}{8}$ as $\kappa_c \to 4q$. We also notice as seen from (48), that the fluctuations of the photons and phonons are approximately the same. Both, the exact numerical and Semiclassical curves confirm that we have transferred the squeezing from the ion to the cavity field.

A qualitative description of how the squeezing is transferred, can be done with a sequence of the processes connecting the energy levels. In order to get steady state light squeezing, one has to have non-vanishing steady state matrix elements of the type:

$$\langle n_v|\langle n_c|\langle i|\rho|i\rangle|n_c + 2\rangle|n_v\rangle,$$

where n_v, n_c labels the Fock state basis for photons and phonons respectively and i denotes the internal ionic levels. To visualize this, we follow the sequence of processes despicted in Fig.[11].

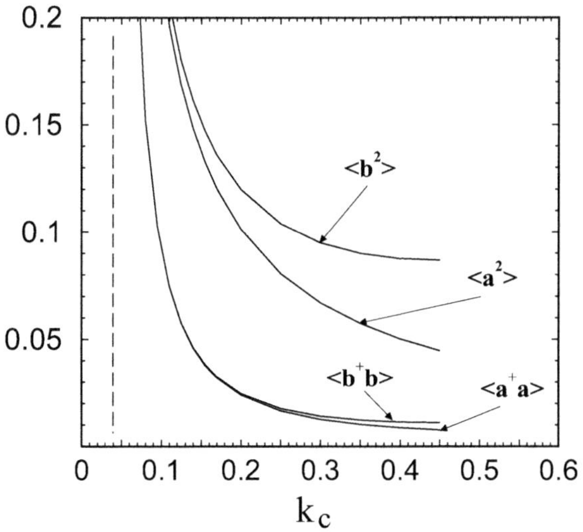

Fig. 9. We have plotted $\langle a^\dagger a\rangle$, $\langle b^\dagger b\rangle$, $\langle a^2\rangle$ and $\langle b^2\rangle$ versus κ_c.These quantities grow as we approach the unstable region $\kappa_c < 4q$, delimited by the dashed vertical line.

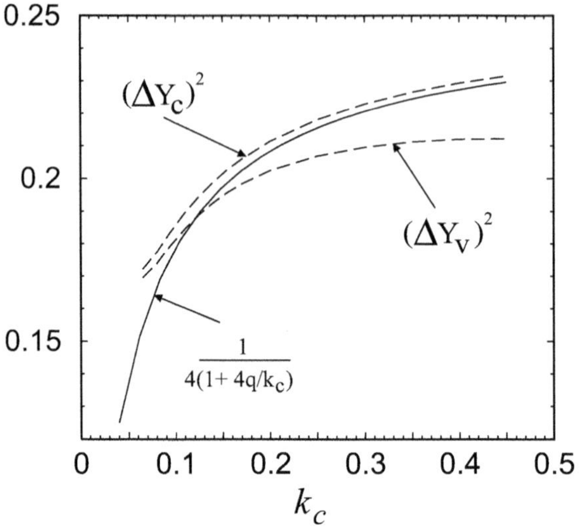

Fig. 10. Fluctuations $(\Delta Y_c)^2$ and $(\Delta Y_v)^2$ versus κ_c. The dashed lines correspond to the numerical integration and the solid lines to the Semiclassical Theory.

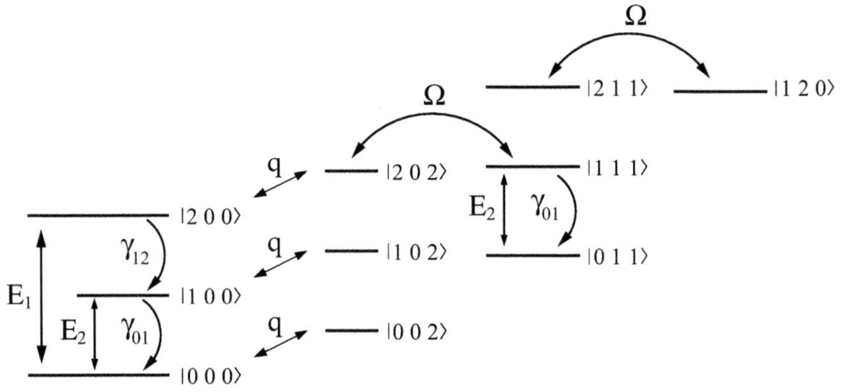

Fig. 11. Sequence of processes to explain how squeezing is tranferred from the ion motion to the cavity field. The arrows labeled with q represent the coupling to the quadratic phonon pump and those with Ω the reversible processes i and ii.

Starting, say from $|i = 0, n_c = 0, n_v = 0\rangle$, we pump coherently via E_1 to get $\mid 2,0,0\rangle$. Then, the quadratic phonon pump will couple $\mid 2,0,0\rangle$ to $\mid 2,0,2\rangle$. With the Hamiltonian H_I, we allow for the absorption of a phonon and emission of a photon through process (i), so we connect $\mid 2,0,2\rangle$ to $\mid 1,1,1\rangle$. Next, we pump coherently from the electronic state $\mid 1\rangle \to\mid 0\rangle$ via E_2 and from $\mid 0\rangle \to\mid 2\rangle$ via E_1, so we end up in the state $\mid 2,1,1\rangle$. Finally, again with the Hamiltonian process (i), we arrive to the state $\mid 1,2,0\rangle$ that populates $\mid 2,2,0\rangle$ and $\mid 0,2,0\rangle$

via the the two pump lasers or the spontaneous emission decay followed by E_1. The difference between the two cases is quite striking.

The existence of the two coherent pumps and the phonon pump q are essential ingredients to produce a non-vanishing $\langle 0,0,0|\rho|0,2,0\rangle$ matrix element. In the absence of E_2, there is no transfer of squeezing. The process connecting the ionic levels $|1\rangle \rightarrow |0\rangle$ only via the spontaneous emission decay γ_{01}, will introduce a random phase noise that produces vanishing off-diagonal density matrix elements in the steady state, thus killing the squeezing. A similar conclusion is obtained if we use incoherent atomic pumps [18]

An experimental realization of an ion-trap laser with a Ca^+ion coupled to an optical cavity has been proposed by Walther et al. In this proposal, the coupling constant is $g = 15MHz$, the spontaneous emission rate ranges from $\gamma = 1MHz$ to $\gamma = 100MHz$ and the cavity decay rate from $\kappa_c = 1MHz$ to $10MHz$. A reasonable value of $\eta = 0.5$ can be obtained for trap parameters as in [8]. Scaling these parameters with $\Omega = g\eta$ we get:$\gamma = 0.14$ to $\gamma = 14$,and $\kappa_c = 0.14$ to 1.4 .Since we want to optimize the system as to get the best results shown in Fig.10, that is for $(\Delta Y_c)^2 = 0.17$, we require a cavity with $\kappa_c = 0.07$, which corresponds to one half of the lowest cavity decay rate used in Walther's proposal.

Aknowledgements. One of the authors (MO) would like to thank FONDECYT (1010777), for partial support of this work.

References

1. J. Javanainen, J.O.S.A.B. **5**, 73 (1988).
2. F. Dietrich, J.C. Bergquist, W.M. Itano, D.J. Wineland, Phys. Rev. Lett. **62**, 403 (1989).
3. W. Nagourney, J. Sandberg, H.G. Dehmelt, Phys. Rev. Lett. **56**, 2797 (1986); T.H. Sauter, W. Neuhauser, R. Blatt, P. Toschek, Phys. Rev. Lett. **57**, 1696 (1986).
4. C. Monroe, D.M. Meekhof, B.E. King, S.R. Jeffers, W.M. Itano, J.M. Wineland, P. Gould, Phys. Rev. Lett. **75**, 4011 (1995); C. Monroe, D.M. Meekhof, B.E. King, W.M. Itano, J.M. Wineland, Phys. Rev. Lett. **75**, 4714 (1995);
5. D. Meschede, H. Walther and G. Muller, Phys. Rev. Lett. **54**, 551 (1985); M. Brune, J.M. Raimond, P. Goy, L. Davidovich, S. Haroche, Phys. Rev. Lett. **59**, 1899 (1987)
6. K. An, J.J. Childs, R.R. Dasari, M.S. Feld, Phys. Rev. Lett. **73** , 3375 (1994)
7. G.M. Meyer, H-J. Briegel and H. Walther, Europhys. Lett. **37**, 317 (1997); G.M. Meyer, M.Löffler and H. Walther, Phys. Rev. A. **56**, R1099, (1997).
8. D.M. Meekhof, C. Monroe, B.E. King, W.M. Itano and D.J. Wineland, Phys. Rev. Lett. **76**, 1796 (1996)
9. The physical discussion of the Paul traps is found in two review papers: Hans Dehmelt, Rev. Mod. Phys. **62**, 525 (1990) and Wolfgang Paul, Rev. Mod. Phys. **62**, 531 (1990).
10. E. Massoni, M. Orszag (To be published in Optics Communications)

11. H. Haken, *Light and Matter*, edited by L. Genzel, Handbuch der Physik Vol. XXv/2c (Springer-Verlag, Berlin, 1970).
12. D.M. Meekhof, C. Monroe, B.E. King, W.M. Itano and D.J. Wineland Phys. Rev. Lett. **76**, 1796 (1996).
13. M. Löffler, G.M. Meyer, H. Walther, Phys. Rev. A **55**, 3923 (1997).
14. G.M. Meyer, H.-J. Briegel and H. Walther, Europhys. Lett. **37**, 317 (1997).
15. G.M. Meyer, M. Löffler and H. Walther, Phys. Rev. A, **56**, R1099, (1997).
16. We have used that the contributions of H_I and $H_I^{(3)}$ are of the order of $\hbar g\eta <\sigma_{21}^\dagger><a><b^\dagger>$ and $\frac{\hbar g\eta^3}{6}<\sigma_{21}^\dagger><a><b^\dagger>^2<b>$ respectively. Since $<b>$ or $<b^\dagger>$ scales as $\sqrt{N}$, the ratio between these contributions is the quantity given by the left hand side of (27).
17. D.J. Heinzen and D.J. Wineland, Phys. Rev. A, **42**, 2977 (1990)
18. E. Massoni, M. Orszag, Opt. Comm. **179**, 315 (2000)

Part II

Quantum Interference, Entanglement, Decoherence and Quantum Computing

Decoherence, Pointer Engineering and Quantum State Protection

A.R.R. Carvalho, P. Milman, R.L. de Matos Filho, and L. Davidovich

Instituto de Física, Universidade Federal do Rio de Janeiro, Caixa Postal 68528, 21945-970 Rio de Janeiro, RJ, Brazil

Abstract. We show that motional states of trapped ions can be protected against decoherence by generating, through transitions involving unstable electronic states, artificial reservoirs for the vibrational motion, which have the states to be protected as their pointer states. We exemplify our procedure with qubits, squeezed states, approximate phase eigenstates and superpositions of coherent states. We show that these states may be efficiently protected against decoherence from thermal reservoirs and random fields.

1 Introduction

It is well known that the interaction of a quantum system with its surrounding environment may lead to quantum entanglement between system and environment, and to an irreversible loss of information on the system. Which set of states ("pointer states") is less sensitive to entanglement depends on the concrete form of the interaction Hamiltonian between system and environment [1]. In the course of the interaction, the reduced density operator of the system becomes rapidly diagonal in the pointer state basis, transforming any initial superposition of these states into a statistical mixture. Furthermore, pointer states minimize the rate of entropy production (due to the coupling between the system of interest and the reservoir) for a dynamic range of times [2]. The decoherence process by which coherent superpositions of pointer states get transformed into statistical mixtures is at the heart of the quantum theory of measurement [3], and plays an essential role in the classical limit of quantum mechanics [4]. It explains why one does not find, in the classical world, coherent superpositions of localized states of macroscopic objects: the decoherence time gets smaller and smaller as the distance between the states increases, and it is extremely small for macroscopic separations [1].

Fighting decoherence has become a major challenge in the last years, motivated by recent progress in the theory of quantum information processing, which relies on the possibility of preserving quantum coherence [5,14]. It is also of interest to high-precision frequency measurements in ion traps [10]. Several strategies have been devised. They include quantum error correction schemes [8], feedback implementations [9,10], the realization of q-bits in symmetric subspaces decoupled from the environment [11], and dynamical decoupling techniques [12].

M. Orszag, J.C. Retamal (Eds.): LNP 575, pp. 65–79, 2001.

In linear ion traps, proposed by Cirac and Zoller [14] as a possible system for implementing a quantum information processor, by far the most important decoherence effect is the one associated with the motional state [13,14]. In the present paper, we show that decoherence in the vibrational motion of a trapped ion can be suppressed by generating, through the techniques of "reservoir engineering" [15], artificial reservoirs associated with properly chosen pointer observables, which have the states to be preserved as their eigenstates, and which dominate over other dissipation processes. We exemplify this procedure by showing how to protect from decoherence several kinds of non-classical states.

A detailed study of the effects of the environment on the dynamics of an open system can be made with the use of master equations. Under the hypotheses of Markovian dynamics and complete positivity [16,17], a master equation describing the reduced dynamics of the system can be written in the Lindblad form

$$\frac{d\hat{\rho}}{dt} = \frac{i}{\hbar}[\hat{H},\hat{\rho}] + \mathcal{L}\hat{\rho}\,, \tag{1}$$

where $\hat{\rho}$ is the reduced density operator of the system in the interaction picture, $-(i/\hbar)[\hat{H},\hat{\rho}]$ is the unitary evolution term, under the action of the Hamiltonian $\hat{H}$, and

$$\mathcal{L}\hat{\rho} \equiv \sum_i (\gamma_i/2)\left(2\,\hat{c}_i\,\hat{\rho}\,\hat{c}_i^\dagger - \hat{c}_i^\dagger\,\hat{c}_i\,\hat{\rho} - \hat{\rho}\,\hat{c}_i^\dagger\,\hat{c}_i\right)\,. \tag{2}$$

The operators $\hat{c}_i$ are closely related to the system operators present in the interaction Hamiltonian and γ_i measures the strength of the system-environment coupling. When the unitary evolution term is neglected, the pointer basis is given by the set of all the eigenstates of the operator $\hat{c}_i$. If all $\hat{c}_i$'s are Hermitian, their eigenstates lead to the vanishing of the non-unitary term in (1). Therefore, these states will not be affected by the environment [11].This will not be the case, however, if the $\hat{c}_i$'s are not Hermitian.

2 Strategy for Quantum State Protection

Our strategy for the protection of a given quantum state against the effects of the environment consists in adding a new term to the right hand side of Eq. (1). More specifically, via adequate external driving of the system, one can engineer the system-environment coupling, so that the net effect of the external driving is the modification of the master equation (1) into an effective master equation of the form

$$\frac{d\hat{\rho}}{dt} = \mathcal{L}\hat{\rho} + (\Gamma_{\mathrm{eng}}/2)\left(2\,\hat{d}\,\hat{\rho}\,\hat{d}^\dagger - \hat{d}^\dagger\,\hat{d}\,\hat{\rho} - \hat{\rho}\,\hat{d}^\dagger\,\hat{d}\right)\,. \tag{3}$$

The operator $\hat{d}$ is chosen so that the state one wants to protect is the *only* steady state of Eq. (3) without the environment term $\mathcal{L}\hat{\rho}$. For $\Gamma_{\mathrm{eng}} \gg \gamma$, the steady state of the new master equation (3) will be very close to the state to be protected (if the state is not unique, the term $\mathcal{L}\hat{\rho}$ could still induce transitions between the steady states). Besides, *any* state of the system will decay into the state chosen to be protected. Therefore, this is also a procedure for preparing quantum states in the presence of decoherence.

3 Application to a Trapped Ion

In the following we apply this method to the one-dimensional motion of the center-of-mass of an ion confined in an electromagnetic trap, described by a harmonic potential. The reservoir engineering process will be implemented by letting the ion interact with several laser beams of adequate frequencies and intensities, oriented along the trap axes, and either resonant with a transition (frequency ω_{21}) between two levels $|1\rangle$ (lower level) and $|2\rangle$ (upper level), or to a sideband, that is, with frequencies $\omega_{21} \pm k\nu$, where ν is the vibrational frequency in the harmonic trap.

3.1 Hamiltonian of the system

The interaction between the trapped ion and the laser beams is described, in the optical rotating-wave approximation, by the Hamiltonian

$$\hat{H}(t) = \hat{H}_0 + \hat{H}_{\mathrm{int}}(t), \tag{4}$$

with

$$\begin{aligned} \hat{H}_0 &= \hbar\omega_1 \hat{A}_{11} + \hbar\omega_2 \hat{A}_{22} + \sum_{i=x,y,z} \hbar\nu_i \hat{a}_i^\dagger \hat{a}_i , \\ \hat{H}_{\mathrm{int}}(t) &= \frac{1}{2} \sum_n \hbar\Omega_n \hat{A}_{21} e^{i(\mathbf{k}_n \cdot \hat{\mathbf{r}} - \omega_n t)} + H.c.\,. \end{aligned} \tag{5}$$

Here, the operators $\hat{a}_i$ $(i = x, y, z)$ are, respectively, the annihilation operators corresponding to the ionic vibrational motion along the directions x, y, and z, with corresponding frequencies ν_i $(i = x, y, z)$. The operators $\hat{A}_{kl}$ $(k, l = 1, 2)$ are electronic operators, defined by $\hat{A}_{kl} = |l\rangle\langle k|$. The n laser fields are characterized by the interaction Rabi frequencies Ω_n, proportional to the field amplitudes, and the wave vectors $\mathbf{k}_n$. The normalized center-of-mass vector operator $\hat{\mathbf{r}}$ is given by

$$\hat{\mathbf{r}} = \sum_{i=x,y,x} \sqrt{\frac{\hbar}{2M\nu_i}} (\hat{a}_i + \hat{a}^\dagger) \mathbf{u}_i \,, \tag{6}$$

where $\mathbf{u}_i$ is the unit vector corresponding to the direction i and M is the ion's mass.

For our purposes it is important to consider the ion to be in the regime of resolved sidebands, given by $\nu \gg \Gamma, \Omega_n$, where Γ is the electronic energy decay rate, ν is the ion vibrational frequency, and Ω_n is the (complex) Rabi frequency corresponding to laser n, tuned to the k-th red vibrational sideband of the ion. We also assume, for the sake of simplicity, that the center-of-mass motion is one-dimensional, as it is the case in linear traps. Under these conditions the interaction Hamiltonian can be written as

$$\hat{\tilde{H}}_{\text{int}}(t) = \sum_n \hat{\tilde{H}}_{\text{int}}{}^{(n)}(t)\,, \tag{7}$$

where $\hat{\tilde{H}}_{\text{int}}{}^{(n)}(t)$ is the interaction term corresponding to laser n, given, in the interaction picture, by [18]

$$\begin{aligned} \hat{\tilde{H}}_{\text{int}}{}^{(n)}(t) &= \hbar g(\hat{A}_{21}\hat{d} + \hat{A}_{12}\hat{d}^\dagger) \\ &= \frac{1}{2}\hbar\Omega_n(i\eta_n)^k \hat{A}_{21} \hat{f}_k(\hat{\tilde{a}}^\dagger\hat{\tilde{a}})\,\hat{\tilde{a}}^k + H.c.. \end{aligned} \tag{8}$$

Here the coupling constant g is taken to be real and

$$\hat{f}_k(\hat{a}^\dagger\hat{a}) = e^{-\eta_n{}^2/2} \sum_{l=0}^{\infty} \frac{(-1)^l \eta_n{}^{2l}}{l!(l+k)!}(\hat{a}^\dagger)^l \hat{a}^l. \tag{9}$$

The operator $\hat{a}$ is the annihilation operator of a quantum of the ionic one-dimensional vibrational motion. The quantity $\eta_n = \sqrt{\hbar(\mathbf{k}_n \cdot \mathbf{u})^2/2M\nu}$ is the Lamb–Dicke parameter with respect to the direction of vibration, fixed by the unit vector $\mathbf{u}$. It is assumed that $\eta_v \ll 1$ for any direction $\mathbf{v}$ orthogonal to $\mathbf{u}$, as it is the case in linear traps.

3.2 Master equation for the center-of-mass motion

In the Born-Markov limit, the time evolution of the vibronic density operator $\hat{\rho}$ for the direction $\mathbf{u}$ (say x) is

$$\begin{aligned} \frac{d\hat{\rho}}{dt} &= -\frac{i}{\hbar}\left[\hat{\tilde{H}}_{\text{int}}, \hat{\rho}\right] + \frac{\Gamma}{2}\left(2\hat{A}_{12}\bar{\hat{\rho}}\hat{A}_{21} - \hat{A}_{22}\hat{\rho} - \hat{\rho}\hat{A}_{22}\right) \\ &+ \mathcal{L}\hat{\rho}, \end{aligned} \tag{10}$$

where the second term corresponds to spontaneous emission with energy relaxation rate Γ, and

$$\bar{\hat{\rho}} = \frac{1}{2}\int_{-1}^{1} ds\, W(s) e^{i\eta(\hat{a}+\hat{a}^\dagger)s}\,\hat{\rho}\, e^{-i\eta(\hat{a}+\hat{a}^\dagger)s} \tag{11}$$

accounts for changes of the vibrational energy along the x direction due to spontaneous emission with angular distribution $W(s)$. A dipole emission pattern symmetric around the z axis corresponds to $W(s) = (3/4)(1 - s^2)$. The last term of (10) describes the coupling of the center-of-mass motion to the environment, and has the general form (1). However, the precise form of this dissipation term is not important for our purposes.

Note that in writing Eq. (10), we have traced the density operator over the directions orthogonal to the one-dimensional motion, using the fact that the corresponding vibrations are weakly coupled to the applied fields. Also, we are using the same Lamb-Dicke parameter η for all fields propagating along the x-direction, since the difference between the parameters corresponding to fields resonant with the electronic or with sideband transitions is much smaller than η itself.

The matrix elements of (10) with respect to the electronic basis yield the equations

$$\dot{\hat{\rho}}_{11} = -ig(\hat{d}^\dagger \hat{\rho}_{21} - \hat{\rho}_{12}\hat{d}) + \Gamma \bar{\hat{\rho}}_{22} + \mathcal{L}\hat{\rho}_{11} , \tag{12}$$

$$\dot{\hat{\rho}}_{22} = -ig(\hat{d}\hat{\rho}_{12} - \hat{\rho}_{21}\hat{d}^\dagger) - \Gamma \hat{\rho}_{22} + \mathcal{L}\hat{\rho}_{22} , \tag{13}$$

$$\dot{\hat{\rho}}_{12} = -ig(\hat{d}^\dagger \hat{\rho}_{22} - \hat{\rho}_{11}\hat{d}^\dagger) - \frac{\Gamma}{2}\hat{\rho}_{12} + \mathcal{L}\hat{\rho}_{12} , \tag{14}$$

We assume now that the decay rate Γ is by far the largest rate in the system, much larger than the coupling g and the decay rate γ. Under this condition, one can eliminate $\hat{\rho}_{12}$ adiabatically, which amounts to setting $\dot{\hat{\rho}}_{12} = 0$ in (14), leading to

$$\hat{\rho}_{12} = -\frac{2ig}{\Gamma}(\hat{d}^\dagger \hat{\rho}_{22} - \hat{\rho}_{11}\hat{d}^\dagger) + \frac{2}{\Gamma}\mathcal{L}\hat{\rho}_{12} . \tag{15}$$

Iterating this equation and taking into account that $\mathcal{L} \propto \gamma$, one gets:

$$\hat{\rho}_{12} = -(2ig/\Gamma)(\hat{d}^\dagger \hat{\rho}_{22} - \hat{\rho}_{11}\hat{d}^\dagger)\left[1 + O\left(\gamma/\Gamma\right)\right] . \tag{16}$$

Replacing (16) into (12) and (13), and adding up these two equations, we get, since the reduced density operator for the vibrational mode is given by $\hat{\rho}_v = \hat{\rho}_{11} + \hat{\rho}_{22}$, and neglecting the correction proportional to γ/Γ in (16):

$$\dot{\hat{\rho}}_v = \frac{2g^2}{\Gamma}\Big[\left(2\hat{d}\hat{\rho}_{11}\hat{d}^\dagger - \hat{d}^\dagger\hat{d}\hat{\rho}_{11} - \hat{\rho}_{11}\hat{d}^\dagger\hat{d}\right) \tag{17}$$
$$+ \left(2\hat{d}^\dagger\hat{\rho}_{22}\hat{d} - \hat{d}\hat{d}^\dagger\hat{\rho}_{22} - \hat{\rho}_{22}\hat{d}\hat{d}^\dagger\right)\Big] - \Gamma(\hat{\rho}_{22} - \bar{\hat{\rho}}_{22}) + \mathcal{L}\hat{\rho}_v .$$

Under the conditions assumed here, the matrix elements of $\hat{\rho}_{22}$ are much smaller than those of $\hat{\rho}_{11}$. Indeed, replacing (16) in (13), and eliminating $\hat{\rho}_{22}$ from (13) adiabatically, one gets $\hat{\rho}_{22} \approx O\left[(g/\Gamma)^2\right]\hat{\rho}_{11}$. We can therefore safely neglect the terms dependent of $\hat{\rho}_{22}$ inside the brackets in (17) and at the same time replace $\hat{\rho}_{11}$ by $\hat{\rho}_v$. We verified numerically that these are indeed excellent approximations. We get then, finally:

$$\dot{\hat{\rho}}_v = \frac{2g^2}{\Gamma}\left(2\hat{d}\hat{\rho}_v\hat{d}^\dagger - \hat{d}^\dagger\hat{d}\hat{\rho}_v - \hat{\rho}_v\hat{d}^\dagger\hat{d}\right) - \Gamma\left(\hat{\rho}_{22} - \bar{\hat{\rho}}_{22}\right)$$
$$+ \mathcal{L}\hat{\rho}_v . \tag{18}$$

We will base our considerations on this equation. The first term on the r.h.s. has the form (1). This is the "engineered reservoir," with a decay constant $\Gamma_{\rm eng} = 4g^2/\Gamma$.

Neglecting terms of $O(\eta^4)$ in the expansion of the second term on the r.h.s. of Eq. (18), one can show that its contribution is $\sim (\eta^2/5)(4g^2/\Gamma)\hat{\rho}_v$, that is, $(2\eta^2/5)$ multiplied by the engineered-reservoir term. For $\eta \approx 0.25$, this yields a factor $\sim 1/40$, a small correction, which is however fully taken into account in our numerical simulations. Therefore, the action of the engineered reservoir will be the dominant one as long as $\Gamma_{\rm eng} \gg \gamma$.

3.3 Effect of random fields

In recent experiments with trapped ions, random fields seem to play an important role in the decoherence process [14]. In the following we will show that their effect may also be described by (1).

We consider a classical random electromagnetic field that has a component in the direction of motion (assumed for simplicity to be one-dimensional) of the following form:

$$E(t) = \mathcal{E}^{(+)}(t)e^{-i\nu t} + \mathcal{E}^{(-)}(t)e^{i\nu t}\,, \tag{19}$$

where ν coincides with the vibration frequency and $\mathcal{E}^{\pm}(t)$ are time-dependent envelopes. We neglect therefore the spatial dependence of the field over the region where the ion is localized (electric dipole approximation).

The Hamiltonian representing the interaction between this classical field and the ion is given by

$$\hat{H} = -e\hat{\boldsymbol{x}} \cdot \boldsymbol{E} = -e\sqrt{\frac{\hbar}{2M\nu}}(\hat{a}^\dagger + \hat{a})E(t)\,, \tag{20}$$

where e is the ionic charge.

Going to the interaction picture and making the rotating-wave approximation we get

$$\hat{H} = -\mu\Big[\mathcal{E}^{(+)}(t)\hat{a}^\dagger + \mathcal{E}^{(-)}(t)\hat{a}\Big]\,, \tag{21}$$

where we have set

$$\mu \equiv e\sqrt{\frac{\hbar}{2M\nu}}\,. \tag{22}$$

The evolution of the density matrix will be given by the Von-Neumann equation $\dot{\hat{\rho}} = -\frac{i}{\hbar}[\hat{H},\hat{\rho}]$. Integrating between t and $t+\Delta t$ and iterating this equation one gets

$$\begin{aligned}\Delta\hat{\rho}(t) = \hat{\rho}(t+\Delta t) - \hat{\rho}(t) &= \frac{1}{i\hbar}\int_t^{t+\Delta t} [\hat{H}(t'),\hat{\rho}(t)]dt' \\ &+ \left(\frac{1}{i\hbar}\right)^2 \int_t^{t+\Delta t} dt' \int_t^{t'} dt'' \Big[\hat{H}(t'),[\hat{H}(t''),\hat{\rho}(t'')]\Big]\,.\end{aligned} \tag{23}$$

As a first approximation let us suppose that Δt is much smaller then the characteristic time T of the evolution of ρ. We can therefore neglect the evolution of $\rho(t)$ in the last term of equation (23) replacing $\rho(t'')$ by $\rho(t)$. The expansion of the commutators with the above approximations yields

$$\begin{aligned}
\Delta\hat{\rho}(t) &= \frac{i\mu}{\hbar}\int_t^{t+\Delta t} dt' \Big[(\mathcal{E}^{(+)}(t')\hat{a}^\dagger + \mathcal{E}^{(-)}(t')\hat{a})\hat{\rho}(t) \\
&- \hat{\rho}(t)(\mathcal{E}^{(+)}(t')\hat{a}^\dagger + \mathcal{E}^{(-)}(t')\hat{a})\Big] \\
&- \frac{\mu^2}{\hbar^2}\int_t^{t+\Delta t} dt' \int_t^{t'} dt'' \mathcal{E}^{(+)}(t')\mathcal{E}^{(+)}(t'') \\
&\times \Big[\hat{a}^{\dagger 2}\rho(t) - 2\hat{a}^\dagger\rho(t)\hat{a}^\dagger + \rho(t)\hat{a}^{\dagger 2}\Big] \\
&+ \mathcal{E}^{(-)}(t')\mathcal{E}^{(-)}(t')\Big[\hat{a}^2\rho(t) - 2\hat{a}\rho(t)\hat{a} + \rho(t)\hat{a}^2\Big] \\
&+ \mathcal{E}^{(+)}(t')\mathcal{E}^{(-)}(t'')\Big[\hat{a}^\dagger\hat{a}\rho(t) - \hat{a}\rho(t)\hat{a}^\dagger - \hat{a}^\dagger\rho(t)\hat{a} + \rho(t)\hat{a}\hat{a}^\dagger\Big] \\
&+ \mathcal{E}^{(+)}(t'')\mathcal{E}^{(-)}(t')\Big[\hat{a}\hat{a}^\dagger\rho(t) - \hat{a}\rho(t)\hat{a}^\dagger - \hat{a}^\dagger\rho(t)\hat{a} + \rho(t)\hat{a}^\dagger\hat{a}\Big] . \qquad (24)
\end{aligned}$$

We introduce now the random character of the classical electric field. From the reality of the field we have $\mathcal{E}^{(\pm)}(t) = \mathcal{E}^{(\mp)^*}(t)$ and we can write its complex amplitudes $\mathcal{E}^{(+)}$ e $\mathcal{E}^{(-)}$ as a sum of the real and imaginary parts

$$\mathcal{E}^{(\pm)}(t) = \mathcal{E}_r(t) \pm i\mathcal{E}_i(t) . \qquad (25)$$

The randomness of the field implies that the equation for the density matrix should be averaged, with the statistical distribution for the field taken into account. This procedure leads to mean values and correlations functions of the electric field appearing in the above master equation. In order to proceed, we need now to make some basic assumptions about the random process. We assume that the electric field has zero mean, and that it is ruled by a second-order stationary process:

i) $\langle\mathcal{E}^{(\pm)}(t)\mathcal{E}^{(\pm)}(t')\rangle$,
ii) $\langle\mathcal{E}^{(+)}(t)\mathcal{E}^{(-)}(t')\rangle = \langle\mathcal{E}^{(-)^*}(t)\mathcal{E}^{(-)}(t')\rangle \equiv D(t'-t)$,
iii) $\langle\mathcal{E}^{(-)}(t)\mathcal{E}^{(+)}(t')\rangle = \langle\mathcal{E}^{(-)^*}(t')\mathcal{E}^{(-)}(t)\rangle \equiv D(t-t') = D^*(t'-t)$.

Expressing these equations in terms of the real and imaginary parts of the complex electric field, one gets

$$\begin{aligned}
D(t'-t) &= \langle\mathcal{E}_r(t)\mathcal{E}_r(t')\rangle + \langle\mathcal{E}_i(t)\mathcal{E}_i(t')\rangle \\
&+ i\Big(\langle\mathcal{E}_r(t)\mathcal{E}_i(t')\rangle + \langle\mathcal{E}_i(t)\mathcal{E}_r(t')\rangle\Big) , \qquad (26)
\end{aligned}$$

and

$$\begin{aligned}
\langle\mathcal{E}^{(\pm)}(t)\mathcal{E}^{(\pm)}(t')\rangle &= \langle\mathcal{E}_r(t)\mathcal{E}_r(t')\rangle - \langle\mathcal{E}_i(t)\mathcal{E}_i(t')\rangle \\
&+ i\Big(\langle\mathcal{E}_r(t)\mathcal{E}_i(t')\rangle + \langle\mathcal{E}_i(t)\mathcal{E}_r(t')\rangle\Big) . \qquad (27)
\end{aligned}$$

The second-order stationarity implies that [19]

$$\begin{aligned}
\langle \mathcal{E}_r(t)\mathcal{E}_r(t')\rangle &= \langle \mathcal{E}_i(t)\mathcal{E}_i(t')\rangle \,, \\
\langle \mathcal{E}_r(t)\mathcal{E}_i(t')\rangle &= -\langle \mathcal{E}_i(t)\mathcal{E}_r(t')\rangle = 0 \,.
\end{aligned}$$

From these equations, one gets

$$\langle \mathcal{E}^{(\pm)}(t)\mathcal{E}^{(\pm)}(t')\rangle = 0 \tag{28}$$

and

$$D(t'-t) = D(t-t') = 2\langle \mathcal{E}_r(t)\mathcal{E}_r(t')\rangle \,. \tag{29}$$

Replacing these results into the master equation and assuming a Markovian process, such that $D(t-t') = 2D\delta(t-t')$, we obtain the master equation in the Lindblad form (1):

$$\dot{\hat{\rho}} = -\frac{\mu^2 D}{\hbar^2}\left[(\hat{a}^\dagger \hat{a}\rho(t) - 2\hat{a}\rho\hat{a}^\dagger + \rho\hat{a}^\dagger\hat{a}) + (\hat{a}\hat{a}^\dagger\rho - 2\hat{a}^\dagger\rho\hat{a} + \rho\hat{a}\hat{a}^\dagger)\right] . \tag{30}$$

This equation contains cooling and heating terms for the vibrational motion of the ion, corresponding respectively to the first and second contributions on the right-hand side, with the same rate $\frac{\mu^2 D}{\hbar^2} \equiv \gamma$. The reservoir obtained by this method corresponds therefore to an infinite temperature thermal reservoir (in fact, it is obtained from a thermal reservoir in the double limit in which the thermal photon number $N_T \to \infty$, and at the same time the dissipation rate $\gamma \to 0$, so that γN_T remains constant). Both random fields and thermal reservoirs will be considered in our simulations.

3.4 Protection of superpositions of Fock states

In order to protect a state $|\psi\rangle$, we look for $\hat{d}$ such that $\hat{d}|\psi\rangle = \lambda|\psi\rangle$ with $\lambda = 0$, and make sure that there is no further eigenstate $|\phi\rangle$ of $\hat{d}$ such that $[\hat{d}, \hat{d}^\dagger]|\phi\rangle = 0$. This implies that $|\psi\rangle$ is the only pure steady state of (3) without $\mathcal{L}\hat{\rho}$. Since $\Gamma_{\text{eng}} \gg \gamma$, this yields a good approximation of the corresponding steady state of (3).

As a first example, we consider the protection (and generation) of the class of states

$$|\psi\rangle = \sum_{n=0}^{N} c_n |n\rangle \,, \tag{31}$$

where $|n\rangle$ is an energy eigenstate of the vibrational motion of the trapped ion and $c_n \neq 0$.

It is easy to see that the operator

$$\hat{d} = \hat{\text{g}}(\hat{n})\,\hat{a} + \hat{\text{h}}(\hat{n}) \tag{32}$$

has $|\psi\rangle$ as its only eigenstate with eigenvalue $\lambda = 0$, provided the eingenvalues of $\hat{\mathrm{g}}(\hat{n})$ and $\hat{\mathrm{h}}(\hat{n})$ fulfill the constraints

$$\mathrm{g}(m) = -\frac{\mathrm{h}(m)}{\sqrt{m+1}} c_m/c_{m+1} \qquad (m = 0, \ldots, N-1)\,, \tag{33}$$

and N is the first zero of $\mathrm{h}(m)$. For this case, $|\psi\rangle$ is the only steady state of (3).

Inspection of Eqs. (12) and (9) shows that the operator $\hat{\mathrm{g}}(\hat{n})\,\hat{a}$ can be engineered by driving the ion with N laser fields, tuned to the first vibrational sideband of the ion. The values of the Rabi frequencies Ω_n of the N lasers are given by the following system of N linear equations ($m = 0, \cdots, N-1$):

$$\sum_{n=1}^{N} e^{-\eta_n^2/2} \eta_n \Omega_n \sum_{l=0}^{m} \frac{(-1)^l \eta^{2l}}{l!(l+1)!} \frac{m!}{(m-l)!} = \frac{i\mathrm{h}(\mathrm{m})}{\sqrt{m+1}} \frac{c_m}{c_{m+1}}, \tag{34}$$

where the Lamb–Dicke parameters η_n depend on the orientation of the laser fields with respect to the x-direction. The operator $\hat{\mathrm{h}}(\hat{n})$ is constructed by driving the ion with two laser fields resonant with the electronic transition, one of them propagating orthogonal to the x-axis (say y-direction, with $\eta_y \ll 1$). The Rabi frequencies of these two lasers are related by $\Omega_y = -\Omega_x L_N(\eta_x^2)$, where $L_N(x)$ is a Laguerre polynomial of order N (η_x, which corresponds to the resonant laser along the x-axis, should not be too large for the first zero of $h(m)$ to occur at $m = N$).

From (34) it follows that our method is sensitive only to relative intensity and phase fluctuations, since state selection is determined by ratios of Rabi frequencies. These quantities are much easier to stabilize than the absolute intensities and phases of the laser beams.

3.5 Protection of a qubit

An important representative of the class of states presented above is the "qubit" state $|\psi\rangle = c_0|0\rangle + c_1|1\rangle$. The discussion above implies that this state can be generated and protected against the action of an external reservoir with just three lasers as shown in Fig. 1: two of them (with Rabi frequencies Ω_1, tuned do the first red sideband, and Ω_x, resonant with the electronic transition) propagate along the x-axis, while a third one (Rabi frequency Ω_y, resonant with the electronic transition) propagates along the y-axis. Their Rabi frequencies must satisfy the following conditions (here η is the Lamb-Dicke parameter corresponding to the lasers propagating along the x axis):

$$\frac{\eta\Omega_x}{i\Omega_1} = -\frac{c_1}{c_0} \quad \text{and} \quad \frac{\Omega_y}{i\Omega_1} = e^{-\eta^2/2} \frac{c_1}{c_0} \frac{1-\eta^2}{\eta}. \tag{35}$$

In this case, $\Gamma_{\text{eng}} = \eta^2 \Omega_1^2/\Gamma$ in Eq. (3). In order for the corresponding reservoir to win over the environment reservoir, one needs $\eta^2 {\Omega_1}^2/\Gamma \gg \gamma$, but at the

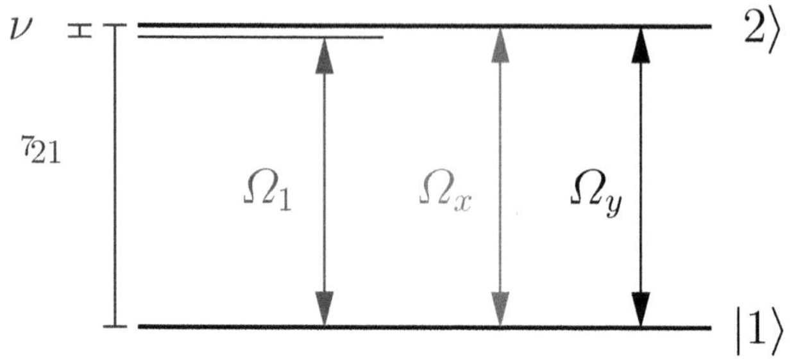

Fig. 1. Level scheme for protection of a qubit: two lasers in $\hat{x}$ direction, one tuned to the first red sideband (Ω_1), other resonant to the electronic transition (Ω_x) and another laser (Ω_y) also resonant but in an orthogonal direction y.

same time $\Gamma, \Omega_1 \ll \nu$ and $\Gamma \gg \eta\Omega_1$. This requirements are satisfied if $\Gamma \approx 4$ MHz, $\Omega_1 \approx 2$ MHz, $\eta = 0.2$, $\nu \approx 20 - 30$ MHz, as long as $\gamma \ll \eta^2 {\Omega_1}^2/\Gamma \approx 40$ kHz. Fig. 2a displays the fidelity $\mathrm{F}(t) = \mathrm{Tr}\{\hat{\rho}(0)\hat{\rho}(t)\}$, with the ion initially in the vibrational state $|\psi\rangle = (|0\rangle + |1\rangle)/\sqrt{2}$ [all our numerical simulations are obtained from Eq. (10), and we always assume the ion to be initially in the electronic ground state]. Both a thermal and a random field reservoir have been considered. As can be seen, the system rapidly reaches a steady state with fidelity very close to unity ($\eta = 0.2$).

One should remark that feedback procedures [10] do not protect states involving superpositions of $|0\rangle$ and $|1\rangle$, since the loss of one photon by the state $|1\rangle$ completely erases any phase information about the original state. Our procedure works however very well in this case.

3.6 Protection of approximate phase eigenstates

Other interesting example is the approximate phase state [20]

$$|\psi\rangle = (1/\sqrt{N+1}) \sum_{n=0}^{N} e^{in\phi}|n\rangle \,, \qquad (36)$$

which can be generated and protected by $N+2$ lasers. Fig. 2b displays the time evolution of the fidelity $\mathrm{F}(t)$, for the approximate phase state $|\psi\rangle = (1/2)\sum_{n=0}^{3}|n\rangle$ (η's in the range $0.1 - 0.2$).

3.7 Superpositions of coherent states

A class of states which is specially fragile against the action of decoherence is the one formed by mesoscopic superpositions of coherent states. Under action of an external reservoir, these states decay to a mixture of coherent states within

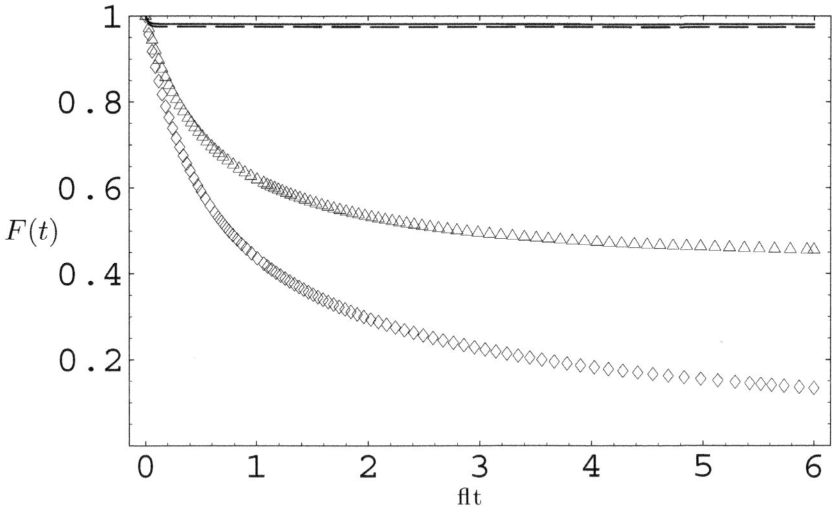

(a)

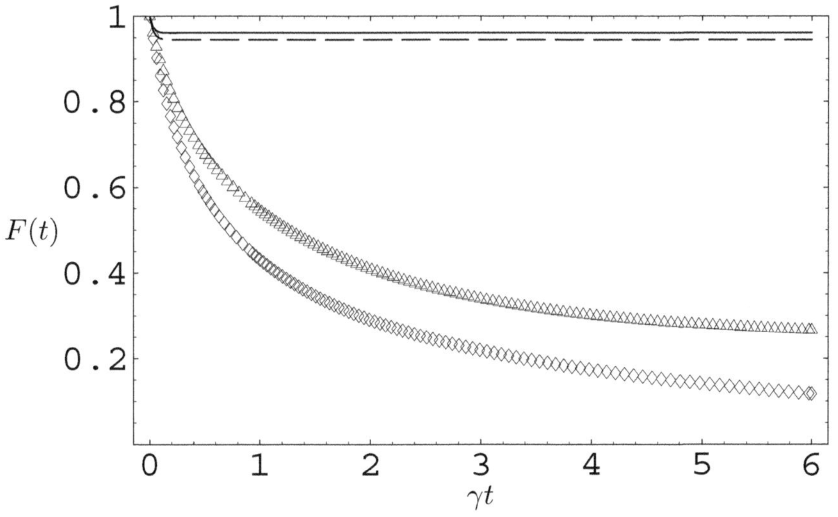

(b)

Fig. 2. Time evolution of the fidelity $F(t)$ for several initial nonclassical states, in the presence of a thermal reservoir with $\langle n_{\rm th}\rangle = 0.5$ (full curves and triangles) and a random field (dashed curves and losangles). For the losangles and triangles, $\Gamma_{\rm eng} = 0$. The full and dashed curves correspond to $\Gamma_{\rm eng} = 40\gamma$. The nonclassical states are: (a) $|\psi\rangle = \frac{1}{\sqrt{2}}(|0\rangle + |1\rangle)$; (b) $|\psi\rangle = \frac{1}{2}\sum_{n=0}^{3}|n\rangle$.

an extremely short time, inversely proportional to the distance between the two states in phase space [4]. Our technique can also be applied to the Schrödinger-cat-like state [21]

$$|\phi_+\rangle = (|\alpha\rangle + i|-\alpha\rangle)/\sqrt{2}. \tag{37}$$

Since $|\phi_+\rangle$ has no "hole" in its number distribution (which coincides with that for the coherent state $|\alpha\rangle$), it can be approximated by one of the states $|\psi\rangle$ discussed above. Consequently, one could, with the use of $N+2$ lasers, generate and protect the state $|\psi\rangle = \sum_{n=0}^{N} c_n|n\rangle$, with the first N coefficients c_n equal to the corresponding coefficients of $|\phi_+\rangle$.

One should notice, however, that it is possible in this case to find directly a Lindblad operator $\hat{d}$, which has the state $|\phi_+\rangle$ as its only eigenstate with zero eigenvalue:

$$\hat{d} = \mathcal{T}\hat{a}\mathcal{T}^\dagger = e^{i\pi\hat{n}}\hat{a} + i\alpha\,. \tag{38}$$

Here $\mathcal{T}$ is the unitary operator

$$\mathcal{T} = \exp[i\pi\hat{n}(\hat{n}-1)/2]\exp(\alpha\hat{a}^\dagger - \alpha^*\hat{a})\,, \tag{39}$$

which yields $|\phi_+\rangle$ when applied to the vacuum. For this choice of $\hat{d}$, we plot in Fig.3a the fidelity $F(t)$, for the initial state $|\phi_+\rangle$ ($\eta = 0.2$). An open problem is how to engineer this operator with a finite number of laser beams.

3.8 Protection of squeezed states

Finally, we describe the protection of a squeezed state. We set $\hat{d} = \hat{a} + \chi\hat{a}^\dagger$, where $\chi = \tanh r$ and r is the squeezing factor. The corresponding setup consists of two lasers along the direction of squeezing, resonant with the first red (laser 1) and the first blue (laser 2) sidebands, and with Rabi frequencies satisfying $\Omega_2/\Omega_1 = \chi$ [15]. The numerical simulation is shown for $r = 0.6$ in Fig.3b, for a realistic set of parameters ($\eta = 0.05$). Higher values of squeezing render our method less effective, since the presence of higher photon numbers would lead to stronger dissipation by the "natural" reservoir.

4 Conclusion

In conclusion, we have suggested a method for protecting quantum states of the vibrational motion of a trapped ion against decoherence by generating artificial reservoirs which have the states to be protected as pointer states. It differs from dynamical decoupling techniques [12] (which require pulses with durations smaller than the reservoir correlation time) and feedback techniques [10] (which have been proposed for the protection of "cat-like" states, but are inefficient for the protection of qubit states). More general pointer states can be generated by applying unitary transformations to the states and operators discussed here.

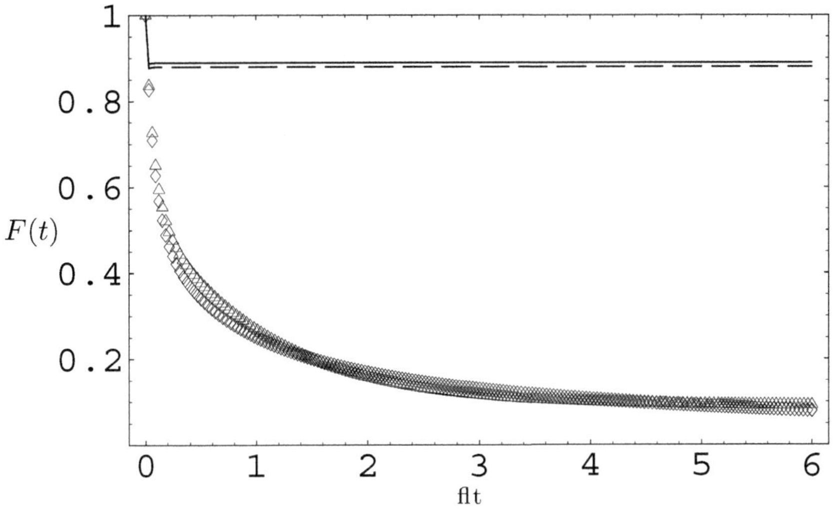

(a)

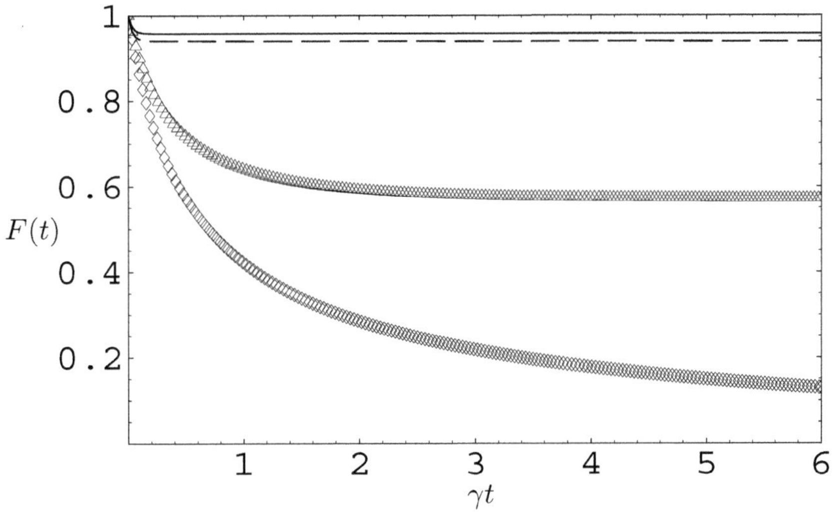

(b)

Fig. 3. Time evolution of the fidelity $F(t)$ for several initial nonclassical states, in the presence of a thermal reservoir with $\langle n_{\rm th}\rangle = 0.5$ (full curves and triangles) and a random field (dashed curves and losangles). For the losangles and triangles, $\Gamma_{\rm eng} = 0$. The full and dashed curves correspond to $\Gamma_{\rm eng} = 150\gamma$ for (a) and $\Gamma_{\rm eng} = 40\gamma$ for (b). The nonclassical states are: (a) $|\psi\rangle = \frac{1}{\sqrt{2}}(|\alpha\rangle + i|-\alpha\rangle)$, with $\alpha^2 = 3$; and (b) vacuum squeezed state with $r = 0.6$.

Indeed, the transformed states would still be the sole steady solutions of the master equation with the transformed operators. This is precisely the mechanism which leads to the protection of the states $|\phi_+\rangle$ above, and also of squeezed states, since they are related by unitary transformations to the vacuum, which is the only steady state for a zero-temperature reservoir (for which $\hat{d} = \hat{a}$). Our method is experimentally robust, since it depends only on the stabilization of relative intensities and phases, and leads to the successful generation and protection of qubits, squeezed states, and approximate phase and Schrödinger cat-like states, even in the deleterious presence of random electric fields, which have been shown to correspond to infinite-temperature thermal reservoirs.

We acknowledge the partial support of Conselho Nacional de Desenvolvimento Científico e Tecnológico (CNPq), Fundação de Amparo à Pesquisa do Estado do Rio de Janeiro (FAPERJ), Fundação Universitária José Bonifácio (FUJB), and Programa de Apoio a Núcleos de Excelência (PRONEX).

References

1. W. H. Zurek, Phys. Rev. D **24**, 1516 (1981); **26**, 1862 (1982); Physics Today **44**, 36 (1991); W. H. Zurek, Phys. Rev. Lett. **82**, 5181 (1999); D. Giulini et al., *Decoherence and the Appearance of a Classical World in Quantum Theory* (Springer Verlag, Berlin Heidelberg, 1996).
2. W.H. Zurek, Prog. Theor. Phys. **89**, 281 (1993); W.H. Zurek, S. Habib, and J.P. Paz, Phys. Rev. Lett. **70**, 1187 (1993).
3. J. Von Neumann, *Die Mathematische Grundlagen der Quantenmechanik* (Springer-Verlag, Berlin, 1932); *Quantum Theory and Measurement,* edited by J. A. Wheeler and W. H. Zurek (Princeton Univ. Press, Princeton, 1983).
4. For the role of decoherence in the classical limit of chaotic systems, see W.H. Zurek and J.P. Paz, Phys. Rev. Lett. **72**, 2508 (1994); for experimental results on decoherence, see M. Brune *et al.*, Phys. Rev. Lett. **77**, 4887 (1996), and C.J. Myatt *et al*, Nature **403**, 269 (2000).
5. For reviews, see D.P. Divicenzo, Science **270**, 255 (1995); A. Ekert and R. Josza, Rev. Mod. Phys. **68**, 733 (1996); J. Preskill, Physics Today **52**, 24 (1999).
6. J.J. Cirac and P. Zoller, Phys. Rev. Lett. **74**, 4091 (1995).
7. D.J. Wineland *et al*, Phys. Rev. A **46**, R6797 (1992); W.M. Itano, Phys. Rev. A **47**, 3554 (1993); S.F. Huelga *et al*, Phys. Rev. Lett. **79**, 3865 (1997); J. Steinbach and C.C. Gerry, Phys. Rev. Lett. **81**, 5528 (1998).
8. P.W. Shor, Phys. Rev. A **52**, 2493 (1995); D. Gottesman, *ibid.* **54**, 1862 (1996); A. Ekert and C. Macchiavello, Phys. Rev. Lett. **77**, 2585 (1996); A.R. Calderband *et al, ibid.* **78**, 405 (1997);
9. H. Mabuchi and P. Zoller, Phys. Rev. Lett. **76**, 3108 (1996).
10. D. Vitali, P. Tombesi, and G. J. Milburn, Phys. Rev. Lett. **79**, 2442 (1997); Phys. Rev. A **57**, 4930 (1998).
11. P. Zanardi and M. Rasetti, Phys. Rev. Lett. **79**, 3306 (1997); D. A. Lidar, I. L. Chuang, and K. B. Whaley, Phys. Rev. Lett. **81**, 2594 (1998); D. Braun, P.A. Braun, and F. Haake, Opt. Comm. **179**, 195 (2000).
12. L. Viola, E. Knill, and S. Lloyd, Phys. Rev. Lett. **82**, 2417 (1999); L. Viola and S. Lloyd, Phys. Rev. A **58**, 2733 (1998).

13. D.J. Wineland *et al* , Journal of Research of the National Institute of Standards and Technology **103**, 29 (1998).
14. D.F.V. James, Phys. Rev. Lett. **81**, 317 (1998).
15. J. F. Poyatos, J. I. Cirac, and P. Zoller, Phys. Rev. Lett. **77**, 4728 (1996); for methods of Hamiltonian engineering, see R.L. de Matos Filho and W. Vogel, Phys. Rev. A.**58**, R1661 (1998).
16. G. Lindblad, Math. Phys. **48**, 119 (1976).
17. R. Alicki and K. Lendi, *Quantum Dynamical Semigroups and Applications*, in Lecture Notes in Physics (Springer-Verlag, Berlin, 1987), No. 286; P. Pechukas, Phys. Rev. Lett. **73**, 1060 (1994); R. Alicki, Phys. Rev. Lett. **75**, 3020 (1995); P. Pechukas, *ibid.*, **75**, 3021 (1995).
18. W. Vogel and R. L. de Matos Filho, Phys. Rev. A **52**, 4214 (1995).
19. L. Mandel and E. Wolf, *Optical Coherence and Quantum Optics* (Cambridge University Press, 1995), Sec. 3.1.3.
20. D.T. Pegg and S.M. Barnett, Europhys. Lett. **6**, 483 (1988).
21. B. Yurke and B. Stoler, Phys. Rev. Lett. **57**, 13 (1986).

High Efficiency in Detection of Photonic Qubits

K.M. Gheri[1] and C. Saavedra[2]

[1] Institut für Theoretische Physik, Universität Innsbruck, Technikerstrasse 25/2, A-6020 Innsbruck, Austria
[2] Departamento de Física, Universidad de Concepción, Casilla 160-C, Concepción, Chile

Abstract. We consider a system of two-coupled cavities as a mechanism for an efficient absorption of a photon wavepacket which is used for encoding a qubit. The system of cavities is composed of a one-mode cavity and a longer cavity with a multi-mode structure. The cavities are coupled by a common dielectric. The long cavity contains an active medium for absorbing the wavepacket, which is switched on when the qubit is inside the cavity. At the same time we connect a source of detuning for the one-mode cavity, which allows trapping the qubit in the long cavity during the absorption process. We also consider a specific example for generating a train of entangled photonic qubits in the context of cavity-QED and show that undesired decoherence effects can be efficiently reduced in the considered scheme.

1 Introduction

Many experiments for testing the EPR hypothesis of local realism in quantum mechanics [1] are based on experiments that utilize a pair of entangled photons as particles in the test. The photons are usually entangled in polarization or momentum, or in both as occurs in spontaneous parametric down-conversion. However, the low efficiency in detecting these entangled photons does not allow to consider the experiments as definitive tests. In particular, a physical parameter that plays an important role here is the so called *single-particle detection efficiency* η_1. For instance, in the case of the GHZ paradox [2] the effect of a non-ideal detector has been studied [3] giving as lower bound for detection efficiency 90.8 %, which has recently been improved [4]. There exists a natural question here and it is: How can we improve the efficiency for detecting these entangled photons? A related problem was studied in connection with a transfer of information between two distant nodes of a quantum network [5]. This scheme, in principle, can be used for detecting each of the components for the entangled photons. However, a practical limitation can appear from the fact that the quantum channel needed to connect the source and detecting cavities needs to be reoriented in the case of entangled wavepackets.

Here, we consider a system of two coupled cavities as a mechanism for an efficient absorption of a photon wavepacket (arising from a given cavity QED source) being used for encoding a quantum bit, qubit. The qubits are codified in the polarization degree of freedom of each generated photon wavepacket. Each

M. Orszag, J.C. Retamal (Eds.): LNP 575, pp. 80–94, 2001.

qubit is stored into an active medium (multi-level atom) embedded in a system of two coupled cavities. These cavities have a common partially transmitting mirror, which allows a direct coupling between them. One of the cavities C_2 is a one-mode cavity which is the entrance cavity for the wavepacket. The other cavity C_1 is much longer than the one-mode cavity C_2. Thus, C_1 has a multi-mode structure in a frequency window of the order of the spectral linewidth of the first cavity. Then the long cavity C_1 behaves as a reservoir with a finite number of modes. The one-mode cavity together with the common dielectric plays the role of a mirror (of intricate structure) for C_1. This feature allows both for the entrance of the incoming photon to the long cavity C_1 by way of the short cavity C_2 and for a later trapping of it by switching on an effective source of detuning within the short cavity C_2. The incoming photon wavepackets can be partially reflected by the system of cavities. However, this can be minimized if the spectral linewidth of the short cavity is wider than that corresponding to the wavepacket.

We also consider a scheme for generating a train of entangled single-photon wavepackets which are well resolved in time [6,7]. This feature permits us to regard them as individual qubits. The scheme is based on well-established techniques from cavity quantum electrodynamics (CQED) [5,8,9]. We consider a nonclassical medium inside a high quality optical resonator. With the use of external pump fields, the coupling of the active medium to the resonator allows a single photon transfer to the resonator and therefrom via cavity decay to the continuum outside. Before the process is repeated the medium is coherently recycled to its initial state, which provides the basis for a build-up of entanglement among subsequently generated photonic qubits. An encoding of quantum information in the one-photon wavepackets could take place by identifying two orthogonal polarization states of the single photon with logical "0" and "1". In a conservative estimation, the number of entangled photon wavepackets that can be generated with high probability are of the order of *ten* entangled photons, for present experimental parameters. To have access to this kind of source represents an important step towards a definitive experimental test when studying violation of Bell inequalities, due to the exponential growing of these violations with the number of entangled particles.

2 Mode Structure of a System of Two-Cavities

Before addressing the question of detecting the photon wavepackets, we shall concentrate on studying the mode structure of a system of two-coupled cavities with a common dielectric. The purpose of having a common dielectric coupling the cavities is to allow a direct transfer of the incoming photon between the cavities. The basic goal of this system is that it can allow a mechanism for *trapping* a photon for a long time compared with the usual decay time of a single optical cavity. The system of cavities is depicted in figure 1. In case the cavity C_1 is much longer than the entrance cavity C_2, we expect that C_1 contains many resonances in a frequency window of the size of one resonance of the cavity C_2. Thus, the following relation holds: $\Delta\omega_1 \ll \Delta\omega_2$, where $\Delta\omega_j$ is the width of a resonance in cavity j. Here, we want to demonstrate that this picture can be implemented.

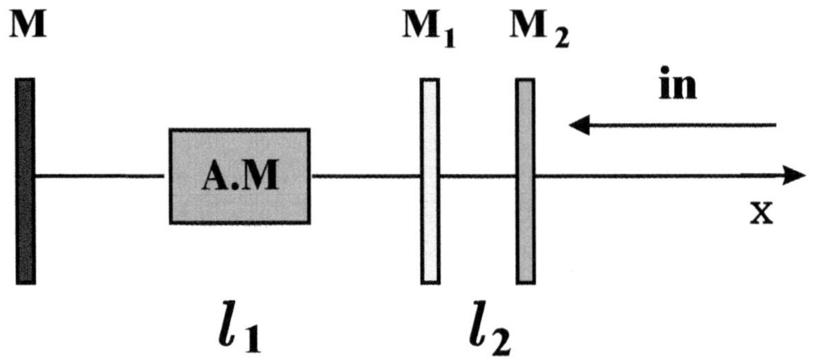

Fig. 1. System of cavities composed of a long cavity C_1 of length l_1, which is the region between a perfectly reflecting mirror M and a partially reflecting mirror M_1. The second cavity C_2 of length l_2 is the region contained between the two partially reflecting mirrors M_1 and M_2. Cavity C_1 contains an active medium composed of an atom with two three-level Λ-configurations. The interaction with the active medium is switched on after the photon wavepacket (qubit) is fully contained in C_1.

The existence of many resonances in the first cavity allows trapping a photon up to times of the order of the *revival time of the wavepacket*. During this time we connect the interaction of an active medium in C_1 for absorbing the wavepacket. This, combined with an additional time dependent large detuning of cavity C_1 with respect to the center frequency of the wavepacket, allows a full photon absorption.

Here, for simplicity in deriving the spectral cavity responses, we assume that both dielectrics have a simple structure with a refractive index n_1 and n_2 for C_1 and C_2 respectively [10]. Of course, this will give origin to a simple expression for reflection and transmittance response coefficients. Furthermore, this allows us to assume a more complicated dielectric composition. The length of a given cavity is denoted by l_j and the width of the corresponding dielectric by d_j, with $j = 1, 2$. From boundary conditions at different interfaces, the mode functions, in the region between dielectrics, can be directly obtained. The result is:

$$\frac{A(k,x)}{A_0} = \begin{cases} \mathcal{T}_1\mathcal{T}_2\left(e^{ikx} - e^{-ikx}\right), & 0 \le x \le l_1 \\ \mathcal{T}_2\left(\frac{\mathcal{T}_1}{\mathcal{T}_1^*}e^{ikx} - e^{-ikx}\right), & l_1 < x \le l_1 + l_2 \\ \frac{\mathcal{T}_1\mathcal{T}_2}{\mathcal{T}_1^*\mathcal{T}_2^*}e^{ikx} - e^{-ikx}, & l_1 + l_2 \le x, \end{cases} \tag{1}$$

where A_0 is a normalization constant. Here, we have explicitly assumed that the incoming field is a plane wave propagating in the $-\hat{x}$ direction. We also have considered that mode functions cancel out at the perfectly reflecting mirror M. Besides, $\mathcal{T}_j$ denotes the spectral response of cavity j, which are defined by

$$\mathcal{T}_1 = \frac{T_1}{(1 + e^{i2kl_1} R_1)},$$
$$\mathcal{T}_2 = \frac{T_2}{\left(1 + \frac{\mathcal{T}_1 e^{2ikl_1}}{\mathcal{T}_1^*} e^{2ikl_2} R_2\right)}. \tag{2}$$

Here R_j and T_j are the reflection and transmission spectral responses of the right mirror of cavity j, respectively. In the limit of thin mirrors, $d_j \ll l_j$, we can make use of the following description for the system of mirrors:

$$n(x) = \theta(x)[1 + n_1\delta(x - l_1) + n_2\delta(x - l_1 - l_2)],$$

where $\theta(x)$ is the Heaviside step function, which describes the existence of a perfect mirror at the origin [11]. Under this assumption the spectral responses can be expressed as:

$$R_j = \frac{ikn_j}{2 - ikn_j}, \qquad T_j = \frac{2}{2 - ikn_j}. \tag{3}$$

The above assumption describes all the physical properties of the system of cavities. We note that the second cavity response $\mathcal{T}_2$ is depending on the response of the first cavity $\mathcal{T}_1$, which appears because of the coupling between cavities. However, in the case of weak transmittance, $|T_1|^2 \ll 1$, it can be shown that $\mathcal{T}_2$ can be approximated by

$$\mathcal{T}_2 = \frac{T_2}{(1 - e^{2ikl_2} R_1 R_2)}. \tag{4}$$

This result shows that the picture of a system of two coupled cavities gives a mechanism for storing a photon wavepacket in a cavity for a time of the order of the photon wavepacket revival time $t_R = 2\pi/\Delta\omega_1$, where $\Delta\omega_1$ is the difference between center frequencies of the resonances in C_1, i.e., $\Delta\omega = \omega_{m+1} - \omega_m = \pi c/l_1$. Thus the revival time is $t_R = 2l_1/c$, which corresponds to the typical round trip flight time inside the first cavity. The frequency range of frequencies for the first cavity is of a few units of the linewidth of the second cavity, so many resonances of the first cavity are interacting with one mode of the cavity C_2. The number of relevant resonances in the first cavity is given by the ratio of the decay constants in the cavities, i.e.,

$$M = \frac{\gamma_1}{\gamma_2} \approx \frac{l_2}{l_1} \frac{|T_1|^2}{(|T_1|^2 + |T_2|^2}. \tag{5}$$

3 Photon Wavepacket Absorption

After having demostrated the existence of the required mode structure in the compound system of cavities, we shall concentrate on the problem of absorbing the qubit encoded in the photon wavepacket. To do this we consider the active medium to be composed of an atom with a double three-level Λ-configuration, each one of these configurations is only interacting with one polarization of the photon. Figure 2 shows one of the Λ -configurations. There is no interaction connecting both polarizations. Thus we will work independently with each one of them.

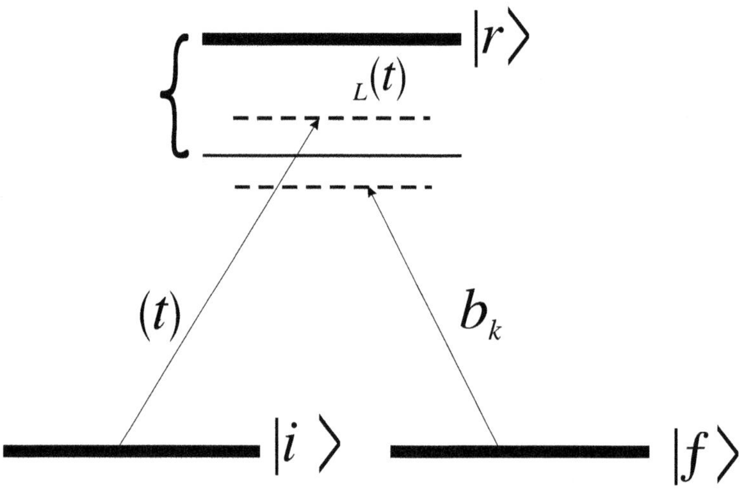

Fig. 2. One of the atomic Λ-configurations of the active medium. The atom is initially prepared in the lower state $|f\rangle$.

The photon wavepackets' source is modeled by a cavity with time dependents decay rate $\kappa_c(t)$ and detuning $\Delta_c(t)$. The Hamiltonian in this case can be written as ($\hbar = 1$):

$$H = H_0 + V,$$
$$H_0 = \omega_c a^\dagger a + \omega_c c^\dagger c + \sum_{k=1}^{M} \omega_k b_k^\dagger b_k + \omega_r |r\rangle\langle r|, \tag{6}$$

where V describes the interaction term, which in an interaction picture $V_I(t) = e^{iH_0 t} V e^{-iH_0 t}$ can be written as:

$$
\begin{aligned}
V_I(t) = i \sum_j & \left[\left(k_j a e^{-i\Delta_j t} + g_j |f\rangle\langle r| e^{-i(\delta-\Delta_j)t} \right) b_j^\dagger + H.c. \right] \\
& -i\Omega(t) \left(|i\rangle\langle r| e^{-i\delta_L(t)t+\phi(t)} - |r\rangle\langle i| e^{i(\delta_L(t)t+\phi(t))} \right) \\
& -i(\kappa_2 + i\Delta_2(t))a^\dagger a - i(\kappa_c(t) + i\Delta_c(t))c^\dagger c \\
& -i\Gamma|r\rangle\langle r| - 2i\sqrt{\kappa_2\kappa_c(t)}a^\dagger c. \qquad (7)
\end{aligned}
$$

These effective non-Hermitian interaction terms contain all the ingredients from the creation of the photon wavepacket up to its absorption by the atom. The wavepacket is generated in a cavity C described by a creation (annihilation) operator $c^\dagger$ (c). This cavity drives in a unidirectional way the system of cavities composed of C_2 and C_1, which is described by the last term of $V_I(t)$. The modes of the long cavity, C_1, are coupled both to the small cavity mode and to the atomic transition $|f\rangle \longrightarrow |r\rangle$. Finally, there is a time dependent classical field of Rabi frequency $\Omega(t)$ coupled to the atomic transition $|i\rangle \longrightarrow |r\rangle$. Even when we work in the large detuning limit we consider a spontaneous emission from the upper atomic level at a rate Γ. Its contribution is small, however the absorption process can be slow and the probability to populate the upper level can not be negligible. The damping term of the cavity C_2 describes both the reflection of the wavepacket during the transfer process and the lost of it due to transmission for later times. The goal is to include a time detuning $\Delta_2(t)$ for cavity C_2 while trapping the photon within C_1 for times longer than $t_R = 2l_1/c$. In the simplest case $\Delta_2(t)$ can be modeled by a stepwise function, which is switched on after the transmission of the wavepacket into C_1. If this detuning is large enough, it will behave like a continuous measurement in C_2 and the photon will be confined into C_1. The physical origin of such a detuning in C_2 can be an atom with a transition of a short decay time interacting with a strong classical resonant field.

In what follows we discuss the dynamics arising from Eq. 7. We shall assume that the system is initially prepared in a number state in cavity C, vacuum state for the other two cavities and the level f for the atom, i.e., in the state $|\alpha\rangle = |1_c, 0_1, 0_j, f\rangle$. Then the evolution is restricted to a subspace with one excitation. The following hansatz is valid for the wavefunction:

$$
\begin{aligned}
|\psi(t) = & A(t)|1_c, 0_1, 0_j, f\rangle + B(t)|0_c, 1_1, 0_j, f\rangle \\
& + \sum_j C_j(t)|0_c, 0_1, 1_j, f\rangle + D(t)|1_c, 0_1, 0_j, i\rangle \\
& + E(t)|0_c, 0_1, 0_j, r\rangle, \qquad (8)
\end{aligned}
$$

where the labels of the elements of this basis are, respectively, photon numbers in cavities C, C_2, and C_1 and the atomic level. The state $|1_j\rangle$ means that the mode j in C_1 contains one photon and all the other modes are in the vacuum state.

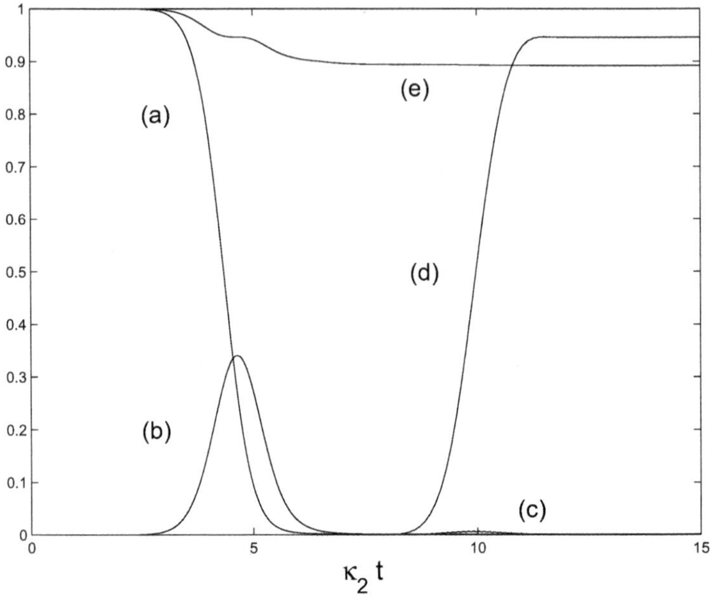

Fig. 3. Evolution of the different probablities to have the initial photon in : (a) cavity C, (b) cavity C_2, (c) atomic level $|r\rangle$ and (d) atomic level $|i\rangle$. Here, we also plot the norm of the wavefunction as a function of time in curve (e). The parameters are: $\delta = \sqrt{50}\kappa_2$, $\kappa_1 = \kappa_2$, $g_j = \sqrt{\delta/7}$.

In Figure 3 we plot the different probabilities involved in the whole process from photon generation to its absorption by the active medium. At time $t^* = 8\kappa^{-1}$ we switch on both the large detuning in cavity C_2 and the classical field $\Omega(t)$ with initial amplitude $\Omega_0 = 1.25\sqrt{\delta/7}$. The optimum process for photon absorption of the wavepacket is a result of the combination of two processes: (i) a linear decreasing of the Rabi frequency of the classical field $\Omega(t)$ (driving the transition $|f\rangle \longrightarrow |r\rangle$) and (ii) a slow frequency change. Besides, we partially compensate the Stark shift arising from the large detuning limit considered in this case, with the phase of the classical field $\phi(t)$. Thus, we assume:

$$\phi(t) = \left(-\frac{\Omega^2(t)}{4\delta} + \frac{g^2}{\delta} \right) (t - t^*). \tag{9}$$

In Figure 3 we observe that the combination of the above mentioned processes allows an almost full photon absorption. We also can observe that for this setup, the upper level $|r\rangle$ is extremely small populated during the evolution, of the order of 1 % during the absorption, and the effect of spontaneous emission can be neglected. Thus, we can work with an effective model of a two-level atom interacting with the cavity modes in C_1. In order to get a simple physical picture for the absorption process of the photon by the active medium, we concentrate during this process only on the interaction between the AM and the cavity modes of C_1. Thus, the multi-mode structure of cavity C_1 can be described by an effective time dependent operator:

$$B(t) = \frac{1}{g'} \sum_{k=1}^{M} g_k b_k e^{-i\Delta_k t} \tag{10}$$

with

$$g'^2 = \sum_{k=1}^{M} g_j^2$$

and $\Delta_k = \omega_k - (2k - M)\Delta\omega/2$

This operator satisfies the usual commutation relations to all times. We know that the wavepacket has a frequency width of few κ_2. Thus, the condition $|\Delta_k| \ll \delta$ holds for each component of the wavepacket. This allows us to study the absorption of the photon as the interaction of an effective two-level atom with a cavity mode $B(t)$:

$$\begin{aligned} V_I(t) = {} & \frac{g'^2}{\delta} B^\dagger(t) B(t) \sigma_{ff} + \frac{\Omega^2(t)}{4\delta} \sigma_{ii} \\ & - i \frac{g' \Omega(t)}{2\delta} \left(\sigma_{if} B(t) e^{i\psi(t)} - B^\dagger(t) \sigma_{fi} e^{-i\psi(t)} \right), \end{aligned} \tag{11}$$

where $\sigma_{ij} = |i\rangle\langle j|$. The atom is assumed to be in its lower level and the "mode" in its state $|1\rangle$ (wavepacket). In this picture, to transfer the wavepacket to the atom we only need to apply a π-pulse interaction, with a parameter α, given by

$$\alpha = \int_{t^*}^{t} du \frac{g' \Omega(u)}{2\delta} e^{i\theta_c(u)}. \tag{12}$$

Thus, the condition for a full absorption is $|\alpha|^2 = \pi^2$. We choose $\Omega(t)$ as

$$\Omega(t) = \begin{cases} \Omega_0 * (T - t) & \text{for } t^* < t \leq T \\ 0 & \text{for others} \end{cases} \tag{13}$$

4 Generation of Photonic Qubits with Three-Level Λ Atoms

In the previous sections we have presented a proposal for detecting photonic qubits. Now, we proceed to describing a scheme for generating entangled photonic qubits in cavity-QED [6]. As an active medium inside the cavity as an optical cavity, we consider a single atom or ion trapped [8,9] as the implementation closest to present day experimental reality. For the atom we assume a double three-level Λ structure in the large detuning limit as depicted in Fig. 4.

The levels $|i_\alpha\rangle$ ($\alpha = 0, 1$) are coupled to the upper levels $|r_\alpha\rangle$ via classical fields $\Omega_\alpha(t)e^{-i(\omega_\alpha t+\phi_\alpha(t))}$, where ω_α are the field center frequencies and the subscript refers to the two polarization states. The external control parameters are the real amplitudes $\Omega_\alpha(t)$ and the phases $\phi_\alpha(t)$. The levels $|f_\alpha\rangle$ are coupled to the upper levels by the cavity modes a_α, with a common frequency ω_c but with orthogonal polarization, with coupling constants g_α. The large detuning (δ) assumption allows us to adiabatically eliminate the upper atomic levels. We are left with two effective two-level systems describable by generalized spin operators $\sigma_{i_\alpha j_\alpha} = |i_\alpha\rangle\langle j_\alpha|$. The center frequencies of the external laser pulses fulfill the Raman resonance condition. Note that any offsets can still be accommodated in the phases $\phi_\alpha(t)$.

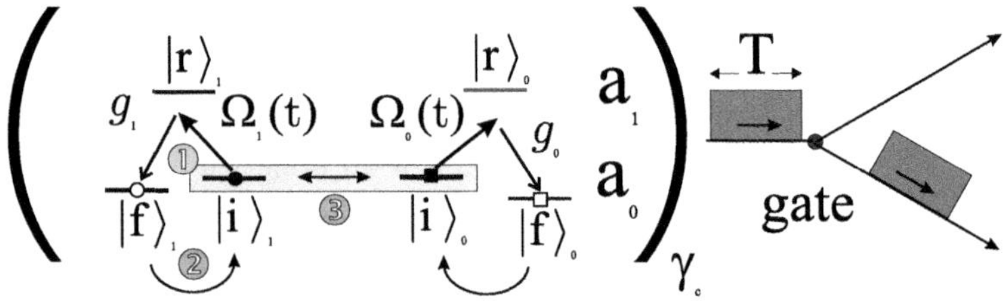

Fig. 4. A single atom with six internal states interacts with two cavity modes of orthogonal polarization a_0, a_1. In a Raman process (step 1) an initial superposition state of levels $|i_0\rangle$ and $|i_1\rangle$ is transformed into an entangled cavity-atom state. Due to cavity leakage the photon will leave the cavity and produce a photon wave packet in the continuum modes outside the resonator. In step 2 the atom is recycled back to $|i_0\rangle$ and $|i_1\rangle$. In between two photon generation sequences levels $|i_0\rangle$ and $|i_1\rangle$ can be coupled (step 3) to tailor the outgoing state.

During the process of photon wavepacket generation there is no interaction connecting the two Λ-systems. Thus, we may independently work with each system corresponding to a single index α, and drop this index for the time being. In this example the total system consists of three building blocks: the continuum outside the resonator, the cavity modes and the internal degrees of freedom of the atom inside the resonator. We switch to an interaction picture with respect to the free dynamics of the compound system. This eliminates the fast optical time scales from the dynamics and leaves us with a simpler effective interaction term:

$$V^I_{\rm mc}(t) = \frac{r^2(t)}{\Delta_s}|i\rangle\langle i| + \Delta_s a^\dagger a|f\rangle\langle f| \\ +ir(t)\left(e^{-i\phi(t)}a^\dagger|f\rangle\langle i| - e^{i\phi(t)}|i\rangle\langle f|a\right), \tag{14}$$

where $r(t) = g\Omega(t)/2\delta$ (Rabi frequency of Raman transition) and $\Delta_s = g^2/\delta$ (cavity induced Stark shift). For simplicity, we have assumed a common detuning parameter between the pump fields and the cavity modes, and real coupling constants. Any additional frequency shifts can be included in the phase of the classical field. The time and intensity dependent terms correspond to dynamical shifts arising from the adiabatic elimination of the upper atomic level. We assume that the atom-cavity system is initially in the state $|\varphi(0)\rangle_{\mathrm{mc}} = |i\rangle|0\rangle_{\mathrm{c}}$. Therefore, the action of the effective evolution operator $U^{\mathrm{eff}}_{\mathrm{mcr}}(t)$ on the chosen initial state is completely characterized by the following expression:

$$U^{\mathrm{eff}}_{\mathrm{mcr}}(t) = C_i(t)e^{-i\theta(t)} + C_f(t)e^{-i\Delta_s t}a^\dagger\sigma_{fi}, \tag{15}$$

where $\theta(t) = \int_0^t dt'|\Omega|^2(t')/4\delta$. The slowly varying functions evolve according to

$$\begin{aligned} \dot{C}_i(t) &= r(t)C_f(t)e^{i\theta_c(t)} \\ C_f(t) &= \int_0^t dt' r(t')C_i(t')e^{-i\theta_c(t')-\kappa_c(t-t')}, \end{aligned} \tag{16}$$

where $\theta_c(t) = \theta(t) + \phi(t) - \Delta_s t$. In the bad cavity limit the main contribution to the integral will stem from time differences $t - t'$ of the order of the intracavity photon lifetime κ_c^{-1}. If both coefficients vary slowly on this timescale, we may approximate the solution of the previous equation by:

$$C_f(t) \simeq \frac{r(t)}{\kappa_c} e^{-\mu(t)-i\theta_c(t)} \tag{17}$$

$$C_i(t) = e^{-\mu(t)}, \tag{18}$$

where $\mu(t)$ is an effective decay constant defined by

$$\mu(t) \equiv \int_0^t dt' \frac{r^2(t')}{\kappa_c}.$$

So far we have found an expression for the state $|\varphi(t)\rangle_{\mathrm{mc}}$. In order to warrant reliable wavepacket creation during a time window T we have to require that $\mu(T) \gg 1$. Note that this sets a lower bound on the size of the pulse area of the externally applied laser fields. Assuming this to be the case we simply can write the state of the system as

$$|\psi(t)\rangle = \int d\omega a G(\omega, T)|\varphi(0)\rangle_{\mathrm{mc}}|1_{\bar{\omega}}\rangle_{\mathrm{r}}, \tag{19}$$

where

$$G(\omega, T) \equiv g(\omega, T)a^\dagger\sigma_{fi} = \sqrt{\frac{\kappa_c}{\pi}} \int_0^T dt e^{i\omega t} U^{\mathrm{eff}}_{\mathrm{mcr}}(t). \tag{20}$$

The operator σ_{fi} maps the state $|i\rangle_{\mathrm{m}}$ to the state $|f\rangle_{\mathrm{m}}$. The spectral envelope of the created wavepacket is given by $g(\omega, T)$ (a mere c-number function). Conservation of the norm of $|\psi\rangle$ implies $\int d\omega|g(\omega, T)|^2 = 1$. In a previous publication [12] it has been shown that a time-dependent decay rate $\kappa_c(t)$ and central frequency $\omega_c(t)$ can be employed to generate a one-photon wavepacket of arbitrary spectral composition. Such an approach may be regarded as a description of the presence of a medium in terms of an effective model. Now, we have to apply the mode destruction operator a to the state $U^{\rm eff}_{\rm mcr}(t)|\varphi(0)\rangle_{\rm mc}$. As we have already argued, doing so will yield a state that is not affected by the interaction Hamiltonian $V^I_{\rm mc}(t)$; in the present case this state is proportional to $|f\rangle|0\rangle_c$. Finally, the spectral envelope as introduced in Eq.(20) can be written as

$$g(\omega,T) = \sqrt{\frac{\kappa_c}{\pi}} \int_0^T dt' \frac{r(t')}{\kappa_c} e^{-\left(\mu(t') - i(\omega - \Delta_s)t' + i\theta_c(t')\right)}.$$

We may now also check whether $|g(\omega)|^2$ is a normalized distribution as it should be in the case of reliable wavepacket generation. We find

$$\int d\omega |g(\omega)|^2 = -\int_0^T dt' \dot{\mu}(t') e^{-\mu(t')} = 1 - e^{-\mu(T)}. \tag{21}$$

If $e^{-\mu(T)}$ is small, which is the underlying assumption, we have created a wavepacket with near certainty. Subsequent wavepackets can be generated by recycling the Λ-system back to state $|i\rangle$ through auxiliary atomic levels immediately before switching on the external light pulse again. Logical qubit states are generated by using a two-mode cavity and two Λ-systems. We simply obtain two wavepackets, each weighted with the appropriate probability amplitude and spectral envelope g_α where $\alpha = 0, 1$ now accounts for the fact that both Λ systems need not be identical in terms of coupling strengths and detunings.

Fig. 5 shows the evolution of the atomic populations for an external driving field with Gaussian pulse shape. The phase of the pump field was chosen in order to compensate the time dependent Stark shift introduced by the external field, i.e., $\theta_c(t) \equiv 0$. The inset displays the spectral envelope of the created 1-photon wavepacket. We observe that the wavepacket has a Gaussian distribution centered around $\omega_c + \Delta_s$. This shift is due to the Stark shift introduced by the cavity field which has not been compensated. The condition $\mu(t) \gg 1$ ensures a full transfer of the atomic state from level $|i\rangle$ to the ancilla level $|f\rangle$ with concomitant photon wavepacket generation in the continuum outside the cavity by way of photon leakage from the cavity.

We have considered an idealized model with no sources of errors present during the generation of the entangled multiphoton wavepackets. A rough estimate of the qubit generation rate our scheme would seem capable of is given by the inverse of the duration of a generation/preparation sequence. For a classical pump field $\Omega(t)$ with a peak frequency $\Omega_0 = 55$ Mhz and a Gaussian pulse shape, $\delta = 1.5$ GHz, $g = 55$ Mhz, and $\kappa_c = 50$ Mhz one-photon pulse durations of around 10 cavity lifetimes are feasible, as can be seen from Fig. 5. Taking into account recycling and reinitialization of the medium, a conservative estimate would yield a generation rate of around 1 MHz. However, the maximum number

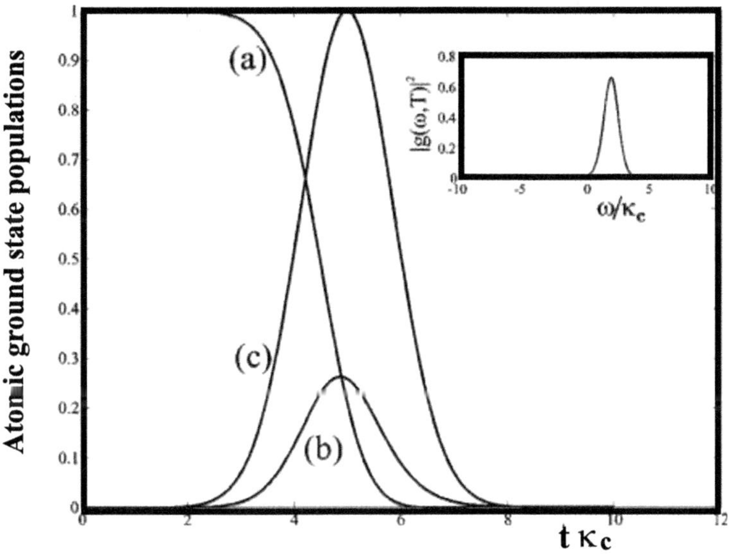

Fig. 5. Temporal evolution of the atomic populations (levels $|i\rangle$ (a) and $|f\rangle$ (b)) of one the of atomic Λ configurations and the time dependence of the applied external pump field (normalised) (c). The parameters are $\Omega_0 = 55$ Mhz and a Gaussian pulse shape, $\delta = 1.5$ GHz, $g = 55$ Mhz, and $\kappa_c = 50$ Mhz. Since the cavity decay constant exceeds all other rates, we are in the overdamped regime, which explains why we do not observe Rabi oscillations. The inset shows the asymptotic spectral envelope of the created photon wavepacket, i.e., $\kappa_c t = 10$.

n of entangled photon wavepackets (qubits) that our scheme can generate with a certain fidelity is limited by technical imperfections. In principle, stabilization of laser phase fluctuations below 1 kHz represents a technical challenge. Amplitude fluctuations in driving fields cause a distortion of the pulse form and lead to incomplete population transfer. An estimate gives that $n \ll I/\Delta I \sim 10^4$, where $\Delta I/I$ are the relative intensity fluctuations. Photon absorption in the cavity mirrors is an essential effect for high–Q optical cavities. In general, it leads to two types of errors: Photon absorption and concomitant destruction of the entanglement. The first type of error is evaded by using postselection schemes, which correspond here to discarding sequences with a number of detected photons smaller than n. State distortion can occur even in the absence of loss of a photon. These kinds of errors vanish for a cavity with equal absorption rates for both polarizations, i.e., $\kappa_0 \simeq \kappa_1$. The most prominent source of decoherence is due to imperfections in the recycling between the ground states $|f_\alpha\rangle \to |i_\alpha\rangle$. The required π-pulses could be ill timed or their effect influenced by the motion of the atom. Moreover, there could be a dephasing between the probability amplitudes of the states $|f_\alpha\rangle$ of the two Λ-systems due to laser fluctuations in the Raman transitions. The correct timing of the π-pulses is critical; we gather that for errors in the 2% range approximately 10 qubits can be created with a fidelity of 90%.

5 Summary and Further Applications

We have presented proposals both for generating and for detecting photonic qubits. The proposals are based on cavity-QED techniques. In both proposals photonic qubits are described as one photon wavepackets, where the quantum information is encoded in the polarization degree of freedom. We observe that the combination of both schemes, in principle, allows implementing definite tests of quantum mechanics, due to the high accuracy in detecting photonic qubits and the possibility of generating high number of maximally entangled particles.

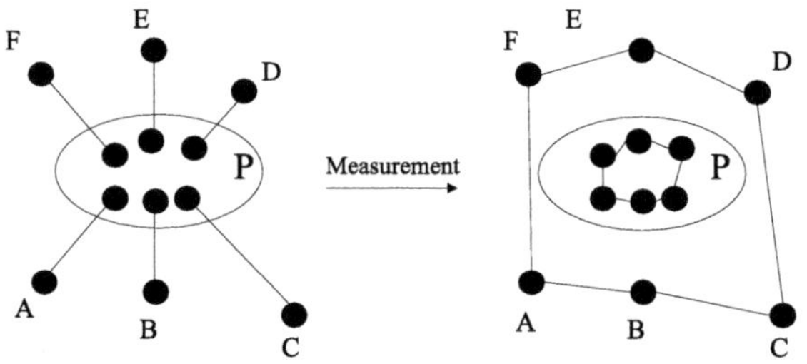

Fig. 6. Topology of a quantum communication network which consists of a single provider of entanglement P and several clients A-F. After a communication channel requirement the provider perform a joint measurement for recognizing generalized Bellś states

In the first portion of this article, we concentrate on giving a proposal for a high efficiency in detecting the photonic qubits. The proposal considers a compound system of two optical cavities, which are very different in length. The first, short, cavity simply behaves as an mirror with an intrincate configuration, which permits a high control of the decay constant of the long cavity after the photon had entered. We end with a proposal for the controlled generation of entangled n-qubit states, where non-overlapping one-photon wavepackets are identified with the logical qubits. We have described a CQED based implementation, where the medium is an atom with two Λ-type level configurations. This implementation seems most promising from the present point of view. However, the theory is quite general and it can be easily adapted to other implementations, for instance quantum dots or single atoms embedded in a host material [13], which may become available in the future as quantum systems with long coherence times.

Our study of decoherence in these models indicates that such proposals are in principle experimentally feasible with state-of-the-art equipment. They could thus form the experimental basis for multi-party communication in future quantum networks. Of particular interest are several recent proposals [14,18] which have suggested making use of many-qubit entangled states in order to share quantum information among several parties. An example is quasi-simultaneous transfer of information to n clients using an n-qubit MES. In this way a key to be used later in an RSA encryption scheme is transferred to n parties in one fell swoop. As has been shown by Molotkov and Nazin [13] that such a procedure not only implements key distribution but also allows one to check efficiently for the presence of an eavesdropper. Another interesting application is a controlled quantum communication network, which consists of a single "provider" (P) who is connected to the clients (A-F) via photonic channels [5]. The provider creates entangled photonic qubits on request and transmits them to the clients, e.g., A and B in Fig. 6. Upon transmission all three parties involved share an entangled 3-qubit state. In order for A or B to teleport the state of an additional qubit between each other the central station P has to measure the state of its qubit (which would be the internal state of the medium) and disclose the result to A and B. The advantage of such a scheme is its relatively simple topology which eliminates the need for photonic channels between all individual nodes A-F. Moreover, no successful transfer of information is possible without the consent of the provider. We have thus a network in which the transfer of information between two parties is always controlled by a third one. This may be of practical importance if the time at which information transfer takes place should be determined by the third party.

All schemes for multi-particle quantum communication assume that, in this case, the provider can perform a joint measurement on his particles, for swapping the entanglement between the provider and parties to entanglement between parties. Actually, the provider needs to perform a measurement for recognizing the maximally entangled states of an n-particle system, which are

$$|\Psi_{\rm mes}\rangle = \frac{1}{\sqrt{2}}(|s_1, s_2, ..., s_n\rangle + |1-s_1, 1-s_2, ..., 1-s_n\rangle), \qquad (22)$$

with $s_j = 0, 1$. This can be implemented if we assume that inside the cavity for generating photonic qubit there are n ions, which can be addressed individually, so that the entangled state is composed of a photon wavepacket and the inner states of a trapped ion. This essentially is the Cirac and Zoller proposal for quantum computing with trapped ions [16]. A joint measurement on ions can be implemented by individual measurement on them after a disentanglement process. Thus, we assume that all the initial pairs are maximally entangled in state

$$|\psi^-_{k,k'}\rangle = \frac{1}{\sqrt{2}}\left(|0_k, 1_{k'}\rangle - |1_k, 0_{k'}\rangle\right), \qquad (23)$$

where the first label refers to the k-th particle and the second label corresponds to a specific ion inside the cavity, i.e., inside the measurement region. Thus, the initial condition of the whole set of particles is given by:

$$|\Psi_{\text{provider,parties}}\rangle = \frac{1}{2^{n/2}} \bigotimes_{k=1}^{n} \left(|0_k, 1_{k'}\rangle - |1_k, 0_{k'}\rangle\right), \tag{24}$$

where k and k' refer to the ion (provider) and photonic qubit (party), respectively. The disentaglement is given by

$$\bigotimes_{k=0}^{n-1} XOR_{n-k+1,n-k}|\Psi_{\text{me}}\rangle = (|s_1\rangle + |1-s_1\rangle) \bigotimes_{k=0}^{n-1} |s_{n-k+1} + s_{n-k}\rangle, \tag{25}$$

where $XOR_{j,j+1}$ denotes a controlled not gate between the j-th ion as a source qubit and the $j+1$-th as a target. The process ends with individual measurements on ions for determining in which logic state they are [17].

Acknowledgments: This work was partially supported by Fundación Andes and FONDECYT 1990838.

References

1. A. Einstein, B. Podolsky, and N. Rosen, Phys. Rev. **47**, 777 (1935).
2. D. M. Greenberger, M. A. Horne, and A. Zeilinger, in *Bell's Theorem, Quantum Theory, and Conceptions of the Universe*, edited by M. Kafatos (Kluwer Academic, Dordrecht, 1989).
3. D. M. Greenberger, M. A. Horne, A. Shimony, and A. Zeilinger, Am. J. Phys. **58**, 1131 (1990).
4. Jan-Ake Larsson, Phys. Rev. A **57**, R3145 (1998).
5. J. I. Cirac, P. Zoller, H. J. Kimble, and H. Mabuchi, Phys. Rev. Lett. **78**, 3221 (1997); T. Pellizzari, *ibid.* **79**, 5242 (1997).
6. K. M. Gheri, C. Saavedra, P. Törmä, I. Cirac, and P. Zoller Phys. Rev. A **58**, **R2267** (1998).
7. C. Saavedra, K. M. Gheri, P. Törmä, I. Cirac, and P. Zoller, Phys. Rev. A **61**, 062311 (2000).
8. Q. A. Turchette, C. J. Hood, W. Lange, H. Mabuchi, and H. J. Kimble Phys. Rev. Lett. **75** 4710 (1995).
9. G. M. Meyer, H.-J. Briegel, and H. Walther, Europhys. Lett. **35**, 317 (1997).
10. Werner Vogel, Dirk-Gunnar Welsch, *Lecture on Quantum Optics* (Akademie Verlag, Berlin 1994)
11. M. Ley and R. Loudon, J. of Mod. Opt. **34**, 227 (1987).
12. K. M. Gheri, K. Ellinger, T. Pellizzari, and P. Zoller, Progress of Physics **46**, 401 (1998).
13. S. N. Molotkov and S. S. Nazin, JEPT Lett. **62**, 956 (1995).
14. M. Hillery, V. Bužek, and A. Berthiaume, Phys. Rev A **59**, 1829 (1999).
15. S. Bose, V. Vedral, and P. L. Knight, Phys. Rev. A, **57**, 822 (1998).
16. J. I. Cirac and P. Zoller, Phys. Rev. Lett **74**, 4091 (1995).
17. J. C. Retamal and C. Saavedra, in preparation.

Macroscopic Entanglement and Relative Phase

G. Nienhuis

Huygens Laborotarium, Universiteit Leiden, Postbus 9504,
2300 RA Leiden, The Netherlands

Abstract. When two bosonic systems are decaying, and the decay products are observed in interference, the relative macroscopic phase between the systems acquires a well-determined value. The spontaneous buildup of this phase is evaluated in the case of an exactly solvable model. The description holds for photons from a cavity, or bosonic atoms from Bose-Einstein condensates. The phase buildup can be viewed as the creation of entanglement between two subsystems by the observation of their decay products.

1 Introduction

Interest in the spontaneous buildup of the relative phase between two systems of identical bosons has regained interest after the realization of Bose-Einstein condensation of bosonic atoms in magneto-optical traps [1,2,3]. When two condensates overlap, interference has been predicted in the overlap region, even when initially their relative phase is undetermined [4,5,6]. Spatial interference between two overlappng condensates has been observed in a number of experiments [7,8]. The concept of phase of a condensate is closely related to the phase of the field in a mode of the radiation field. However, there are also important differences. The phase of a mode field is a classical quantity, and the same is true for the relative phase between two modes. However, for two condensates, a well-determined value of their relative phase requires indeterminacy of the difference in particle number. This implies a quantummechanical entanglement between the two condensates.

Two quantum systems in a pure state are entangled when the state of the combined system cannot be factorized into product states pertaining to the separate systems. Entanglement is one out of several fundamental features of quantum mechanics, with no classical counterpart. It arises naturally from the superposition principle. It is commonly believed that for two systems to be in an entangled state it is necessary that they share a common past. The simplest possibility is that they arise from a common source, such as two photons created by parametric downconversion, or that they have been interacting with each other, such as two atoms separating after a collision. A slightly more involved possibility is that each system of an entangled pair has been interacting with a common partner system. More fancy possibilities has been recently realized, where two entangled pairs of photons separate, in such a way that two photons,

M. Orszag, J.C. Retamal (Eds.): LNP 575, pp. 95–109, 2001.

one out of each pair, are brought to interfere, leading to entanglement of the remaining two photons [9].

In this paper we analyse the creation of entanglement between two decaying systems by the detection of their decay products in interference. The systems are modelled as boson modes, and can be thought of as two single-mode radiation cavities, or two Bose-Einstein condensates. Particles emitted from the systems enter the two input ports of a beam splitter, and detectors are attached to the output ports. A sketch of the setup is indicated in Fig. 1. Finite detection

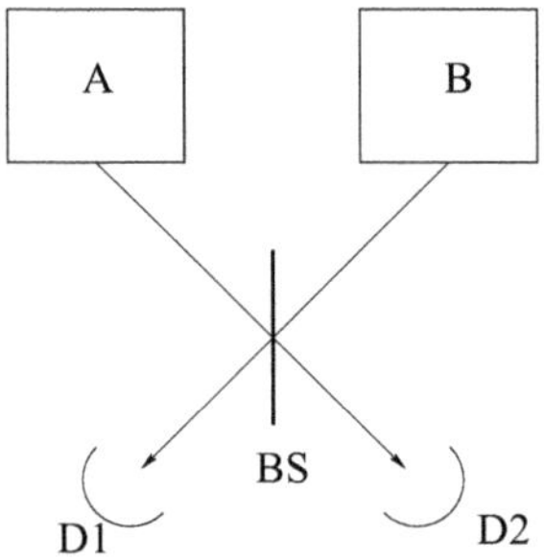

Fig. 1. Sketch of setup. Two decaying boson modes A and B emit particles, that pass a beam splitter BS, and are detected at D1 and D2

efficiencies are allowed for. In the case of Bose-Einstein condensates, this is a situation where spontaneous symmetry breaking gives rise to a relative phase between the condensates. We consider a class of initial states where the buildup of the phase during a detection history can be solved analytically. We give a simple decription of such histories by generalizing the idea of quantum trajectories to include imperfect detection efficiencies.

2 Single Histories with Arbitrary Detection Efficiency

2.1 Perfect detection efficiency

A decaying boson mode is viewed as an open quantum system, whose evolution can be described by a master equation for its density matrix $\hat{\rho}$. The simplest form of this equation is

$$\frac{d}{dt}\hat{\rho} \equiv \mathcal{L}\hat{\rho} = -\frac{i}{\hbar}[\hat{H}, \hat{\rho}] - \frac{1}{2}\Gamma(\hat{a}^\dagger\hat{a}\hat{\rho} + \hat{\rho}\hat{a}^\dagger\hat{a}) + \Gamma\hat{a}\hat{\rho}\hat{a}^\dagger \ , \tag{1}$$

where the operator $\hat{a}$ represents a quantum jump, and Γ is a measure of the jump rate. The last term in eq. (1) describes the gain in the final state after quantum jumps. In the prototype case of spontaneous emission the operator $\hat{a}$

describes the transition of an excited atom to a lower state. In the present case of decaying boson modes, quantum jumps describe the loss of a boson from the mode. For a mode of the radiation field in a cavity, $\hat{a}$ represents the annihilation of a photon from the mode. Master equations for open quantum systems are valid provided that the time scale of the evolution is slow compared with the correlation time of the outside world (the 'reservoir' [10]).

Even when the system is initially in a pure state, it cannot be expected to remain in a pure state, due to the loss of information to the outside world. However, the density matrix $\hat{\rho}$ can be represented as an ensemble of pure states, where each realization corresponds to the precise specification of the instants of time at which the quantum jumps occurred. This can be clarified by separating the evolution operator $\mathcal{L}$ in eq. (1) as

$$\mathcal{L} = \mathcal{L}_0 + \mathcal{L}_1 \ , \tag{2}$$

with $\mathcal{L}_1\hat{\rho} = \Gamma a\hat{\rho}\hat{a}^\dagger$ the gain term. Then eq. (1) can be expressed as an integral equation

$$\hat{\rho}(T) = e^{\mathcal{L}_0 T}\hat{\rho}(0) + \int_0^T dt e^{\mathcal{L}_0(T-t)}\mathcal{L}_1\hat{\rho}(t) \ , \tag{3}$$

which after iteration leads to a formal solution of the master equation in the form of an expansion in $\mathcal{L}_1$. The integrand in each term can be viewed as a possible history of the system with given instants of the jumps. The first term $\hat{\tilde{\rho}}_0(T) \equiv e^{\mathcal{L}_0 T}\hat{\rho}(0)$ in (3) represents the case that no jump occurred in the interval $[0,T]$, the second term describes all histories with the last jump at time t. The strength of the first term $p_0(T) = \mathrm{Tr}\hat{\tilde{\rho}}_0(T)$ has the physical significance of the probability that no jump occurred during the interval $[0,T]$. The pure-state single histories are convenient for numerical simulation of single runs of the evolution of open quantum systems [11,12,13]. They are commonly referred to as quantum trajectories.

The statistics of the jumps can also be derived from (3). When precisely N jumps have occurred in the count interval $[0,T]$, at the successive time instants $t_1 \leq t_2 \leq \cdots \leq t_N$, the corresponding contribution to the total density matrix at time T is

$$\hat{\tilde{\rho}}_n(t_1, t_2, \ldots, t_n; T) = e^{\mathcal{L}_0(T-t_k)}\mathcal{L}_1 e^{\mathcal{L}_0(t_k - t_{k-1})} \ldots \mathcal{L}_1 e^{\mathcal{L}_0 t_1}\hat{\rho}(0) \ . \tag{4}$$

The strength of this contribution determines the probability density for precisely N jumps in the count interval $[0,T]$, at the indicated time instants, so that

$$w_N(t_1, t_2, \ldots, t_N; T) = \mathrm{Tr}\ \hat{\tilde{\rho}}_N(t_1, t_2, \ldots, t_N; T) \ . \tag{5}$$

The probability $p_N(T)$ for N jumps in the count interval is obtained after integration of w_n over the ordered time instants $t_1 \leq t_2 \leq \cdots \leq t_N$. Finally, the ratio $\hat{\tilde{\rho}}_N/w_N$ of (4) and (5) represents the normalized conditional density matrix, given that N particle detections have occurred at the instants $t_1, t_2, \ldots, t_N$.

2.2 Imperfect detection efficiency

In practice, the ejected bosons are detected with a limited detection efficiency. In order to evaluate the conditional state of the modes for a given detection history, we generalize the concept of quantum trajectories. The treatment is related to earlier work on the statistics of detected jumps [14], or single histories of the evolution of the system [15] in the case of imperfect detection efficiency. When the jumps are detected with the efficiency η with $0 \leq \eta \leq 1$, we separate the evolution operator $\mathcal{L}$ as in eq. (2), with $\mathcal{L}_1 = \eta\Gamma\hat{a}\hat{\rho}\hat{a}^\dagger$. The complementary part of the gain term is included in $\mathcal{L}_0$. With this redefinition of the partial evolution operators the integral equation (3) remains valid, and it leads to single histories with a specific number of *detected* jumps. In this case, eqs. (4) and (5) give the density matrix and the probability density for a given number of detected jumps at the indicated time instants. Undetected jumps contribute to $\mathcal{L}_0$ to an amount $1 - \eta$, with the result that single histories are no longer represented by pure states. In the limiting case of full detection efficiency $\eta = 1$, the standard pure-state trajectories are recovered. In the opposite limiting case of $\eta = 0$, there is no difference between the solution of the full master equation (1) and the density matrix $\hat{\tilde{\rho}}_0(t)$ during a detection-free period, since $\mathcal{L} = \mathcal{L}_0$.

3 Single Boson Mode

3.1 Arbitrary state

Before turning to the case of two decaying boson modes, we first sketch the general solution of a single boson mode, thereby fixing the notation. Particles are leaking to the outside world, where they have a chance η to be detected. The description is equally valid for a mode of the radiation field, or for a Bose-Einstein condensate. We are interested in the conditional density matrix corresponding to a specific detection history. A single decaying boson mode is described by the master equation (1), where the Hamiltonian takes the simple form $\hat{H} = \hbar\omega\hat{a}^\dagger\hat{a}$. The jump operator $\hat{a}$ is the annihilation operator of a boson from the mode. The initial density matrix $\hat{\rho}(0)$ can be expanded in coherent states $|\alpha>$ in terms of the Glauber-Sudarshan P function as [16,17,10]

$$\hat{\rho}(0) = \int d_2\alpha \, |\alpha >< \alpha| P(\alpha) \, , \tag{6}$$

where the integration extends over the complex α-plane. In order to find P we introduce the characteristic function

$$\chi(\lambda) = < e^{\lambda\hat{a}^\dagger} e^{-\lambda^*\hat{a}} > = \mathrm{Tr}\hat{\rho}(0) \, e^{\lambda\hat{a}^\dagger} e^{-\lambda^*\hat{a}} \, , \tag{7}$$

so that knowledge of χ as a function of λ determines the density matrix $\hat{\rho}$. Then the distribution function P defined as the two-dimensional Fourier transform of χ

$$P(\alpha) = \frac{1}{\pi^2} \int d_2\lambda \, \chi_A(\lambda) \, e^{\lambda^* \alpha - \lambda \alpha^*} \, . \tag{8}$$

determines $\hat{\rho}(0)$ as in (6). The merit of this representation is that the evolution both of the total density matrix and of the state during a detection-free interval is easily found, since it is sufficient to evaluate the evolution of an initially coherent state. The function P is real and normalized. On the other hand, P is not necessarily positive definite, so one cannot interpret it as a probability distribution function. In fact, when $\hat{\rho}(0)$ is a number state, the function χ is a polynomial, so that P contains derivatives of a delta function. This possible singular behavior of P presents no problems in the present paper.

The solution of the master equation with the initial condition (6) is equal to

$$\hat{\rho}(T) = \int d_2\alpha \, |\alpha(T) >< \alpha(T)| P(\alpha) \, , \tag{9}$$

with $\alpha(T) = \alpha \exp[-(i\omega + \Gamma/2)T]$ a time-dependent coherent-state index. For a detection efficiency η, the contribution to the density matrix corresponding to a detection-free interval is

$$\hat{\tilde{\rho}}_0(T) \equiv e^{\mathcal{L}_0 T} \hat{\rho}(0) = \int d_2\alpha \, |\alpha(T) >< \alpha(T)| P(\alpha) \exp\Big(-\eta |\alpha|^2 (1 - e^{-\Gamma T}) \Big) \, . \tag{10}$$

The trace of (10) gives the zero-detection probability

$$p_0(T) = \int d_2\alpha \, P(\alpha) \exp\Big(-\eta |\alpha|^2 (1 - e^{-\Gamma T}) \Big) \, . \tag{11}$$

The effect of the detection of a particle as described by $\mathcal{L}_1$ can simply be accounted for, since by definition the coherent states $|\alpha >$ are eigenstates of the annihilation operator $\hat{a}$ with eigenvalue α. The contribution (4) to the density matrix corresponding to N detections in the interval $[0, T]$ at the instants $t_1 \leq t_2 \leq \cdots \leq t_N$ is now easily evaluated. The result is

$$\hat{\tilde{\rho}}_N(t_1, t_2, \ldots, t_N; T) = \int d_2\alpha \, |\alpha(T) >< \alpha(T)| \, P(\alpha) \\ \times \exp\Big(-\eta |\alpha|^2 (1 - e^{-\Gamma T}) \Big) \prod_{i=1}^{N} \Big(\Gamma \eta |\alpha|^2 e^{-\Gamma t_i} \Big) \, , \tag{12}$$

The trace of (12) gives the N-fold probability density $w_N(t_1, t_2, \ldots, t_N : T)$ for precisely N detections at the given instants of time. The ratio $\hat{\tilde{\rho}}_N / w_N$ gives the conditional density matrix, given that N particles have been detected. After an N-fold integration of w_N over the ordered time instants, one obtains for the probability distribution for precisely N particles in the interval $[0, T]$

$$p_N(T) = \int d_2\alpha \, P(\alpha) \frac{1}{N!} \Big(\eta |\alpha|^2 (1 - e^{-\Gamma T}) \Big)^N \exp\Big(-\eta |\alpha|^2 (1 - e^{-\Gamma T}) \Big) \, . \tag{13}$$

Notice that $\sum_N p_N(T) = 1$, as it should. The distribution (33) has the form of the average of a Poisson distribution, with the P function as the probability distribution over the coherent-state index α. The mean value is

$$\overline{N} = \int d_2\alpha \; P(\alpha)\eta|\alpha|^2(1 - e^{-\Gamma T}) \,, \tag{14}$$

which is determined by the time integral of the detection rate $\eta\Gamma|\alpha(t)|^2$. Since the function $P(\alpha)$ is not necessarily positive definite, the actual probability distribution p_N can be sub-Poissonian. The difference between the variance and the mean value of N is found to be

$$\Delta N^2 - \overline{N} = \eta^2(1 - e^{-\Gamma T})^2 \times \Big(\int d_2\alpha \; P(\alpha)|\alpha|^4 - (\int d_2\alpha \; P(\alpha)|\alpha|^2)^2 \Big) \tag{15}$$

When this result is negative, the distribution of the number of detections is sub-Poissonian. Evidently, this is the case when also the initial distribution of the number of particles in the mode is sub-Poissonian.

3.2 Fixed amplitude

The situation simplifies in the special case that $P(\alpha)$ is nonzero only for a single value of $|\alpha|$, so that the initial state is specified by the distribution over the phase of $\alpha = r\exp(-i\phi_A)$. In eqs. (6)-(33) we can make the replacement $\int d_2\alpha P(\alpha) \to \int d\phi_A g_A(\phi_A)$, with g_A a normalized distribution over the phase ϕ_A. In this case, the n-fold distribution function takes the simple form

$$w_N(t_1, t_2, \ldots, t_N; T) = \exp\Big(- \eta r^2(1 - e^{-\Gamma T})\Big) \prod_{i=1}^{N} \Big(\Gamma\eta r^2 e^{-\Gamma t_i}\Big) \,, \tag{16}$$

independent of the phase distribution. The corresponding contribution to the density matrix (12) takes the factorized form

$$\hat{\tilde{\rho}}_N(t_1, t_2, \ldots, t_N; T) = w_N(t_1, t_2, \ldots, t_N) \int d\phi_A g_A(\phi_A) \; |\alpha(T) >< \alpha(T)| \,. \tag{17}$$

The integral in (17) is the conditional density matrix after precisely N detections. This is identical to the unconditioned density matrix (9) in this case. It is remarkable that this conditional density matrix is not affected by the detections. This is due to the fact that the detection probability is independent of the phase ϕ_A. This also implies that subsequent particle detections are uncorrelated, so that the distribution (33) reduces to a Poisson distribution. The mean value of the number of detections is $\overline{N} = \eta r^2(1 - e^{-\Gamma T})$. The simplest example is an initial coherent state $|r\exp(-i\phi_0)>$, so that g_A is effectively a delta function. Then the state remains a pure state at all times, given by the coherent state $|r\exp(-i\phi_0 - i\omega T - \Gamma T/2)>$.

In a radiation field the number of photons is not conserved, and it is quite common to have states that are not diagonal in the particle number. Howeve, the number of bosonic atoms is conserved, and no coherent superposition of states with different particle number can exist. This implies that for an isolated Bose-Einstein condensate, the density matrix must be diagonal in the particle number. This requires that the phase ϕ_A is uniformly distributed, so that in eqs. (6)-(33) we can make the replacement $\int d_2\alpha P(\alpha) \to \int d\phi_A/2\pi$. The time-dependent state (9) is then a statistical mixture of number states with a Poissonian distribution, and with mean value $r^2 \exp(-\Gamma T)$ of the number of particles in the mode.

4 Master Equation for Two Boson Modes

4.1 Two representations

Now we consider two independently decaying boson modes A and B. The combined density matrix obeys the master equation

$$\begin{aligned}\frac{d}{dt}\hat{\rho} = &- i\omega[\hat{a}^\dagger\hat{a}, \hat{\rho}] - \frac{1}{2}\Gamma(\hat{a}^\dagger\hat{a}\hat{\rho} + \hat{\rho}\hat{a}^\dagger\hat{a}) + \Gamma\hat{a}\hat{\rho}\hat{a}^\dagger \\ &- i\omega[\hat{b}^\dagger\hat{b}, \hat{\rho}_2] - \frac{1}{2}\Gamma(\hat{b}^\dagger\hat{b}\hat{\rho} + \hat{\rho}\hat{b}^\dagger\hat{b}) + \Gamma\hat{b}\hat{\rho}\hat{b}^\dagger \ . \end{aligned} \tag{18}$$

When the density matrix $\hat{\rho}(0)$ at time 0 can be expressed as a product $\hat{\rho}_A\hat{\rho}_B$ of terms corresponding to the separate systems, also the combined Glauber-Sudarshan function $P(\alpha, \beta)$ can be written as a product $P_A(\alpha)P_B(\beta)$. Then the state of the modes is unentangled. Since the two modes evolve independently, the time-dependent solution of the master equation (18) is just the product of two solutions of the form (9). Hence, the two modes remain unentangled at all times. Also when emitted particles are detected, while the products from both modes are distinguishable, no entanglement can arise.

The situation is different when identical particles emitted from both modes are observed in interference. As sketched in Fig. 1, we assume that particles emitted from both sources are combined in a beam splitter. Detection at the output ports D1 and D2 are described by the two orthogonal detection operators

$$\hat{c}_\pm = (\hat{a} \pm \hat{b})/\sqrt{2} \, . \tag{19}$$

This same detection scheme has been analyzed in Ref. [5] in the case of two condensates. The operators $\hat{c}_\pm$ obey the standard commutation rules $[\hat{c}_\pm, \hat{c}_\pm^\dagger] = 1$, $[\hat{c}_\pm, \hat{c}_\mp] = [\hat{c}_\pm, \hat{c}_\mp^\dagger] = 0$. When we express the operators $\hat{a}$ and $\hat{b}$ in terms of $\hat{c}_\pm$, the master equation (18) takes the fully equivalent form

$$\begin{aligned}\frac{d}{dt}\hat{\rho} = &- i\omega[\hat{c}_+^\dagger\hat{c}_+, \hat{\rho}] - \frac{1}{2}\Gamma(\hat{c}_+^\dagger\hat{c}_+\hat{\rho} + \hat{\rho}\hat{c}_+^\dagger\hat{c}_+) + \Gamma\hat{c}_+\hat{\rho}\hat{c}_+^\dagger \\ &- i\omega[\hat{c}_-^\dagger\hat{c}_-, \hat{\rho}_2] - \frac{1}{2}\Gamma(\hat{c}_-^\dagger\hat{c}_-\hat{\rho} + \hat{\rho}\hat{c}_-^\dagger\hat{c}_-) + \Gamma\hat{c}_-\hat{\rho}\hat{c}_-^\dagger \equiv \mathcal{L}\hat{\rho} \ . \end{aligned} \tag{20}$$

4.2 Correlations created by observation

Detection histories with imperfect efficiencies can be described as a direct generalization of the one-mode case discussed in Sec. 3. A detection history is now described by specifying for each detection within a time interval $[0, T]$ both the time instant and the detection channel. The evolution operator $\mathcal{L}$ in eq. (20) is separated as in (2), with $\mathcal{L}_1\hat{\rho} = \Gamma(\eta_+\hat{c}_+^\dagger\hat{\rho}\hat{c}_+ + \eta_-\hat{c}_-^\dagger\hat{\rho}\hat{c}_-) \equiv \mathcal{L}_{1+} + \mathcal{L}_{1-}$, with η_1 and η_2 the two detection efficiencies. These two terms describe two types of quantum jumps, corresponding to the two detection channels. Two-mode coherent states $|\alpha, \beta >$ are eigenfunctions of $\hat{a}$ and $\hat{b}$, and therefore also of $\hat{c}_\pm$, with eigenvalues $\gamma_\pm = (\alpha \pm \beta)/\sqrt{2}$. When we express the factorized initial state in terms of the combined P-function as

$$\hat{\rho}(0) = \int d_2\alpha d_2\beta \, |\alpha, \beta >< \alpha, \beta| P_A(\alpha) P_B(\beta) \, , \tag{21}$$

the contribution to the time-dependent density matrix corresponding to zero detections in $[0, T]$ is

$$\hat{\tilde{\rho}}_0(T) \equiv e^{\mathcal{L}_0 T}\hat{\rho}(0) = \int d_2\alpha d_2\beta \, |\alpha(T), \beta(T) >< \alpha(T), \beta(T)| \\ \times P_A(\alpha) P_B(\beta) \exp\Big(- (\eta_+|\gamma_+|^2 + \eta_-|\gamma_-|^2)(1 - e^{-\Gamma T})\Big) \, , \tag{22}$$

where the time dependence of the coherent states $|\alpha(T), \beta(T) >$ is the same as defined in Sec. 3. The probability for zero detections is the trace of (22), so that

$$p_0(T) = \int d_2\alpha d_2\beta P_A(\alpha) P_B(\beta) \exp\Big(- (\eta_+|\gamma_+|^2 + \eta_-|\gamma_-|^2)(1 - e^{-\Gamma T})\Big) \, . \tag{23}$$

Remarkably, when the two efficiencies $\eta_\pm$ are unequal, and the distributions P_A and P_B have a finite width, the last line in (22) does not factorize in the variables α and β. This indicates that entanglement has been created, by the simple fact that a detection experiment has been performed with a null outcome. In particular, entanglement is created when a detector is attached only to the output port $+$. Then after an interval with zero detection, the entangled state is described by the ratio of eq. (22) and (23), with the substitution $\eta_- = 0$. This situation is reminiscent of a measurement scheme discussed in Ref. [18], where an entangled subradiant state of two atoms is created by a null result of a photon measurement leaking out of a cavity.

Formal expressions can be given for the contributions to the density matrix corresponding to any detection history $t_1, s_1; t_2, s_2; \ldots; t_N, s_N$, with t_i the time instant, and $s_i = \pm$ the channel of the ith detection. Each subsequent detection has the tendency to enhance the entanglement between the two modes, even though they share no common past. This will be illustrated with an exactly solvable special case in the next section.

5 Initial States with Fixed Amplitudes

5.1 Separation of total number and relative phase

We evaluate the full detection statistics and the corresponding entangled states in the symmetric case that $P_A(\alpha)$ and $P_B(\beta)$ are nonzero only for the single value $|\alpha| = |\beta| = r$, while the efficiencies $\eta_+ = \eta_- = \eta$ are equal. Then we can write $\alpha = r\exp(-i\phi_A)$, $\beta = r\exp(-i\phi_B)$, so that the integrations effectively extend only over the phases, weighed by a factorized two-phase distribution $g_A(\phi_A)g_B(\phi_B)$. The multiplication factors corresponding to a detection are proportional to $|\gamma_+|^2 = 2r^2\cos^2(\phi/2)$, $|\gamma_-|^2 = 2r^2\sin^2(\phi/2)$, which depend only on the relative phase $\phi = \phi_A - \phi_B$ between the two modes. The unconditioned density matrix, which solves the master equation (18) (or (20)) is then

$$\hat{\rho}(T) = \int d\phi_A d\phi_B |\alpha(T), \beta(T) >< \alpha(T), \beta(T)| g_A(\phi_A) g_B(\phi_B) \,. \tag{24}$$

The contribution to the density matrix corresponding to a given detection history can be written down as a direct generalization to the single-mode case, with the result

$$\begin{aligned}
&\hat{\tilde{\rho}}_N(t_1, s_1; t_2, s_2; \ldots; t_N, s_N; T) = \exp\Big(- 2\eta r^2 (1 - e^{-\Gamma T})\Big) \\
&\times \prod_{i=1}^{N} \Big(2\Gamma\eta r^2 e^{-\Gamma t_i}\Big) \int d\phi_A d\phi_B |\alpha(T), \beta(T) >< \alpha(T), \beta(T)| \\
&\times g_A(\phi_A) g_B(\phi_B) \cos^{2n}(\phi/2) \sin^{2m}(\phi/2) \,.
\end{aligned} \tag{25}$$

Here n and m are the number of detections in each channel, with $n + m = N$ the total number. The r-dependent factor preceding the integral in (25) has the physical significance of the N-fold probability distribution function, regardless of the detection channel. Its expression is very similar to the N-fold distribution function (16), the only difference being that now the detection rate is doubled, corresponding to the presence of two decaying modes. It follows that the statistics of the total number of detections, regardless of the channel, is given by a Poisson distribution with mean value $2\eta r^2(1 - e^{-\Gamma T})$, as one would expect from the single-mode results.

The distribution function for the indicated detection history is just the trace of (25), which takes the factorized form

$$\begin{aligned}
&w_N(t_1, s_1; t_2, s_2; \ldots; t_N, s_N; T) = \\
&F(n, m) \exp\Big(- 2\eta r^2 (1 - e^{-\Gamma T})\Big) \prod_{i=1}^{N} \Big(2\Gamma\eta r^2 e^{-\Gamma t_i}\Big) \,.
\end{aligned} \tag{26}$$

We introduced the coefficients

$$F(n, m) = \int d\phi \; g(\phi) \cos^{2n}(\phi/2) \sin^{2m}(\phi/2) \,, \tag{27}$$

with $g(\phi) = \int d\phi_B g_A(\phi_B + \phi) g_B(\phi_B)$ the normalized distribution function over the relative phase ϕ in the initial state. The physical significance of the coefficients $F(n, m)$ is the probability that the first N detections occur in an ordered sequence $s_1, s_2, \ldots, s_N$ of the detection channels. It turns out that this probability depends only on the total number of detections in each channel, not on their time order nor on the initial particle number r^2. This makes it easy to obtain the probability distribution over the possible partitions (n, m) over the two detection channels, with $m = N - n$, in the form

$$p(n, m) = \binom{N}{n} F(n, m) . \tag{28}$$

Obviously, for a given value of N these numbers add up to one, as they should. The probability distribution (28) over the detection channels is independent of the initial mean occupation number r^2 in the modes. Moreover, the distribution depends only on the initial distribution of the relative phase ϕ, not on the absolute phases of the two modes.

It is interesting to notice that the full statistics $F(n, m)$ can be easily expressed in terms of the values $F(n, 0)$ with $m = 0$. This follows after introducing the branching ratios

$$f_+(n, m) = F(n+1, m)/F(n, m) \ , \ f_-(n, m) = F(n, m+1)/F(n, m) \tag{29}$$

that specify the branching over the two channels for the next detection after a series (n, m). From the expression (27) it follows that $f_+ + f_- = 1$. Knowledge of all $F(n, 0)$ also determines the branching ratios $f_+(n, 0)$, and thereby also all $f_-(n, 0)$. Moreover, one proves the identity

$$f_+(n, m) f_-(n+1, m) = f_-(n, m) f_+(n, m+1) \ , \tag{30}$$

since both sides equal the probability that the first two detections after a series (n, m) consist of one in the channel $+$, and one in the channel $-$. Hence, eq. (30) for $m = 0$ also determines the branching ratios $f_+(n, 1)$, and thereby all values $f_-(n, 1)$. Likewise, eq. (30) for $m = 1$ produces the values of $f(n, m)$ for $m = 2$, etc. Once the branching ratios are known, the values of $F(n, m)$ are given as products of branching ratios corresponding to the sequence of the particle detections.

From the coefficients $F(n, m)$ we can also derive expressions for the conditional statistics of the number of detections, following a given initial number of n_0 and m_0 detections in the two channels. Since $F(n, m)$ represents the probability for a specific order of the detections, the conditional probability that the next N detections follow a specific sequence $s_1, s_2, \ldots, s_N$ is given by

$$F(n, m | n_0, m_0) = F(n_0 + n, m_0 + m)/F(n_0, m_0) \ , \tag{31}$$

which again depends only on the partition (n, m) of the conditional sequence. For the conditional probability distribution over the partitions (n, m) we find

$$p(n, m | n_0, m_0) = \binom{N}{n} F(n_0 + n, m_0 + m)/F(n_0, m_0) . \tag{32}$$

The normalized conditional density matrix, given a detection history (n, m) in both output channels, is given by

$$\hat{\rho}(n, m; T) = \int d\phi_A d\phi_B |\alpha(T), \beta(T) >< \alpha(T), \beta(T)| \times g_A(\phi_A) g_B(\phi_B) \cos^{2n}(\phi/2) \sin^{2m}(\phi/2)/F(n, m) \ . \tag{33}$$

It is equal to the ratio of (25) and (26), it is fully specified by the detection numbers n and m, and it does not depend on the instants or the ordering of the detections. Equation (33) shows as the effect of each detection that the phase distribution function is multiplied by $\cos^2(\phi/2)$ or $\sin^2(\phi/2)$, apart from normalization.

5.2 Coherent states

A trivial special case of Sec. 5.1 arises when both modes are in a coherent state. This is a natural situation when the particles are photons leaking out of cavities, but it is not possible for two condensates. In this case the phase distribution function is $g(\phi) = \delta(\phi - \phi_0)$ with ϕ_0 the difference of the phases of α and β. In this case, the branching ratios (29) are $f_+ = \cos^2(\phi_0/2)$ and $f_- = \sin^2(\phi_0/2)$ for all values of n and m. The distribution (28) over the two channels is binomial, with average numbers of detections $N \cos^2(\phi/2)$ and $N \sin^2(\phi/2)$ in both channels. When the phase difference is $\phi_0 = \pi/2$, the branching ratios are equal to 1/2, and the probability distribution (28) over the channels is

$$p(n, m) = \frac{1}{2^N} \binom{N}{n} \ , \tag{34}$$

just as the conditional distribution (32). Likewise, the conditional density matrix (33) after (n, m) detections is equal to the unconditioned density matrix (24), so that the density matrix is not affected by the detection. The measurement process is highly classical, and no entanglement is created. The average value is $\overline{n} = N/2$, and the standard deviation is equal to $\Delta n = \sqrt{N}/2$. In the limit of large detection numbers n, the distribution (34) is well approximated by a Gaussian, specified by these numbers. Hence, the relative width decreases proprtional to $\sqrt{N}$.

5.3 Uniform phase distribution

The opposite extreme case occurs when the initial phase distributions of both modes are completely uniform, so that both modes contain a Poissonian mixture of number states. The unconditional state that solves the master equation is given by (24), with $g_A = g_B = 1/2\pi$. This state is unentangled at all times. This uniform-phase state has been considered for the analysis of the buildup of a fixed relative phase between two Bose-Einstein condensates in a continuous measurement [4]. In fact, the detection statistics depends only on the distribution function of the relative phase, which is also uniform, so that $g(\phi) = 1/2\pi$. Initially, there is no preference for either channel, and the two branching ratios $f_+(0,0) = f_-(0,0) = 1/2$ are equal. However, after a number of detections, a preference is created for the channel that had already the most detections. This effect has been discussed for Bose condensates in initial number states in the large-number limit [5]. Here we derive exact expressions for the statistical distribution, while accounting for the limited detection efficiency, and the decay of the modes during the detection interval.

From the definitions (27) and (29) of F and f_+, with a uniform phase distribution, it is an easy matter to obtain the identity

$$f_+(n,0) = \frac{2n+1}{2n+2} .$$

By the method sketched in Section 5.1, we can then obtain expressions for all branching ratios, with the result

$$f_+(n,m) = \frac{2n+1}{2(N+1)} , \; f_-(n,m) = \frac{2m+1}{2(N+1)} , \tag{35}$$

with $N = n + m$. For the coefficients F this gives the result

$$F(n,m) = \frac{(2n)!(2m)!}{2^{2N} N!n!m!} , \tag{36}$$

so that the probability distribution (28) of N detected particles over the two channels is expressed by

$$p(n,m) = \frac{1}{2^{2N}} \binom{2n}{n} \binom{2m}{m} . \tag{37}$$

Indeed, these probabilities (37) for a fixed value of $N = n + m$ add up to one. Whereas the binomial distribution (34) has maximal values for $n = m = N/2$, the distribution (37) is maximal for $n = 0$ or $m = 0$, so that the detected particles tend to bunch in one channel. For the most probable detection histories, with most particles in one channel, the phase distribution becomes very narrow. The conditional density matrix after a detection sequence (n, m) with $n \ll m$ (or $n \gg m$) is given by eq. (33), with $g_A g_B = 1/4\pi^2$. After a typical detection history, this state is characterized by a phase-dependent function that strongly peaks at $\phi = 0$ (or at $\phi = \pi$). On the other hand, the phases of each one of the modes are still uniformly distributed. Provided that the remaining number of particles per mode $r^2 \exp(-\Gamma T)$ is still large, the state of the modes is highly entangled.

The distribution (37) cannot be approximated by a Gaussian distribution, even in the limit of large detection numbers $N = n + m$. This illustrates the strong correlation between successive detections. For not too small values of n, the binomial coefficients occurring in (37) can be approximated by

$$\binom{2n}{n} \approx 2^{2n}\frac{1}{\sqrt{\pi n}} \, .$$

In the limit of large N values, the distribution (37) is therefore well represented by the continuous normalized distribution

$$q(\nu) = \frac{1}{\pi\sqrt{\nu(1-\nu)}} \tag{38}$$

over $\nu = n/N$, with $0 \leq \nu \leq 1$. This distribution can be understood by considering the situation that two classical fields with amplitudes $\propto \exp(-i\phi_A)$ and $\propto \exp(-i\phi_B)$ enter the input port of the beam splitter. The relative intensities of the two output channels are then $\nu = \cos^2(\phi/2)$ and $\mu = \sin^2(\phi/2)$. A uniform distribution over the relative phase $\phi = \phi_A - \phi_B$ reproduces the distribution $q(\nu)$. A plot of the limiting distribution (38) is plotted in Fig. 2.

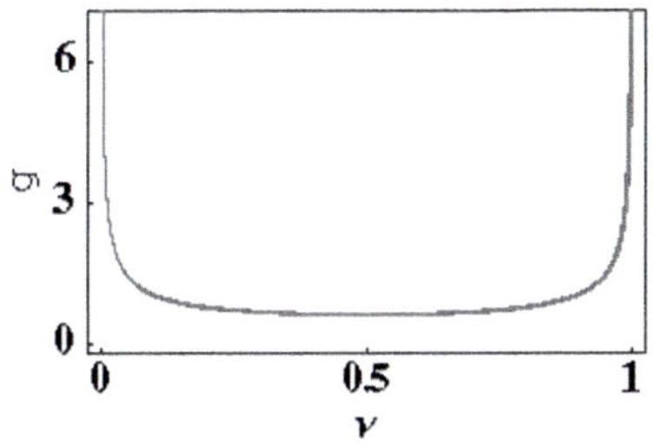

Fig. 2. Plot of distribution (38), which is equivalent to the distribution (37) in the limit of large N

The distribution of ν (or μ), resulting from the distribution of ϕ is illustrated in Fig. 3. The indicated complex amplitudes a and b correspond to the superposition of two fields with equal amplitudes and a relative phase ϕ.

The conditional probability distribution over the partitions (n, m) of N detections following an initial series containing n_0 and m_0 detections in both channels is given by eq. (32). When n_0 is much larger than m_0, the distribution has its maximal value around $n = n_0 N/N_0$. Then the state of the two modes is characterized by a narrow distribution over the relative phase ϕ, which implies that the distribution (32) approaches a binomial distribution with average value $\overline{n} = n_0 N/N_0$.

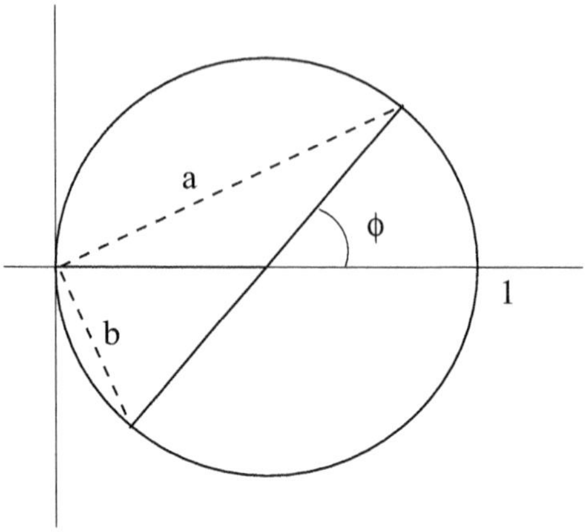

Fig. 3. Geometrical illustration of distribution of relative intensity ν in the complex plane. A uniform distribution over the phase difference ϕ corresponds to the distribution (38) over ν. The values of $\nu = |a|^2$ and $\mu = |b|^2$ are equal to the squares of the lengths of the indicated dashed lines, in a circle with diameter 1

6 Conclusions

We describe the decay of two boson modes, detected in interference after passing the two decay channels through a beam splitter. The initial states are represented in terms of a Glauber-Sudarshan quasiprobability function P. By applying quantum trajectories with a finite detection probability, we obtain expressions for the conditional state after a given detection history, and for the full detection statistics. Exact solutions are obtained for the conditional density matrix following a given detection history. In the case that the initial state is a mixture of coherent states with the same average number of particles in both modes, the distribution over the phases separates from that over the amplitudes, and the detection statistics factorizes into a product of probabilities for the total number and a relative distribution over the two detection channels. The description applies just as well to the case of bosonic atoms leaking out of a Bore-Einstein condensate, as to the situation of decaying modes of two cavities.

We point out that the detection of he decay products in interference gives rise to entanglement of the two modes. This entanglement arises between two systems that have no common past. In the special case that initially the phases of both modes are uniform, the final state after a most probable detection history has a well-defined value of the relative phase, while the phase distributions of the two separate modes remains uniform. Conversely, the state remains diagonal in the total number of particle, while the difference in particle number becomes highly indeterminate.

References

1. M.H. Anderson, J.R. Ensher, M.R. Matthews, C.E. Wieman, E.A. Cornell: Science **269**, 198 (1995)
2. C.C. Bradley, C.A. Sackett, J.J. Tollett, R.G. Hulet: Phys. Rev. Lett. **75**, 1687 (1995)
3. K. Davis, M.O. Mewes, M.R. Andrews, N.J. van Druten, D.S. Durfee, D.M. Kurn, W. Ketterle: Phys. Rev. Lett. **75**, 3969 (1995).
4. J.I. Cirac, C.W. Gardiner, M. Naraschewski, P. Zoller: Phys. Rev A **54**, R3714 (1996)
5. Y. Castin, J. Dalibard: Phys. Rev. A **55**, 4330 (1997)
6. S.M. Yoo, J. Ruostekoski, J. Javanainen: J. Mod. Optics **44**, 1763 (1997)
7. M. R. Andrews, C. G. Townsend, H.-J. Miesner, D. S. Durfee, D. M. Kurn, W. Ketterle: Science **275**, 637 (1997)
8. D. S. Hall, M. R. Matthews, C. E. Wieman, C. A. Cornell: Phys. Rev. Lett. **81**, 1543 (1998)
9. J.-W. Pan, D. Bouwmeester, H. Weinfurtner, A. Zeilinger: Phys. Rev. Let. **80**, 3891 (1998)
10. L. Mandel, E. Wolf: *Optical Coherence and Quantum Optics* (Cambridge University Press, 1995)
11. H.J. Carmichael: *An Open System Approach to Quantum Optics* (Springer-Verlag, Berlin, 1993)
12. J. Dalibard, Y Castin, K. Mølmer: Phys. Rev. Lett. **68**, 580 (1992)
13. R. Dum, P. Zoller, H. Ritsch: Phys. Rev. A **45**, 4879 (1992)
14. G. Nienhuis: J. Stat. Phys. **53**, 417 (1988)
15. G. C. Hegerfeldt, M. B. Plenio. Phys. Rev. A **47**, 449 (1993)
16. R.J. Glauber: Phys. Rev. **131**, 2766 (1963)
17. E.C.G. Sudarshan: Phys. Rev. Lett. **10**, 277 (1963)
18. M.B. Plenio, S.F. Huelga, A. Beige, P.L. Knight: Phys. Rev. A **59**, 2468 (1999)

Decoherence Effects of Motion Induced Radiation

P.A. Maia Neto[1] and D.A.R. Dalvit[2]

[1] Instituto de Física, UFRJ, Caixa Postal 68528
21945-970 Rio de Janeiro, Brazil
[2] T-6, Theoretical Division, MS B288, Los Alamos National Laboratory
Los Alamos, NM 87545, USA

Abstract. The radiation pressure coupling with vacuum fluctuations gives rise to energy damping and decoherence of an oscillating particle. Both effects result from the emission of pairs of photons, a quantum effect related to the fluctuations of the Casimir force. We discuss different alternative methods for the computation of the decoherence time scale. We take the example of a spherical perfectly-reflecting particle, and consider the zero and high temperature limits. We also present short general reviews on decoherence and dynamical Casimir effect.

1 Introduction and Brief Summary of Decoherence Theory

The understanding of the quantum-to-classical transition has been the subject of extensive research[1,2]. The core of the problem is that the Hilbert space of allowed states of a quantum system is huge, whereas the set of states with associated classical properties is a tiny subset of the whole Hilbert space. Some questions that naturally arise are the following: which mechanism is responsible for the classical appearance of macroscopic and mesoscopic quantum systems? How are those few classical states selected from the huge Hilbert space? The common wisdom is that classicality is an emergent property induced on subsystems by their environment.

The interaction between a system S and its environment E creates entanglement (i.e. non reducible correlations) between the states of the system and those of the environment. Imagine that at a given time (say $t = 0$) the state $|\Psi(t)\rangle$ of S+E is a product state, that is, there are no initial correlations. We have $|\Psi(t=0)\rangle = |s\rangle|\epsilon\rangle$, where the first ket corresponds to the system state (assumed for simplicity to be in a pure state), and the second one to the environmental state, also assumed pure. When the two parts begin to interact, entanglement is generally produced. This means that at a later time t, the state $|\Psi(t)\rangle$ will be given by a linear superposition of the form

$$|\Psi(t)\rangle = |s_1\rangle|\epsilon_1\rangle + |s_2\rangle|\epsilon_2\rangle + \dots \tag{1}$$

M. Orszag, J.C. Retamal (Eds.): LNP 575, pp. 110–124, 2001.

where $\{|s_i\rangle\}$ and $\{|\epsilon_i\rangle\}$ are states of the system and environment Hilbert spaces, respectively. If the interaction is such that the states $|\epsilon_i\rangle$ become approximately orthogonal ($\langle\epsilon_n|\epsilon_m\rangle \approx \delta_{nm}$), then interference between the system states $|s_n\rangle$ and $|s_m\rangle$ will not be observed. These set of states $\{|s_i\rangle\}$ usually have classical properties. Any quantum superposition of them is a non-classical state, and quickly decays away into a statistical mixture of the states. The coherence of the phase relation between the components of the superposition is lost, and this process is accordingly known as *decoherence*. In other words, the environment monitors the different classical alternatives for the system (the different states $|s_i\rangle$), thereby providing which-way information, even though such information is usually unaccessible to the observer. The set of states $\{|s_i\rangle\}$ are called *pointer states* [3], and they are the states within the huge Hilbert space of the system that become less entangled with the environment. Perfect pointer states are those that produce no entanglement at all, so that an initial product state of S+E will remain a product state throughout the interaction time, which means that those states are robust and stay unperturbed by the interaction. All this will be illustrated in Section 3 in the particular case where the environment is the radiation field at zero temperature (vacuum field) and radiation pressure is responsible for the coupling between system (a mirror) and the environment.

A possible method to identify pointer states is called the 'predictability sieve criterion' [4], which is based on the fact that pointer states are the ones that produce least entropy and remain most pure. Let us explain these concepts. The evolution of the closed combined system S+E is unitary, so that the purity of the whole state $|\Psi\rangle$ is preserved, i.e. $P(\rho_{S+E}) = \mathrm{Tr}\rho^2_{S+E} = 1$ for all times. However, the purity of S is not preserved. To show it one needs to calculate the reduced density matrix of the subsystem S by tracing out the environmental degrees of freedom, $\rho_S = \mathrm{Tr}_E\rho_{S+E}$, and then $P(\rho_S) < 1$. The loss of purity can be associated with a loss of information about the system state. When no measurement involving the environment is made, the density matrix ρ_S contains the state of knowledge of an observer about the system, and purity is a measure of that knowledge. Initially, there is full knowledge of the system state, which is described by a single ket state. Subsequently, the interaction with the environment produces entanglement, and part of the information about the system S is lost to the environment, causing a decrease in the purity of the system. One can also measure the information content of ρ_S through the von Neumann entropy, $S(\rho_S) = -\mathrm{Tr}\rho_S \log \rho_S$. Initially one has full information, and entropy is identically zero; as time goes on, information is lost and entropy is produced. Pointer states are least affected by the environment, so their information content is preserved and hence they produce the least entropy. The idea is then to take every state in the Hilbert space of S, calculate the von Neumann entropy at time t produced via interaction with the environment starting from the given state of the system, and order the initial states in a tower of increasing entropy. The states that lie at the bottom of such a tower are candidates for pointer states. Finally, one must check that those states remain at the bottom when the time t when entropy is calculated is changed, so that the states are robust pointers.

If one can satisfy these conditions, one gets the pointer states of the system S. If not, there are no pointer states for the problem. When these ideas are put into practice, it is more convenient to work with the linear entropy, defined as $S(\rho_S) = 1 - P(\rho_S) = 1 - \mathrm{Tr}\rho_S^2$.

The determination of pointer states for a given problem depends both on the system and environment dynamics, as well as on their interaction. There are three different regimes: 1) When the system Hamiltonian is irrelevant, pointer states are given by the eigenstates of the system's operators contained in the interaction Hamiltonian. A typical example is that of a measuring apparatus (the system) that has no internal dynamics, measuring an external reservoir (the environment), say a photocurrent. 2) When the system dynamics as well as the interaction are relevant, pointers come from an interplay between the two. The most thoroughly studied example is that of quantum Brownian motion (QBM), in which a particle is coupled to a set of harmonic oscillators in a thermal state, thereby suffering decoherence [5]. Although the coupling between the system and the environment is of the type position-position, pointer states are not position eigenstates of the system because the self dynamics of the harmonic oscillator interchanges position and momentum every quarter of a cycle. It turns out that the interplay between interaction and self dynamics leads to coherent states as pointers [6]. This second case is also the relevant one for this paper. 3) Finally, when the environment evolves much slower than the system, pointers may correspond to energy eigenstates of the system's Hamiltonian [7].

In the above we have ignored the information contained in the environmental state, and that is the reason why one traces over the environmental degrees of freedom in order to find the reduced density matrix of the system. However, the information lost to the environment could be, in principle, intercepted and recovered. Performing measurements on the environment one may extract information about the decohering system. In [8] it is shown that the preferred pointer states remain unchanged, even when that information is kept and modifies the dynamical evolution of the system.

Another related way to study the dynamical process underlying in the quantum-to-classical transition is via phase space representations of the reduced quantum dynamics for the system S. Among the many possible representations, one of particular interest is the Wigner function $W(x,p)$, which is defined as a Fourier transform of the reduced density matrix. $W(x,p)$ is a pseudo probability distribution in phase space, and encapsulates the quantum coherence of the system in interference fringes that take both positive and negative values. Imagine that one starts with an initial state for S which is highly non classical, such as a cat state $|\mathrm{cat}\rangle = 1/\sqrt{2}(|\alpha\rangle + |-\alpha\rangle)$, where $|\alpha\rangle$ is a coherent state with large amplitude ($|\alpha| \gg 1$). The corresponding Wigner function will have interference fringes, showing the quantum nature of the state. However, when the system is put in contact with the environment and each component of the state becomes entangled with almost orthogonal states of the environment, the interference fringes will be washed out. In the end the Wigner function becomes positive defined, with two peaks at the values corresponding to $\pm\alpha$, as a true probability distribution. Decoherence transforms the initial pure state into a mixture of the two coherent states $| \pm \alpha\rangle$.

Until not very long ago the ideas of decoherence were restricted to the theoretical domain. Recent experimental developments have succeeded in studying in real time the process of decoherence in the laboratory, and have tested the predictions of the theory. Here we shall mention a few experiments that have been a hallmark in those developments. First, in the field of cavity QED, superposition states of photons and Rydberg atoms have been created within high-Q microwave cavities. Cat states of around 3 photons have also been produced, and it has been studied how they decay due to decoherence [9]. The coherence of the state was monitored with the help of a measurement of correlations between two consecutive atoms crossing the cavity [10]. Second, in the field of ion trapping, methods for creating superposed motional states of ions were developed, as well as schemes of environment engineering to protect those states from decoherence [11]. Finally, it has been possible to push the size of the cat states further into the macroscopic realm by generating a mesoscopic cat inside a rf-SQUID. The two components of the cat correspond to superconducting currents moving either clockwise or counterclockwise, each containing around 10^9 Cooper pairs [12].

In what we have discussed so far decoherence has been portraited as a "good" effect, in the sense that is responsible for the quantum-classical transition and the appearance of our classical world. Decoherence can also have a "bad" role in the field of quantum computation and quantum information processing. There one performs logical operations making use of the superposition states of quantum mechanics. For such operations to be successful it is very important to maintain the relative phase between the components of the superpositions all along the operations. If decoherence acts, it produces quantum errors that must be somehow corrected. Several methods have been proposed to minimize the effects of decoherence (see [13] for an example in nuclear magnetic resonance).

The prototype calculation of environment induced decoherence is the heuristic position-position interaction Hamiltonian for describing quantum Brownian motion, where the environment is taken to be a collection of harmonic oscillators. Although such a model is quite useful for studying many physical processes associated to dissipation and decoherence of a quantum system, the results that follow from it do not apply to every situation. That is, it is necessary to perform a case by case analysis in order to compute physical observables, such as decoherence and damping rates, how they scale with the parameters of the system, the environment, and their coupling, etc. For usual environments (thermal atoms, thermal light, phonons, etc.) it is in principle possible to design engineering schemes to protect the state of the system from decoherence, for example by reducing the coupling to the environment.

Then, the following question naturally arises: is it possible, at least in principle, to have arbitrarily weak decoherence? In this paper, we consider a fundamental source of decoherence that cannot be 'turned off': the radiation pressure coupling with the vacuum field [14]. As reviewed in Sec. 2, any particle not completely transparent unavoidably scatters vacuum field fluctuations. This type of coupling is responsible for the Casimir effect. More generally, photons are created out of the vacuum field when moving boundaries are considered, an effect known as dynamical Casimir effect or motion-induced radiation. In Sec. 3, we show how the dynamical Casimir effect engenders decoherence. Our emphasis is on the basic physical ideas, and most of the calculations are referred to [15], but we also briefly discuss a model alternative to the one employed in this reference.

2 Dynamical Casimir Effect

The Casimir effect is perhaps the simplest and most striking effect of the quantum vacuum field (see [17] for reviews). The essential idea is that the boundary conditions modify the spectrum of the radiation field, and thereby its zero-point energy. This modification has direct physical consequences, leading, for example, to an attractive force between two parallel perfectly-reflecting plates (of surfaces A) and at a distance L, given by [18]

$$F = \frac{\pi^2}{240}\frac{\hbar c}{L^4}A.$$

A series of recent experiments [19] reported precision measurements of the Casimir force in agreement with the predictions of Quantum Electrodynamics, although more complete theoretical calculations, taking into account corrections due to finite temperature and conductivity as well as to roughness and geometry of the surfaces are partially yet to be done [20].

The Casimir force may also be computed by taking the average of the Maxwell stress tensor over the field vacuum state [21]. This method suggests that the Casimir force is itself a fluctuating quantity, as noted by Barton. Its fluctuations were first computed for plane perfectly reflecting mirrors [22], and later for spherical and spheroidal particles [23]. More generally, any particle scattering the radiation field is under the action of a fluctuating radiation pressure force exerted by the vacuum field, even in the situations where the *average* force vanishes (for example a single plane mirror at rest). The coupling responsible for those fluctuations also gives rise to a *dissipative* force, when the particle is moving in vacuum. Dissipation of the mirror's mechanical energy is needed to enforce energy conservation, since the motion induces the emission of pairs of photons (for reviews see [27][28]). Because of their common physical origin, fluctuations and dissipation are related by a very general result [29], whose most known application is the Einstein relation between diffusion and friction coefficients for a Brownian particle in the high-temperature limit. This connection provides a very useful tool for deriving the response to an external small perturbation from the fluctuations in the unperturbed case. Linear response theory was employed by Jaekel and Reynaud to infer the vacuum radiation pressure force on partially-reflecting moving mirrors in the one-dimensional (1D) case [30]. For a single perfect mirror (position $x(t)$) the force is given by

$$F = \frac{\hbar}{6\pi c^2} \frac{d^3 x}{dt^3}, \tag{2}$$

a result first obtained by solving the boundary conditions of a moving mirror in the long wavelength approximation, and assuming the effect of the motion to be a small perturbation [27]. Eq. (2) was also derived as the $n \to \infty$ limit of a moving half-space of refractive index n [32]. It also corresponds to the nonrelativistic approximation of the exact result (for a perfect mirror) derived with the help of a conformal coordinate transformation to the co-moving frame [33].

Since the wave equation in three dimensions is not invariant under a general conformal transformation, only approximated methods are used in this case. The dissipative force on a plane mirror was computed within the long wavelength approximation for a scalar [27] and electromagnetic [34] field models. The angular and frequency distributions of the emitted radiation were also computed for a single plane moving mirror [35], a moving dielectric half-space [36] [37] and two parallel plane mirrors [38]. Linear response theory was employed to derive the dissipative force on moving spheres [39]. Small but otherwise arbitrary time-dependent deformations of an initially plane surface were analyzed with the help of different approaches: linear response theory [40], long wavelength approximation [41], and path integrals [42].

The magnitude of the dynamical Casimir effect may be illustrated with the following example, which we shall discuss in detail in Sec. 3. We consider that the 'mirror' is a particle of mass M in a 1D harmonic potential, such that the oscillation frequency is ω_0. From Eq. (2), the equation of motion reads

$$\frac{d^2 x}{dt^2} = -\omega_0^2 x + \frac{\hbar}{6\pi M c^2} \frac{d^3 x}{dt^3}. \tag{3}$$

For any situation of physical interest, the zero point energy is much smaller than the rest mass energy: $\hbar\omega_0 \ll Mc^2$. In this case, (3) has solutions corresponding to oscillations damped at the rate

$$\Gamma = \frac{\hbar\omega_0}{12\pi M c^2}\omega_0 \ll \omega_0, \tag{4}$$

showing that the dynamical Casimir effect provides a tiny perturbation of the free oscillations.

As could be expected, a larger effect takes place when field modes of a cavity resonator are coupled to the moving boundaries, mainly when the mechanical frequency lies close to a given cavity eigenfrequency. Moore considered a scalar 1D field inside a cavity where one of the mirrors follows a prescribed motion [11]. The field modes were formally built in terms of the solution of a functional equation. This method was later developed [44] and extended to the case where the two mirrors are set in motion [45]. The case of partially-transmitting mirrors was also calculated, allowing for a reliable estimation of the orders of magnitude for the rate of transmitted photons and the number of photons inside the cavity at steady-state [46]. So far, few three-dimensional (3D) calculations along these lines have been reported. A rectangular cavity made of perfectly-reflecting moving mirrors [47] [48], and a spherical bubble with time-dependent radius [49] were analyzed, the latter motivated by the problem of sonoluminescence.

In this article we only consider a single scatterer, so that no resonant enhancement takes place. In this section, we have shown that the radiation pressure coupling gives rise to energy damping of a particle scattering vacuum fluctuations. In the next section, we show that it also destroys the quantum coherence of the particle.

3 Decoherence and the Casimir Effect

Most treatments of the dynamical Casimir effect consider the particle that scatters the vacuum field (the 'mirror') to follow a prescribed motion (an exception is Ref. [50], which considers fluctuations of position of a particle driven by vacuum radiation pressure). In this article, however, we want to focus on the particle as the dynamical degree of freedom of interest. More specifically, we analyze how the radiation pressure coupling destroys the quantum coherence of an initial superposition state of the particle.

We consider as before that the particle is in a harmonic potential well, corresponding to a frequency of oscillation ω_0. The connection with the previous approaches, where the (classical) particle is assumed to follow a prescribed oscillation, is made by taking a coherent quasi-classical state $|\alpha\rangle$ for the particle, so that the combined particle-field state at $t = 0$ is

$$|\Psi(t=0)\rangle = |\alpha\rangle|0\rangle, \tag{5}$$

where we have assumed that the field is initially in the vacuum state $|0\rangle$. The oscillation gives rise to the emission of photon pairs at time t at the field modes λ_1 and λ_2, with probability amplitudes $b(\lambda_1, \lambda_2, t)$:

$$|\Psi(t)\rangle = |\alpha\rangle \left(B(t)|0\rangle + \sum_{\lambda_1,\lambda_2} b(\lambda_1, \lambda_2, t)|\lambda_1, \lambda_2\rangle \right), \tag{6}$$

where $B(t)$ is such that this state is normalized. As discussed in Sec. 2 (see, in particular, Eq. (4)), the energy damping associated to the dynamical Casimir effect is very small. This effect, and more generally the recoil of the particle, is neglected in (6), where the particle state is assumed not to be modified. Even at this level of approximation, there is decoherence, as we show by taking the initial state of the particle to be the cat state $|\text{cat}\rangle = (|\alpha\rangle + |-\alpha\rangle)/\sqrt{2}$, an example already mentioned in Sec. 1. It corresponds to the coherent superposition of two wavepackets oscillating out-of-phase in the harmonic potential well. The amplitudes $b(\lambda_1, \lambda_2, t)$ depend on the phase of the oscillation, so that they have an opposite sign when we take the state $|-\alpha\rangle$. Since the evolution operator is linear, the complete state at time t is the superposition of the r.-h.-s. of (6) with the analogous state for $|-\alpha\rangle$. It turns out to be an entangled state of the form discussed in (1):

$$|\Psi(t)\rangle = |\alpha\rangle|\epsilon^{(+)}(t)\rangle + |-\alpha\rangle|\epsilon^{(-)}(t)\rangle, \tag{7}$$

with $|\epsilon^{(\pm)}(t)\rangle = B(t)|0\rangle \pm \sum_{\lambda_1,\lambda_2} b(\lambda_1, \lambda_2, t)|\lambda_1, \lambda_2\rangle$. These field states work as tags for the particle states, providing which-way information about the phase of the oscillation. As time goes on, the information gets better defined, since

$$\langle\epsilon^{(-)}(t)|\epsilon^{(+)}(t)\rangle = 1 - 2\sum_{\lambda_1,\lambda_2} |b(\lambda_1, \lambda_2, t)|^2 \tag{8}$$

decreases as the probability for photon emission increases. When the emitted photons are not detected, all the relevant information about the particle is contained in the reduced matrix $\rho(t) = \mathrm{Tr_F}(|\Psi(t)\rangle\langle\Psi(t)|)$, where the trace is taken over the field states. Since the interference term is gradually washed out as a consequence of the photon emission effect and the corresponding entanglement with the field, $\rho(t)$ decays into the statistical mixture $\rho_m = (|\alpha\rangle\langle\alpha| + |-\alpha\rangle\langle-\alpha|)/2$. The corresponding time scale t_d may be computed [15] from Eq. (7), and turns out to be proportional to the energy damping time $1/\Gamma$, which is related to the two-photon probabilities by energy conservation:

$$t_d = \frac{1}{4|\alpha|^2}\frac{1}{\Gamma}. \tag{9}$$

Eq. (9) also holds when the coupling with the environment is described by a heuristic master equation in the Lindblad form (derived with the help of the rotating-wave approximation) [51], as well as in the case of position-position coupling to a zero-temperature environment of harmonic oscillators, and has a very simple interpretation [16]: if $1/\Gamma$ is the time needed to damp the energy $2|\alpha|^2\hbar\omega_0$, it corresponds to the emission of $2|\alpha|^2$ pairs of photons (each pair has a total energy equal to $\hbar\omega_0$). On the other hand, coherence is much more delicate than energy, since a single photon provides which-way information that destroys the quantum phase of the cat state. Hence the decoherence time is the time scale for the emission of a single photon. Since $4|\alpha|^2$ photons are emitted during the time interval $1/\Gamma$, the time for a single photon scales as in the r.-h.-s. of (9).

Eq. (9) only holds when $|\alpha| \gg 1$. In this limit, decoherence is much faster than damping, justifying the approach of neglecting the decay of the amplitude α of the coherent states in (6) and (7). This is of course in line with the idea that in the 'macroscopic' limit weird non-classical states are extremely fragile and difficult to observe. For truly macroscopic systems t_d is so short that no experimental monitoring of the decoherence process is possible. However the validity of Eq. (9) is restricted by the additional condition that decoherence is slower than the free oscillation (this condition is fulfilled by the experiments [9] [11] discussed in Sec. 1). In this regime, the particle oscillates several times in the potential well before coherence is lost, and the r.-h.-s. of Eq. (9) may be written in terms of the distance $\Delta x = 2\sqrt{2\hbar/M\omega_0}|\alpha|$ between the two wavepackets when they are at their turning points (M is the mass of the particle):

$$t_d = 4\left(\frac{\Delta x_0}{\Delta x}\right)^2 \frac{1}{\Gamma}, \tag{10}$$

where $\Delta x_0 = \sqrt{\hbar/(2M\omega_0)}$ is the position uncertainty of the oscillator ground state. Eq. (10) shows more explicitly that the decoherence rate scales as the squared distance in phase space between the two components of the cat state. In Eq. (9), the distance is expressed in terms of the squared difference $\Delta\alpha = 2\alpha$ between the amplitudes of the two coherent states $|\pm\alpha\rangle$. Such dependence, already experimentally observed in [9], was fully verified in [11]. Thus, the decoherence rate is directly connected to the quality of which-way information, for the possibility of resolving the two wavepackets is quantified by the distance between them divided by their width Δx_0.

The second factor entering in the r.-h.-s. of (9) is the damping coefficient Γ. Rather than a phenomenological constant, here Γ quantifies the strength of the radiation pressure coupling to the vacuum field, and is calculated from first principles. As discussed in Sec. 2, it may be obtained directly from the expression for the dissipative radiation pressure force on the particle. In the 1D case, Γ is given by Eq. (4), which jointly with Eq. (9) yields

$$t_d = \frac{3}{(v/c)^2}\frac{2\pi}{\omega_0}, \tag{11}$$

where $v = \sqrt{2\hbar\omega_0/M}\,|\alpha|$ is the velocity of the wavepackets at the moment they cross the bottom of the potential well. Therefore, in the nonrelativistic limit considered in this paper, decoherence is much slower than the free oscillation. The ratio between the two time scales is even larger when considering the real 3D case. If we take a spherical perfectly-reflecting particle of radius R smaller than the range of oscillation, then $\omega_0 R/c < v/c \ll 1$. Since the relevant field modes have frequencies of the order of ω_0, in this limit the particle is much smaller than typical wavelengths (Rayleigh scattering regime), and hence is weakly coupled to the field. The dissipative force in this regime was calculated in [39]; the resulting damping coefficient scales as the squared polarizability of the sphere, leading to an additional factor $(\omega_0 R/c)^6$:

$$t_d = \frac{324}{(v/c)^2}\left(\frac{c}{\omega_0 R}\right)^6 \frac{2\pi}{\omega_0} \gg \left(\frac{c}{v}\right)^8 \frac{2\pi}{\omega_0}. \tag{12}$$

It is also possible to analyze the decoherence effect in a more complete theoretical framework, where the dynamical radiation pressure coupling between particle and field is fully taken into account. This approach also accounts properly for damping of the particle's energy, as well as for additional effects resulting from the coupling with the field. Moreover, it allows us to analyze decoherence in the more general case of an arbitrary temperature of the field. An ab-initio Hamiltonian model for the particle-field system was derived from first principles in Ref. [26]. This model was the starting point for the discussion of decoherence in Refs. [14] and [15]. The field scattering corresponds to frequency dependent reflection and transmission coefficients that satisfy the passivity requirements discussed in [52]. This means that the dynamics of the particle does not suffer from the instabilities associated to the model of a perfect mirror (as well known from classical electron theory, Eq. (3) is plagued with 'runaway' solutions).

Here we describe the radiation pressure coupling with the alternative, more intuitive model, where the interaction Hamiltonian corresponds to the energy transfer between field and particle:

$$H_{\text{int}} = -xF, \tag{13}$$

where F is the radiation pressure force on the particle, and x its position. This type of model was extensively employed in several contexts associated to the dynamical Casimir effect [27]. Here we focus on the limit where the particle perfectly reflects the (1D) field, but a discussion of partially-reflecting mirrors along these lines is also possible. As shown below, it leads to results for the decoherence and damping rates in agreement with those found in Ref. [15].

Starting from (13), we derive a master equation for the reduced density matrix of the particle. It is similar to the master equation for QBM, derived from the position-position interaction Hamiltonian. Technically, the essential difference arises from the fact that the force operator F in (13) is quadratic in the field operators, which leads to a damping coefficient that depends on the state (and hence temperature) of the field (reservoir). Although the formalism relies on a 1D model, the final results may be generalized to the 3D case.

We write the master equation in terms of the Wigner function $W(x,p,t)$:

$$\partial_t W = -\frac{p}{M}\partial_x W + M{\omega_0^*}^2 x\partial_p W + 2\Gamma\partial_p(pW) + D_1\frac{\partial^2 W}{\partial p^2} - D_2\frac{\partial^2 W}{\partial x\partial p}. \tag{14}$$

The first two terms in (14) correspond to the harmonic oscillation in the potential well, with a frequency $\omega_0^* = \omega_0 + \delta\omega$ modified by the coupling with the field (on the other hand, when the interaction Hamiltonian is linear in the momentum of the particle, a mass correction appears [26]). The remaining terms describe non-unitary evolution. The damping as well as the diffusion coefficients D_1 and D_2 are time dependent and given in terms of correlation functions of the force operator. The diffusion coefficients are related to the symmetric correlation function:

$$\sigma_{FF}(t) = \langle\{F(t), F(0)\}\rangle, \tag{15}$$

where the brackets denote the anticommutator, and the average is taken over the field state (thermal equilibrium, temperature T).

The term proportional to D_2 in (14) yields a negligible contribution, so that we focus on D_1 :

$$D_1(t) = \frac{1}{2}\int \frac{d\omega}{2\pi}\sigma_{FF}[\omega]\,\mathrm{sync}_t(\omega), \tag{16}$$

where $\sigma_{FF}[\omega]$ is the Fourier transform of $\sigma_{FF}(t)$ and

$$\mathrm{sync}_t(\omega) = \frac{\sin[(\omega-\omega_0)t]}{\omega-\omega_0}$$

is a function peaked around $\omega = \omega_0$ of width $2\pi/t$. Clearly, for a time t long enough, the function $\mathrm{sync}_t(\omega)$ is so sharply peaked that $\sigma_{FF}[\omega]$ is approximately constant over the short frequency interval that contributes in the integral in Eq. (16), and hence may be replaced by its value at $\omega = \omega_0$. In this case, we find

$$D_1(t\to\infty) = \frac{1}{4}\sigma_{FF}[\omega_0]. \tag{17}$$

A sufficient (and also necessary at $T = 0$) condition for the validity of (17) is $\omega_0 t \gg 1$. In other words, for times much longer than the period of oscillation, the field fluctuations at frequency ω_0 provide the dominant contribution to diffusion.

The damping coefficient is likewise connected to the average value of the *commutator* of the force operator taken at different times (anti-symmetric correlation function). When the interaction Hamiltonian is linear in the operators of the environment, as in the position-position model, the commutator is a c-number times a delta function (in time), and as a consequence, the damping coefficient has a constant value that does not depend on the state of the environment. As already mentioned, this is not the case for radiation pressure coupling. In particular, the damping coefficient depends on the temperature of the field, as could be expected having in mind the Stefan-Boltzmann law. At zero temperature, we recover the result given by Eq. (4).

We calculate the pointer states using the predictability sieve criterion, discarding all information about the environment, as discussed in Sec. 1. We start from the master equation, and evaluate the rate of change of linear entropy, assuming an initial pure state. It is straightforward to show that the entropy is minimized for minimum uncertainty Gaussian states, hence the pointer states are the coherent states. This result agrees with the well-known fact that coherent states provide the closest possible realization of a classical state of oscillation, given the constraint imposed by the Heisenberg uncertainty relation. In short, coherent states remain approximately pure because they do not entangle with field states, at least for times shorter than the damping time $1/\Gamma$, as shown by Eq. (6).

On the opposite extreme in Hilbert space, superpositions of coherent states are highly nonclassical and cannot last when the distance between the two components is large. This may be analyzed in detail from Eq. (14). The coherence of the initial state is imprinted on the Wigner function in the form of an interference term W_{int} that oscillates in phase space. When the two state components are spatially separated by a distance Δx, the oscillation is along the axis of momentum: $W_{\text{int}}(x,p) \sim \cos(\Delta x\, p/\hbar)$. Thus, according to Eq. (14), diffusion washes out this oscillatory term, the faster the larger the value of Δx. With an additional factor of 2 to take into account the average over several free rotations of the state in phase space, we find

$$t_d = 2\frac{\hbar^2}{D_1(\Delta x)^2}. \tag{18}$$

To derive the decoherence time when the field is in the vacuum sate, we compute the correlation function $\sigma_{FF}[\omega_0]$ at zero temperature. When replacing the result for D_1 as given by (17) into (18), we obtain the same result already derived in this Section by a more elementary method.

For finite temperatures, the spectrum is approximately constant at low frequencies, so that (17) also holds when $\omega_0 \ll kT/\hbar$ (k is the Boltzmann constant), including the free particle limit $\omega_0 = 0$, provided that the entire frequency interval around ω_0 is contained in the low frequency part of the spectrum, which corresponds to the condition $1/t \ll kT/\hbar$. The damping coefficient may calculated in the high temperature limit as well, and the results are in agreement with Einstein relation

$$D_1 = 2MkT\Gamma. \tag{19}$$

More generally, we may derive a relation between diffusion and damping coefficients valid for arbitrary values of temperature [15], including $T = 0$, starting from the general relation between symmetric and anti-symmetric correlation functions (fluctuation-dissipation theorem).

The decoherence time for high T is derived by replacing (19) into (18). As we discuss below, usually in this limit decoherence is faster than the free oscillation, so that, contrary to the $T = 0$ case, there is no average over many oscillations in this case. To describe the decoherence process, we must evaluate the diffusion coefficient at a time t much shorter than t_d. Hence, we are allowed to use its asymptotic value as given by the Einstein relation (19) only if we assume that $t_d \gg \hbar/(kT)$. The resulting expression is very general [2], and also holds in the free particle case:

$$t_d = \frac{\lambda_T^2}{(\Delta x)^2}\frac{1}{\Gamma}, \tag{20}$$

where $\lambda_T = \hbar/\sqrt{2MkT}$ is the de Broglie wavelength of a particle of mass M in thermal equilibrium. Eq. (20) has a form similar to (10), except that now the reference of distance is set by thermal fluctuations instead of zero point fluctuations.

In order to complete the evaluation of the decoherence time, we need to evaluate the damping coefficient Γ in the high temperature limit. We consider as before a sphere of radius R, which is usually much larger than typical field wavelengths, which are of the order of $\hbar c/(kT)$ (except for very low temperatures or very small spheres). In this short-wavelength regime, the radiation pressure force may be calculated by replacing the surface of the sphere by a collection of tangent planes, and the final result reads

$$F = -\frac{4\pi^3}{45}\frac{(kT)^4}{\hbar^3 c^4}R^2\frac{dx}{dt}. \tag{21}$$

The force scales with the surface or cross section of the sphere, and is proportional to T^4, in agreement with Stefan-Boltzmann law. As opposed to the vacuum case, here we have a true friction force, i.e. proportional to the velocity of the particle and not to higher-order time derivative as in Eq. (2) (the thermal field is not Lorentz invariant).

In the free case ($\omega_0 = 0$), Γ is simply the coefficient multiplying the velocity in Eq. (21) divided by M. Then, with the help of (20) we find

$$t_d = \frac{45}{8\pi^3}\frac{\hbar^5 c^4}{(kT)^5 R^2 (\Delta x)^2}. \tag{22}$$

Eq. (22) shows that the decoherence time depends strongly on temperature (the same temperature dependence was found in Ref. [54]). Even at the temperature corresponding to the cosmic background radiation, $T = 2.7$K, radiation pressure is a very efficient source of decoherence. As an example, for $R = 1$cm, we have $t_d = 2.7 \times 10^{-21}/(\Delta x[\mathrm{m}])^2$ s, which is in the nanosecond range for a separation $\Delta x = 1\mu$m.

4 Conclusion

The master equation provides a complete description of the particle dynamics when no measurement on the field is made. It accounts for the renormalization of the oscillation frequency, damping, and diffusion and the associated decoherence effect. It also allows for the determination of the pointer states, and all that for any temperature T. On the other hand, the decoherence time scale at $T = 0$ may be calculated by a simpler approach, in which we follow the evolution of the complete particle-field state, calculated with the help of the superposition principle, and trace over the field at the very end. This approach explicitly shows that decoherence results from entanglement between particle and field states.

The decoherence induced by radiation pressure coupling with vacuum fluctuations is a very slow effect, when compared with the the time scale of the free evolution. Yet, it is remarkable, from a conceptual point-of-view, that classical behavior of a macroscopic system emerges from the formalism of Quantum Mechanics itself, even though in very long time scale, provided that the quantum vacuum radiation field is taken into account.

P.A.M.N. thanks M.T. Jaekel, A. Lambrecht and S. Reynaud for discussions, and CNPq, PRONEX and FAPERJ for partial financial support.

References

1. D. Giulini *et al*: *Decoherence and the Appearance of a Classical World in Quantum Theory* (Springer, Berlin 1996) H.D. Zeh: Found. Phys. **3**, 109 (1973)
2. W.H. Zurek: Phys. Today **44**, No. 10, 36 (1991)
3. W.H. Zurek: Phys. Rev. D **24**, 1516 (1981)
4. W.H. Zurek: Prog. Theor. Phys. **89**, 281 (1993)
5. W.G. Unruh and W.H. Zurek: Phys. Rev. D **40**, 1071 (1989) B.L. Hu, J.P. Paz and Y. Zhang: Phys. Rev. D **45**, 2843 (1992)
6. W.H. Zurek, S. Habib, and J. P. Paz: Phys. Rev. Lett. **70**, 1187 (1993)
7. W.H. Zurek and J. P. Paz: Phys. Rev. Lett. **82**, 5181 (1999)
8. D.A.R. Dalvit, J. Dziarmaga and W.H. Zurek, to appear in Phys. Rev. Lett.
9. M. Brune *et al*: Phys. Rev. Lett. **77**, 4887 (1996) A. Rauschenbeutel *et al.*: Science **288**, 2024 (2000)
10. L. Davidovich *et al*: Phys. Rev. A **53**, 1295 (1996)
11. C.J. Myatt *et al*: Nature **403**, 269 (2000) C. A. Sackett *et al*: Nature **404**, 256 (2000)
12. J. Friedman *et al*: Nature **406**, 43 (2000) C.H. van der Wal *et al*: Science **290**, 773 (2000)
13. D.G. Cory *et al*: Phys. Rev. Lett. **81**, 2152 (1998)
14. D.A.R. Dalvit and P.A. Maia Neto: Phys. Rev. Lett. **84**, 798 (2000)
15. P.A. Maia Neto and D.A.R. Dalvit: Phys. Rev. A **62**, 042103 (2000)
16. A.O. Caldeira and A.J. Leggett: Phys. Rev. **A** 31, 1059 (1985)
17. S.K. Lamoreaux: Am. J. Phys. **67**, 850 (1999) G. Plunien, B. Müller and W. Greiner: Phys. Rep. **134**, 87 (1986) V.M. Mostepanenko and N.N. Trunov: *The Casimir Effect and its Applications* (Clarendon, London, 1997)
18. H.B.G. Casimir: Proc. K. Ned. Akad. Wet. **51**, 793 (1948)
19. S.K. Lamoreaux: Phys. Rev. Lett. **78**, 5 (1997) U. Mohideen and A. Roy: Phys. Rev. Lett. **81**, 4529 (1998)
20. A. Lambrecht and S. Reynaud: Eur. Phys. J. D **8**, 309 (2000) C. Genet, A. Lambrecht and S. Reynaud: Phys. Rev. A **62**, 012110 (2000) M. Bordag *et al*: Phys. Rev. Lett. **85**, 503 (2000)
21. L.S. Brown and G.J. Maclay: Phys. Rev. **184**, 1272 (1969)
22. G. Barton: J. Phys. A: Math. Gen. **24**, 991 (1991) J. Phys. A: Math. Gen. **24**, 5533 (1991)
23. C. Eberlein: J. Phys. A: Math. Gen. **25**, 3015 (1992) J. Phys. A: Math. Gen. **25**, 3039 (1992)
24. M.T. Jaekel and S. Reynaud: Quantum Opt. **4**, 39 (1992)
25. M.T. Jaekel and S. Reynaud: Phys. Lett. A **172**, 319 (1993)
26. G. Barton and A. Calogeracos: Ann. Phys. (NY) **238**, 227 (1995) A. Calogeracos and G. Barton: Ann. Phys. (NY) **238**, 268 (1995)
27. M.T. Jaekel and S. Reynaud: Rept. Prog. Phys. **60**, 863 (1997)
28. M. Kardar and R. Golestanian: Rev. Modern Phys. **71**, 1233 (1999)
29. H.B. Callen and T.A. Welton: Phys. Rev. **83**, 34 (1951) R. Kubo: Rep. Progr. Phys. **29**, 255 (1966)
30. M.T. Jaekel and S. Reynaud: Quantum Opt. **4**, 39 (1992)
31. L.H. Ford and A. Vilenkin: Phys. Rev. D **10**, 2569 (1982)
32. G. Barton and C. Eberlein: Ann. Phys. (N.Y.) **227**, 222 (1993)
33. S.A. Fulling and P. Davies: Proc. R. Soc. London Ser. A **348**, 393 (1976)
34. P.A. Maia Neto: J. Phys. A: Math. Gen. **27**, 2167 (1994)

35. L.A.S. Machado and P.A. Maia Neto: Phys. Rev. A **54**, 3420 (1996)
36. G. Barton and C. North: Ann. Phys. (N.Y.) **252**, 72 (1996)
37. R. Gütig and C. Eberlein: J. Phys. A: Math. Gen. **31**, 6819 (1998)
38. D.F. Mundarain and P.A. Maia Neto: Phys. Rev. A **57**, 1379 (1998)
39. P.A. Maia Neto and S. Reynaud: Phys. Rev. A **47**, 1639 (1993)
40. G. Barton: 'New Aspects of the Casimir Effect'. In: *Cavity Quantum Electrodynamics, Supplement: Advances in Atomic, Molecular and Optical Physics*, ed. by P.R. Berman (Academic Press, New York 1993)
41. P.A. Maia Neto and L.A.S. Machado: Brazilian J. Phys. **25**, 324 (1995)
42. R. Golestanian and M. Kardar: Phys. Rev. Lett. **78**, 3421 (1997) R. Golestanian and M. Kardar: Phys. Rev. A **58**, 1713 (1998)
43. G.T. Moore: J. Math. Phys. **11**, 2679 (1970)
44. D.A.R. Dalvit and F.D. Mazzitelli: Phys. Rev. A **57**, 2113 (1998) C.K. Cole and W.C. Schieve: Phys. Rev. A **52**, 4405 (1995) V.V. Dodonov, A.B. Klimov and D.E. Nikonov: J. Math. Phys. **34**, 2742 (1993).
45. D.A.R. Dalvit and F.D. Mazzitelli: Phys. Rev. A **59**, 3059 (1999)
46. A. Lambrecht, M.T. Jaekel and S. Reynaud: Phys. Rev. Lett. **77**, 615 (1996)
47. V.V. Dodonov and A.B. Klimov: Phys. Rev. A **53**, 2664 (1996)
48. M. Crocce, D.A.R. Dalvit and F.D. Mazzitelli: *xxx archives* quant-ph/0012040 (2000)
49. C. Eberlein: Phys. Rev. A **53**, 2772 (1996) Phys. Rev. Lett. **76**, 3842 (1996)
50. M.T. Jaekel and S. Reynaud: J. Phys. France **I 3**, 1 (1993)
51. D.F. Walls and G.J. Milburn: Phys. Rev. A **31**, 2403 (1985)
52. M.T. Jaekel and S. Reynaud: Phys. Lett. A **167**, 227 (1992) A. Lambrecht, M.T. Jaekel and S. Reynaud: Phys. Lett. A **225**, 188 (1997)
53. M.T. Jaekel and S. Reynaud: J. Phys. France **I 2**, 149 (1992)
54. E. Joos and H.D. Zeh: Z. Phys. B **59**, 223 (1985)

Control of Cold Atomic Collisions by Multiparticle Entanglement and a Modified Vacuum in Cavity QED

J.I. Kim, R.B.B. Santos, and P. Nussenzweig

Instituto de Física, Universidade de São Paulo,
Caixa Postal 66318, CEP 05315-970, São Paulo, SP,
Brazil.

Abstract. We show how the indistinguishable nature of the interaction of a collection of cold atoms with the electromagnetic field in an optical cavity leads to entanglement and a very fast radiative decay dynamics. This acceleration of spontaneous emission has important consequences in the process known as radiative escape, one of the most important sources of trap loss due to cold collisions. In this process, a pair of atoms excited at some internuclear separation R_C picks up kinetic energy on the long range excited state potential. The spontaneous emission time is a crucial parameter in this dynamics for it will determine whether or not the pair gains enough kinetic energy to escape from the trap. Considering realistic experimental parameters we can predict a very large suppression of radiative escape trap loss in a sample of cold rubidium atoms.

1 Introduction

The current developments of cooling and trapping techniques have allowed the creation of very cold gases of atoms, down to the quantum degeneracy regime [1,2,3,4], and understanding collisions of such cooled neutral trapped atoms has become of increasing importance [5]. In this context, the use of light fields provides a powerful tool to analyze collisional properties, like scattering lengths, interatomic potentials or atomic wavefunctions [6,7]. This is illustrated, e.g., by photoassociation spectroscopy, optical shielding, or measurements of light-induced trap-losses [5]. In these situations, the outcome of a collisional encounter can be said to be strongly driven by the *presence* of photons interacting with the atoms. A notable feature to be pointed out, in contrast, is that some collisions can also be manipulated in the complete *absence* of photons. More precisely, the presence of a quantized vacuum of a cavity can be responsible for the fate of a collisional encounter. This curious effect can be understood from the fact that the spontaneous decay time of atomic excited states, for sufficiently slow atoms, becomes of the same order of magnitude as the collision time. Atoms can then pick up kinetic energy during a collision before spontaneous decay takes place. From the dependence of potential energy curves on the internal electronic states, this decay "turns off" the atom–atom interaction. By manipulating the

M. Orszag, J.C. Retamal (Eds.): LNP 575, pp. 125–137, 2001.

spontaneous decay time [8], one determines the amount of energy gained and thereby controls collisional dynamics. An essential ingredient that comes into play is the entanglement between several pairs of colliding atoms that the cavity can allow for, even though the average distance between pairs is too long for them to interact directly. The resulting multiparticle entangled state shows a highly increased collective decay probability akin to the phenomenon of superradiance [9,10,11,12]. The colliding pairs end up influencing one another and the collisional evolution as well. This is made possible by preparing the gas sample within the cavity and serves to improve the cavity effect on collisions up to the point of being experimentally observable. Cavity QED effects on cold atoms have been recently investigated in the context of mechanical forces exerted by light fields on the atomic motion [13,14], and for high-Q cavities in the context of cold collisions [15,16].

We present first some basic aspects of cold collisions and Cavity QED, in § 2. Then we proceed in § 3 to demonstrate a cavity-induced manipulation of cold collisions, using a relatively simple model, that has proved adequate in some collisional experiments, with suitable modifications. We end this section by calculating orders of magnitude which indicate that the effect should be observable within realistic experimental conditions. In view of the scope of this paper, we suppressed some technical details which are presented elsewhere [17].

2 Cold Collisions and Cavity QED

Cold collisions can be of several types: in presence of light (excited state collisions) or in the "dark" (ground state collisions), with light blue-detuned from an atomic resonance (leading, e.g., to optical shielding) or red-detuned (as in photoassociation spectroscopy) [5]. We single out one particular type of excited state collisions between alkali neutral ^{85}Rb atoms, the so-called radiative escape collisions, in order to explore the effect of Cavity QED. The description of cold collisions given below follows some approximations introduced in [18]. A discussion of its validity can be found in [17] and is based on [19,20].

2.1 Radiative escape collisions

In the Born-Oppenheimer approximation, an internuclear configuration is fixed and the electronic wavefunctions and eingenvalues are solved, assuming the electrons are much faster than the heavy nuclei. These eigenvalues become then part of the potential for the internuclear degrees of freedom. As a consequence, the dynamics of the atomic nuclei depends strongly on the electronic state. In cold collisions, this has a profound effect on the collisional outcome, since the slowness of the collision allows for transitions between electronic states during the very collision time.

One of the first identified types of collisions between cold atoms in the presence of light is that which leads to the so-called radiative escape from traps [5].

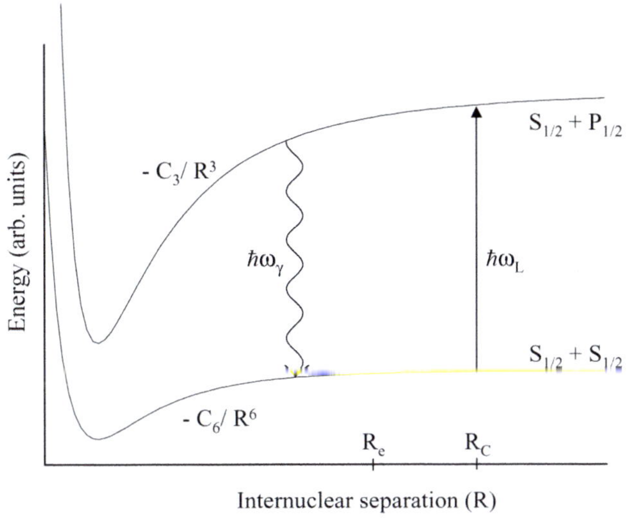

Fig. 1. Excited state long range dipole-dipole potential $U = -C_3/R^3$ and the ground state van der Waals attractive potential $-1/R^6$, which is here neglected and taken as the zero of energy. Its distance to the asymptote of U is the atomic separation $\hbar\omega_A$ between $5S_{1/2}$ and $5P_{1/2}$.

A colliding pair in their ground state is excited by a laser of frequency ω_L at a large internuclear separation R_C. The choice of an appropriate excited state determines the potential that is "turned on". For our case, we deal with an attractive long-range dipole–dipole potential. The atoms are then accelerated towards each other by

$$U(R) = -\frac{C_3}{R^3} , \tag{1}$$

where C_3 is a constant that depends on the chemical species under consideration. If the spontaneous emission time is long, the pair may gain enough kinetic energy $\hbar\omega_L - \hbar\omega_\gamma$ to escape from the trap by emitting a photon with energy $\hbar\omega_\gamma$ smaller than that of the absorbed photon, $\hbar\omega_L$. The interaction potentials are sketched in Fig. 1 for the ground state $5S + 5S$ and the excited state $5S + 5P$ of a pair of ^{85}Rb atoms, with energy difference $\hbar\omega_A$. Two regions can be defined: $R < R_e$ and $R_e < R < R_C$, where R_C (the so-called Condon point, chosen by tuning ω_L) is the internuclear separation at which a weakly bound molecule is excited. The separation R_e is the smallest one for which the kinetic energy gained by spontaneous emission does not overcome the trapping potential depth and therefore does not lead to trap loss.

A colliding pair of atoms is treated in the two-level approximation (which is justified in [17]) with a ground state $|g\rangle$ and an excited state $|e\rangle$ connected to the $5S_{1/2}+5S_{1/2}$ and $5S_{1/2}+5P_{1/2}$ asymptotic states, respectively. Here, the pair in state $|g\rangle$ is not bound (i.e., we neglect the van der Waals interaction arising from

$\sim -1/R^6$) but, in state $|e\rangle$ the atoms form a weakly bound, unstable molecule. For simplicity, we will refer to these pairs as quasimolecules, independent of their state ($|e\rangle$ or $|g\rangle$). For a given R, the energy separation between $|g\rangle$ and $|e\rangle$ is (Fig. 1)

$$\hbar\omega_R = \hbar\omega_A - \frac{C_3}{R^3} , \tag{2}$$

so that the Condon point R_C is given by the resonance condition

$$\omega_L = \omega_R , \text{ or } \delta = U(R_C) , \tag{3}$$

where $\delta = \omega_L - \omega_A < 0$ is the detuning between the excitation laser frequency and the atomic resonance. The detuning range is of the order of hundreds of MHz, therefore well below (in absolute value) the GHz range in which the discrete vibrational levels of the potential U are resolved in photoassociation spectroscopy. Accordingly, the relative nuclear motion is approximated by localized wave packets following classical trajectories [5,17]. This semiclassical treatment, however, is restricted to low excitation laser intensities and detunings not too small compared to the atomic excited state decay constant Γ_A [5].

Each quasimolecule interacts with the electromagnetic field with a dipole moment $(\sigma_i + \sigma_i^\dagger)\, \boldsymbol{d}_i$, where σ_i and $\sigma_i^\dagger$ are Pauli operators acting in the subspace spanned by $|e\rangle$ and $|g\rangle$. The dipole $\boldsymbol{d}_i$, determined by the molecular axis [23], is randomly oriented in space. Its magnitude is $|\boldsymbol{d}_i| = \sqrt{2}\, d_A$, where d_A is the atomic dipole moment, with a resulting decay constant $\Gamma = 2\Gamma_A$ [5]. This doubled decay rate is a simple example of two-atom superradiance in which the interatomic separation R_C is smaller than the wavelength of the emitted optical radiation [24,25].

2.2 Cavity QED

The modification of atomic radiative properties was one of the first effects to be demonstrated in Cavity QED [8]. Spontaneous emission enhancement [26,27] and inhibition [28] were demonstrated in the 80's. Radiative level shifts, such as a cavity-induced Lamb shift, were also demonstrated in this context [27,29]. However, spontaneous emission for single atoms in the optical domain was not significantly enhanced, owing to the relatively small solid angle encompassed by a centimeter-sized Fabry-Perot cavity. It is, nevertheless, possible to achieve a large enhancement of spontaneous emission when we consider a sample of many identical atoms coupled to the same cavity mode. Quantum interference will thence be important in the process of spontaneous emission into the cavity, in a process known as superradiance [9,10,11,12].

In our case, the atoms are replaced by quasimolecules composed each of a pair of colliding atoms. Except for a phase factor, the different excited quasimolecules are then indistinguishable when interacting with the cavity field. Since, for most experiments in cold collisions, the typical separation between different quasimolecules is greater than an optical wavelength, this interference will only be constructive if the quasimolecules are excited into a multiparticle entangled state by a laser beam injected into the cavity mode [10]. One also has the possibility of sending the laser beam in the normal direction to the cavity axis and bypass the cavity effect, exciting the quasimolecules individually.

The electric field operator in the cavity can be written as a sum of modes, all of which can be chosen to have the same frequency [30]

$$\boldsymbol{E}(\mathbf{r}) = \sum_{\alpha} \mathcal{E}_\alpha(\omega) f_\alpha(\boldsymbol{r}) \boldsymbol{\epsilon}_\alpha \, a_\alpha^\dagger + \text{h.c.} \,, \tag{4}$$

where $a_\alpha^\dagger$ is the creation operator of the mode α, $\mathcal{E}_\alpha(\omega) = (2\pi\hbar\omega/V_\alpha)^{1/2}$ is the field strength (or field per photon), V_α being the effective mode volume, $\boldsymbol{\epsilon}_\alpha$ is the polarization, and $f_\alpha(\boldsymbol{r})$ is the spatial profile (obtained from the time-independent Maxwell Equations). Of the various modes a cavity can sustain, we may choose the fundamental TEM_{00} mode for the excitation field, characterized by the frequency ω_L, polarization $\boldsymbol{\epsilon}_L$ and a gaussian profile $f_c(\boldsymbol{r})$. In obtaining spontaneous decay rates, other higher order modes TEM_{nm} should be taken into account as well.

In the high-Q regime of Cavity QED, the cavity photons can survive long enough to be reabsorbed and emitted several times before being dissipated. Some consequences of such Rabi oscillations for cold collisions are examined elsewhere [16]. We work here in the low-Q regime, where an emitted photon can not be reabsorbed.

3 Collisional Dynamics in a Cavity

Let us now bring together the concepts developed in the previous section by considering the implications of an altered spontaneous decay rate on cold collisions. As seen above, the excitation procedure will be of great importance to achieve multiparticle entanglement. Therefore, in order to observe the effect, one should consider traps in which the atoms remain in the ground state. Atoms can then be excited by an independent probe laser. Good candidates would be Far-Off Resonant Traps (FORT) [21] or Magnetic Traps [22]. For the discussion below, we chose to consider a FORT since it is well-adapted to the cavity geometry.

3.1 Multiparticle entanglement

A pencil shaped gas sample is supposed to be trapped at the center of the cavity and along its axis of symmetry, with all atoms in their ground state (Fig 2). After one excitation is injected *into* the cavity mode, the interaction between the atoms and the field is modeled by collecting all atomic pairs, or quasimolecules, whose

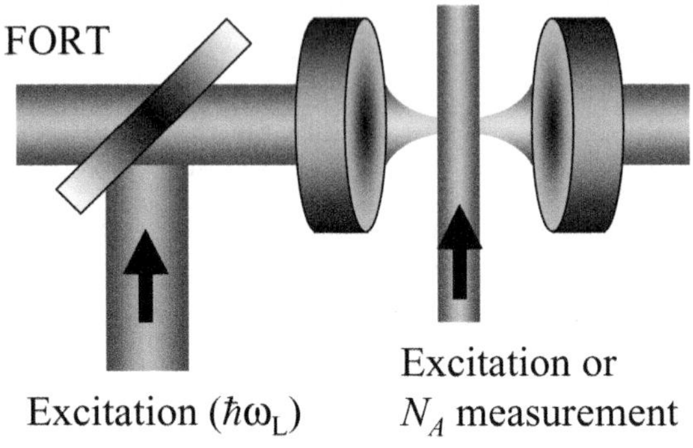

Fig. 2. Sketch of a possible experiment to probe Cavity QED-modified cold collisions. The N_A atoms are held near the center of a Fabry-Perot cavity by a FORT laser. The excitation (probe) laser, of frequency ω_L, can be sent either perpendicular to the cavity axis or via the cavity mode.

separation R allows them to be resonant with the cavity mode frequency ω_L. In the dipole and rotating wave approximations, the hamiltonian is given by

$$H = \frac{\hbar\omega_R}{2}\sum_{i=1}^{N}(\sigma_{zi}+1) + \sum_{i=1}^{N}\left(\frac{\hbar\Omega_i}{2}\,\sigma_i^\dagger\, a + \text{h.c.}\right) + \hbar\omega_L a^\dagger a \;, \tag{5}$$

where only the excitation mode is included, and $\Omega_i = 2\mathcal{E}(\omega_L) f_c(\boldsymbol{r}_i)\boldsymbol{\epsilon}_L \cdot \boldsymbol{d}_i/\hbar$ is the single quasimolecule Rabi frequency. Since ω_L and ω_R are in the optical range, the variation of the field along the quasimolecules' positions $\boldsymbol{r}_i$ can not be neglected. The total number N of quasimolecules depends on the detuning δ, the number N_A of atoms in the gas sample and their density n_A, as estimated below. The initial state of the total system is

$$|G,1\rangle \equiv |gg\cdots g\rangle|1\rangle \;, \tag{6}$$

with all quasimolecules in state $|g\rangle$ and one photon in the cavity mode. It can be seen that this state couples to the collective excited state

$$|E,0\rangle = \sum_i \frac{\Omega_i}{\tilde{\Omega}}\,|i,0\rangle \;, \tag{7}$$

where $|1,0\rangle \equiv |eg\cdots g\rangle|0\rangle$, $|2,0\rangle \equiv |ge\cdots g\rangle|0\rangle$ and so forth, with the collective Rabi frequency given by

$$\tilde{\Omega} = (\sum_i |\Omega_i|^2)^{1/2} \;. \tag{8}$$

The respective eigenvalues are $E_g = \hbar\omega_L$ and $E_e = \hbar\omega_R$, and these two states form a closed subspace of the hamiltonian H.

The total number N of quasimolecules can be estimated by counting all pairs of atoms whose separation R is such that $\omega_R = \omega_L$, allowing for a spread ΔR about R_C, determined by the linewidth Γ of state $|e\rangle$, namely,

$$\begin{aligned} N &\sim \frac{1}{2} N_A n_A \, 4\pi R_C^2 \Delta R \\ &\sim \frac{1}{2} N_A n_A \, 4\pi R_C^2 \frac{\Gamma}{|d\omega_R/dR|_{R_C}} \\ &\sim N_A n_A \frac{2\pi C_3}{3\hbar\Gamma} \left(\frac{\Gamma}{\delta}\right)^2 , \end{aligned} \tag{9}$$

having used that $C_3/R_C^3 = \hbar|\delta|$. For each of the N_A atoms, we've counted the pairs it can form with all others located in a spherical shell of width ΔR and radius R_C.

In the case of a single pair of colliding atoms in a light field, there is a probability that the excitation towards $|e\rangle$ takes place as the atoms approach the Condon point R_C [5]. Under conditions of low excitation intensity (met here) and our detuning range, a Landau-Zener calculation [20,31,32] may be used to estimate the probability $1 - \mathrm{e}^{-2\pi\Delta}$, where $\Delta = \hbar\Omega^2/v_\infty|U'(R_C)|$, Ω being an average coupling (Rabi frequency) between $|g\rangle$ and $|e\rangle$, and v_∞ (dependent on the atomic gas temperature), the asymptotic relative velocity with which the nuclei approach each other. Analogously, the probability P_E that $|E,0\rangle$ is excited from $|G,1\rangle$ may be estimated by

$$P_E = 1 - \mathrm{e}^{-2\pi\tilde{\Delta}}, \qquad \tilde{\Delta} = \frac{\hbar\tilde{\Omega}^2}{v_\infty|U'(R_C)|} , \tag{10}$$

where the collective coupling $\tilde{\Omega}$ replaces Ω.

The fact that each quasimolecule in the excited state $|E,0\rangle$ is *quantum-mecha-*
nically entangled with all the others implies that their evolution is not separable. This is of fundamental importance for the manipulation of the collisional dynamics in the cavity vacuum.

3.2 Control of cold collisions by a modified vacuum

Let us consider the potential curves in Fig. 1, and the region $R_e < R < R_C$. The cavity mode has a finite spectral width γ_c that can be made "to fit" to this region, in such a way that a quasimolecule in the excited state $|e\rangle$ with an internuclear separation within this region is at resonance with the cavity mode and practically off-resonance outside this region. Therefore, right after excitation at R_C has taken place, the spontaneous decay from $|e\rangle$ can be enhanced while in this resonant region. The collision is altered and if the quasimolecule remains excited below R_e, it is no longer subject to the cavity, evolving as if it were in free space. The chances to reach the trap-escape region below R_e can then be significantly reduced. This would be observed as an inhibition of trap-loss probabilities compared to a situation with no cavity. The role of the quantum entanglement is to increase this effect since the decay rate of $|E, 0\rangle$ is greatly enhanced if N is large.

3.3 Collective decay rate

In order to calculate the spontaneous decay rate Γ_c of $|E, 0\rangle$ on resonance with the cavity mode, we need to consider the distribution of quasimolecules along the cavity axis and their random orientations. From perturbation theory *à la* Fermi's golden rule,

$$\Gamma_c = \frac{1}{\Delta t} \sum_{\boldsymbol{k}\lambda} \left| \langle G, 1_{\boldsymbol{k}\lambda} | \frac{1}{\mathrm{i}\hbar} \int_0^{\Delta t} dt \, H_{\mathrm{int}}(t) \, |E, 0\rangle \right|^2 , \tag{11}$$

where the summation is over all modes $\boldsymbol{k}$ with polarization λ (including those into free space, away from the solid angle of the cavity mirrors), with the corresponding meaning of the state $|G, 1_{\boldsymbol{k}\lambda}\rangle$. The interaction hamiltonian (in the interaction picture) is

$$H_{\mathrm{int}}(t) = \sum_{i=1}^{N} \sum_{\boldsymbol{k}\lambda} \left[\mathcal{E}(\omega_{\boldsymbol{k}}) f_{\boldsymbol{k}}(\boldsymbol{r}_i) \; \boldsymbol{\epsilon}_{\boldsymbol{k}\lambda} \cdot \boldsymbol{d}_i \right] \, a_{\boldsymbol{k}\lambda} \mathrm{e}^{-\mathrm{i}\omega_{\boldsymbol{k}} t} \sigma_i^{\dagger} \mathrm{e}^{\mathrm{i}\omega_R t} + \mathrm{h.c.} \, , \tag{12}$$

with $\omega_R \approx \omega_L$ (since the quasimolecules are on resonance with the cavity) and definitions analogous to those of (4). Substituting (12) back into (11), going to the continuum limit $\sum_{\mathbf{k}} \to V \int dk \, k^2 \, d\Omega_{\mathbf{k}}/(2\pi)^3$, and using $[\sin(\frac{1}{2}\omega\Delta t)/\frac{1}{2}\omega]^2 \sim 2\pi \, \Delta t \, \delta(\omega)$ yields

$$\frac{\Gamma_c}{2\pi} = \int d\omega_{\boldsymbol{k}} d\Omega_{\boldsymbol{k}} \; \rho(\omega_{\boldsymbol{k}}) \delta(\omega_{\boldsymbol{k}} - \omega_L) \, \frac{\sum_\lambda |\boldsymbol{D}_{\boldsymbol{k}} \cdot \boldsymbol{\epsilon}_{\boldsymbol{k}\lambda}|^2}{\hbar^2} \, , \tag{13}$$

where ρ is the spectral density, $\boldsymbol{D}_{\boldsymbol{k}} \equiv \sum_i \mathcal{E}(\omega_{\boldsymbol{k}}) f_{\boldsymbol{k}}(\boldsymbol{r}_i)(\Omega_i^*/\tilde{\Omega}) \, \boldsymbol{d}_i$ is a collective dipole moment and $|\boldsymbol{D}_{\boldsymbol{k}} \cdot \boldsymbol{\epsilon}_{\boldsymbol{k}\lambda}|$ is the absolute value of the collective coupling between the quasimolecules and the mode $\boldsymbol{k}\lambda$ (matrix element of H_{int}). Taking into account that only emission into the solid angle $\Delta\Omega_c$ encompassed by the cavity mirrors is affected by the enhanced spectral density

$$\rho(\omega_{\boldsymbol{k}}) = \rho_0(\omega_{\boldsymbol{k}}) \Lambda(\omega_{\boldsymbol{k}}) \, , \tag{14}$$

where $\rho_0(\omega_{\boldsymbol{k}}) = V\omega_{\boldsymbol{k}}^2/(2\pi)^3 c^3$ is the free-space spectral density and $\Lambda(\omega)$ is the cavity line-shape function [27,30], we separate its contribution from emission into the rest of free space

Since $\Delta\Omega_c$ is small, the contribution of emission into free space can be estimated by integrating (13) over the full solid angle with $\rho = \rho_0$. We suppose a thin gas sample of length L along the cavity z-axis,

$$f_{\boldsymbol{k}}(\boldsymbol{r}_i) = \mathrm{e}^{-\mathrm{i}\boldsymbol{k}\cdot\boldsymbol{r}_i} \sim \mathrm{e}^{-\mathrm{i}k_z z_i} ,$$

and the N quasimolecules equally spaced by $\sim L/N$, so that the summation over i is a discrete one-dimensional one. Using the polarization summation formula

$$\sum_{\lambda} |\boldsymbol{\epsilon}_{\boldsymbol{k}\lambda} \cdot \boldsymbol{D}_{\boldsymbol{k}}|^2 = \boldsymbol{D}_{\boldsymbol{k}}^* \cdot \boldsymbol{D}_{\boldsymbol{k}} - \frac{|\boldsymbol{k} \cdot \boldsymbol{D}_{\boldsymbol{k}}|^2}{k^2} ,$$

it turns out that to leading order in $1/N$

$$\left(\frac{\Gamma_c}{2\pi}\right)_{\text{incoher}} \approx \frac{\Gamma}{2\pi} . \tag{15}$$

This result means that the collective decay rate into free space differs little from the individual molecular decay rate. In fact, the emitted radiation has an optical wavelength much smaller than the average separation L/N (for N not too large) between quasimolecules, and hence the indistinguishability between excited quasimolecules no longer holds (it is possible, in principle, to determine which quasimolecule emits radiation), washing out the collective effect.

The contribution of emission into the cavity solid angle $\Delta\Omega_c$ has a complicated structure. If only the fundamental mode TEM_{00} (used to excite the quasimolecules) is taken into account, this contribuition tends to be very small due to the smallness of the effective solid angle $\Delta\Omega_{00}$ of this fundamental mode. This can be remedied, in turn, by collecting other cavity TEM_{nm} modes, as represented in (4), with larger solid angles $\Delta\Omega_{nm}$. The modes for which $\Delta\Omega_{nm} > \Delta\Omega_c$ are neglected on the basis of high losses by diffraction. Each mode has its own profile f_{nm}, and line-shape function Λ_{nm} [30]. In (13), the integral $d\Omega_{\boldsymbol{k}}$ over the cavity solid angle $\Delta\Omega_c$ is then substituted by a summation over discrete higher order modes TEM_{nm}. Defining an effective solid angle by

$$\Delta\Omega_{\text{eff}} \equiv \sum_{nm} \Delta\Omega_{nm} \frac{\Lambda_{nm}}{\Lambda_{00}} \frac{|\, \frac{1}{N}\sum_i |\boldsymbol{\epsilon}_L \cdot \hat{\boldsymbol{d}}_i|^2 f_{nm}^*(\boldsymbol{r}_i) f_c(\boldsymbol{r}_i)\,|^2}{\frac{1}{N}\sum_i |\boldsymbol{\epsilon}_L \cdot \hat{\boldsymbol{d}}_i|^2 |f_c(\boldsymbol{r}_i)|^2} , \tag{16}$$

where the nm summation can stop as $\Delta\Omega_{nm} \sim \Delta\Omega_c$, the contribution from coherent collective emission into the cavity adds to the incoherent one to give the total decay rate of $|E, 0\rangle$ on resonance

$$\Gamma_c \approx \left(1 + \frac{3}{2}\frac{\Delta\Omega_{\text{eff}}}{4\pi} N \Lambda_{00}\right) \Gamma , \tag{17}$$

with $|\omega_L - \omega_A| = |\delta| \ll \omega_L$. The point to be noted is that Γ_c can be significantly higher than Γ for large N and Λ_{00}, despite the fact that the quasimolecules are randomly oriented and distributed along the cavity axis.

This enhanced decay rate applies as far as the quasimolecules are on resonance with the cavity $\omega_R \approx \omega_L$, when the internuclear separation is in the region $R_e < R < R_C$, as discussed in §3.2. If the state $|E,0\rangle$ survives and the quasimolecules remain excited as the separation R decreases down below R_e, the decay rate should shift from Γ_c to the value calculated without a cavity, which is given in (15)[1].

We discuss below a strategy to observe the Cavity QED effect on Cold Collisions based on measurements of trap-loss probabilities. The nuclear relative motion will be allowed to perform round trips along the potential curve U, between $R = 0$ and R_C. The decay rate of $|E,0\rangle$ will then be taken as either Γ_c or Γ, according to the quasimolecules' internuclear distance R. Conditional probabilities will be associated to these round trips until $|E,0\rangle$ decays in the region $R < R_e$, leading to the escape of an atomic pair from the confining trap.

3.4 Trap-loss probabilities

Let t_e and t_c denote the time intervals spent by an excited quasimolecule in the regions $0 < R < R_e$ and $R_e < R < R_C$, respectively, as the atomic nuclei are attracted to each other by the potential $U(R)$. These time intervals can be obtained from the classical energy conservation condition [18]

$$\frac{\mu \dot{R}^2}{2} + U(R) = \text{const.}\,, \tag{18}$$

where μ is the reduced mass. The probability that *any* of the N quasimolecules in the state $|E,0\rangle$ remains excited *and* decays in the region $0 < R < R_e$ (where the gained kinetic energy is enough to overcome the trap depth) is constructed as

$$\begin{aligned} l_1 &= \sum_{i=1}^{N} \left| \frac{\Omega_i}{\tilde{\Omega}} \right|^2 \mathrm{e}^{-t_c \Gamma_c} \left(1 - \mathrm{e}^{-2t_e \Gamma}\right) \\ &= \mathrm{e}^{-t_c \Gamma_c} \left(1 - \mathrm{e}^{-2t_e \Gamma}\right) . \end{aligned} \tag{19}$$

This refers to the decay probability in the first passage through the region $0 < R < R_e$. It is possible that the quasimolecules remain excited and keep themselves vibrating between R_C and $R = 0$ before emission takes place, that is, they may pass several times across R_e while still in $|E,0\rangle$. Summing over these multiple passages [19] and composing them with conditional probabilities,

[1] Actually, this value should not include the emission towards the cavity solid angle $\Delta\Omega_c$ (inhibition of spontaneous emission) for the emitted radiation would be off-resonance, but $\Delta\Omega_c$ is very small and can be neglected.

the probability $\mathcal{L}_c$ for a pair of atoms to be ejected at any time is then

$$\begin{aligned}\mathcal{L}_c &= l_1 + l_2 + l_3 + \cdots \\ &= \sum_{n=1}^{\infty} \left[\sum_{i=1}^{N} \left| \frac{\Omega_i}{\tilde{\Omega}} \right|^2 (1 - \mathrm{e}^{-2\Gamma t_e}) \mathrm{e}^{-(2n-1)\Gamma_c t_c - 2(n-1)\Gamma t_e} \right] \\ &= \frac{\sinh\,(t_0 - t_c)\Gamma}{\sinh\,[t_0 + (\Gamma_c/\Gamma - 1)t_c]\Gamma} \equiv p\,\mathcal{L}_0 \,, \end{aligned} \tag{20}$$

where $t_0 = t_e + t_c$ is the total time interval and the *cavity-free* loss-probability is given by [19]

$$\mathcal{L}_0 \equiv \frac{\sinh\,(t_0 - t_c)\Gamma}{\sinh t_0 \Gamma} \,. \tag{21}$$

As pointed out in §2.2, comparison of losses with and without cavity could be possible by exciting the quasimolecules either via a cavity mode or by a laser beam propagating orthogonally to the cavity axis. In order to measure $p = \mathcal{L}_c/\mathcal{L}_0$, it would be necessary then to adjust the laser intensities so as to have the same fraction of excited state quasimolecules in both situations: in this case the excitation probability factors out.

3.5 Orders of magnitude

Since the collision is slow, the total time $t_0 \sim \Gamma^{-1}$. Indeed, for a typical detuning of $\delta = \omega_L - \omega_A = -2\pi \times 100$ MHz and taking $C_3 = 11.4 \times 10^{-11}$ erg Å^3 [33], the integration of (18) from R_C to $R = 0$ gives $t_0 = 3.0 \times 10^{-8}$ s, neglecting the initial velocity $\dot{R}$ at the Condon point $R_C \approx 556$ Å (see (3)). Relative to the atomic transition in ^{85}Rb with $\lambda_A = 2\pi c/\omega_A = 795$ nm and $\Gamma_A/2\pi = 6$ MHz, one can approximate $\sinh(x) \approx e^x/2$ for $x \geq t_0\Gamma = 2.3$, so that

$$\mathcal{L}_c/\mathcal{L}_0 = p \approx \mathrm{e}^{-(\Gamma_c - \Gamma)t_c} \tag{22}$$

is simply limited by the ratio between the survival probabilities, in the first passage through the region $R_e < R < R_C$, with and without cavity. The value of t_c is $t_c = 0.74 t_0$.

From (17), the trap-loss suppression can be very large depending primarily on N. Assuming $N_A \sim 10^6$ atoms at a density $n_A \sim 10^{12}$ cm^{-3}, we have from (9) $N \simeq 45$ quasimolecules. Even though this is a relatively crude approximation, attaining a number of this order should be feasible, since even larger numbers N_A of atoms in a FORT were recently reported by Corwin *et al.* [34]. Moreover, if the experiment were to be performed in a magnetic trap, N_A could be larger by a few orders of magnitude.

We consider a (quasi-)confocal cavity with mirrors of diameter $2b = 1.0$ cm and reflectivities $r = 0.97$, separated by $\ell = 2.9$ cm. The excitation field, with circular polarization, matches a TEM_{00} mode. The cavity spectral width is then $\gamma_c/2\pi \approx 200$ MHz (see §3.2) and the lineshape function is $\Lambda_{00} = 66$, with a corresponding finesse $\mathcal{F} = 103$. A numerical simulation of (16), sorting out the positions $\boldsymbol{r}_i$ and dipole orientations $\boldsymbol{d}_i$, gives an effective solid angle $\Delta\Omega_{\text{eff}}/4\pi \approx 7.4 \times 10^{-4}$, implying $\Gamma_c \approx 4.3\Gamma$ and thus a great reduction $p = \mathcal{L}_c/\mathcal{L}_0 \sim 3 \times 10^{-3}$ in the trap-loss probability.

The above analysis has assumed that not more than one single quasimolecule is excited, although any of them has a probability amplitude of being the excited one. In superradiance [9,10,11,12], the fastest decay rates, proportional to N^2, originate from collective states ("Dicke states") with several excited atoms entangled to each other. If we extend such conclusions to the case of cold collisions, despite the much more elaborate theoretical description required when several quasimolecules get excited and entangled by the cavity field, an even stronger effect can be expected on the collisional dynamics.

4 Conclusion

The fact that cold collisions can be driven by electromagnetic field photons is shown to have a counterpart in that an electromagnetic *vacuum* can be used to control the outcome of cold collisions as well. This is demonstrated theoretically considering radiative escape collisions. The life-time of the excited state involved in these collisions can be significantly reduced when the atoms are placed in the center of an optical resonator and excited via the cavity mode. The resulting multiparticle entangled state of the colliding atoms shows, even with a single excitation, a large collective decay rate owing to the coherent effect of many (quasi-) molecules being able to emit, and to the increased density of cavity modes. The normal course of the collision is thus interrupted by such fast cavity-induced decay, preventing the atoms to gain enough kinetic energy to escape from the trap. Within realistic experimental parameters, we predict a strong, observable, suppression of collisional trap-loss.

References

1. M. H. Anderson, J. R. Ensher, M. R. Matthews, C. E. Wieman, E. A. Cornell: Science **269**, 198 (1995)
2. C. C. Bradley, C. A. Sackett, J. J. Tollett, R. G. Hulet: Phys. Rev. Lett. **75**, 1687 (1995)
3. K. B. Davies, M. O. Mewes, M. R. Andrews, N. J. van Druten, D. S. Durfee, D. M. Kurn, W. Ketterle: Phys. Rev. Lett. **75**, 3969 (1995)
4. See related articles in the present volume
5. J. Weiner, V. S. Bagnato, S. Zilio, P. S. Julienne: Rev. Mod. Phys. **71**, 1 (1999)
6. P. S. Julienne: J. Res. Natl. Inst. Stand. Technol. **101**, 487 (1996)
7. K. Burnett, P. S. Julienne, K. -A. Suominen: Phys. Rev. Lett. **77**, 1416 (1996)

8. S. Haroche, in: *Les Houches, Session XXXVIII, 1982, Vol. I*, ed. by G. Grynberg and R. Stora (North-Holland, Amsterdam 1984), pp. 193–309
9. R. H. Dicke: Phys. Rev. **93**, 99 (1954)
10. R. Bonifacio, P. Schwendimann, F. Haake: Phys. Rev. A **4**, 302 (1971); *ibid.* **4**, 854 (1971)
11. M. F. H. Schuurmanns, Q. H. F. Vrehen, D. Polder, H. M. Gibbs: Adv. At. Mol. Phys. **17**, 167 (1981)
12. M. Gross, S. Haroche: Phys. Rep. **93**, 301 (1982)
13. M. Gangl, H. Ritsch: Phys. Rev. A **61**, 043405 (2000)
14. V. Vuletić, S. Chu: Phys. Rev. Lett. **84**, 3787 (2000)
15. B. Deb, G. Kurizki: Phys. Rev. Lett. **83**, 714 (1999)
16. J. I. Kim, P. Nussenzveig: submitted to J. Mod. Opt.
17. J. I. Kim, R. B. B. Santos, P. Nussenzveig: to appear in Phys. Rev. Lett. (2001) ; available at quant-ph/0008001
18. A. Gallagher, D. E. Pritchard: Phys. Rev. Lett. **63**, 957 (1989)
19. M. G. Peters, D. Hoffmann, J. D. Tobiason, T. Walker: Phys. Rev. A **50**, R906 (1994)
20. K. -A. Suominen, Y. B. Band, I. Tuvi, K. Burnett, P. S. Julienne: Phys. Rev. A **57**, 3724 (1998)
21. S. Chu, J. E. Bjorkholm, A. Ashkin, A. Cable: Phys. Rev. Lett. **57**, 314 (1986); J. D. Miller, R. A. Cline, D. J. Heinzen: Phys. Rev. A **47**, R4567 (1993)
22. V. S. Bagnato, G. P. Lafyatis, A. C. Martin, E. L. Raab, R. Ahmad-Bitar, D. E. Pritchard: Phys. Rev. Lett. **58**, 2194 (1987)
23. R. Napolitano, J. Weiner, P. S. Julienne: Phys. Rev. **A55**, 1191 (1997)
24. E. A. Power: J. Chem. Phys. **46**, 4297 (1967)
25. D. Kleppner: Physics Today **48** (7), 11 (1995)
26. P. Goy, J. M. Raimond, M. Gross, S. Haroche: Phys. Rev. Lett. **50**, 1903 (1983)
27. D. J. Heinzen, J. J. Childs, J. E. Thomas, M. S. Feld: Phys. Rev. Lett. **58**, 1320 (1987); D. J. Heinzen, M. S. Feld, *ibid.* **59**, 2623 (1987)
28. H. G. Hulet, E. S. Hilfeer, D. Kleppner: Phys. Rev. Lett. **55**, 2137 (1985)
29. M. Brune, P. Nussenzveig, F. Schmidt-Kaler, F. Bernardot, A. Maali, J. M. Raimond, S. Haroche: Phys. Rev. Lett. **72**, 3339 (1994)
30. A. Yariv: *Quantum Electronics*, 3rd edn. (John Wiley & Sons, New York 1989)
31. C. Zener: Proc. R. Soc. **A137**, 696 (1932)
32. L. Landau, E. Lifchitz: *Mécanique Quantique*, 3ème édition (Éditions Mir, Moscou 1980)
33. B. Bussery, M. Aubert-Frécon: J. Chem. Phys. **82**, 3224 (1985)
34. K. L. Corwin, S. J. M. Kuppens, D. Cho, C. E. Wieman: Phys. Rev. Lett. **83**, 1311 (1999)

Decoherence Evolution of a Harmonic Oscillator

J.C. Retamal

Departamento de Física, Universidad de Santiago de Chile
Casilla 307, correo 2, Santiago, Chile

Abstract. We consider the problem of decoherence for a harmonic oscillator evolving under a generalized dynamics. In particular we study the behavior of a decoherence measure around the initial time when the system is initially prepared in a pure state. In addition we calculate exacts analytical solutions for the linear entropy in some particular cases.

1 Introduction

The understanding of the boundary between the microworld and the macroworld has remained a central problem in Quantum Mechanics[1]. In the microscopic world a system can exist in a superposition of different quantum states, does exhibiting interference. On the other hand, these superpositions apparently do not manifest themselves in the classical world. This problem was already discussed by Schrödinger [2] in a famous paper in 1935.

In the last years there has been important theoretical insights that the interaction of a large system with its environment plays a major role in the disappearance of quantum coherence in the classical domain [3]. Recent progress in microwave cavities and electromagnetic traps for ions, led to experimental access to study this fascinating problem in the microscopic and mesoscopic level. The monitoring of decoherence, due to the effects of dissipation, in a superposition of two coherent states[4], was recently reported in a cavity QED microwave experiment [5]. Also the generation of controlled superpositions of states of the vibrational motion of the center of mass of the ion[6], as well as entangled states of internal states of two ions[7,8] and four ions[8] have already been achieved. The use of laser action in these systems opens the possibility of engineering several kinds of dissipative quantum dynamics [14], that may be interpreted as due to different kinds of interactions with reservoirs. This engineering of reservoirs allows the study of the process of quantum decoherence under several situations. Recently authors in reference [10] have presented an experimental study of decoherence in a high temperature amplitude reservoir, in a zero temperature amplitude reservoir and in a high temperature phase reservoir.

In this paper we will be mostly concerned with the study of the evolution of quantum mechanical coherence of a harmonic oscillator interacting with a reservoir. We consider the problem of the stability of quantum coherence for quantum states evolving under generalized reservoir dynamics. In this sense we

M. Orszag, J.C. Retamal (Eds.): LNP 575, pp. 138–157, 2001.

wants to answer the following questions: (a) which are the states that maintain its coherence (remain pure states) under the time evolution of an open system; (b) given an initial pure state, what is the scale of time where the decoherence process occurs. The evolution of quantum coherence can not be exactly known for an arbitrary dynamics, because analytical solutions are available only in some particular cases. We consider in addition some cases of interest where the evolution of quantum mechanical coherence can be exactly evaluated.

In section II we present the conditions for stability under dissipation. In section III we discuss the onset of decoherence. In section IV we present some exact analytical calculation for a harmonic mode intereracting with a reservoir at finite temperature and a nonclassical reservoir. Finally we present our final remarks.

2 Stable Quantum States

The stability of quantum coherence may be understood as the process where quantum coherence of the state of a physical system is preserved along its time evolution. In this sense we say that a initial pure quantum state, described by a density operator ρ, is "stable" if $\mathrm{tr}\rho^2 = 1$, for all times. One way to measure decoherence, or the stability of an initial pure state is to use the linear order entropy

$$s(t) = 1 - tr\rho^2. \tag{1}$$

A classical example of stability under dissipation is given by the time evolution of an initial coherent state of a quantum field mode, inside a cavity at zero temperature. In the Markov approximation, the dissipation may be described through the master equation [11] :

$$\dot{\rho} = \frac{\gamma}{2}(2c\rho c^\dagger - c^\dagger c\rho - \rho c^\dagger c), \tag{2}$$

where ρ is the density operator in the interaction representation, c and $c^\dagger$ are the annihilation and creation operators of the field cavity mode and γ is the field decay rate. Consider as the initial state a pure coherent state vector:

$$|\alpha\rangle_{\mathrm{coh}} = e^{-\frac{|\alpha|^2}{2}} \sum \frac{\alpha^n}{\sqrt{n!}} |n\rangle \,, \tag{3}$$

where α is a complex amplitude and $|n\rangle$ is the number state. It is easy to show that the state at a later time t is still a coherent state given by $| \ \alpha e^{-\gamma t/2}\rangle_{\mathrm{coh}}$. That is, coherent states remain pure states and we may say that they are "stable" under the dynamics given by Eq. (2). In fact, as we will show below, they are the only class of states with this property. This is not true anymore if the cavity is in equilibrium with a reservoir at a finite temperature $T > 0$. In this case the dynamics is described by

$$\dot{\rho} = \frac{\gamma}{2}(\bar{n}+1)(2c\rho c^\dagger - c^\dagger c\rho - \rho c^\dagger c) + \frac{\gamma}{2}\bar{n}(2c^\dagger \rho c - cc^\dagger \rho - \rho cc^\dagger), \tag{4}$$

where $\bar{n}$ is the equilibrium mean photon number at temperature T. For $T \neq 0$, it can be shown that there is no "stable state" and that the time of decoherence scales as $1/(\gamma\bar{n})$.

The aim of the present work is to analyze the stability of quantum coherence and the time scales in which a given quantum state became unstable. It possible to show that, under the assumptions of Markovian dynamics, complete positivity and initial decoupling between the system and the reservoir[12], the evolution of the system in the interaction picture is described by a Lindblad form

$$\dot{\rho} = \sum_{ij} \frac{\gamma_{ij}}{2}(2f_i\rho f_j^\dagger - f_i^\dagger f_j\rho - \rho f_i^\dagger f_j). \tag{5}$$

Here ρ is the reduced density operator in the interaction representation, and f_i are operators that act on the system. This structure is particularly interesting in the case of trapped ions. Poyatos, Cirac and Zoller [14] have modeled different reservoirs, corresponding to several kind of combinations of annihilation operator a and creation operator $a^\dagger$, associated to the vibrational motion, by properly adjusting the interaction of the ions with external laser fields. They have shown how to engineer dissipative dynamics of the form of Eq.(5) for a single operator f, such as a^k, $(a-\alpha)(a-\beta)$, $\mu a + \nu a^\dagger$, $a^\dagger a$,etc. Motivated by these ideas we will consider in our analysis, without loss of generality, that the dynamics is governed by Eq. (5), for a single operator f.

We first show that the only states that could remain pure under the dynamics described by Eq. (5) are the eigenstates of the operator f. Consider the rate of change in the linear entropy, $s(t)$, by calculating its first derivative, sometimes called the "entropy production":

$$\dot{s}(t) = 2\gamma[\text{ tr } f^\dagger f\rho_t^2 - \text{tr } f^\dagger \rho_t f \rho_t]. \tag{6}$$

If at an arbitrary time t_0 the system is in a pure state $|\Psi(t_0)\rangle$, this means that $\rho_{t_0}^2 = \rho_{t_0} = |\Psi(t_0)\rangle\langle\Psi(t_0)|$ and we may write

$$\begin{aligned}\dot{s}(t) &= 2\gamma(\text{tr } f^\dagger f\rho_t - \text{tr } f^\dagger\rho_t \text{tr} f\rho_t) \\ &= 2\gamma(\langle f^\dagger f\rangle - \langle\, f^\dagger\rangle\langle f\rangle).\end{aligned} \tag{7}$$

Therefore, $\dot{s}(t_0) = 0$ *only* if $|\Psi(t_0)\rangle$ is an eigenstate of f. We immediately exclude the possibility of dynamical stability when f does not have an eigenstate, as for example when $f = (a^\dagger)^k$. In these cases, which describe gain in the system, the entropy always increases. JI Il Kim it et al. [13] analyzed the coherence loss in terms of a perturbation expansion of the reduced density operator and found that for any initial factorized density operator, $s(t)$ increases at least as t^2, which seems to be in contradiction with the above result. The Lindblad equation, however, is obtained after a coarse grain derivative over intervals of time much greater than the correlation time of the reservoir.

Trivial examples of stable states are the eigenstates of Hermitean operators and normalizable eigenstates associated with the eigenvalue zero of any operator f[14,14]. These eigenstates remain constant in time as it can be easily shown by evaluating directly the right hand side of Eq. (5) to find that $\dot{\rho} = 0$. As examples we mention the Fock states $|n\rangle$ if $f = a^\dagger a$, and the states

$$\sum_{n=0}^{k-1} c_n |e^{2\pi i n/k}\beta\rangle_{\rm coh} \,, \tag{8}$$

where c_n are arbitrary complex numbers, if $f = a^k - \beta$.

When $f = a - \lambda$. it is easy to show that if the initial state is the coherent state $|\alpha + \lambda\rangle_{\rm coh}$, it evolves in time to the coherent state $|\alpha e^{-\gamma t/2} + \lambda\rangle_{\rm coh}$, that is an eigenstate of $a - \lambda$ with eigenvalue $\alpha e^{-\gamma t/2}$. In fact, for $f = a - \lambda$, we have

$$\dot{\rho} = \frac{\gamma}{2} D(\lambda)^\dagger (2a\hat{\rho}a^\dagger - a^\dagger a\hat{\rho} - \hat{\rho}a^\dagger a) D(\lambda) \tag{9}$$

where $D(\lambda) = e^{\lambda^* a - \lambda a^\dagger}$ is a displacement operator and $\hat{\rho} = D(\lambda)^\dagger \rho D(\lambda)$. As λ does not depend on t, it is clear that

$$\begin{aligned} \rho &= D(\lambda)|\alpha e^{-\gamma/2t}\rangle_{\rm coh}\langle \alpha e^{-\gamma/2t}|_{\rm coh} D(\lambda)^\dagger \\ &= |\alpha e^{-\gamma/2t} + \lambda\rangle_{\rm coh}\langle \alpha e^{-\gamma/2t} + \lambda|_{\rm coh} \,, \end{aligned} \tag{10}$$

is the solution of Eq. (9) since $|\alpha e^{-\gamma/2t}\rangle_{\rm coh}\langle \alpha e^{-\gamma/2t}|_{\rm coh}$ is a solution of Eq. (2). Another example are the eigenstates of the operator $S(re^{i\theta})aS^\dagger(re^{i\theta}) = \cosh(r)a + e^{i\theta}\sinh(r)a^\dagger$. If the initial state is an eigenstate with eigenvalue α

$$S(z)|\alpha\rangle_{\rm coh} = e^{\frac{1}{2}(z^* a^2 - z a^{\dagger 2})}|\alpha\rangle_{\rm coh}, \tag{11}$$

where $z = re^{i\theta}$,. Using a similar procedure as in the above case for the operator $a - \lambda$., it is easy to show that it evolves to the squeezed state $S(z)|\alpha e^{-\gamma/2t}\rangle_{\rm coh}$,.

3 The Onset of Unstabilities

To find out if the eigenstates of f are the only states that remain pure under the dynamics described by Eq.(5) and, simultaneously, the time scale of the onset of instabilities we will consider the change in $s(t)$, by calculating their derivatives.

Assume that we have an initial state which is an eigenstate of f, so that the entropy $s(0) = 0$ and the entropy production $\dot{s}(0) = 0$. To show that $s(t) = 0$ for any time, the derivatives of any order, should be zero (we are assuming that $s(t)$

is an analytical function of t). Let us denote by s_n and ρ_n, the nth derivative of $s(t)$ and $\rho(t)$ calculated at $t = 0$, respectively. The first few derivatives of the entropy can be easily written as:

$$\begin{aligned} s_1 &= -2\mathrm{tr}\rho\rho_1, \\ s_2 &= -2\mathrm{tr}(\rho_1^2 + \rho\rho_2), \\ s_3 &= -2\mathrm{tr}(3\rho_1\rho_2 + \rho\rho_3), \\ s_4 &= -2\mathrm{tr}(3\rho_2^2 + 4\rho_1\rho_3 + \rho\rho_4). \end{aligned} \quad (12)$$

The explicit form of these derivatives as a function of the operators ρ, f and $f^\dagger$ is, in general, very complicated. Below we give the expressions for the first two derivatives:

$$\begin{aligned} s_1 &= -2\gamma\mathrm{tr}(\rho f^\dagger \rho f - \rho^2 f^\dagger f), \\ s_2 &= -\gamma^2\mathrm{tr}[4\rho f^\dagger f \rho f^\dagger f + 2\rho^2(f^\dagger f)^2 - 3\rho f \rho f^{\dagger 2} f \\ &\quad - 3\rho f^\dagger \rho f^\dagger f^2 + 2\rho f^2 \rho f^{\dagger 2} - \rho f^\dagger \rho f f^\dagger f - \rho f \rho f^\dagger f f^\dagger]. \end{aligned} \quad (13)$$

Using the fact that the initial state is a pure state and is an eigenstate of $\hat{f}$, we set $\rho^2(0) = \rho(0)$, and $f \mid \Psi_f(t_0)\rangle = \beta \mid \Psi_f(t_0)\rangle$in Eq.(13). In this case the first and second derivatives are zero and the third derivative may be obtained after a lengthy but straightforward calculation:

$$s_3 = \gamma^3 \mid \beta \mid^2 (\langle A^2 \rangle - \langle A \rangle^2), \quad (14)$$

where $A = [f, f^\dagger]$. That is, the third derivative of the idempotency defect, s, will vanish at $t = 0$ either if $\beta = 0$ or if the variance , $\Delta A^2 = \langle A^2 \rangle - \langle A \rangle^2$, of the commutator A is zero. If $\beta \neq 0$ the initial state is a simultaneous eigenfunction of A and f. A sufficient condition for this to be possible is that $[A, f] = 0$, which happens either when f is Hermitean, in which case $A = 0$, or when $f = \mu a + \nu a^\dagger + \lambda I$, which implies that $A = 2\,\mathrm{Re}(\mu\nu^*)I$. It is easy to show that for all "stable" states discussed above we have either $\beta = 0$ or $\Delta A^2 = 0$ as it should be, since all derivatives $s_n(0)$ are zero. For other operators f we do not have stable states with eigenvalues $\beta \neq 0$. However if we start with an initial eigenstate of one of these operators, the idempotency defect increases, for small t, at most as t^3 :

$$s(t) \approx \frac{1}{3!}(\frac{t}{t_d})^3. \quad (15)$$

We should remark that this short time behavior of the linear entropy is a non trivial property of the Lindblad structure we are concerned in this work. The value of $t_d = (1/s_3(0))^{1/3}$ gives a time scale for the onset of decoherence for states that initially are eigenstates of f. For example, let us consider the case of $f = a^2$. The eigenstates of a^2 belonging to the eigenvalue $\beta = \alpha^2$ are any linear combination of $|\alpha\rangle$ and $|-\alpha\rangle$:

$$|\psi(0)\rangle = c_+|\alpha\rangle_{\mathrm{coh}} + c_-|-\alpha\rangle_{\mathrm{coh}}, \quad (16)$$

where

$$|c_+|^2 + |c_-|^2 + 2Re(c_+c_-^*)e^{-2|\alpha|^2} = 1. \tag{17}$$

In this case the third derivative of $s(t)$, at $t = 0$ is given by

$$\begin{aligned} s_3 = 16\gamma^3 \mid \alpha \mid^6 [1 - 4\,\mathrm{Re}(c_+^* c_-)e^{-2|\alpha|^2} + \\ 8 \mid \alpha \mid^2 \mathrm{Re}(c_+^* c_-)e^{-2|\alpha|^2}(1 - 2\,\mathrm{Re}(c_+^* c_-)e^{-2|\alpha|^2})], \end{aligned} \tag{18}$$

so that, for $|\alpha|^2 \gg 1$,

$$t_d = \left(\frac{1}{16}\right)^{1/3} \frac{1}{\gamma|\alpha|^2}. \tag{19}$$

Thus for large $|\alpha|$ the values of t_d for any combination of $|\alpha\rangle$ and $|-\alpha\rangle$ coincide. In particular this shows, that for this kind of reservoir, t_d is the same for a coherent state $|\alpha\rangle_{\mathrm{coh}}$ ($c_- = 0$) and for an even or an odd coherent state:

$$\frac{|\alpha\rangle_{\mathrm{coh}} \pm |-\alpha\rangle_{\mathrm{coh}}}{(2 \pm 2e^{-2|\alpha|^2})^{1/2}}. \tag{20}$$

This means that for times smaller than t_d the entropy increases in the same way regardless of the distance between the two "hills " in the Wigner representation of these states. It does not mean, however, that for longer times the behavior of $s(t)$ will be the same. In the cases where $|\psi(0)\rangle = |\alpha\rangle$ and $|\psi(0)\rangle = c(|\alpha\rangle + |-\alpha\rangle)$, the linear entropies differ at large times, the even coherent state evolving to the vacuum state while the coherent evolves to a statistical mixture of the vacuum $\mid 0\rangle$ and the state $\mid 1\rangle$. This may be explained by the fact that the even coherent state is a linear combination of Fock states with even numbers of quanta while the coherent state $|\alpha\rangle$ is a linear combination of Fock states with even and odd numbers of quanta. As the system can exchange only even numbers of quanta with the reservoir, the even coherent state will evolve to the vacuum, while the coherent state will evolve to a statistical mixture, with almost equal weights of $n = 1$ and $n = 0$ Fock states.

The time scale for decoherence of any initial eigenstate,

$$\sum_{n=0}^{k-1} c_n |e^{2\pi i n/k}\alpha\rangle_{\mathrm{coh}}, \tag{21}$$

of $f = a^k$, can also be easily found when $|\alpha| \gg 1$. In this case

$$A = \sum_{\ell=0}^{k-1} \frac{k!^2}{\ell!^2 (k-\ell)!} a^{\dagger \ell} a^{\ell}. \tag{22}$$

Therefore, to leading order in $|\alpha|$ we have, from Eq.(14),

$$\begin{aligned} s_3 \approx \gamma^3 \mid \alpha \mid^{2k} k^4 (\langle a^{\dagger(k-1)} a^{(k-1)} a^{\dagger(k-1)} a^{(k-1)} \rangle \\ - \langle a^{\dagger(k-1)} a^{(k-1)} \rangle^2). \end{aligned} \tag{23}$$

From this expression it is not difficult to show that for $|\alpha|^2 \gg 1$

$$t_d \approx \frac{1}{(k^4(k-1)^2)^{1/3}} \frac{1}{\gamma|\alpha|^{2(k-1)}}, \tag{24}$$

which shows that t_d has a behavior that depends on the nonlinearity of the interaction. The greater is the nonlinearity the smaller is the decoherence time.

Our results above refer to a situation where initially we have a single eigenstate of the operator f. The decoherence time scale may be completely different if the initial state is a linear combination of two eigenstates of the operator f belonging to different eigenvalues. In this case $\dot{s}(0) \neq 0$ and the decoherence time scale is given by $\tau = 1/\dot{s}(0)$. As an example consider $f = a^k$, and an initial state $|\psi(0)\rangle = N(|\alpha\rangle_{\rm coh} + |\beta\rangle_{\rm coh})$. In this case it is easy to show that

$$\dot{s}(0) \approx \frac{\gamma|\alpha^k - \beta^k|^2}{2}, \tag{25}$$

when $|\alpha - \beta|^2 >> 1$. Until now we were mostly concerned with operators that would have normalizable eigenstates, so that there would be a chance of obtaining stable states.

It is worth to mention that for other operators we may also obtain decoherence rates easily. Let us consider again the finite temperature reservoir of Eq.(4). The rate of linear entropy change for a initial coherent state $|z\rangle$ going to a thermal state [15] can be obtained by directly calculation using Eq.(12)

$$\dot{s}(0) = 2\gamma\bar{n}. \tag{26}$$

As we will see in the next section we can also obtain this time scale from the exact analytical result for $s(t)$. This can be obtained by considering the time evolution of a coherent state in a thermal reservoir.

Another interesting situation studied in the literature is the squeezed vacuum state reservoir. In this case the Lindblad form is given by[16]:

$$\begin{aligned}\dot{\rho} = &\frac{\gamma}{2}(\bar{N}+1)(2a\rho a^\dagger - a^\dagger a\rho - \rho a^\dagger a) + \\ &\frac{\gamma}{2}\bar{N}(2a^\dagger \rho a - aa^\dagger \rho - \rho aa^\dagger) \\ &\frac{\gamma}{2}\bar{M}(2a^\dagger \rho a^\dagger - a^\dagger a^\dagger \rho - \rho a^\dagger a^\dagger) + \\ &\frac{\gamma}{2}\bar{M}^*(2a\rho a - aa\rho - \rho aa),\end{aligned} \tag{27}$$

where $\bar{M}$ and $\bar{N}$ are the squeezed reservoir parameter parameters[16] satisfying the inequality $|\ \bar{M}\ |^2 \leq \bar{N}(\bar{N}+1)$. In this case it is not difficult to prove that

$$\begin{aligned}\dot{s}_t = &\, 2\gamma(2\bar{N}+1)(\mathrm{tr}a^\dagger a\rho^2 - \mathrm{tr}a\rho a^\dagger \rho) - 2\gamma\bar{N}s_t + 2\gamma\bar{N} + \\ &2\ \gamma\bar{M}(\mathrm{tr}a^2\rho^2 - \mathrm{tr}a\rho a\rho) + 2\gamma\bar{M}^*(\mathrm{tr}a^{\dagger 2}\rho^2 - \mathrm{tr}a^\dagger \rho a^\dagger \rho).\end{aligned}$$

For a coherent state the last two terms are zero, so that $\dot{s}_t$ has a behavior similar to that of a thermal reservoir, the thermal number of photon being replaced by the excitation number $\bar{N}$ of the squeezed reservoir. For an initial even coherent state the entropy production is given by

$$\dot{s}_0 = 2\gamma \mid \alpha \mid^2 [(2\bar{N}+1)\frac{(1-e^{-2|\alpha|^2})}{(1+e^{-2|\alpha|^2})} + 2\,\mathrm{Re}\,\bar{M})] + 2\gamma\bar{n}. \tag{28}$$

For $\bar{M} = |\bar{M}|e^{i\pi}$ and $\mid \alpha \mid^2 \gg 1$, this last equation may be written as

$$\dot{s}_0 = 2\gamma \mid \alpha \mid^2 (2(\bar{N}-\bar{M})+1) + 2\gamma\bar{n}. \tag{29}$$

which coincides with the result obtained in reference[17].

We have chosen the particular functional operator $s(t) = 1 - tr\rho^2$ to characterize the decoherence evolution. This is a choice that depends only on the system in consideration. Other invariants, like $I_n(t) = 1 - tr\rho^n$, could have also been used to characterize the tendency for a system to evolve to a statistical mixture. If the initial state is an eigenstate of the operator $\hat{f}$, it easy to show that the first and second derivatives of $I_n(t)$ are zero while the third derivative is given by:

$$\frac{d^3 I_n(0)}{dt^3} = \frac{n}{2}\gamma^3 \mid \beta \mid^2 (\langle A^2\rangle - \langle A\rangle^2). \tag{30}$$

That is, a result $n/2$ times larger than that obtained in the case of $n = 2$. Of course, these choices of measuring decoherence, although basis independent, are arbitrary.

4 Analytical Solutions for the Linear Entropy

An interesting question is related to the possibility to calculate the exact evolution of the decoherence process. Unfortunately this can not be made for an arbitrary dynamics such as given by Eq. 5. Even in the case of a single operator the possibility to solve the master equation in a simple way depends on the form of the operator $\hat{f}$. An analytical solution exist for $\hat{f} = a$ [11]. In other cases of interest, it is hard to calculate closed solutions of master equations, because of the existence of complicated couplings between different matrix elements of the density matrix. For example, a central problem is a dissipative cavity mode interacting with a thermal reservoir. Several approaches to solve this problem are available in the literature. The solution can be calculated through the resolution of a partial differential equation for a quasiprobability distribution[15,18]. An exact solution is given as an infinite series expressed in terms of the matrix elements of the initial conditions[18]. Another interesting case is that of dissipative mode in the presence of a squeezed reservoir[17,19]. In this case a solution has been found for the Wigner function. In a more recent work the problem of an oscillator in a dissipative environment has been revisited. It has been shown that for an initial coherent state the temporal evolution can be expressed in a remarkably simple way, by a displacement of a thermal state with a time dependent number of thermal excitations[20]. In that follows we present an analytical calculation for the dynamics of a cavity mode oscillator interacting with a finite reservoir of harmonic modes. This method allows us to solve the problem of an oscillator in a reservoir at finite temperature as well as in a squeezed reservoir. As we will see the difference between both cases is in the choice of the initial state of the reservoir.

Let us consider the Hamiltonian describing the interaction of a harmonic mode and a finite reservoir is

$$H = \hbar\omega a^\dagger a + \sum_{k=1}^{N} \hbar\omega_k b_k^\dagger b_k + \sum_{k=1}^{N} \hbar g_k (a b_k^\dagger + b_k a^\dagger) \tag{31}$$

In order to study this problem we introduce a collective mode operator

$$\Phi = \begin{pmatrix} b_0 \\ b_1 \\ b_2 \\ \cdot \\ \cdot \\ b_N \end{pmatrix} \tag{32}$$

where the system operator is given by $b_0 = a$. In addition we define the Hermitian matrix

$$G = \begin{pmatrix} \omega & g_1 & g_2 & \cdot\cdot & g_N \\ g_1 & \omega_1 & & & \\ g_2 & & \omega_2 & & \\ \cdot & & & & \\ \cdot & & & & \\ g_N & & & & \omega_N \end{pmatrix} \tag{33}$$

These definitions allow us to write the Hamiltonian in the form

$$H = \Phi^\dagger G \Phi \tag{34}$$

The matrix G can be diagonalized so that the Hamiltonian is written as

$$H = \varphi^\dagger D \varphi \tag{35}$$

where $\varphi = S\Phi$, with S being the similarity transformation which diagonalizes G. The transformed vector φ is defined as

$$\varphi = \begin{pmatrix} c_0 \\ c_1 \\ \cdot \\ \cdot \\ c_N \end{pmatrix} \tag{36}$$

Where operators c_i , $i = 0, 1..N$ are boson operators which satisfy the conmutations $[c_i, c_j] = \delta_{ij}$. In this representation the evolution operator for the total system (harmonic mode plus reservoir) is given by

$$U(t) = e^{-i \sum_{\ell} \nu_\ell t c_\ell^\dagger c_\ell} \tag{37}$$

Where ν_ℓ are the eigenvalues of G. This expression is useless if we do not know how a general initial state of our system transforms into the new basis . For most purposes our harmonic mode system could be assumed to be in a coherent state and the bath modes in the vacuum, which represents a reasonable assumption. A more interesting situation arise for other states of the reservoir, for example, a reservoir at finite temperature, or a reservoir in a perfect squeezed vacuum.

In order to illustrate this solution can be used, let us consider as a first step the well known case of a reservoir at zero temperature with all modes in an ordinary vacuum state, that is:

$$\mid \psi\rangle = \mid \alpha\rangle \prod_{k=1}^{N} \mid 0_k\rangle \tag{38}$$

A first observation is that this state is an eigenstate of operator a and the operators b_k (for each k). The question now is: how the transformed operators c_k operate on these states?. We observe that

$$c_\ell \mid \psi\rangle = \sum_{\ell'} S_{\ell\ell'} b_{\ell'} \mid \psi\rangle \tag{39}$$

Our choice for the initial state tells us that $\mid \psi\rangle$ is an eigenstate of c_ℓ with eigenvalue $S_{0\ell}\alpha$ so that we can write

$$c_\ell \mid S_{\ell 0}\alpha\rangle = S_{\ell 0}\alpha \mid S_{\ell 0}\alpha\rangle \tag{40}$$

This result allows us to write the initial state chosen above as follows

$$\mid \psi\rangle = \prod_{\ell=0}^{N} \mid S_{\ell 0}\alpha\rangle \tag{41}$$

where $\mid S_{\ell 0}\alpha\rangle$ is a coherent state associated with the transformed mode c_ℓ. It is readily apparent that the temporal evolution in the transformed basis is given by

$$\mid \psi(t)\rangle = \prod_{\ell=0}^{N} \mid S_{\ell 0}\alpha e^{-i\nu_\ell t}\rangle \tag{42}$$

where we still have a tensor product of coherent states. But it is not obvious how to separate explicitly the system coordinates from bath coordinates, so that we need to go back to the original basis. Let us note that

$$\begin{aligned} b_k \mid \psi(t)\rangle &= \sum_{k'} S^*_{k'k} c_{k'} \prod_{\ell=0}^{N} \mid S_{\ell 0}\alpha e^{-i\nu_\ell t}\rangle \\ &\sum_{k'} S^*_{k'k} S_{k'0}\alpha e^{-i\nu_{k'} t} \prod_{\ell=0}^{N} \mid S_{\ell 0}\alpha e^{-i\nu_\ell t}\rangle \\ &\sum_{k'} S^*_{k'k} S_{k'0}\alpha e^{-i\nu_{k'} t} \mid \psi(t)\rangle \end{aligned} \tag{43}$$

We learn that $\mid \psi(t)\rangle$ is a coherent state associated with b_k with eigenvalue

$$\sum_{k'} S^*_{k'k} S_{k'0}\alpha e^{-i\nu_{k'} t} \tag{44}$$

Then we can write the general state of the coupled system reservoir in the original basis as

$$\mid \psi(t)\rangle = \prod_{k=0}^{N} \mid \sum_{k'} S^*_{k'k} S_{k'0} e^{-i\nu_{k'} t}\alpha\rangle \tag{45}$$

We can now express explicitly the state of the coupled system separating the harmonic mode state from the bath state in the form

$$\mid \psi(t)\rangle = \mid \alpha_t\rangle \mid \boldsymbol{\beta}_t\rangle \tag{46}$$

where we use the following notation

$$\begin{aligned} \mid \alpha_t\rangle &= \mid \sum_{k'} S^*_{k'0} S_{k'0} e^{-i\nu_{k'} t}\alpha\rangle \\ \mid \boldsymbol{\beta}_t\rangle &= \prod_{k=1}^{N} \mid \sum_{k'} S^*_{k'k} S_{k'0} e^{-i\nu_{k'} t}\alpha\rangle \end{aligned} \tag{47}$$

Normalization are ensured because they are coherent states at all times, as we have previously found. From this result it is particularly simple to study the evolution of special states of the harmonic mode, for example a Schrödinger cat. Let us consider a initial state for the harmonic system bath mode interaction as follows

$$\mid \psi\rangle = \frac{1}{\sqrt{2(1+e^{-2|\alpha|^2})}} (\mid \alpha\rangle + \mid -\alpha\rangle) \prod_{k=1}^{N} \mid 0_k\rangle \tag{48}$$

By following the previous discussion we find the state of the coupled system at an arbitrary time is given by

$$\mid \psi(t)\rangle = \frac{1}{\sqrt{2(1+e^{-2|\alpha|^2})}} (\mid \alpha_t\rangle \mid \boldsymbol{\beta}_t\rangle + \mid -\alpha_t\rangle \mid -\boldsymbol{\beta}_t\rangle) \tag{49}$$

For simplicity we choose to study linear entropy $s(t) = 1 - \mathrm{tr}\rho^2$ to characterize the loss of quantum coherence of a Schrodinger cat. From the expression above we find the harmonic mode density matrix is given by

$$\begin{aligned} \rho(t) &= N_+(\mid \alpha_t\rangle\langle\alpha_t \mid + \mid -\alpha_t\rangle\langle-\alpha_t \mid \\ +& \quad \mid \alpha_t\rangle\langle-\alpha_t \mid \langle-\boldsymbol{\beta}_t \mid \boldsymbol{\beta}_t\rangle + \mid -\alpha_t\rangle\langle\alpha_t \mid \langle\boldsymbol{\beta}_t \mid -\boldsymbol{\beta}_t\rangle) \end{aligned} \tag{50}$$

where we defined $N_+ = [2(1+e^{-2|\alpha|^2})]^{-1/2}$. We can easily see that

$$\langle\boldsymbol{\beta}_t \mid -\boldsymbol{\beta}_t\rangle = e^{-2F(t)|\alpha|^2} \tag{51}$$

where we defined the quantity

$$F(t) = \sum_{k=1}\sum_{\ell\ell'} S^*_{\ell'0} S_{\ell' k} S^*_{\ell k} S_{\ell 0} e^{i(\nu_{\ell'}-\nu_\ell)t} \tag{52}$$

Then we can write the system density matrix as follows

$$\begin{aligned} \rho(t) &= N_+^2(\mid \alpha_t\rangle\langle\alpha_t \mid + \mid -\alpha_t\rangle\langle-\alpha_t \mid \\ +& \quad \mid -\alpha_t\rangle\langle\alpha_t \mid e^{-2F(t)|\alpha|^2} + \mid \alpha_t\rangle\langle-\alpha_t \mid e^{-2F(t)|\alpha|^2}) \end{aligned} \tag{53}$$

Normalization of the density matrix is apparent from these expression considering the following fact

$$tr \mid -\alpha_t\rangle\langle\alpha_t \mid = e^{-2\xi(t)|\alpha|^2} \tag{54}$$

where we defined

$$\xi(t) = \sum_{\ell\ell'} S^*_{\ell'0} S_{\ell' 0} S^*_{\ell 0} S_{\ell 0} e^{i(\nu_{\ell'}-\nu_\ell)t} \tag{55}$$

It is not difficult to show that

$$\begin{aligned} \xi(t) + F(t) &= \sum_{k=0}\sum_{\ell\ell'} S^*_{\ell'0} S_{\ell' k} S^*_{\ell k} S_{\ell 0} e^{i(\nu_{\ell'}-\nu_\ell)t} \\ &= 1 \end{aligned} \tag{56}$$

From these expression it follows that the reduced density operator for the harmonic system calculated above is normalized. It is now possible to calculate linear entropy which, after some calculations, is given by

$$s(t) = \frac{(1-e^{-4\xi(t)|\alpha|^2})(1-e^{-4(1-\xi(t))|\alpha|^2})}{2(1+e^{-2|\alpha|^2})^2} \tag{57}$$

Is worth to mention that at zero temperature this solution can be easily found from the exact solution for the master equation describing the dissipative harmonic mode[11].In that case we replace in this expression the quantity $\xi(t)$ by its corresponding limit in the large reservoir case $\xi(t) \to e^{-\gamma t}$.

4.1 A reservoir at a finite temperature

In the previous analysis we made a very reasonable assumption to characterize the state of the reservoir as the vacuum state. A limitation of this assumption arises because it ignores thermal excitations, which are included in the usual master equation approach. This problem has been considered to some extent in a recent work to study the decoherence of excitons in a leaky cavity[22]. In this section we address this problem in order to find the general evolution of a coherent state projector in reservoir at finite temperature. This evolution allows us to find analytically some closed expression for the decoherence evolution of a coherent state and a Schrodinger cat state. Th same expression could be used for a more general state which can be expanded in a coherent state basis. From the point of view we are concerned with in this work, we can include finite temperature effects considering the representation for a thermal vacuum in a coherent state basis:

$$\rho_{th} = \int P_{th}(\alpha, \alpha^*) \mid \alpha\rangle\langle\alpha \mid d^2\alpha \tag{58}$$

where

$$P_{th} = \frac{1}{\pi\bar{n}} e^{-|\alpha|^2/\bar{n}} \tag{59}$$

where $\bar{n}$ represent the number of thermal excitations of the reservoir. Then we assume a distribution like this for each mode of the reservoir. As is done in the procedure to obtain the master equation at finite temperature, we assume here that each mode of the reservoir has $\bar{n} = \bar{n}(\omega_0)$, ω_0 being the frequency of the system. This last assumption is well justified in the Markov approximation.

In this case it is not possible to write a wave function for the coupled system, however, we can write the density matrix. Let us consider an arbitrary coherent projector $\mid z\rangle\langle\omega \mid$ for the harmonic system, so that the initial condition for the harmonic mode and the reservoir is written as

$$\begin{aligned}\hat{\rho}_0 &= \mid z\rangle\langle\omega \mid \prod_{k=1}^{N} \int P_k(\alpha_k, \alpha_k^*) \mid \alpha_k\rangle\langle\alpha_k \mid d^2\alpha_k \\ &= \int \mathcal{D}^2\boldsymbol{\alpha} P(\boldsymbol{\alpha}, \boldsymbol{\alpha}^*) \mid z\rangle \mid \boldsymbol{\alpha}\rangle\langle\boldsymbol{\alpha} \mid \langle\omega \mid\end{aligned} \tag{60}$$

where we define

$$\mathcal{D}^2\boldsymbol{\alpha} = \prod_{k=1}^{N} d^2\alpha_k \tag{61}$$

$$P(\boldsymbol{\alpha}, \boldsymbol{\alpha}^*) = \frac{1}{(\pi\bar{n})^N} \prod_{k=1}^{N} P_k(\alpha_k, \alpha_k^*) \tag{62}$$

$$\mid \boldsymbol{\alpha}\rangle = \prod_{k=1}^{N} \mid \alpha_k\rangle \tag{63}$$

Now following the same procedure leading to equation (47), we obtain for the evolution of the global density matrix the expression

$$\hat{\rho}_t = \int \mathcal{D}^2\boldsymbol{\alpha} P(\boldsymbol{\alpha},\boldsymbol{\alpha}^*) \mid z_t\rangle \mid \boldsymbol{\alpha}_t^z\rangle\langle\boldsymbol{\alpha}_t^\omega \mid \langle\omega_t \mid$$

where

$$\begin{aligned}
\mid z_t\rangle &= \mid \sum_j a_{0j} z_j\rangle & z_0 &= z & z_{j\geq 1} &= \alpha_j \\
\mid \omega_t\rangle &= \mid \sum_j a_{0j} \omega_j\rangle & \omega_0 &= \omega & \omega_{j\geq 1} &= \alpha_j \\
\mid \boldsymbol{\alpha}_t^z\rangle &= \prod_{k=1}^{N} \mid \sum_j a_{kj} z_j\rangle & z_0 &= z & z_{j\geq 1} &= \alpha_j \\
\mid \boldsymbol{\alpha}_t^\omega\rangle &= \prod_{k=1}^{N} \mid \sum_j a_{kj} \omega_j\rangle & \omega_0 &= \omega & \omega_{j\geq 1} &= \alpha_j
\end{aligned} \tag{64}$$

We introduce the shorter notation

$$a_{kj} = \sum_{k'=0}^{N} S_{k'k}^* S_{k'j} e^{-i\nu_{k'} t} \tag{65}$$

We notice that this definition allows us to write a simple summation rule

$$\sum_{k=0}^{N} a_{kj}^* a_{kj'} = \delta_{jj'} \tag{66}$$

Then, the harmonic mode density matrix reads as follows

$$\rho_t = \int \mathcal{D}^2\boldsymbol{\alpha} P(\boldsymbol{\alpha},\boldsymbol{\alpha}^*) \mid z_t\rangle\langle\boldsymbol{\alpha}_t^\omega \mid \boldsymbol{\alpha}_t^z\rangle\langle\omega_t \mid \tag{67}$$

This is the formal solution for the time dependent global density matrix for the arbitrary initial coherent projector $\mid z\rangle\langle\omega \mid$. This expression is completely general and it allows to calculate the evolution of any state which can be expanded in a coherent state basis. In that follows we use this formula to calculate the evolution of the linear entropy for a coherent state and a Schrodinger cat.

4.2 Finite temperature entropy for a coherent state

For a initial coherent state with $z = \omega$ the density matrix for the harmonic mode is obtained from Eq. 67

$$\rho_t = \int \mathcal{D}^2\boldsymbol{\alpha} P(\boldsymbol{\alpha},\boldsymbol{\alpha}^*) \mid z_t\rangle\langle z_t \mid$$

This expression has a particularly simple structure which allows us to calculate the linear entropy. Let us calculate $\mathrm{tr}\rho^2$.

$$tr\rho^2 = \int \mathcal{D}^2\boldsymbol{\alpha}\mathcal{D}^2\boldsymbol{\eta} P(\boldsymbol{\alpha},\boldsymbol{\alpha}^*) P(\boldsymbol{\eta},\boldsymbol{\eta}^*) \langle\bar{z}_t \mid z_t\rangle\langle z_t \mid \bar{z}_t\rangle \tag{68}$$

It is not difficult to get convinced that this last expression involves only the integration of quadratic forms, and then it is exactly solvable. Let us note that

$$\langle z_t \mid \bar{z}_t\rangle = e^{-\frac{1}{2}|z_t|^2-\frac{1}{2}|\bar{z}_t|^2+z_t^*\bar{z}_t} \tag{69}$$

which, in accordance to Eq. (64) can be equivalently expressed as

$$\langle z_t \mid \bar{z}_t\rangle = e^{-\frac{1}{2}\sum_{kk'=0}^{N} a_{0k}^* a_{0k'}(z_k^* z_{k'}+\bar{z}_k^*\bar{z}_{k'}-2z_k^*\bar{z}_{k'})} \tag{70}$$

we remember that the variables to be integrated run from $k, k' = 1, ..N$, so that we separate explicitly the variables in the exponent

$$\langle z_t \mid \bar{z}_t\rangle = e^{-\frac{1}{2}\mathbf{v}^*(\boldsymbol{\alpha}-\boldsymbol{\eta})-\frac{1}{2}\mathbf{v}(\boldsymbol{\eta}^*-\alpha^*)-\frac{1}{2}\boldsymbol{\alpha}^*\mathbf{A}\boldsymbol{\alpha}-\frac{1}{2}\boldsymbol{\eta}^*\mathbf{A}\boldsymbol{\eta}+\boldsymbol{\alpha}^*\mathbf{A}\boldsymbol{\eta}} \tag{71}$$

where we defined in addition

$$\mathbf{v}_k = z a_{00} a_{0k}^* \tag{72}$$

$$\mathbf{A} = \sum_{kk'=1}^{N} a_{0k}^* a_{0k'} |k\rangle\langle k'| \tag{73}$$

we observe that

$$|\langle z_t \mid \bar{z}_t\rangle|^2 = e^{-\boldsymbol{\alpha}^*\mathbf{A}\boldsymbol{\alpha}-\boldsymbol{\eta}^*\mathbf{A}\boldsymbol{\eta}+\boldsymbol{\alpha}^*\mathbf{A}\boldsymbol{\eta}+\boldsymbol{\eta}^*\mathbf{A}\boldsymbol{\alpha}} \tag{74}$$

because the matrix $\mathbf{A}$ is hermitean. Then we finally attack the integral which can be expressed as

$$tr\rho^2 = (\frac{1}{\pi\bar{n}})^{2N}\int \mathcal{D}^2\boldsymbol{\alpha}\mathcal{D}^2\boldsymbol{\eta}e^{\mathbf{Q}(\boldsymbol{\alpha},\boldsymbol{\alpha}^*,\boldsymbol{\eta},\boldsymbol{\eta}^*)} \tag{75}$$

where we define

$$\mathbf{Q} = -\boldsymbol{\alpha}^*(\bar{n}^{-1}+\mathbf{A})\boldsymbol{\alpha}-\boldsymbol{\eta}^*(\bar{n}^{-1}+\mathbf{A})\boldsymbol{\eta}+\boldsymbol{\alpha}^*\mathbf{A}\boldsymbol{\eta}+\boldsymbol{\eta}^*\mathbf{A}\boldsymbol{\alpha} \tag{76}$$

This integral can be carried out by changing variables to a basis where $\mathbf{A}$ is diagonal. However, considering that $\mathbf{A}$ is a matrix arising from a tensor product of a vector by itself, it has only one eigenvalue different from zero, so that the integral simplifies itself into

$$tr\rho^2 = \frac{1}{(\pi\bar{n})^2}\int d^2\beta d^2\bar{\beta}e^{-(\frac{1}{\bar{n}}+\lambda)|\beta|^2-(\frac{1}{\bar{n}}+\lambda)|\bar{\beta}|^2+2\lambda(\beta^*\bar{\beta}+\bar{\beta}^*\beta)} \tag{77}$$

after some calculations this integral is given by the expression

$$tr\rho^2 = \frac{1}{1+2\bar{n}\lambda} \tag{78}$$

This result is simplicity itself, however it means nothing unless we can infer the expression for the eigenvalue λ. It is worth noticing that the temporal dependence is precisely in this eigenvalue. In order to guess the meaning of λ we can think as follows. If the vector $\boldsymbol{a}$ with components a_{0k} were normalized, the eigenvalue would be equal to unity. But actually it is not normalized, because of Eq. 66,so that we can write for the eigenvalue λ

$$\lambda = \sum_{k=1}^{N} |a_{0k}|^2 = 1 - |a_{00}|^2 \tag{79}$$

now we have by definition $|a_{00}|^2 = \xi(t)$ and considering the continuous limit of the model, $\xi(t)$ should converge to $\xi(t) = e^{-\gamma t}$ and the expression for the entropy could finally be written as

$$s(t) = 1 - \frac{1}{1 + 2\bar{n}(1 - e^{-\gamma t})} \tag{80}$$

We can see which are the limits of the entropy, for example if $t \gg \gamma^{-1}$ then

$$s(t) = \frac{2\bar{n}}{1 + 2\bar{n}} \tag{81}$$

If $t \lesssim (2\bar{n}\gamma)^{-1}$the decay law of quantum coherence is given by

$$s(t) = 2\bar{n}\gamma t \tag{82}$$

The rate of decoherence is then given by $s^{-1}(0) = 2\bar{n}\gamma$. This expression agrees with that previously by authors in reference[15].

4.3 Finite temperature entropy for a Schrodinger cat

Following a line of reasoning similar to that of the previous analysis, let us consider the initial condition for the coupled system

$$\begin{aligned}\hat{\rho}_0 = N_+^2 \textstyle\int \mathcal{D}^2\boldsymbol{\alpha} P(\boldsymbol{\alpha}, \boldsymbol{\alpha}^*)(\mid z\rangle\langle z \mid + \mid -z\rangle\langle -z \mid \\ + \mid z\rangle\langle -z \mid + \mid -z\rangle\langle z \mid) \mid \boldsymbol{\alpha}\rangle\langle\boldsymbol{\alpha} \mid\end{aligned} \tag{83}$$

From equation (67) the temporal evolution of the density operator for the system is given by

$$\begin{aligned}\rho_t = N_+^2 \textstyle\int \mathcal{D}^2\boldsymbol{\alpha} P(\boldsymbol{\alpha}, \boldsymbol{\alpha}^*)(\mid z_t\rangle\langle z_t \mid \langle \boldsymbol{\alpha}_t^z \mid \boldsymbol{\alpha}_t^z\rangle + \mid \bar{z}_t\rangle\langle \bar{z}_t \mid \langle \boldsymbol{\alpha}_t^{\bar{z}} \mid \boldsymbol{\alpha}_t^{\bar{z}}\rangle \\ + \mid z_t\rangle\langle \bar{z}_t \mid \langle \boldsymbol{\alpha}_t^{\bar{z}} \mid \boldsymbol{\alpha}_t^z\rangle + \mid \bar{z}_t\rangle\langle z_t \mid \langle \boldsymbol{\alpha}_t^z \mid \boldsymbol{\alpha}_t^{\bar{z}}\rangle)\end{aligned} \tag{84}$$

where $\bar{z} = \sum a_{0k}\bar{z}_k$ so that $\bar{z}_0 = -z$, $z = \sum a_{0k} z_k$ so that $z_0 = z$. In this case the idea follows in the same way as in the previous case of a coherent state projector, but a few more calculations are involved. Again, only integration of quadratic forms has to be done. The final expression for the linear entropy is given by:

$$s(t) = 1 - 2N_+^4 \frac{1}{1+2\bar{n}\lambda}(1 + e^{-4z^2a_{00}^2 + 8z^2a_{00}^2\lambda\frac{\bar{n}}{1+2\bar{n}\lambda}} \\ +4e^{-2z^2} + e^{-4z^2} + e^{-4z^2(1-a_{00}^2) - 8z^2a_{00}^2\lambda\frac{\bar{n}}{1+2\bar{n}\lambda}}) \tag{85}$$

Where it is understood that in the continuous limit $a_{00} \longrightarrow e^{-\gamma t/2}$, and $\lambda \longrightarrow 1-e^{-\gamma t}$. This expression converges to the right result in the regime of $\bar{n} = 0$. From this expression we can find the behavior of the linear entropy in some special regimes, such as in the case of the coherent state, for example, for $t \gg \gamma^{-1}$ the decoherence rate given through the linear entropy by $\dot{s}(0)^{-1}$. After some algebra we find

$$s(t) = \frac{2\bar{n}}{1+2\bar{n}} \tag{86}$$

this expression coincides with that obtained previously for a coherent state projector. For $t \lesssim (2z^2\gamma)^{-1}$ we have

$$\dot{s}(0) = 2\bar{n}\gamma + 2z^2\gamma(2\bar{n}+1)\frac{(1-e^{-2z^2})}{(1+e^{-2z^2})} \tag{87}$$

an expression which coincides with previous expressions obtained from calculation based on quasiprobability distributions[17], and or simply expanding the linear entropy around the initial time [21], as we have done in section II.

The approach we considered in this work allowed us to obtain some analytical expressions for useful quantities associated with the dynamics of the harmonic mode. Averages of energy or quadrature fluctuations can also be easily calculated. In addition to these results, we can obtain an analytical formula for the time dependent density matrix of a coherent state projector. Let us go back to the equation (67). We should note that

$$\begin{aligned} | z_t\rangle &= |a_{00}z + \boldsymbol{a}\cdot\boldsymbol{\alpha}\rangle \\ &= e^{-i\,\mathrm{Im}(a_{00}z\boldsymbol{a}\cdot\boldsymbol{\alpha})}D(a_{00}z)|\boldsymbol{a}\cdot\boldsymbol{\alpha}\rangle \end{aligned} \tag{88}$$

where $\boldsymbol{a}_k = a_{0k}$. In this way the expression for the density matrix of the coherent state projector can be written as

$$\rho_t = D(a_{00}z)\int \mathcal{D}^2\boldsymbol{\alpha}P(\boldsymbol{\alpha},\boldsymbol{\alpha}^*)|\boldsymbol{a}\cdot\boldsymbol{\alpha}\rangle\langle\boldsymbol{\alpha}_t^{\omega} \mid \boldsymbol{\alpha}_t^{z}\rangle\langle\boldsymbol{a}\cdot\boldsymbol{\alpha}|e^{-i\,\mathrm{Im}(a_{00}z-a_{00}\omega)\boldsymbol{a}\cdot\boldsymbol{\alpha}}D^\dagger(a_{00}\omega) \tag{89}$$

from here it is particularly simple to obtain a closed expression for the case of $z=\omega$, considering that in the continuous limit $a_{00} \longrightarrow e^{-\gamma t/2}$, and $\lambda \longrightarrow 1-e^{-\gamma t}$. After doing an integration over $N-1$ reservoir modes, we obtain the expression

$$\rho_t = D(e^{-\gamma t/2}z)\rho_{th}(\bar{n}(1-e^{-\gamma t}))D^\dagger(e^{-\gamma t/2}z) \tag{90}$$

This formula is in agreement with that obtained by authors in reference [20]. The main difference in our treatment is the explicit consideration of the reservoir modes in the whole procedure. This formula allows us to calculate a variety of quantities related to an initial coherent state evolving in a reservoir at finite temperature. In particular, one quantity of interest which is of common use is Von Neumann entropy $S(t) = -\mathrm{tr}\rho \ln \rho$. In this case, it is particularly simple to show that

$$S(t) = -\sum_{n=0}^{\infty} \left(\frac{\bar{n}(t)}{\bar{n}(t)+1} \right)^n \ln \left(\frac{\bar{n}(t)}{\bar{n}(t)+1} \right)^n \tag{91}$$

where $\bar{n}(t) = \bar{n}(1-e^{-\gamma t})$. After some calculations we obtain for the Von Neumann entropy the expression

$$S(t) = -\bar{n}(t) \ln \frac{\bar{n}(t)}{\bar{n}(t)+1} + \ln(\bar{n}(t)+1) \tag{92}$$

As a final remark to emphasize the usefulness of the master formula 67 we can write a compact expression for the evolution of a Schrodinger cat and visualize directly from the general expression the scale of decoherence as can be done for the case of a reservoir in a vacuum state. In fact, the temporal evolution of a Schrodinger cat $\mid \psi\rangle = N_+(\mid x\rangle + \mid -x\rangle)$ is given by the expression

$$\rho = N_+^2 (\rho_{xx} + \rho_{-x-x} + \rho_{x-x} + \rho_{-xx}) \tag{93}$$

where ρ_{xx} represents the evolution of diagonal terms and ρ_{-xx} represents the evolution of off-diagonal terms. Quantum decoherence can be read from ρ_{-xx}. In our case the expression for this term is given by:

$$\rho_{-xx} = e^{-2x^2(1-a_{00}^2)-x^2\bar{n}(t)a_{00}^2} D(a_{00}x(\bar{n}(t)+1))\rho_{th}(\bar{n}(t))D^\dagger(a_{00}x(\bar{n}(t)-1)) \tag{94}$$

From here we can obtain an estimation of the scale of decoherence as $t_d = 1/(2x^2\gamma + x^2\bar{n}\gamma)$.

Analytical formulas for the decoherence evolution can also be calculated in the case of a harmonic mode evolving in a squeezed vacuum. In this case the difference with the previous calculations is in the choice for the representation of the squeezed vacuum state in a coherent state basis, given by

$$\mid s\rangle = \left(\frac{a(a+1)}{\pi^2} \right)^{\frac{1}{4}} \int_{-\infty}^{\infty} dy e^{-ay^2} \mid e^{i\phi} y\rangle \tag{95}$$

It not difficult to show that this state can be obtained from the vacuum by the action of the squeeze operator $S(z) = e^{\frac{1}{2}(za^{\dagger 2} - z^* a^2)}$ where $z = re^{i\phi}$, being the complex squeezing parameter such that the width of the gaussian weight is related to the parameter r through the relation $\coth r = 2a + 1$. In this case, after a similar procedure we arrives to the following expression for the linear entropy of a initial coherent state:

$$s(t) = 1 - \frac{1}{\sqrt{1 + 4\sinh r^2 \lambda(1-\lambda)}} \tag{96}$$

where λ is same as given in Eq. (79).

5 Conclusions

Summarizing, we have studied the stability of quantum states under dissipation in a class of Lindblad dissipative dynamics, by considering the behavior of first order entropy around the initial time. Stable states may be characterized through the stability of linear entropy. We have found the nontrivial result that the instabilities of $s(t)$ are at least of third order in time, if the initial state is an eigenstate of the operator f. In addition we have presented a detailed calculation of the dynamics of a dissipative harmonic mode in the presence of a reservoir. We have considered a model of a reservoir composed of a finite number of harmonic modes with frequencies distributed around the central frequency of the harmonic mode system. Through this model we aimed to calculate closed expressions for quantities of interest to the study of quantum mechanical decoherence. The main attraction that the present calculations have shown is that the same procedure can be applied for different reservoirs, characterized through the initial state of reservoir modes, the vacuum, the thermal reservoir, the nonclassical reservoir. Certainly the modification in this case is only on the initial state we consider for the reservoir modes, and the possibility to have a representation in a coherent state basis.

6 Acknowledgments

This work was partially supported by the Fondecyt (Chile) under grant 1990838, DYCYT Universidad de Santiago de Chile and the Nucleo Cientifico Milenio de Materia Condensada

References

1. For a review see W. H. Zurek Physics Today **44**, 36 (1991).
2. E. Schrödinger, Naturwissenschaften **23**, 807(1935).
3. See, for example, Roland Omnès in *The Interpretation of Quantum Mechanics*, Princeton University Press, Princeton, 1994.
4. L. Davidovich, M. Brune, J.M. Raimond and S. Haroche, Phys. Rev. A **53**, 1295 (1996)
5. M. Brune et. al. Phys. Rev. Lett. **77**, 4887 (1996).

6. C. Monroe, D. M. Meekhop, B. E. King and D. J. Wineland, Science, **272**, 1131 (1996).
7. Q. A. Turchette, C. S. Wood, B. E. King, C. J. Myatt, D. Leibfried, W. M. Itano, C. Monroe, and D. J. Wineland, Phys. Rev. Lett. **81**, 3631 (1998).
8. C. A. Sackett, D. Kielpinski, B. E. King, C. Langer, V. Meyer, C. J. Myatt, M. Rowe, Q. A. Turchette, W. M. Itano, D. J. Wineland and C. Monroe, Nature **404**, 256 (2000).
9. J. F. Poyatos, J.I. Cirac and P. Zoller Phys. Rev. lett. **77**, 4728 (1996)
10. Q. A. Turchette, C. J. Myatt, B. E. King, C. A. Sackett, D. Kielpinski, W. M. Itano, C. Monroe, D. Wineland, Phys. Rev. A **62**, 053807 (2000)
11. See, for example, M. Sargent, M. Scully, W. Lamb, " Laser Physics"(Adison-Wesley, Reading, MA, 1974)
12. See for example, R. Alicki and K. Lendi, in Quantum Dynamical Semigroups and Applications in Lecture Notes in Physics (Springer Verlag, Berlin, 1987), $N^0$286.
13. Ji Il Kim, M.C. Nemes, A.F.R. de Toledo Piza, and H.E. Borges, Phys. Rev. Lett. **77,** 207(1996).
14. D. A. Lidar, I. L. Chuang, K.B. Whaley, Phys. Rev. Lett. **81**, 2594 (1998).
15. M. S. Kim and V. Buzek Phys. Rev. A **46**, 4239-4251 (1992)
16. D. F. Walls, G. J. Milburn, *Quatum Optics,* Springer-Verlag, Berlin, Heidelberg, 1994.
17. M. S. Kim and V. Buzek Phys. Rev. A **47**, 610-619 (1993)
18. M. Orszag Quantum Optics (Springer Verlag, Berlin 2000).
19. T. A. B. Kennedy and D. F. Walls, Phys. Rev. A 37, 152 (1988).
20. Hiroki Saito and Hiroyuki Hyuga Journal of the Physical of Japan **65**, 1648 (1996)
21. J. C. Retamal and N. Zagury Phys. Rev. A **63**, 321 (2001).
22. Yu-xi Liu, C. P. Sun and S. X. Yu Preprint quant.ph/0006059

Part III

Non-linear Optics, Matter Waves

Atomic Squeezing and Entanglement from Bose–Einstein Condensates

H. Pu[1], M.G. Moore[2], and P. Meystre[1]

[1] Optical Sciences Center, University of Arizona, Tucson, AZ 85721, USA
[2] ITAMP, Harvard-Smithonian Institute for Astrophysics, Cambridge, MA, USA

Abstract. We review several nonlinear atom optical approaches to the generation of entangled particle beams. The systems being considered, which involve both atomic pairs and atom-photon pairs, may find applications in high-precision spectroscopy, atomic interferometry and quantum information processing. All schemes deal with strictly ground atomic states, a considerable advantage, since the resulting entanglement and squeezing is therefore robust against decoherence and immune from the quantum fluctuations caused by the electromagnetic vacuum field modes.

1 Introduction

Quantum entanglement lies at the heart of the profound difference between quantum mechanics and classical physics[1]. The entanglement between the states of space-like separated particles is the fundamental reason for the violation of Bell inequality, and causes many of the "paradoxes" of quantum physics. In recent years,the discussions of entanglement have largely drifted away from the foundations of quantum mechanics and to "applications" in the emerging field of quantum information processing. Much attention has been paid to quantum correlated atomic systems, particularly nonclassical multi-atom states[2,3,4,5], as they have important applications in quantum measurement beyond the "standard quantum limit" as well as in quantum computation. There have been several proposals to create entangled atomic ensembles[6,7] and two of them have recently been demonstrated experimentally[8,9]. In addition, a number of applications would benefit from the combined advantages presented by the easy storage of massive particles and the fast propagation of photons, leading to the need for correlated atom-photon pairs[10,11].

In this paper, we review several proposed approaches to the generation of quantum entanglement that rely on *nonlinear atom optics* techniques for the realization of entangled particle beams. We first discuss how spin-exchange collisions in a three-component condensate can be used to produce a beam of entangled atoms, and then turn to a two-component system to produce arbitrary Dicke states and samples of maximally entangled atoms. Finally, we outline a discussion of the use of nonlinear mixing between optical and matter waves to generate entangled atom-photon pairs. These systems have obvious advantages

M. Orszag, J.C. Retamal (Eds.): LNP 575, pp. 161–176, 2001.

for quantum information applications: photons are well suited for transmission over long distances but are difficult to store at a fixed location, while the reverse is true for atoms. Creating an entangled pair of particles where one member is readily used to couple to a distant system while the other is stored by the sender thus seems ideal for use in quantum cryptography and teleportation related devices.

Much of the early work on nonlinear atom optics was developed in the framework of the Hartree approximation, in which the ultracold atomic sample is described by a Gross-Pitaevskii nonlinear Schrödinger equation. This approach makes important implicit assumptions about the statistical properties of the field — essentially, it is treated as a number state or as a coherent state, depending on the point of view adopted. While describing many effects very accurately, the semiclassical description fails in more subtle situations, where the quantum statistical properties of the field are tantamount. Inspired by the techniques of theoretical quantum optics, one can use an alternative method, based on the selection of "essential modes", to carry out a treatment of nonlinear matter-wave mixing going past these approaches. It allows for a full quantum mechanical treatment of the quantum statistical properties of the matter-wave field, at the expense of replacing a full spatial description of the system by just a few modes.

2 Entangled Atomic Beams

In this section we show that by taking advantage of coherent spin-exchange ultracold collisions, one can generate macroscopic atomic Einstein-Podolsky-Rosen (EPR) states[12] from a spinor Bose-Einstein condensate without the need of nonclassical light fields[13]. Our scheme is illustrated in Fig. 1. A spinor Bose-Einstein condensate consisting of a dilute $F = 1$ atomic sample is initially polarized such that only the spin-0 hyperfine ground state is populated at time $t = 0$. Binary spin-exchange interaction then convert the spin-0 atoms into pairs of spin-(± 1) atoms. The irreversibility of the process is provided by shifting the energy of the spin-0 state above that of the spin-(± 1) states. As a result of this detuning, the phase-matching condition, i.e., conservation of momentum and energy, ensures that the resultant atoms in the pair move in opposite directions away from each other and escape the trap. Quantum entanglement results from our ignorance about which of the two escaping atoms is in the spin-($+1$) state and which has spin-(-1).

We assume that at $t = 0$, a condensate of N_0 spin-0 atoms is confined in an optical dipole trap. An additional off-resonant optical field is used to shift the energy of the spin-0 state above those of the spin-(± 1) states by an amount $\hbar\delta$ (see Fig. 1). The spatial wave function $\varphi(\mathbf{r})$ of the condensate is determined by the stationary Gross-Pitaevskii equation.

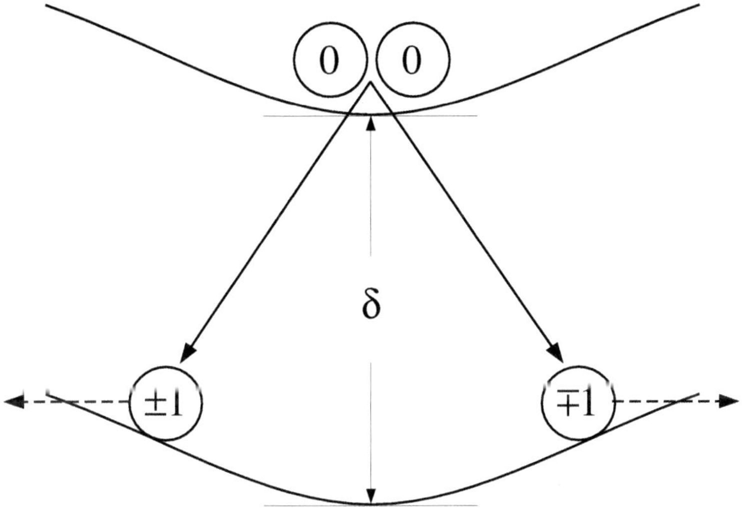

Fig. 1. Entanglement scheme: A spin-0 condensate is initially prepared. Spin-exchange interaction creates spin-(±1) atom pairs whose energy level is shifted below that of spin-0 atom by an amount of $\hbar\delta$. This excess energy is transferred into the kinetic energies of spin-(±1) atoms which escape the trap.

At $t > 0$, the spin-(±1) states start being populated by the spin-exchange interaction

$$H = \lambda_a \int d\mathbf{r}\, \hat{\psi}^\dagger_{+1}(\mathbf{r},t)\hat{\psi}^\dagger_{-1}(\mathbf{r},t)\hat{\psi}_0(\mathbf{r},t)\hat{\psi}_0(\mathbf{r},t) + h.c. \tag{1}$$

where λ_a is a constant related to the s-wave scattering lengths associated with the hyperfine levels involved, and $\hat{\psi}_\alpha$ is the boson annihilation operator for spin-α atoms. The effect of atomic recoil during this process is to transfer the excess energy $\hbar\delta$ into the kinetic energy of the spin-(±1) atoms. Therefore, for the short time scale where the propagation of (±1) atoms can be neglected, we expand the boson field operators as

$$\hat{\psi}_0(\mathbf{r},t) \quad = \varphi(\mathbf{r})\, e^{-i\delta t}\, \hat{c}_0(t) \tag{2}$$

$$\hat{\psi}_{\pm1,\mathbf{q}}(\mathbf{r},t) = \varphi(\mathbf{r})\, e^{i(\mathbf{q}\cdot\mathbf{r}-\omega_q t)}\, \hat{c}_{\pm1,\mathbf{q}}(t), \tag{3}$$

where $\omega_q \equiv \hbar|\mathbf{q}|^2/(2m)$, and the operators $\{\hat{c}_\mu\}$ obey the boson commutation relations $[\hat{c}_\mu, \hat{c}^\dagger_\nu] = \delta_{\mu,\nu}$. With these expansions, Hamiltonian (1) can be reexpressed as

$$H = \kappa \int\!\!\int d\mathbf{q}\, d\mathbf{q}'\, \rho(\mathbf{q},\mathbf{q}')e^{i\Delta_{q,q'}t}\, \hat{c}^\dagger_{+1,\mathbf{q}}\hat{c}^\dagger_{-1,\mathbf{q}'}\hat{c}_0\hat{c}_0 + h.c., \tag{4}$$

where $\Delta_{q,q'} \equiv (\omega_q + \omega_{q'} - 2\delta)$, $\kappa \equiv \lambda_a V^2/(2\pi)^6$ with V being the quantization volume, and

$$\rho(\mathbf{q},\mathbf{q}') = \int d\mathbf{r}\, |\varphi(\mathbf{r})|^4\, e^{-i(\mathbf{q}+\mathbf{q}')\cdot\mathbf{r}}. \tag{5}$$

Eq. (4) is reminiscent of the Hamiltonian describing parametric down conversion processes in nonlinear and quantum optics. As is well known, these processes lead to squeezing and to the generation of entangled photon pairs.

For short enough interaction times, the population of the sidemodes (± 1) remain small compared to N_0. In this regime, we neglect the depletion of the spin-0 state and treat $\hat{c}_0$ as a c-number c_0 such that $|c_0|^2 = N_0$. We can furthermore neglect those terms in the Hamiltonian that describe atom-atom interactions involving only the spin-(± 1) states. Under these assumptions, the Heisenberg dynamics of the operators $\hat{c}_{\pm 1,\mathbf{q}}$ is completely determined by Eq. (4) with $\hat{c}_0$ replaced by c_0, and the equations of motion for $\hat{c}_{\pm 1,\mathbf{q}}$ may be written as:

$$\frac{d}{dt}\hat{c}_{+1,\mathbf{q}} = -i\kappa \int d\mathbf{q}'\, \rho(\mathbf{q},\mathbf{q}')\, e^{i\Delta_{q,q'}t}\, \hat{c}^\dagger_{-1,\mathbf{q}'} c_0^2, \tag{6}$$

$$\frac{d}{dt}\hat{c}_{-1,\mathbf{q}} = -i\kappa \int d\mathbf{q}'\, \rho(\mathbf{q},\mathbf{q}')\, e^{i\Delta_{q,q'}t}\, \hat{c}^\dagger_{+1,\mathbf{q}'} c_0^2. \tag{7}$$

These equations can be solved in the Markov approximation to give

$$\hat{c}_{+1,\mathbf{q}}(t) = \mathcal{G}_\mathbf{q}(t)\, \hat{c}_{+1,\mathbf{q}}(0) + \int_0^t d\tau\, \mathcal{G}_\mathbf{q}(\tau)\, \hat{f}^\dagger_\mathbf{q}(t-\tau) \tag{8}$$

$$\hat{c}_{-1,\mathbf{q}}(t) = \mathcal{G}_\mathbf{q}(t)\, \hat{c}_{-1,\mathbf{q}}(0) + \int_0^t d\tau\, \mathcal{G}_\mathbf{q}(\tau)\, \hat{g}^\dagger_\mathbf{q}(t-\tau) \tag{9}$$

where $\mathcal{G}_\mathbf{q}(t) \equiv \exp(N_0^2 G_\mathbf{q} t/2)$, and we have defined the gain parameter

$$G_\mathbf{q} = 2\pi\kappa^2 \int d\mathbf{q}'\, |\rho(\mathbf{q},\mathbf{q}')|^2\, \delta(\Delta_{q,q'}) \tag{10}$$

and the noise operator

$$\hat{f}^\dagger_\mathbf{q}(t) = -i\kappa c_0^2 \int d\mathbf{q}'\, \rho(\mathbf{q},\mathbf{q}')\, e^{i\Delta_{q,q'}t}\, \hat{c}^\dagger_{-1,\mathbf{q}'}(0),$$

$$\hat{g}^\dagger_\mathbf{q}(t) = -i\kappa \int d\mathbf{q}'\, \rho(\mathbf{q},\mathbf{q}')\, e^{i\Delta_{q,q'}t} \hat{c}^\dagger_{+1,\mathbf{q}'}(0)\, c_0^2.$$

whose correlation functions are given in the Markov approximation by

$$\langle \hat{f}^\dagger_\mathbf{q}(t)\, \hat{f}_\mathbf{q}(t')\rangle = 0,$$

$$\langle \hat{f}_\mathbf{q}(t)\, \hat{f}^\dagger_{\mathbf{q}'}(t')\rangle = N_0^2 G_\mathbf{q}\, \delta(\mathbf{q}-\mathbf{q}')\, \delta(t-t'),$$

and similarly for $\hat{g}^{\dagger}_{\mathbf{q}}(t)$. It is these noise operators that trigger the populating of spin-(±1) state from quantum fluctuations. ¿From this result we can calculate the population in modes $\{\pm1, \mathbf{q}\}$:

$$N_{\pm1,\mathbf{q}} = \langle \hat{c}^{\dagger}_{\pm1,\mathbf{q}} \hat{c}_{\pm1,\mathbf{q}} \rangle = \exp(N_0^2 G_{\mathbf{q}} t) - 1$$

It is also straightforward to calculate the correlation function

$$\begin{aligned} \mathcal{C}_{\mathbf{q},\mathbf{q}'} &\equiv \langle \hat{c}_{-1,\mathbf{q}}\, \hat{c}_{+1,\mathbf{q}'} \rangle \\ &= -i\kappa\, \mathcal{G}_{\mathbf{q}}(t)\, c_0^2\, \rho_{\mathbf{q}}(\mathbf{q}') \frac{\mathcal{G}_{\mathbf{q}'}(t) - e^{i\Delta_{q,q'}t}}{N_0^2 G_{\mathbf{q}'}/2 - i\Delta_{q,q'}} \end{aligned} \tag{11}$$

The fact that the ±1 modes are correlated implies that the two spin states (±1) are entangled. From the definition (5) of $\rho(\mathbf{q}, \mathbf{q}')$, we conclude that as long as the spatial size of the condensate wave function is much larger than the reciprocal length $1/|\mathbf{q}|$ and $1/|\mathbf{q}'|$, $\rho(\mathbf{q}, \mathbf{q}')$ is approximately proportional to a delta-function, $\rho(\mathbf{q}, \mathbf{q}') \longrightarrow \delta(\mathbf{q} + \mathbf{q}')$. In other words, under this condition the two correlated atoms resulting from a spin-exchanging collision move in opposite directions. Additionally, the particle momenta $|\mathbf{q}|$ and $|\mathbf{q}'|$ have to satisfy the conservation of energy condition $\omega_q + \omega_{q'} - 2\delta \approx 0$, a condition that can be met for a large light shift $\hbar\delta$. We observe that in addition, a large energy shift $\hbar\delta$ is also required to produce spin-(±1) atoms with sufficiently large kinetic energy to escape the trap, which is also required to prevent them from undergoing a collision resulting in a pair of spin-0 atoms.

In general, it is not sufficient to just produce entangled atomic pairs. Rather, one would like to subsequently store them, e.g. in a dipole trap. It is desirable for this purpose to achieve a high degree of directionality in the generated atoms, so that they have a narrow enough angular distribution. To see how this can be achieved, let us take a closer look at the gain parameter $G_{\mathbf{q}}$ appearing in Eq. (10). Fig. 2 illustrates the gain along different directions for the case of a cylindrically symmetric condensate, for various light shifts $\hbar\delta$ and condensate aspect ratios. For simplicity, we choose $q = |\mathbf{q}| = \sqrt{2m\delta/\hbar}$, and assume that the condensate has a Gaussian shape. Fig. 2 illustrates quite clearly that a smaller angular distribution of emitted atoms is obtained for larger aspect ratios and larger q. Thus for a strongly elongated cigar-shaped condensate, the matter-wave modes along the long axis, which have the largest gain coefficient $G_{\mathbf{q}}$, will typically deplete all the condensate atoms before the population of the off-axis modes can significantly build up. As a consequence of mode competition, the emission of the spin-(±1) atoms is therefore largely confined to two narrow cones at the two ends of the cigar-shaped condensate.

¿From this discussion we conclude that in order to experimentally realize the proposed scheme, one should first create an elongated spin-0 condensate with a large light shift $\hbar\delta$. Spin-exchange interactions then generate pairs of spin-(±1) atoms flying in opposite directions along its long axis. These atoms can be subsequently captured and stored by two traps located at opposite sides of the original trap.

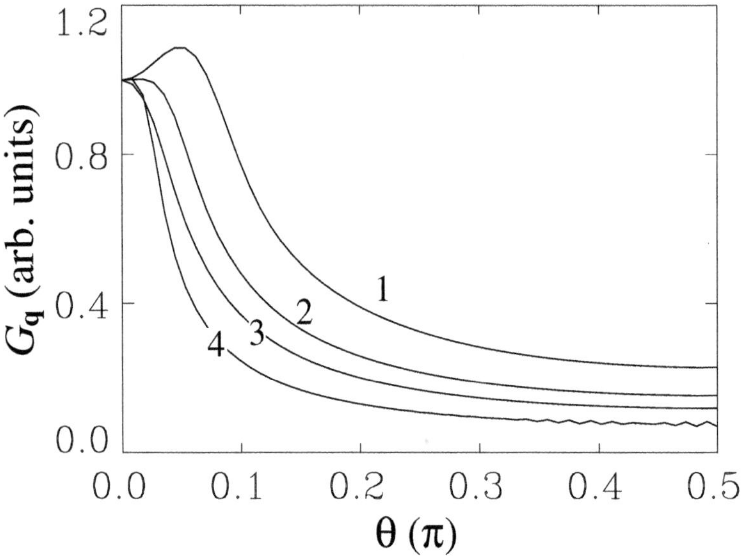

Fig. 2. The gain parameter $G_\mathbf{q}$ versus θ, for a Gaussian and cylindrically symmetric condensate wave function of the form $\varphi(\mathbf{r}) \propto \exp[-z^2/(2\sigma_z^2) - (x^2+y^2)/(2\sigma_\perp^2)]$. θ is the angle between $\mathbf{q}$ and the z-axis. In the calculation, we set $q = |\mathbf{q}| = \sqrt{2m\delta/\hbar}$. Curve 1: $\sigma_z = 10$, $q = 10$; curve 2: $\sigma_z = 10$, $q = 20$; curve 3: $\sigma_z = 10$, $q = 40$; curve 4: $\sigma_z = 20$, $q = 40$. The units for σ_z and q are $\sigma_\perp$ and $1/\sigma_\perp$, respectively.

The spin-(±1) atoms being created in pairs, we know for sure that taken together, the two ensembles must contain equal number of spin-(+1) and spin-(−1) atoms — although how many spin-(+1) and spin-(−1) atoms are in each ensemble is unknown. In the Schrödinger picture, such a state may be expressed as

$$|\Psi\rangle = \sum_{m=-N/2}^{N/2} a_m \left|\frac{N}{2}, m\right\rangle_l \left|\frac{N}{2}, -m\right\rangle_r , \tag{12}$$

where $N/2$ is the total number of atoms in each of the two "left" and "right" side traps, labeled by l and r, respectively. The integers m and $-m$ represent the difference in numbers of atoms in the spin states (+1) and (−1) in each of the two traps.

Introducing the z-component of the quasi-spin operator

$$\hat{L}_z^{(i)} = \hat{N}_{+1}^{(i)} - \hat{N}_{-1}^{(i)},$$

where $\hat{N}_{\pm 1}^{(i)}$ is the number operator for state-(±1) in ensemble i and $i = l, r$, we have that

$$\hat{L}_z^{(i)} \left|\frac{N}{2}, m\right\rangle_i = m \left|\frac{N}{2}, m\right\rangle_i .$$

Since the explicit expressions of the coefficients a_m in (12) are unknown, so are the expectation value and variance for $\hat{L}_z^{(i)}$. However, a simple calculation shows that

$$\begin{aligned} \langle \hat{L}_z \rangle &= 0, \\ (\Delta \hat{L}_z)^2 &= 0, \end{aligned}$$

where $\hat{L}_z \equiv \hat{L}_z^{(l)} + \hat{L}_z^{(r)}$ is the z-component of the *total* quasi-spin operator. Hence, although the variance of the $\hat{L}_z^{(i)}$ may be large for the individual ensembles, the variance for the whole system vanishes. In other words, taken as a whole the two ensembles represent a maximally spin-squeezed state. This should be contrasted to the case of N independent atoms in the state $(|+1\rangle + |-1\rangle)^N$, for which one finds $(\Delta \hat{L}_z)^2 = N/4$.

If we randomly pick one atom each from the two side-traps for an atomic ensemble prepared in state (12), then their degree of entanglement is only of order $1/N$. This is because although the atoms are created in pairs, we cannot tell which particular pairs of atoms are entangled. It is only through the collective spin measurement that the quantum entanglement can be revealed. The observation of such an macroscopic entanglement can be carried out with the spectroscopic technique described in Refs. [8,14]. In practice, the state (12) has to be averaged over the statistical distribution of the total particle number N. However, as noted in Ref. [6], such fluctuations do not affect the entanglement significantly for large numbers of atoms.

3 Dicke States

We now turn to a scheme to generate arbitrary atomic Dicke states in condensates[15]. Consider a two-component condensate with N_i atoms in component i, and $N_1 + N_2 = N$. A general state of the system can be written as a superposition of number difference states

$$|\Psi\rangle = \sum_{N_2=0}^{N} c_{N_2} |N_1, N_2\rangle. \tag{13}$$

Alternatively, we may introduce the effective angular momentum quantum number $j = (N_1 + N_2)/2 = N/2$ and the z-projection quantum number $m = (N_2 - N_1)/2$ and reexpress the state $|\Psi\rangle$ in terms of angular momentum states $|j, m\rangle$. In particular, we have the correspondence

$$|j, -j\rangle = |N_1 = N, N_2 = 0\rangle \tag{14}$$

a state which can be thought of as the "ground state" of coherent spin states. Our goal is to construct arbitrary Dicke states of the system, $j, m\rangle$, from this ground state $|j, -j\rangle$.

In a landmark paper, Kitagawa and Yamamoto [16] showed that photons could be number-squeezed by making use of a Kerr nonlinear medium. Their procedure consisted of splitting a coherent beam, passing one of its components through a Kerr nonlinear medium, and interfering the resulting light with the other coherent beam in a Mach-Zehnder interferometer. Kitagawa and Ueda [3,17] showed that it is possible to likewise produce squeezed spin states (SSS's) by making use of spin Hamiltonians quadratic in the spin operators. Our proposed scheme relies similarly on the existence of such a term in the Hamiltonian describing the dynamics of a coupled two-component condensate[20].

We consider for simplicity a two-mode model which neglecting all modes except the condensate modes. It is described by the second-quantized Hamiltonian

$$\begin{aligned} H = E_1 b_1^\dagger b_1 + E_2 b_2^\dagger b_2 + \frac{u_{11}}{2}(b_1^\dagger b_1^\dagger b_1 b_1) + \frac{u_{22}}{2}(b_2^\dagger b_2^\dagger b_2 b_2) \\ + u_{12}(b_1^\dagger b_1 b_2^\dagger b_2) - \frac{1}{2}(\Omega b_1 b_2^\dagger + \Omega^* b_2^\dagger b_1), \end{aligned} \tag{15}$$

where b_i is the annihilation operator for component i,

$$E_i = \int d^3r\, \phi_i^*(\mathbf{r}) \left[\hat{\mathbf{p}}^2/(2m) + V_i(\mathbf{r})\right] \phi_i(\mathbf{r}) \tag{16}$$

is the single-particle energy of mode i, $V_i(\mathbf{r})$ is the trapping potential, and

$$u_{ij} = \frac{4\pi\hbar^2 a_{ij}}{m} \int d^3r\, |\phi_i(\mathbf{r})|^2 |\phi_j(\mathbf{r})|^2 \tag{17}$$

describes the two-body collisions in the condensate in the s-wave scattering approximation. Here a_{ij} is the scattering length for a two-body collision between an atom of ith component and that of jth component and $\phi_i(\mathbf{r})$ represents the condensate wave function for mode i. Finally

$$\Omega = \Omega_0 \int d^3r\, \phi_1(\mathbf{r})\phi_2^*(\mathbf{r}) \tag{18}$$

gives the strength of the linear coupling between components and arises from the application of external fields.

The analysis of Eq. (15) is greatly simplified by the introduction of the angular momentum operators

$$\hat{J}_x = \frac{1}{2}(b_1^\dagger b_2 + b_2^\dagger b_1)\,, \tag{19}$$

$$\hat{J}_y = \frac{i}{2}(b_1^\dagger b_2 - b_2^\dagger b_1)\,, \tag{20}$$

$$\hat{J}_z = \frac{1}{2}(b_2^\dagger b_2 - b_1^\dagger b_1)\,, \tag{21}$$

in terms of which Hamiltonian (15) may be rewritten as

$$H = \kappa \hat{J}_z^2 - \Omega_x \hat{J}_x - \Omega_y \hat{J}_y, \tag{22}$$

where we have introduced the effective nonlinear coupling

$$\kappa = \frac{1}{2}(u_{11} + u_{22}) - u_{12}, \tag{23}$$

and the real Rabi couplings $\Omega_x = Re(\Omega), \Omega_y = Im(\Omega)$. Some control of the nonlinear parameter κ can be achieved through the proper engineering of the trapping potential, and hence of the condensate wave functions $\phi_i(\mathbf{r})$. In deriving (22), we have assumed that

$$E_1 - E_2 + (N - 1/2)(u_{11} - u_{22}) = 0,$$

a condition that can always be achieved by shifting the energy levels of the condensate components.

With the Hamiltonian (22) at hand, we now demonstrate how an appropriate choice of the external coupling fields allows one to generate arbitrary Dicke states from the ground state $|j, -j\rangle$. We first observe that a general coherent spin state (CSS) $|\theta, \varphi\rangle$ can be created by rotating $|j, -j\rangle$ by the angle θ about the axis $\boldsymbol{n} = (\sin\varphi, -\cos\varphi, 0)$[21],

$$|\theta, \varphi\rangle = \exp[-i\theta(\hat{J}_x \sin\varphi - \hat{J}_y \cos\varphi)]\,|j, -j\rangle \tag{24}$$

Starting from state $|j, -j\rangle$, switching on the coupling pulses and properly adjusting the strength of Ω_x and Ω_y, we can prepare any CSS. For the CSS $|\theta, \varphi\rangle$, the expectation value and the variance of operator $\hat{J}_z$ are

$$\begin{aligned} \langle \hat{J}_z \rangle &= -j \cos\theta\,, \\ \langle \Delta \hat{J}_z^2 \rangle &= \frac{j}{2} \sin^2\theta\,. \end{aligned}$$

(Note that neither $\langle \hat{J}_z \rangle$ nor $\langle \Delta \hat{J}_z^2 \rangle$ will change after the coupling fields are turned off.) Following the usual convention we say that a j-spin state is *squeezed along* $\hat{J}_z$ if the state has the same $\langle \hat{J}_z \rangle$ but reduced $\langle \Delta \hat{J}_z^2 \rangle$ compared to the CSS $|\theta, \varphi\rangle$. Alternatively, we can define a squeezing parameter [18]

$$\xi_z = \frac{2j\langle \Delta \hat{J}_z^2 \rangle}{j^2 - \langle \hat{J}_z \rangle^2}\,. \tag{25}$$

in terms of which a state is said to be squeezed along $\hat{J}_z$ if $\xi_z < 1$.

Squeezed states with arbitrary prescribed value of $\langle \hat{J}_z \rangle = m_0$ can be obtained as follows: first, a coupling pulse of appropriate strength is applied to the ground state $|j, -j\rangle$, creating an initial CSS with $\langle \hat{J}_z \rangle = m_0$. Assume without loss of generality that that pulse is along the x-axis. Following this pulse, we then apply a coupling field along the negative x-axis. During the subsequent time evolution, both $\langle \hat{J}_z \rangle$ and $\langle \Delta \hat{J}_z^2 \rangle$ start to oscillate. But at certain times when $\langle \hat{J}_z \rangle$ returns back to its initial value m_0, $\langle \Delta \hat{J}_z^2 \rangle$ comes close to a local minimum which is less

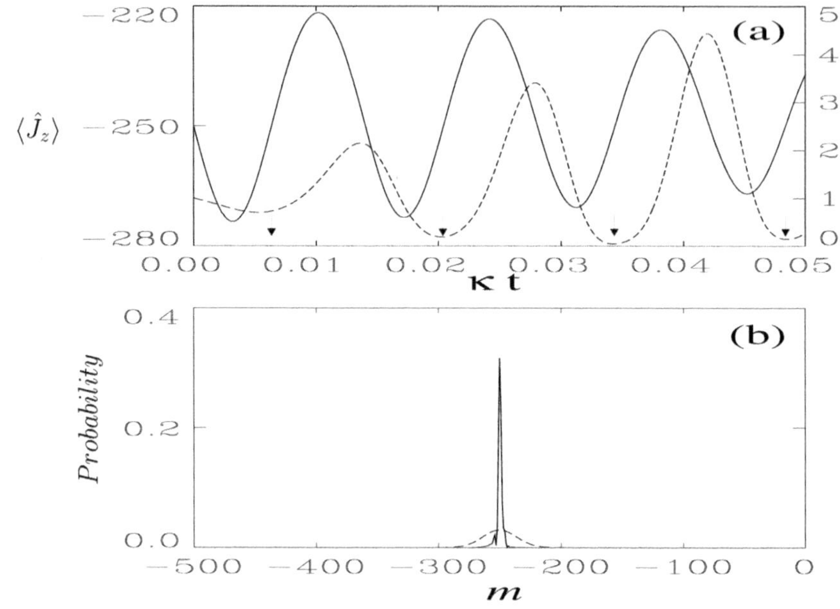

Fig. 3. (a) Time evolution of $\langle \hat{J}_z \rangle$ (left panel) and ξ_z (right panel). The initial CSS is $|\theta = \pi/3, \varphi = 0\rangle$ with $j = 500$. After the initial CSS is created, the system evolves under the Hamiltonian $H = \kappa \hat{J}_z^2 - \Omega_x \hat{J}_x$ with $\Omega_x/\kappa = -30$. At times indicated by the arrows, $\langle \hat{J}_z \rangle$ reaches the initial value (-250) while ξ_z is reduced. (b) Probability distribution of the azimuthal number m. Dashed curve: the initial CSS $|\theta = \pi/3, \varphi = 0\rangle$; solid curve: the spin squeezed state at $\kappa t = 0.0344$.

than the initial variance, i.e., $\xi_z < 1$. Provided that the coupling field is turned off at these precise times, a state squeezed along $\hat{J}_z$ with $\langle \hat{J}_z \rangle = m_0$ will be created. For large enough squeezing, the state thus prepared can be regarded as an approximation of the Dicke state $|j, m_0\rangle$ which is the spin analog of the Fock state. A typical time evolution of $\langle \hat{J}_z \rangle$ and ξ_z is shown in Fig. 3(a) where the initial CSS has $j = 500$ and $\langle \hat{J}_z \rangle = -250$. In this example, the optimal squeezing occurs at $\kappa t = 0.0344$ where the variance $\langle \Delta \hat{J}_z^2 \rangle$ is reduced by a factor of more than 20. Fig. 3(b) shows the probability distribution against the azimuthal number m at the optimal squeezing.

Procedures to achieve spin squeezing in the equatorial plane $\langle \hat{J}_z \rangle = 0$ was previously proposed by Kitagawa and Ueda[17,19], and more recently by Law *et al.* [22]. The essential point of both schemes is that they do not allow for the creation of a squeezed state with any prescribed $\langle \hat{J}_z \rangle = m_0 \neq 0$. By using our procedure however, this limitation can be lifted.

Squeezing and entanglement are oftentimes closely related. In particular, it is well known that the quadratic Hamiltonian $H_{spin} = \kappa \hat{J}_z^2$ also produces entanglement in a collective spin system. In particular, starting from the ground state $|j,-j\rangle_x$, H_{spin} generates at time $t^* = \hbar\pi/(2\kappa)$ the state

$$|\Psi(t^*)\rangle = \exp\left(-i\frac{\pi}{2}\hat{J}_z^2\right)|j,-j\rangle_x$$
$$= \frac{1}{\sqrt{2}}\left[e^{-i\pi/4}|j,-j\rangle_x + (-1)^j e^{i\pi/4}|j,+j\rangle_x\right]. \quad (26)$$

This state is maximally entangled, because if one individual spin is found to be aligned along the negative x-axis or the positive x-axis, so are all other spins. However, the degree of entanglement degrades after t^*, a consequence of the fact that the state (26) is not an eigenstate of the Hamiltonian H_{spin}. To preserve the entanglement, a $\pi/2$ pulse along the y-direction can be applied at $t = t^*$. This converts the state $|j,\pm j\rangle_x$ into $|j,\pm j\rangle$. The resultant state, still maximally entangled, it now an eigenstate of H_{spin}, hence it remains maximally entangled during the subsequent time evolution. This state is also a Schrödinger cat state since it is a coherent superposition of two macroscopically distinct states — one state has all the population in component 1 and the other in component 2.

4 Atom-Photon Entanglement

We conclude this brief review with a third nonlinear atom optical scheme, this time to generate atom-photon pairs whose internal degrees of freedom are maximally entangled[11].

Much effort has recently been devoted to the transfer of quantum coherence and correlations between optical and atomic systems, so as to be able to combine the advantages of atoms as a storage media and of light as a carrier of quantum information [23,24]. The use of parametric down conversion to generate entangled atom-photon pairs is a natural and important step towards achieving this goal.

In recent work [25,26] we have shown that the interaction between a Bose Einstein condensate and an off-resonant light field can take the form of a cubic nonlinearity, and hence leads to simultaneous parametric amplification of optical and atomic fields. In this case a pump photon and a condensate atom may be 'absorbed' simultaneously and a correlated atom-photon pair spontaneously 'emitted'. In this section, we show that if the atomic species in the condensate has multiple ground hyperfine states, then it is possible to generate atom-photon pairs whose internal states — ground hyperfine level and photon polarization — are maximally entangled.

There are two reasons why the high phase-space densities available in condensates is advantageous in this process: first, with a subrecoil velocity spread, an entangled atom moving at the recoil velocity is readily resolved in momentum space from the remaining condensate atoms, and second, a small spatial extent of the atom source allows the recoiling atom to quickly reach an empty region of space. This condition thus minimizes the occurrence of decoherence due to collisions with condensate atoms. Taken in combination, these two requirements are best satisfied by a minimum-uncertainty-state atom source such as a BEC [27].

In the electric-dipole and rotating-wave approximations the atom-field interaction operator is given by

$$\hat{\mathcal{V}}_{int} = -\hbar \sum_{Fm}^{(e)} \sum_{F'm'}^{(g)} \sum_{\mathbf{k},\lambda} \int d^3r\, \hat{\Psi}_e^\dagger(Fm\mathbf{r}) g_{\mathbf{k}\lambda}(FmF'm') \times e^{i\mathbf{k}\cdot\mathbf{r}} \hat{b}_{\mathbf{k}\lambda} \hat{\Psi}_g(F'm'\mathbf{r}) + H.c., \tag{27}$$

where $\hat{\Psi}_e(Fm\mathbf{r})$ is the annihilation operator for excited atoms with hyperfine quantum numbers F and m, $\hat{\Psi}_g(F'm'\mathbf{r})$ is the annihilation operator for ground-state atoms, and $\hat{b}_{\mathbf{k}\lambda}$ is the annihilation operator for photons with wave vector $\mathbf{k}$ and polarization $\epsilon_{\mathbf{k}\lambda}$, $\lambda = 1, 2$ corresponding to any two orthogonal polarization vectors for a given $\mathbf{k}$. The notation $\sum^{(e,g)}$ indicates a summation over excited or ground hyperfine levels, respectively. The coupling coefficient $g_{\mathbf{k}\lambda}(FmF'm')$ is given by

$$\hbar g_{\mathbf{k}\lambda}(FmF'm') = \mathcal{E}_{\mathbf{k}} \sum_q (\epsilon_q^* \cdot \epsilon_{\mathbf{k}\lambda}) \langle Fm|e\mathbf{r}\cdot\epsilon_q|F'm'\rangle, \tag{28}$$

where $\mathcal{E}_{\mathbf{k}}$ is the 'electric field per photon' at wave vector $\mathbf{k}$ and the three unit vectors ϵ_q, $q = \{-1, 0, +1\}$, are the usual polarization vectors for σ_-, π, and σ_+ polarized photons. They are given explicitly by $\epsilon_{\mp 1} = \pm(\hat{\mathbf{x}} \mp i\hat{\mathbf{y}})/\sqrt{2}$ and $\epsilon_0 = \hat{\mathbf{z}}$.

The dipole matrix elements $\langle Fm|e\mathbf{r}\cdot\epsilon_q|F'm'\rangle$ can be decomposed in the standard way as

$$\langle Fm|e\mathbf{r}\cdot\epsilon_q|F'm'\rangle = -(-1)^{-I-J'-F}\sqrt{(2F'+1)(2J+1)} \times \mathcal{D}\langle Fm|F'1m'q\rangle \begin{Bmatrix} I & J' & F' \\ 1 & F & J \end{Bmatrix}, \tag{29}$$

where I is the nuclear spin quantum number, J and J' are the angular momentum quantum numbers of the excited and ground states respectively, $\mathcal{D}$ is the reduced dipole moment for the $J' \leftrightarrow J$ transition, $\langle Fm|F'1m'q\rangle$ is a Clebsch-Gordan coefficient and the quantity in brackets is a Wigner 6-j symbol. For sodium we have $I = 3/2$, $J = 3/2$, and $J' = 1/2$.

We assume that the atomic system is driven by a strong off-resonant pump laser of frequency ω_L far detuned from any ground-excited transition. The excited state field operator can then be adiabatically eliminated, leaving an interaction operator which describes two-photon transitions between various ground atomic states. This operator is given by

$$\hat{\mathcal{V}}_{eff} = \sum_{FmF'm'}^{(g)} \sum_{\mathbf{k}\lambda\mathbf{k}'\lambda'} \int d^3r\, \hat{\Psi}_g^\dagger(Fm\mathbf{r}) e^{-i(\mathbf{k}-\mathbf{k}')\cdot\mathbf{r}} \times \hbar G_{\mathbf{k}\lambda\mathbf{k}'\lambda'}(FmF'm') \hat{b}_{\mathbf{k}\lambda}^\dagger \hat{b}_{\mathbf{k}'\lambda'} \hat{\Psi}_g(F'm'\mathbf{r}), \tag{30}$$

where we have introduced

$$G_{\mathbf{k}\lambda\mathbf{k}'\lambda'}(FmF'm') = \sum_{F''m''}^{(e)} g^{*}_{\mathbf{k}\lambda}(F''m''Fm) \times \frac{g_{\mathbf{k}'\lambda'}(F''m''F'm')}{\Delta(F',m')} \tag{31}$$

as the effective coupling constant, and where the detuning $\Delta(F',m')$ is given by

$$\Delta(F',m') \approx \omega_L - \omega_a + \delta(F',m'). \tag{32}$$

Here ω_a is the atomic resonance frequency and $\delta(F',m')$ is the frequency separation between ground hyperfine level F',m' and the lowest ground hyperfine level, including any AC Stark shifts due to driving laser fields. In deriving Eq. (32) we have neglected the difference in the kinetic energies of the excited and ground state atomic field due to recoil, as well as the excited state hyperfine splittings. These latter terms are neglected under the assumption that the far off-resonant condition $\Delta(F',m') \gg \Gamma$, the spontaneous decay rate, is satisfied and that the neglected frequency shifts are comparable to or less than Γ. It is clear that by scattering a photon into a different polarization it is possible for an atom to make a transition from one ground hyperfine sublevel into another. It is these Raman transitions that can be employed to generate entangled atom-photon pairs in a controlled manner from a BEC.

The initial state of the atomic field for the case of a nearly pure condensate is well approximated by assuming that a fixed number of atoms N_0 occupy the mode $\varphi_0(Fm\mathbf{r})$, with all other modes taken to be in the vacuum state. Similarly, the initial state of the optical field is assumed to consist of a single pump mode in the coherent state β_L, all other modes being in the vacuum. For times short enough for the fraction of scattered atoms to remain negligible we neglect the depletion evolution of the condensate and pump laser. They can then be factored out of the Hilbert space, in which case the interaction operator acting on the space of initially unoccupied states is found by making the substitutions

$$\begin{aligned} \hat{\Psi}_g(Fm\mathbf{r}) &\to \varphi_0(Fm\mathbf{r})\sqrt{N_0} + \hat{\psi}(Fm\mathbf{r}), \\ \hat{b}_{\mathbf{K}1} &\to \beta_L, \end{aligned} \tag{33}$$

where $\hat{\psi}(Fm\mathbf{r})$ is the annihilation operator for atoms in the subspace orthogonal to $\varphi_0(Fm\mathbf{r})$, and $\mathbf{K}$ is the wave vector of the pump laser.

We now focus on the terms in (30) which describe the spontaneous creation of correlated atom photon pairs. We express the operator for photon scattering in the $\mathbf{k}$-direction as

$$\hat{\mathcal{V}}_{\mathbf{k}} = \sum_{\lambda} \hbar\sqrt{N_0}\beta_L \mathcal{N}_{\mathbf{k}\lambda} \hat{b}^{\dagger}_{\mathbf{k}\lambda} \hat{c}^{\dagger}_{\mathbf{k}\lambda}, \tag{34}$$

so that $\hat{V}_{int} = \sum_{\mathbf{k}} \hat{V}_{\mathbf{k}}$. In deriving Eq. (34) we have introduced the operator $\hat{c}^{\dagger}_{\mathbf{k}\lambda}$ which creates an atom in the state

$$\varphi_{\mathbf{k}\lambda}(Fm\mathbf{r}) = \mathcal{N}_{\mathbf{k}\lambda}^{-1} \sum_{F'm'}^{(g)} G_{\mathbf{k}\lambda\mathbf{K}1}(FmF'm') e^{-i(\mathbf{k}-\mathbf{k}_L)\cdot\mathbf{r}} \times \varphi_0(F'm'\mathbf{r}), \tag{35}$$

$\mathcal{N}_{\mathbf{k}\lambda}$ being a normalization constant.

The frequency of the scattered light in a given direction is determined by energy conservation. Taking into account the recoil energy of the atom, it follows that a unique photon frequency corresponds to each scattering direction. In addition, if the atoms change their ground hyperfine F quantum number one also needs to account for the associated hyperfine splitting. From Eq. (34) we therefore see that Raman processes result in the creation of atom-photon pairs in which the spin state of the atom is entangled with the polarization state and frequency of the photon. We note, however, that the atomic spin states correlated with orthogonal photon polarizations need not be themselves orthogonal.

As a specific example now consider an $F = 1$, $m = 0$ condensate driven by a pump laser which propagates along the $\hat{\mathbf{y}}$ axis and is polarized along the $\hat{\mathbf{z}}$ axis. The condensate spinor is therefore

$$\varphi_0(Fm\mathbf{r}) = \phi_0(\mathbf{r})\delta_{F1}\delta_{m0}, \tag{36}$$

and the pump polarization is given by $\epsilon_{\mathbf{K}1} = \epsilon_0$. A scattered photon with polarization $\epsilon_{\mathbf{k}\lambda}$ is correlated with an atom in state

$$\varphi_{\mathbf{k}\lambda}(Fm\mathbf{r}) = \mathcal{N}_{\mathbf{k}\lambda}^{-1} G_{\mathbf{k}\lambda\mathbf{K}1}(Fm10) e^{-i(\mathbf{k}-\mathbf{K})\cdot\mathbf{r}} \phi_0(\mathbf{r}). \tag{37}$$

The internal state $|S_{\mathbf{k}\lambda}\rangle$ of the scattered atom is now completely specified by

$$\begin{aligned} \langle Fm|S_{\mathbf{k}\lambda}\rangle &= \mathcal{N}_{\mathbf{k}\lambda}^{-1} G_{\mathbf{k}\lambda\mathbf{K}1}(Fm10) \\ &= \mathcal{N}_{\mathbf{k}\lambda}^{-1} \frac{\mathcal{E}^*_{\mathbf{k}}\mathcal{E}_{\mathbf{K}}}{\hbar^2 \Delta(1,0)} \sum_q \sum_{F''m''}^{(e)} (\epsilon_q \cdot \epsilon^*_{\mathbf{k}\lambda}) \\ &\times \langle F''m''|e\mathbf{r}\cdot\epsilon_q|Fm\rangle^* \langle F''m''|e\mathbf{r}\cdot\epsilon_0|10\rangle, \end{aligned} \tag{38}$$

which can be evaluated explicitly by making use of the matrix elements (29).

The pump laser consists of π-polarized light, hence the atoms are excited from the $F = 1$, $m = 0$ ground state into the $F = 0$, $m = 0$ and $F = 2$, $m = 0$ excited states states. If we then set up our photon detector along the $\hat{\mathbf{z}}$ axis we guarantee that only σ_+ and σ_- photons will be detected. By taking the basis for the scattered photon polarization to be $\epsilon_{\mathbf{z}1} = \epsilon_{+1}$ and $\epsilon_{\mathbf{z}2} = \epsilon_{-1}$ we then find that the resulting entangled atom-photon state $|\psi_e\rangle$ is given by

$$\begin{aligned} |\psi_e\rangle = \frac{1}{2\sqrt{2}} \big[&|\omega_1, \sigma_+\rangle|1,-1\rangle + \sqrt{3}|\omega_2, \sigma_+\rangle|2,-1\rangle \\ &+ |\omega_1, \sigma_-\rangle|1,1\rangle - \sqrt{3}|\omega_2, \sigma_-\rangle|2,1\rangle \big], \end{aligned} \tag{39}$$

where the photon state is specified by its frequency and polarization and the atomic hyperfine state is specified in terms of $|F, m\rangle$ states. The frequency ω_1 of photons emitted during the $F = 1 \to F = 1$ atomic ground-state transition differs from the frequency ω_2 of photons emitted during the $F = 1 \to F = 2$ transition by the ground-state hyperfine splitting frequency, which in the case of sodium is of the order of GHz.

If the photodetectors are preceded by narrow spectral filters such that only photons of the shorter wavelength are detected, then they are entangled only with atoms in the $F = 1$ state, in which case the atom-photon state, specified by the photon polarization and the atomic m quantum number, is then

$$|\psi_e\rangle = \frac{1}{\sqrt{2}}(|\sigma_+, -1\rangle + |\sigma_-, 1\rangle), \tag{40}$$

the desired maximally entangled state. It can be shown that for realistic parameters, this scheme can produce as many as 10^5 maximally entangled atom-photon pairs, in a time of about 100 sec.

5 Conclusion

In this brief review, we have discussed several schemes to generate entangled particle pairs with the help of Bose-Einstein condensates. A first scheme created a macroscopic EPR-correlated atomic state which possesses a nonlocal entanglement and is maximally squeezed We showed that an elongated spinor condensate with large aspect ratio and large energy difference between spin-0 and spin-(± 1) states is the best candidate to create such a state. We have also shown that an arbitrary collective spin squeezing and entanglement of a two-component spinor condensate can be readily controlled by the coupling fields between the two components. Finally, we have proposed a realistic scheme to generate entangled atom-photon pairs via off-resonant light scattering from a Bose-Einstein condensate.

Squeezed or entangled spin states will find applications in high-precision spectroscopy, atomic interferometry and quantum information, and spinor condensates are attractive candidates to create such states. All schemes presented here deal with strictly ground atomic states. This is of considerable advantage, since the resulting entanglement and squeezing is therefore robust against decoherence and immune from the quantum fluctuations caused by the electromagnetic vacuum field modes.

This work is supported in part by the U.S. Office of Naval Research under Contract No. 14-91-J1205, by the National Science Foundation under Grant No. PHY98-01099, by the U.S. Army Research Office, and by the Joint Services Optics Program.

References

1. See, for example, J. S. Bell, *Speakable and unspeakable in quantum mechanics* (Cambridge University Press, 1987).
2. D. J. Wineland, J. J. Bolinger, W. M. Itano, F. L. Moore and D. J. Heinzen, Phys. Rev. A **46**, R6797 (1992).
3. M. Kitagawa and M. Ueda, Phys. Rev. A **47**, 5138 (1993).
4. A. Kuzmich, N. P. Bigelow and L. Mandel, Europhys. Lett. **42**, 481 (1998).
5. C. A. Sackett *et al.*, Nature **404**, 256 (2000).
6. E. S. Polzik, Phys. Rev. A **59**, 4202 (1999).
7. M. D. Lukin, S. F. Yelin and M. Fleischhauer, Phys. Rev. Lett. **84**, 4232 (2000).
8. J. Hald, J. L. Sørensen, C. Schori and E. S. Polzik, Phys. Rev. Lett. **83**, 1319 (1999).
9. A. Kuzmich, L. Mandel and N. P. Bigelow, Phys. Rev. Lett. **85**, 1594 (2000).
10. E. Hagley, X. Maitre, G. Nogues, C. Wunderlich, M. Brune, J. M. Raimond and S. Haroche, Phys. Rev. Lett. **79**, 1 (1997).
11. M. G. Moore and P. Meystre, Phys. Rev. Lett. **85**, 5026 (2000).
12. A. Einstein, B. Podolsky and N. Roson, Phys. Rev. **47**, 777 (1935).
13. H. Pu and P. Meystre, Phys. Rev. Lett. bf85, 3987 (2000); L. M. Duan, A. Sørensen, J. I. Cirac and P. Zoller, Phys. Rev. Lett. **85**, 3991 (2000).
14. J. L. Sørensen, J. Hald and E. S. Polzik, Phys. Rev. Lett. **83**, 3487 (1998).
15. S. Raghavan, H. Pu, P. Meystre and N. P. Bigelow, preprint cond-mat/0010140.
16. M. Kitagawa and Y. Yamamoto, Phys. Rev. A **34**, 3974 (1986).
17. M. Kitagawa and M. Ueda, Phys. Rev. Lett. **67**, 1852 (1991).
18. We note that this definition of ξ_z differs from those used previously in the literature. We prefer it because of its utility in evaulating the Dicke states discussed here.
19. M. Kitagawa and M. Ueda, Phys. Rev. A **47**, 5138 (1993).
20. A. Sørensen, L.-M. Duan, J. I. Cirac and P. Zoller, e-print quant-ph/0006111.
21. J. Klauder, Ann. Phys. (N.Y.) **11**, 123 (1960); J. M. Radcliffe, J. Phys. A **4**, 313 (1971); F. T. Arecchi, E. Courtens, R. Gilmore, and H. Thomas, Phys. Rev. A **6**, 2211 (1972).
22. C. K. Law, H. T. Ng and P. T. Leung, preprint quant-ph/0007056.
23. X. Maistre et al, Phys. Rev. Lett. **79** 769 (1997).
24. A. S. Parkins, and H. J. Kimble, J. Opt. B-Quantum Semiclass. Opt. **1** 496 (1999).
25. M. G. Moore, P. Meystre, Phys. Rev. A **59**, R1754 (1999).
26. M. G. Moore, O. Zobay, P. Meystre, Phys. Rev. A **60**, 1491 (1999).
27. J. Stenger et al, Phys. Rev. Lett. **82**, 4569 (1999).

Atomic Coherence Effects in Doppler-Broadened Three-Level Systems with Standing-Wave Drive

F. Silva[1], J. Mompart[2], V. Ahufinger[2], and R. Corbalán[2]

[1] Departament d'Òptica, Universitat de València, E-46100 Burjassot, Spain
[2] Departament de Física, Universitat Autònoma de Barcelona, E-08193, Bellaterra, Spain

Abstract. We study atomic coherence effects (e.g., electromagnetically induced transparency, EIT, and amplification without inversion, AWI) for a probe travelling-wave (TW) laser field in closed Doppler-broadened three-level systems driven by a standing-wave (SW) laser field of moderate intensity (its Rabi frequencies are smaller than the Doppler width of the driven transition). We show that probe windows of transparency occur only for values of the probe to drive field frequency ratio R close to half integer values. For optical transitions and typical values of Doppler broadening for atoms in a vapor cell, we show that for $R > 1$ a SW drive field is appreciably more efficient than a TW drive in inducing probe transparency. We show that folded (cascade) schemes driven by a suficiently detuned SW field can exhibit AWI for odd (even) values of R. Results for AWI with frequency up-conversion ratios $R = 3$ and 4 are presented and compared to those obtained with a TW drive.

1 Introduction

Linear and nonlinear optical properties of an atomic medium can be controlled and modified by applying a strong drive laser field that induces atomic coherence. Atomic coherence manifests itself in a rich variety of phenomena that are a subject of current research interest, such as Electromagnetically Induced Transparency, EIT; Amplification and Lasing Without Inversion, AWI and LWI (for reviews see, for instance [1,2,3,4]); high refractive index without absorption [5]; steep dispersion and either ultraslow group velocity [6] or gain-assisted superluminal light propagation [7]; and enhanced nonlinear optical susceptibility [8]. Here we consider EIT and AWI, which are among the most basic of these phenomena.

In EIT, an otherwise absorbing medium is made transparent to a weak coherent field on resonance with a certain atomic transition by applying an intense coherent "drive" field to an adjacent transition. There has been a large number of theoretical contributions [1,2,3] and EIT was demonstrated experimentally [9,10,11,12,13,14].

In AWI, a weak coherent field coupled to a non-inverted atomic transition, but with some population in its upper level, is amplified by applying an intense coherent drive field to an adjacent transition. There has been also a large number

M. Orszag, J.C. Retamal (Eds.): LNP 575, pp. 177–194, 2001.

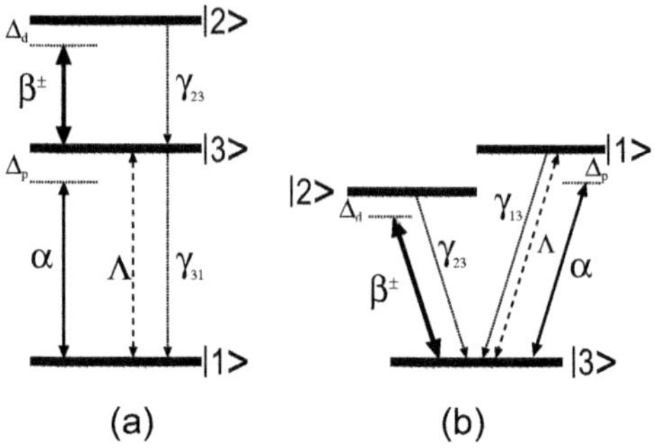

Fig. 1. Closed three-level systems under investigation: **(a)** $p-$, **(b)** $V-$ schemes.

of theoretical papers on AWI [1,2,3,4], which has been demonstrated experimentally [15]. Several recent proof-of-principle experiments have demonstrated also actual lasing without inversion [16]. The main motivation in these studies is the pursuit of lasing at a wavelength substantially shorter than that of the drive laser (the so-called frequency up-conversion regime), which would open a new way toward uv and x-ray lasers. In the above experiments [16] lasing occurred at a frequency either smaller or slightly larger than that of the drive field. Only a few theoretical papers have addressed recently the frequency up-conversion regime [17], pointing out that one of the main problems in this challenging regime is Doppler broadening.

Up to now, most theoretical papers dealing with atomic coherence effects in gas media have considered only the case of a traveling-wave (TW) drive field. In this paper we study EIT [18] and AWI [19] in Doppler-broadened three-level media with standing-wave (SW) drive and focus our analysis on the advantages of using a SW drive configuration instead of a TW one in the frequency up-conversion regime. In a related context, it has been demonstrated both theoretically [20] and experimentally [21] that SW pumping of far infrared lasers leads to gain values larger than those obtained for a TW pumping.

The case of a SW drive intuitively appears as promising in the frequency up-conversion regime. Let us consider, for instance, a Doppler-broadened three-level ladder scheme (Fig. 1a) driven on transition $|2\rangle \leftrightarrow |3\rangle$ by a SW laser field and probed on transition $|1\rangle \leftrightarrow |3\rangle$ by a weak collinear TW laser field. The wavenumber and the detuning from atomic resonance are (k_d, Δ_d) and (k_p, Δ_p) for drive and probe fields, respectively. The resonance conditions for the processes contributing to probe amplification or absorption can be easily traced if the drive-induced ac Stark effect is neglected. Since the probe is weak, only processes involving one probe-photon must be considered. Multi-photon processes involving one probe-photon and several pump-photons (see Fig. 2), with alternating absorption and emission from the two counterpropagating pump components, occur for velocities satisfying:

$$\Delta_p - k_p v = q k_d v, \qquad q = 0, \pm 2, ... \tag{1a}$$

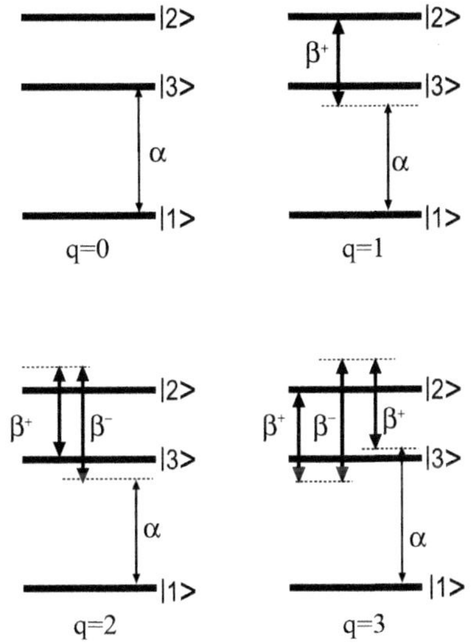

Fig. 2. Lowest order processes participating in the probe absorption/amplification, with q being the number of drive photons involved. Similar processes, interchanging β^+ by β^- and β^- by β^+ correspond to $q < 0$.

$$\Delta_p - k_p v = -\Delta_d + q k_d v, \qquad q = \pm 1, \pm 3, ... \tag{1b}$$

where v is the atomic velocity along the probe beam and $|q|$ is the number of pump-photons involved in the process. For a TW drive counterpropagating with the probe, only processes (1a) with $q = 0$ and (1b) with $q = -1$ occur. The resonance condition for this second process can be made $v-$ independent only if $k_p = k_d$. This is the well-known result that with TW fields Doppler-free resonances only occur for equal frequency drive and probe fields. In contrast, a SW drive allows also Doppler-free resonances at $\Delta_p = 0$ with $k_p/k_d = |q| = 2, 4, ...$, and at $\Delta_p + \Delta_d = 0$ with $k_p/k_d = |q| = 1, 3, ...$, and it is therefore interesting for frequency up-conversion AWI. On the other hand, in the case of EIT (i.e., without population in the upper level of the probed transition) there will be a strong probe absorption under conditions of Doppler-free resonances, since all the atoms, irrespective of their velocities, contribute to the absorption. Therefore, one expects to find windows of transparency for $k_p/k_d \simeq |q| + \frac{1}{2}$, i.e., approximately midway between consecutive Doppler-free multiphoton resonances. These main conclusions will be confirmed by the detailed analysis presented below.

Section 2 is devoted to the semiclassical density matrix approach to the problems at hand. In Section 3 we briefly review the dressed-atom analysis for the SW excitation of a Doppler-broadened transition. Results for EIT are presented in Section 4, while AWI is discussed in Section 5. Section 6 summarizes our main conclusions.

2 Semiclassical Density Matrix Analysis

To be specific, let us consider the closed cascade three-level system shown in Fig. 1a, although similar results can be obtained for the folded (V and Λ) schemes. A TW laser field $E_p(z,t) = \frac{1}{2}A_p \exp\left[-i(\omega_p t - k_p z)\right] + c.c.$ to which transparency will be induced or which is to be amplified probes the lower transition $|3\rangle$-$|1\rangle$, while the upper transition $|2\rangle$-$|3\rangle$ is driven by a collinear strong SW laser field $E_d(z,t) = \sum_{\mu=\pm1} \frac{1}{2}A_d^{\mu} \exp\left[-i(\omega_d t - \mu k_d z)\right] + c.c.$ The coupling of the atoms with the laser fields is characterized by the detunings $\Delta_p = \omega_p - \omega_{31}$ and $\Delta_d = \omega_d - \omega_{23}$ from their respective resonance, and by the three Rabi frequencies $2\alpha \equiv \mu_{31}A_p/\hbar$ and $2\beta^{\pm} \equiv \mu_{23}A_d^{\pm}/\hbar$, where μ_{31} and μ_{23} are electric dipole moments between the corresponding states. Without loss of generality the three Rabi frequencies will be taken to be real. Due to electric dipole selection rules the states $|2\rangle$ and $|1\rangle$ must have the same parity, therefore $\mu_{21} = 0$. The three-level system is also subjected to incoherent pumping and relaxation processes, which will be described by means of a Liouville operator R acting on the density matrix $\rho(v,z,t)$, whose evolution is given by:

$$i\hbar(\frac{\partial}{\partial t} + v\frac{\partial}{\partial z})\rho = [H,\rho] + R\rho \tag{2}$$

where the Hamiltonian H includes the internal energy of the three-level atoms and the atom-laser coupling in the standard rotating wave approximation, and v is the atomic velocity along the z axis. The operator R contains the rate Λ of the incoherent pump process used to bring some population to the upper level of the probed transition. In EIT problems $\Lambda = 0$ and R contains only relaxation terms. For simplicity, we consider that there are only terms γ_{31} and γ_{23} describing the population decay rates of the middle and upper levels, respectively, and terms Γ_{ij} describing the dephasing rates of the ρ_{ij} off-diagonal elements. In the so-called radiative limit when there are no dephasing collisions one has, $\Gamma_{21} = \Gamma_{12} = (\Lambda + \gamma_{23})/2$; $\Gamma_{32} = \Gamma_{23} = (\gamma_{23} + \Lambda + \gamma_{31})/2$; and $\Gamma_{31} = \Gamma_{13} = (2\Lambda + \gamma_{31})/2$.

It is convenient to work in the interaction picture to eliminate the fast dependences in the time evolution equations. Thus, we define new density matrix elements σ_{ij} by $\rho_{23}(v,z,t) \equiv \sigma_{23}(v,z,t)\exp(-i\omega_d t)$, $\rho_{31}(v,z,t) \equiv \sigma_{31}(v,z,t)$ $\exp(-i\omega_p t)$, and $\rho_{21}(v,z,t) \equiv \sigma_{21}(v,z,t)\exp\left[-i(\omega_d + \omega_p)t\right]$. The resulting equations of motion are:

$$(\frac{\partial}{\partial t} + v\frac{\partial}{\partial z})\rho_{22} = -\gamma_{23}\rho_{22} + \left[i(\beta^+ e^{ik_d z} + \beta^- e^{-ik_d z})\sigma_{32} + c.c.\right] \tag{3a}$$

$$(\frac{\partial}{\partial t} + v\frac{\partial}{\partial z})\rho_{11} = -\Lambda d_{13} + \gamma_{31}\rho_{33} - (i\alpha e^{ik_p z}\sigma_{13} + c.c.) \tag{3b}$$

$$(\frac{\partial}{\partial t} + v\frac{\partial}{\partial z})\sigma_{23} = -R_{23}\sigma_{23} - id_{23}(\beta^+ e^{ik_d z} + \beta^- e^{-ik_d z}) - i\alpha e^{-ik_p z}\sigma_{21} \tag{3c}$$

$$(\frac{\partial}{\partial t} + v\frac{\partial}{\partial z})\sigma_{31} = -R_{31}\sigma_{31} + id_{13}\alpha e^{ik_p z} + i(\beta^+ e^{-ik_d z} + \beta^- e^{ik_d z})\sigma_{21} \tag{3d}$$

$$(\frac{\partial}{\partial t} + v\frac{\partial}{\partial z})\sigma_{21} = -R_{21}\sigma_{21} + i(\beta^+ e^{ik_d z} + \beta^- e^{-ik_d z})\sigma_{31} - i\alpha e^{ik_p z}\sigma_{23} \quad (3e)$$

Where $d_{13} = (\rho_{11} - \rho_{33})$, $d_{23} = (\rho_{22} - \rho_{33})$, $R_{23} = (\Gamma_{23} - i\Delta_d)$, $R_{31} = (\Gamma_{31} - i\Delta_p)$, and $R_{21} = [\Gamma_{21} - i(\Delta_p + \Delta_d)]$. Since we are considering a closed system, the normalization condition $\rho_{11}+\rho_{22}+\rho_{33} = 1$ must be added to Eqs. (3). The corresponding equations for the $V-$ scheme of Fig. 1b are obtained, apart from obvious changes in the incoherent pump and relaxation terms, by simply replacing Δ_p by $-\Delta_p$ in Eqs. (3).

Even in the steady-state, to which we restrict in the following, these equations do not have general analytical solution due to the simultaneous presence in the right-hand sides of the explicit z dependences $\exp(\pm ik_d z)$, obviously linked to the SW nature of the drive field. The solution must therefore be obtained numerically, in contrast with the well known case of a TW drive which can be treated analytically. In the case of weak probe and strong drive fields (i.e. $\alpha \ll \gamma_{31}$, $\beta^\pm \gg \gamma_{23}$) one first solves Eqs. (3) to zeroth-order in the probe field and to all orders in the drive field. This solution is well known from the theory of a high-intensity gas laser [22,23]: The density matrix elements are expanded in spatial Fourier components with a fundamental frequency k_d. One obtains a set of coupled difference equations in the Fourier coefficients c_n which may be solved subject to appropriate boundary conditions on c_n for large n. The solution can be expressed in terms of continued fractions [22,23]. This zeroth-order solution is then used to solve Eqs. (3) to first order in the probe field [24]. In some cases of practical interest it may be necessary to consider also a probe field of arbitrary intensity. The steady-state solution of Eqs. (3) for this case can be obtained introducing a two-index Fourier expansion for the density matrix elements with k_d and k_p fundamental spatial frequencies. However, this expansion can be more easily handled if we define, without loss of generality, two integers a and b such that a/b is a rational number and $k_d = ak$, $k_p = bk$, where k is a reduced wavenumber [25]. Then the steady-state solution of Eqs. (3) becomes the one-index expansion:

$$\rho_{jj} = \sum_{q=-\infty}^{+\infty} y_{jj}^q(v)\exp(iqkz); \quad \sigma_{jk} = \sum_{q=-\infty}^{+\infty} y_{jk}^q(v)\exp(iqkz) \quad (4)$$

with $j,k = 1,2,3$. The set of coupled difference equations in the Fourier coefficients y_{jj}^q and y_{jk}^q can be recast in a form identical to that derived in [22] for complex numbers, but involving now vectors and matrices of dimension 4 [25]. The solution can therefore be obtained in terms of continued fractions of 4x4 matrices [25]. The continued fractions are evaluated numerically by truncation to a finite number of terms in the denominator and it is a standard procedure in the calculations to check if the addition of one term in the denominator modifies the numerical value within the required accuracy. We have also checked that for $\alpha \ll \gamma_{31}$ the matrix continued-fraction solution reproduces the Feldman and Feld's solution [24].

The absorption and dispersion characteristics of the atomic medium with the $p-$ or ladder-configuration shown in Fig. 1a are derived in the usual way from the components of the polarization $P(z,t) = \int N(v)dv[\mu_{23}\rho_{32}(v,z,t) + \mu_{31}\rho_{13}(v,z,t)] + c.c.$, whose temporal and spatial oscillation matches the coupled field. Here $N(v)dv$ is the number of atoms per unit volume with velocity v and we take for $N(v)$ the usual Maxwellian distribution with a most probable velocity u. Thus, for instance, the intensity absorption coefficient for the probe is given by $a_p = \frac{2\hbar\omega_p\alpha}{I_p}\int dvN(v)\,\mathrm{Im}\,y_{31}^b(v)$, where $I_p = \varepsilon_0 cA_p^2/2$ is the probe intensity.

We will consider here the following drive field configurations: (i) a symmetric SW drive (i.e., $\beta \equiv \beta^+ = \beta^-$), (ii) a TW drive copropagating with the probe (i.e., $\beta \equiv \beta^+$ and $\beta^- = 0$), and (iii) a counterpropagating TW drive (i.e., $\beta \equiv \beta^-$ and $\beta^+ = 0$). In all cases the drive will be strong enough to saturate the driven transition, i.e., $\beta \gg \gamma_{23}$, although with a Rabi frequency smaller than the Doppler broadening $\Delta\omega_D$ of the driven transition, i.e., $\beta < \Delta\omega_D = 2\sqrt{\ln 2}\frac{u}{c}\omega_d$. Let us emphasize that Doppler broadening is one of the main difficulties in observing and using atomic coherence effects in atomic gases, in particular in the case of a large frequency up-conversion ratio $R = \omega_p/\omega_d$. It is also well known that a way to mitigate the negative influence of Doppler broadening is to use a high intensity TW drive with $\beta \geq \Delta\omega_D$, which usually requires pulsed lasers. In contrast, we here explore the opposite regime, i.e. $\beta < \Delta\omega_D$, but use a SW drive.

3 Dressed-Atom Analysis

The dressed-atom analysis [26] is very useful for the physical interpretation of numerical results obtained by solving the density matrix Eqs. (3). Consider again the ladder scheme of Fig. 1a, although generalization to any other three-level configuration is straightforward. We use a quantum description of the laser fields and denote by $|i;N^+,N^-;n_p\rangle$ the states of the uncoupled atom+laser fields, with the atom in state $i = 1,2,3$, in the presence of $n_p$probe photons and N^+(N^-) photons of the TW drive component copropagating (counterpropagating) with the probe field. For a TW drive it is either $N^+ = 0$ or $N^- = 0$. In the standard rotating-wave approximation, and considering only processes involving one probe photon, due to the weakness of the probe field, one has the following infinite chain of couplings between the unperturbed states:

$$\begin{array}{c} \ldots \overset{\beta^+}{\leftrightarrow} |2;N^+,N^- - 1;n_p\rangle \overset{\beta^-}{\leftrightarrow} |3;N^+,N^-;n_p\rangle \overset{\beta^+}{\leftrightarrow} |2;N^+ - 1,N^-;n_p\rangle \overset{\beta^-}{\leftrightarrow} \ldots \\ \updownarrow \alpha \\ |1;N^+,N^-;n_p+1\rangle \end{array} \quad (5)$$

Where we have indicated the matrix elements of the atom-laser interaction Hamiltonian between the levels connected by the arrows, in the semiclassical approximation. In this approximation, the choice of N^+, N^- and n_p is unrelevant, therefore we introduce the following shortened label:

$$\left|3; N^{+} - m, N^{-} + m; n_p\right\rangle \equiv |3; q = 2m, 0\rangle \tag{6a}$$

$$\left|2; N^{+} - m, N^{-} + m - 1; n_p\right\rangle \equiv |2; q = 2m - 1, 0\rangle \tag{6b}$$

$$\left|1; N^{+}, N^{-}; n_p + 1\right\rangle \equiv |1; 0, 1\rangle \tag{6c}$$

with $m = 0, \pm 1, \pm 2, \ldots$ If the energy of the state $|3; 0, 0\rangle$ is taken as the origin, the energies of the above unperturbed states become (with $\hbar = 1$):

$$E^{u}_{3;q} = q k_d v, \qquad q = \text{even} \tag{7a}$$

$$E^{u}_{2;q} = -\Delta_d + q k_d v, \qquad q = \text{odd} \tag{7b}$$

$$E^{u}_{probe} = \Delta_p - k_p v \tag{7c}$$

that represented as a function of $k_d v$ (see Fig. 3) appear as two manifolds of straight lines, one of them formed by lines with even slope passing through the origin, and the other by lines with odd slope passing through the point with coordinates $(0, -\Delta_d)$. The probe level is represented by a straight line with ordinate at the origin Δ_p and slope $-\frac{k_p}{k_d} = -R$. Notice that the crossing points between the probe level and the other levels determine the (unperturbed) resonant velocities at which various multiphoton resonances occur. Thus, the crossings of E^{u}_{probe} with the levels of slopes $q = 0, 1, 2,$ and 3 determine the resonant velocities for the four processes shown in Fig. 2.

For a TW drive copropagating with the probe, only the following three states are involved: $|3; q = 0, 0\rangle, |2; q = 1, 0\rangle, |1; 0, 1\rangle$. If the fields are counterpropagating the levels involved are: $|3; q = 0, 0\rangle, |2; q = -1, 0\rangle, |1; 0, 1\rangle$.

The laser-atom coupling can be considered in two steps owing to the weak-probe assumption. First one considers the coupling with the strong SW drive field. By diagonalizing the infinite Hamiltonian associated to the first row in Eq. (5) one obtains the perturbed energies, E_n, and eigenstates, $|x_n\rangle$, of the so-called "SW-dressed two-level atom" [27]. The $|x_n\rangle$ states can be written in the form $|x_n\rangle = \sum_{q'=-\infty}^{+\infty} C_{q',n} \left|j; q', 0\right\rangle$, with $j = 3$ (2) for q' even (odd). These dressed states are labelled by the integer $-\infty < n < \infty$, whose choice is completely arbitrary. The coupling with the probe field is treated as a perturbation on the basis formed by the $|x_n\rangle$ states and the probe state $|1; 0, 1\rangle$ and its coupling elements depend on the $|3; 0, 0\rangle$ term in the expansion of $|x_n\rangle$. Within the weak probe assumption the influence of the coupling with the probe on the energies E_n and E^{u}_{probe} is negligible. This coupling produces a transition rate $P_{|1;0,1\rangle \to |x_n\rangle} \propto \alpha^2 \left|C_{0,n}\right|^2$, with the probe resonance occurring at the velocity where E_n and E^{u}_{probe} are degenerate.

An exact solution of the SW-dressed two-level atom problem can be obtained analytically when the drive is on resonance, while numerical treatment is required off resonance [27]. Figure 3 shows typical energy diagrams for these

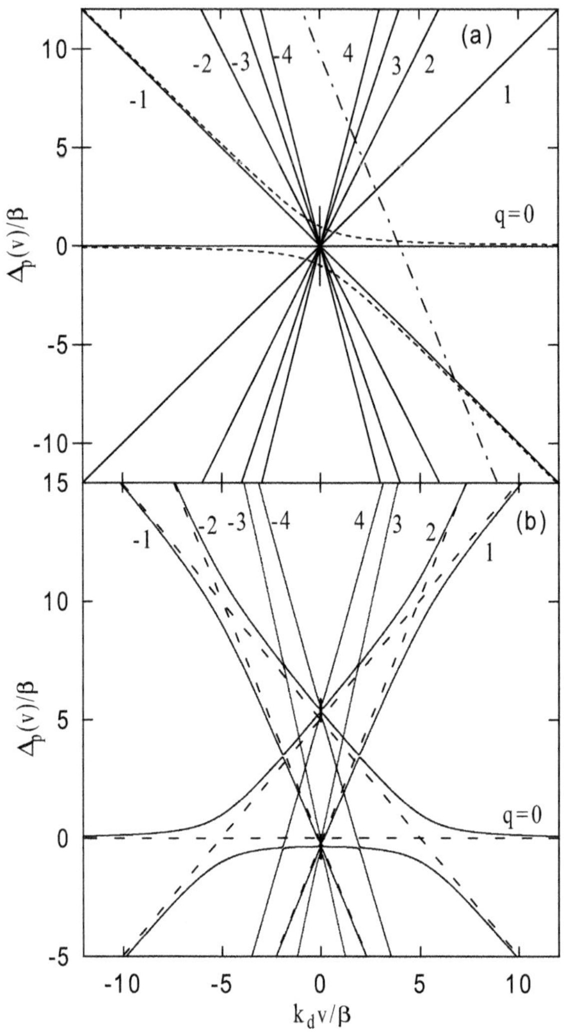

Fig. 3. Energy diagram of the SW-dressed atom (solid lines) for **(a)** $\Delta_d = 0$ and **(b)** $\Delta_d = -5\beta$. Dashed lines represent unperturbed energies. Dotted lines in (a) correspond to the case of a counterpropagating TW drive. The dotted-dashed line in (a) represents the probe level for $R = 2$, $\Delta_p = 10\beta$.

two cases [27]. For the resonant case (Fig. 3a), at $v \neq 0$ the energy levels are *identical* to the unperturbed ones, however the associated eigenstates are linear combinations of the unperturbed states. At $v = 0$, the energies cover a finite interval between -2β and 2β due to the spatially inhomogeneous ac-Stark splitting suffered by the stationary atoms. The level anticrossings (see Fig. 3b) centered at $k_d v \sim \pm\Delta_d$, $\pm\Delta_d/3$, ... and the two vertical intervals, at $v = 0$, manifest the ac-Stark effect induced by a detuned SW drive. Notice that in the resonant SW drive case (Fig. 3a), for $\Delta_p = 0$ and $R = \frac{k_p}{k_d} = 1, 2, 3, ...$ there is a complete superposition between the probe level and a dressed-state level, i.e. a complete cancelation of the Doppler effect, since all the atomic velocity classes contribute simultaneously to the probe resonance. For the detuned SW case (Fig. 3b), due to the presence of anticrossings, only a partial cancelation of Doppler effect can occur, for appropriate Δ_p values.

The TW drive case is much simpler. One obtains the two perturbed energies, E_n, and eigenstates, $|x_n\rangle$, by diagonalizing the 2×2 matrix corresponding to the coupling of $|3; q = 0, 0\rangle$ and $|2; q = 1, 0\rangle$ (or of $|3; q = 0, 0\rangle$ and $|2; q = -1, 0\rangle$). The two E_n are the branches of a hyperbola whose asymptotes are the unperturbed energy levels (see dotted lines in Fig. 3a).

On the other hand, for the case of a TW drive, it has been shown [28] that AWI in closed V and Λ schemes is due to two-photon (Raman) processes ($|q| = 1$ in our notation) while one-photon processes contribute to probe absorption. On the contrary, one-photon processes ($q = 0$ in our notation) are responsible for AWI in closed cascade schemes while two-photon processes contribute to probe absorption. In the SW case, one expects therefore, that the multi-photon processes shown in Fig. 2 with $|q|$ odd (even) will contribute to AWI in V and Λ schemes (cascade schemes). In the resonant drive case (Fig. 3a) to enhance the contribution of multi-photon processes with $|q|$ even, which are those contributing to probe amplification in the cascade case, one has to select $\Delta_p = 0$ and $\frac{k_p}{k_d} = |q|$ even, but since all the energy levels pass through the origin one also gets the contribution to probe absorption of the $v = 0$ atoms via the multi-photon processes with $|q|$ odd. For $\Delta_d \neq 0$ lines with q even pass also close to the origin and those with q odd are displaced vertically. Therefore, in the detuned SW case, by simply changing the nominal probe detuning Δ_p, one can enhance the contribution of either multi-photon processes with q odd ($\Delta_p \approx -\Delta_d$) or with q even ($\Delta_p \approx 0$).

To end this Section let us remark that here we have considered only probe resonances involving the drive-dressed levels, however, it is well known [28] that in AWI problems these resonances always contribute to probe absorption, while quantum-interference-induced probe gain appears either between two absorption dressed-state resonances or at the sidebands of these absorbing resonances.

4 Electromagnetically Induced Transparency

For the numerical results shown in this Section, we have considered parameters for a real cascade system (Fig. 1a) in atomic barium [29] with a resonant SW drive of $\beta = 350$ MHz coupling the upper transition at $\lambda_d = 821$ nm and a weak field with $\alpha = 0.035$ MHz probing the lower transition at $\lambda_p = 554$ nm. Note that the probe to drive frequency ratio is in this case $R = 1.48$ which, as we shall see, makes this configuration very convenient to observe EIT with a SW drive. The spontaneous emission rates of lower and upper transitions read, respectively, $\gamma_{31} = 19$ MHz and $\gamma_{23} = 3.5$ MHz. Here we set $\Lambda = 0$. For a

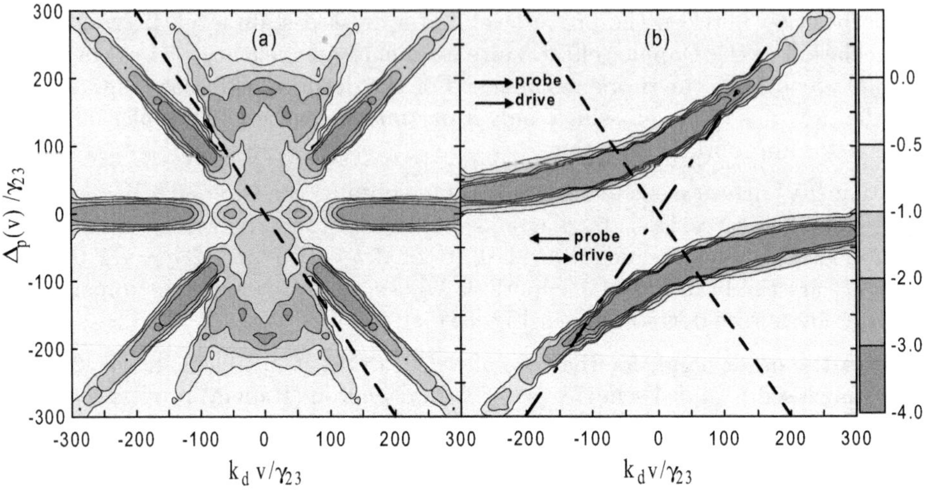

Fig. 4. Probe absorption contour map in the parameter plane ($k_d v$, $\Delta_p(v)$) for **(a)** resonant SW drive, **(b)** resonant TW drive. For parameters and discussion see text.

Maxwellian atomic velocity distribution corresponding to barium atoms in a vapor cell at $T = 500$ K one has $k_d u = 930$ MHz. For a resonant drive field ($\Delta_d = 0$), we plot in Fig. 4 a contour map of $N(v)\,\mathrm{Im}\, y_{31}^b(v)$ in the parameter plane $\Delta_p(v) = \Delta_p - k_p v = \Delta_p - R k_d v$ versus $k_d v$ for (a) a SW drive ($\beta^+ = \beta^- = 350$ MHz) and (b) a copropagating TW drive ($\beta^+ = 350$ MHz , $\beta^- = 0$). Regions with dark gray (light gray) correspond to high (low) probe absorption. A comparison of Figs. 4a and 4b reveals that the absorption characteristics of the atomic medium are dramatically different for SW and TW driving. While for a TW drive (see Fig. 4b) the regions of high probe absorption occur close to the two branches of a hyperbola whose asymptotes have slopes 0 and 1, for a SW drive high probe absorption occurs about straight lines passing through the origin with slopes 0, ± 1, ± 2,... In Fig. 4a the regions with slopes 0 and ± 1 are clearly visible, those with slopes ± 2 can be appreciated clearly only far from the origin. Regions with higher slopes overlap and cannot be seen individually. These results clearly reflect the atom's dressed-level structure discussed above. To show this structure more clearly we have assumed in Fig. 4 a flat velocity distribution (i.e., $k_d u \to \infty$). The probe absorption domains in Fig. 4 are due to probe resonances involving state $|1;0,1\rangle$ and the dressed-states built by the drive field. The discussion in Section 3 then clarifies the physical origin of the probe absorption shown in Fig. 4a and shows that the main features in this figure are generic, i.e., with a resonant SW drive, probe absorption will *always* occur about straight lines through the origin, with integer slope, in diagrams of the type shown in Fig. 4a.

One can already "read" in Fig. 4 the advantages of using resonant SW drive instead of TW drive when $R > 1$. To obtain the probe absorption coefficient a_p

one must perform an integration over the velocity distribution $N(v)$ in order to obtain the contributions of all the atoms. If also $\Delta_p = 0$, the points representing different atomic velocity classes in the diagrams of Fig. 4 lie on a straight (dashed) line through the origin with slope -R (or R in the case of counterpropagating probe and TW drive). To get a_p one must add the contributions to probe absorption found along this dashed line. In Fig. 4b, with $R > 1$ there is always partial overlap between the dashed lines and the two absorbing regions. (Notice, however, that with counterpropagating drive and probe fields there is no absorption, i.e., complete transparency, for $0 < R < 1$, which corresponds to frequency down-conversion). For the SW drive case, we see in Fig. 4a that one gets small absorption not only when $0 < R < 1$ but also for $R \simeq n + 1/2$ with $n = 1, 2, 3...$. These features of the probe absorption coefficient are clearly seen in Fig. 5, where we plot the probe absorption coefficient normalized to its value in the absence of the drive field, $a_p/a_p(\beta = 0)$, as a function of R (and other parameters as in Fig. 4) for both resonant SW drive and TW drive. Although this figure is not very interesting from a physical point of view, since one cannot explore in practice all the R values with the same set of parameters (selected here for a cascade system in Barium atoms with $R = 1.49$), we plot it because it clearly exhibits the advantages of using resonant SW drive instead of TW drive in the frequency up conversion regime, i.e., when $R > 1$. For SW drive we observe strong absorption resonances at integer values of the probe to drive frequency ratio, i.e., at $R = 1, 2, 3....$ Between these resonances there are probe windows of electromagnetically induced transparency at $R \simeq n + 1/2$ ($n = 1, 2, 3...$) whose depth decreases as R increases. It is also appreciated that up to $R = 1$ a counterpropagating TW drive is the most convenient configuration in order to produce probe transparency. Nevertheless, for $R > 1$ such that $R \simeq n + 1/2$ ($n = 1, 2, 3...$), the resonant SW drive configuration is the most advantageous configuration to observe EIT or even the unique one in which it occurs, e.g., in Fig. 5 there is almost no EIT with a TW drive for $R \geq 2$ while there are appreciable EIT windows with a SW drive up to $R = 9.5$.

Again for the parameters of the real cascade system in atomic barium we have obtained the probe spectra shown in Fig. 6a for both SW and TW drive. Clearly, the maximum induced probe transparency is observed for a SW drive configuration and, as a characteristic feature of this case, three different peaks of transparency appear. In particular, at probe field line center, i.e., $\Delta_p = 0$, a_p reduces to about $0.1a_p\,(\beta = 0)$ (i.e., it reduces by $\sim 90\%$).

Finally, we have considered the probe propagation through a vapor cell of length $L = 4$ cm with an atomic density of $N = 10^{15}$ atoms/cm^3. Assuming an electric dipole moment for the probed transition of $\mu \sim 10^{-27}$ C·cm and neglecting drive field depletion through the vapor cell, we have obtained the probe transmission spectrum shown in Fig. 6b. Probe transmission at line center increases from the value $\exp(-10.6)$ in the absence of the drive field to a value above 50% with the SW drive, but remains almost negligible with a co- or a counter-propagating TW drive field. To obtain the same resonant probe transmission peak with a counterpropagating TW drive one should use $\beta \simeq 1680$

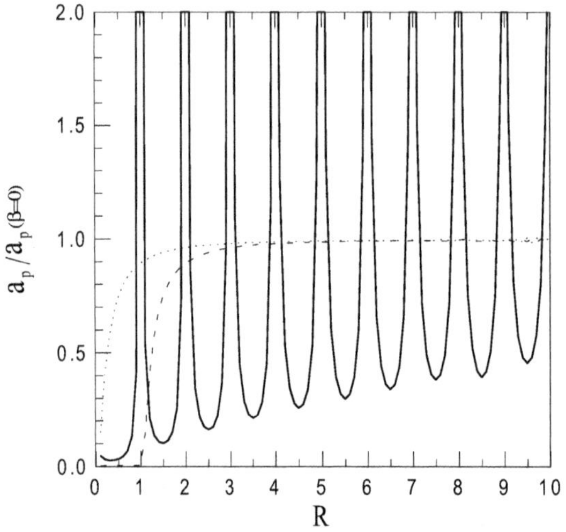

Fig. 5

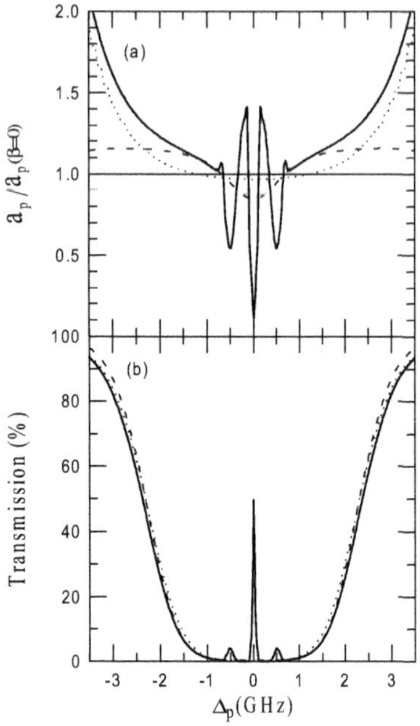

Fig. 6

Fig. 5. Normalized probe absorption coefficient as a function of the probe to drive frequency ratio R for $\Delta_p = \Delta_d = 0$, and other parameters as in Fig. 4. Three different cases are represented: (i) SW drive (solid line), (ii) copropagating TW drive (dotted line), and (iii) counterpropagating TW drive (dashed line).

Fig. 6. (a) Normalized probe absorption coefficient as a function of the probe detuning Δ_p for (i) SW drive (solid line), (ii) copropagating TW drive (dotted line), and (iii) counterpropagating TW drive (dashed line). **(b)** Corresponding probe transmission through a vapor cell of barium atoms of length $L = 4$ cm and atomic density of $N = 10^{15}$ atoms/cm^3. Other parameters as in Fig. 4.

MHz, which means an intensity 23 times that of one of the TW components of the SW drive used here. We have computed also transmission spectra for increasing probe strength. The peak at $\Delta_p = 0$ does not change appreciably for α up to $\alpha = 3.5$ MHz $= 0.18\gamma_{31}$. For $\alpha \sim 2\gamma_{31}$ there is still a resonant peak above 50% but differences between the cases of TW and SW drive are less marked.

5 Amplification Without Inversion

To investigate whether it is advantageous to use a SW drive instead of a TW one for getting AWI in the frequency up-conversion regime, we consider next $V-$ and $p-$ schemes (Fig. 1). We do not refer here to real atomic configurations, instead, for simplicity, we take identical parameters for both schemes, except that the

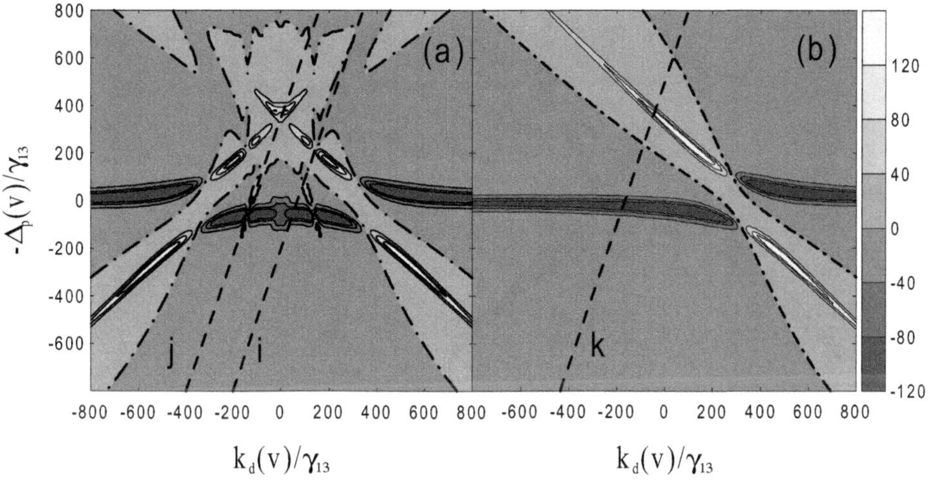

Fig. 7. Contour map of the probe response in the parameter plane ($k_d v$,$\Delta_p(v)$) for the V-scheme for detuned **(a)** SW drive, **(b)** TW drive. $\beta = 100\gamma_{13}$, $\Delta_d = -300\gamma_{13}$. The dashed-dotted lines separate the absorption (dark gray) regions from the amplification (light gray) ones. Dashed lines with slope $R = 3$ represent $-\Delta_p(v) = -\Delta_p + Rk_d v$ for nominal detunings Δ_p corresponding to several features in Fig. 9a with the same labels.

ratio between probe and drive frequencies is $R = \frac{k_p}{k_d} = 3$ for the V-scheme and $R = 4$ for the ladder scheme. Specifically, we take $\gamma_{13} = \gamma_{31}$, $\gamma_{23} = 5\gamma_{31}$, $\Lambda = 20$ $\gamma_{13}, \beta \equiv \beta^+ = \beta^- = 100$ γ_{13} , $\alpha = 0.01\gamma_{13}$ and $k_d u = 300$ γ_{13}. We have studied first these two cases with a resonant SW drive and found that one does not get gains larger than with a TW drive [19]. This is due to probe absorption at $\Delta_p = 0$ by $v = 0$ atoms placed at the nodes of the SW drive, where they are unperturbed.

Fig. 7 shows, for the $V-$ scheme driven by a detuned field with $\Delta_d = -300$ γ_{13}, a contour map of $N(v)\,\mathrm{Im}\, y^b_{13}(v)$ in the parameter plane $-\Delta_p(v) = -\Delta_p + k_p v = -\Delta_p + Rk_d v$ versus $k_d v$ for (a) a SW drive and (b) a counterpropagating TW drive ($\beta^- = 100\gamma_{13}$, $\beta^+ = 0$). The corresponding results for the $p-$ scheme with $\Delta_d = -200$ γ_{31} and $\Delta_p(v) = \Delta_p - Rk_d v$ are shown in Fig. 8. As for Fig. 4, we take in Figs. 7 and 8 a flat velocity distribution. We again see that the SW and TW results are dramatically different. Inversionless amplification with TW drive and probe fields has been analyzed in detail previously (for instance, see [30,31,32]). Well known features that can be appreciated in Figs. 7b and 8b are the following:

(i) On resonance, in the V -scheme (i.e., at $k_d v = -\Delta_d, \Delta_p(v) = 0$), since $\gamma_{23} > \gamma_{13}$ and $\Lambda > \gamma_{13}^2/(\gamma_{23} - \gamma_{13})$, there is a symmetrical gain domain about $\Delta_p(v) = 0$, between the two absorbing dressed-states resonances. On increasing drive detuning $|\Delta_d(v)| = |-\Delta_d - k_d v|$, the domain of gain becomes asymmetric, its width increases and its maximum appears close to the two-photon (Raman)

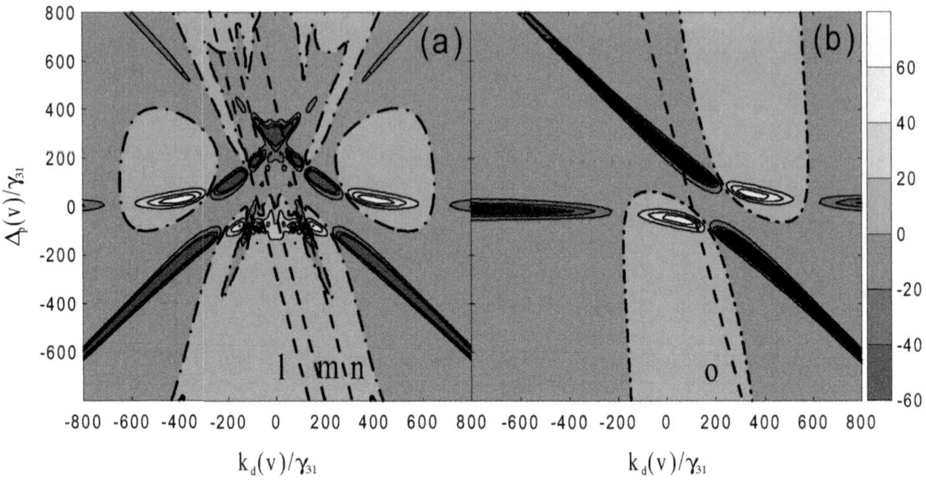

Fig. 8. The same as in Fig. 7 but for the p-scheme with $\beta = 100\gamma_{31}$, $\Delta_d = -200\gamma_{31}$, and $R = 4$. Dashed lines represent now $\Delta_p(v) = \Delta_p - Rk_dv$.

resonance (thus, the AWI domain in Fig. 7b has a global slope -1); this gain peak first increases until it reaches a maximum and then it decreases with further increase of drive detuning. At small drive detuning this gain appears without population inversion either at the one photon transition ($\rho_{11} < \rho_{33}$) or at the two photon transition ($\rho_{11} < \rho_{22}$), while there is Raman inversion ($\rho_{11} > \rho_{22}$) at large detunings.

(ii) In the cascade scheme (Fig. 8b) there is no on-resonance gain: For $k_dv = -\Delta_d$ there are symmetrical domains of gain (for $|\Delta_p(v)| \geq \beta$) at either side of the resonance region determined by the two absorbing dressed state resonances. With increasing drive detuning $\Delta_d(v)$ the domains of gain become asymmetric and the peak with larger (smaller) gain increases (decreases) and moves towards (away from) resonance. Gain takes place in this configuration without population inversion at the one- or at the two-photon transition ($\rho_{11} < \rho_{33}$, ρ_{22}).

With a SW drive, as expected, gain appears in the V-scheme (see Fig. 7a) associated with dressed levels with odd slope: large domains of gain with slope $q = \pm 1$ are clearly visible; smaller gain domains with $q = \pm 3$ can also be appreciated at both sides of $k_dv = 0$. Probe absorption is associated with even slope dressed levels as can be clearly seen for $q = 0, \pm 2$ in Fig. 7a. In contrast, the reverse is true for the cascade scheme shown in Fig. 8a: domains of gain with slopes $q = 0, \pm 2, \pm 4$ are clearly visible as well as domains of probe absorption with slopes $q = \pm 1, \pm 3$. These results clearly suggest that the folded schemes (V and Λ) are appropriate for frequency up-conversion AWI with $\frac{k_p}{k_d} = 3, 5, \ldots$, while cascade schemes could be useful with $\frac{k_p}{k_d} = 2, 4, \ldots$.

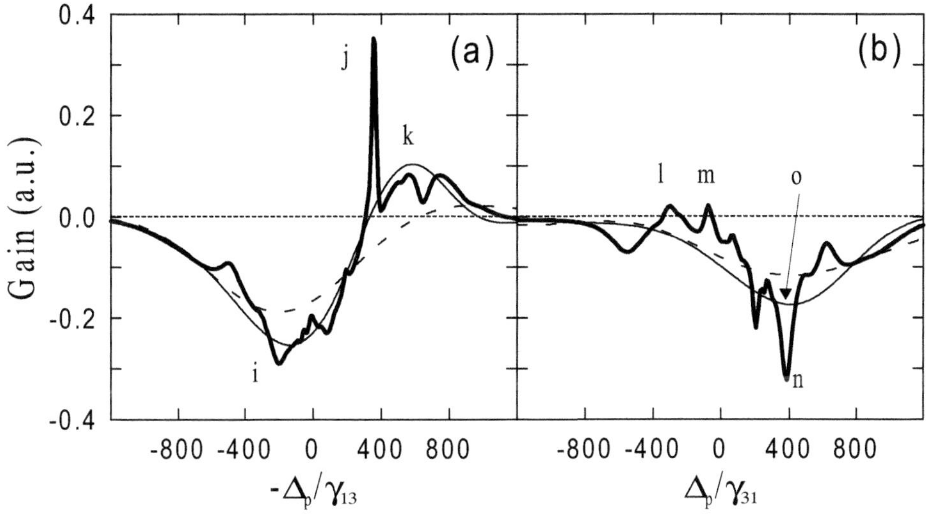

Fig. 9. (a) Doppler integrated probe gain for the parameter values corresponding to the V-scheme of figure 7 with $R = 3$ and $k_d u = 300\gamma_{13}$. The wide solid line corresponds to the SW drive case while the light solid and the dashed lines correspond to the copropagating ($\beta^+ = 100\gamma_{13}, \beta^- = 0$) and counterpropagating ($\beta^+ = 0, \beta^- = 100\gamma_{13}$) TW drive cases, respectively. **(b)** The same for the p-scheme with the parameter values of Fig. 8 with $R = 4$ and $k_d u = 300\gamma_{31}$.

On Fig. 7 (Fig. 8) the dashed lines with slope 3 (-4) represent $-\Delta_p(v) = \Delta_p + Rk_d v$ ($\Delta_p(v) = \Delta_p - Rk_d v$) lines for the nominal Δ_p values corresponding to the most salient features which appear on the velocity-integrated probe gain/absorption spectra shown in Fig. 9 for V-scheme with $\Delta_d = -300\gamma_{13}$ and $\frac{k_p}{k_d} = 3$ (a), and for the cascade-scheme with $\Delta_d = -200\gamma_{31}$ and $\frac{k_p}{k_d} = 4$ (b). The solid line represents the SW drive case, the thin and dashed lines refer to a TW drive copropagating or counterpropagating with the probe beam, respectively. In Fig. 9a the maximum gain in the SW case is 3.4 times larger than that in the TW case. In both cases gain occurs with Raman inversion ($\rho_{11} > \rho_{22}$). Notice in Fig. 9b that AWI (without inversion either at the one-photon or two-photon transitions) only occurs in the SW case. All the features in these spectra can be easily explained by means of the corresponding $\mp\Delta_p(v)$ straight lines in Figs. 7 and 8, as we have done for the peaks and dips numbered in Fig. 9.

In order to show that the advantages of the SW driving over the TW driving do not restrict to a limited domain of parameters, we plot in Fig. 10 the contour map of the ratio between the maximum gain of the SW and TW cases in the parameter plane (β, Δ_d) for a $V-$ scheme with $\frac{k_p}{k_d} = 3$. Clearly, for small drive detunings (below the dashed line) the TW drive configuration is favorable, as already pointed out at the beginning of this Section, while for large enough detunings (above the dashed line) the SW drive case is considerably better.

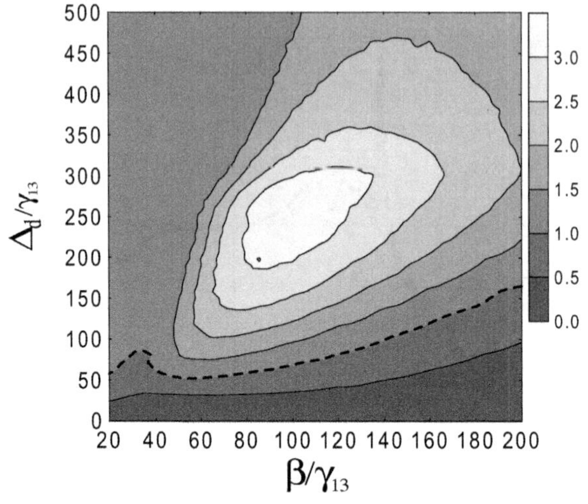

Fig. 10. Contour map of the ratio between the maximum gain in the probe gain spectrum for the SW and TW cases in the parameter plane (β,Δ_d) for the V-scheme with other parameters as in Fig. 7.

Conclusions

In this paper we have analyzed atomic coherence effects for a probe TW field in closed Doppler-broadened three-level systems driven by a SW field. Electromagnetically induced probe transparency windows appear close to half integer values of the probe to drive field frequency ratio R. These results have been compared to those obtained for a TW drive proving that for $R > 1$ a SW drive field is appreciably more efficient than a TW drive in inducing probe transparency. In particular, we have considered parameters for a real closed cascade scheme in barium atoms with $R \simeq 1.5$ showing probe transmission resonances with values above 50% when the transmission is almost negligible either without drive field or with only one of the TW components of the SW drive.

We have also investigated here amplification without inversion in closed $V-$ and $p-$ schemes, although similar results to those of the V-scheme ($p-$ scheme) can be obtained for the $\Lambda-$scheme ($h-$scheme or ladder system driven on the lower transition). Our results are compared to those of the well-known case of a travelling-wave drive. A distinguishing feature of the SW case is the contribution to probe gain of multi-photon processes involving one probe-photon and several pump-photons, with alternating absorption and emission from the two counterpropagating pump components. Depending on the type of three-level system (folded or cascade), multi-photon processes involving an even (odd) number of drive photons can contribute to probe amplification (absorption) or viceversa. These processes are only weakly affected by Doppler broadening for probe to drive frequency ratios R equal to integer numbers. Examples of inversionless amplification with frequency up-conversion ratios $R = 3$ and 4 have been presented. Detailed contour maps of the probe field absorption/amplification in the parameter plane ($k_d v$, $\Delta_p(v)$) have been presented and discussed with the help of the dressed-atom approach.

Acknowledgments

Support from the DGESIC (Spanish Government) under contract PB98-0935-C03 and from the DGR (Catalan Government) under contract 1999SGR00096 is acknowledged.

References

1. M. O. Scully and M. S. Zubairy, *Quantum Optics*, (Cambridge Univ. Press., 1997) Chap. 7, pp. 220-245.
2. E. Arimondo, *Coherent population trapping in laser spectroscopy*, in Progress in Optics XXXV, Ed. E. Wolf (Elsevier, Amsterdam 1996) Chap. V, pp. 258-354.
3. S. E. Harris, *Electromagnetically Induced Transparency*, Phys. Today **50** (7), 36 (1997).
4. J. Mompart and R. Corbalán, *Lasing without inversion,* J. Opt. B: Quantum Semiclass. Opt. **2**, R7-R24 (2000).
5. M.O. Scully et al. Phys. Rev. Lett. **67**, 1855 (1991); A.S. Zibrov et al. Phys. Rev. Lett. **76**, 3935 (1996).
6. L.V. Hau et al. Nature, **397**, 594 (1999); M.M. Kash et al. Phys. Rev. Lett. **82**, 5229 (1999).
7. L.J. Wang et al. Nature, **406**, 277 (2000).
8. K. Hakuta et al. Phys. Rev. Lett. **66,** 596 (1991); G. Z. Zhang et al. Phys. Rev. Lett. **71,** 309 (1993); Y. Li and M. Xiao, Opt. Lett. **21**, 1064 (1996).
9. G. Alzetta et al. Lett. Nuovo Cimento **17**, 333 (1976).
10. H. R. Gray et al. Opt. Lett. **3**, 218 (1978).
11. K.-J. Boller et al. Phys. Rev. Lett. **66**, 2593 (1991); J. E. Field et al. Phys. Rev. Lett. 67, 3062 (1991).
12. K. Hakuta et al. Phys. Rev. Lett. **66**, 596 (1991).
13. M. Xiao et al. Phys. Rev. Lett. **74**, 666 (1995).
14. J. R. Boon et al. Phys. Rev. A **57**, 1323 (1998).
15. A. Nottelmann et al. Phys. Rev. Lett. **70**, 1783 (1993); E. S. Fry et al. Phys. Rev. Lett. **70**, 3235 (1993); W. E. van der Veer et al. Phys. Rev. Lett. **70**, 3242 (1993). Y. Zhu and J. Lin, Phys. Rev. A, **53**, 1767 (1996); Y. Zhu et al. Optics Comm. **128**, 254 (1996); J. A. Kleinfeld, and A. D. Streater, Phys. Rev. A **53**, 1836 (1996); P. B. Sellin et al. Phys. Rev. A **54**, 2402 (1996); C. Fort et al. Optics Comm. **139**, 31 (1997); J. Kitching and L. Hollberg, Phys. Rev. A **59**, 4685 (1999).
16. A. S. Zibrov et al. Phys. Rev. Lett. **75**, 1499 (1995); G. G. Padmabandu et al. Phys. Rev. Lett. **76**, 2053 (1996); F. B. de Jong et al. Phys. Rev. A **57**, 4869 (1998).

17. M. D. Lukin et al. Laser Physics **6**, 436 (1996). S. F. Yelin et al. Phys. Rev. A **57**, 3858 (1998); J. Mompart et al. Opt. Commun. **147**, 299 (1998); J. R. Boon et al. Phys. Rev. A **58**, 2560 (1998); V. Ahufinger et al. Phys. Rev. A **60**, 614 (1999); V. Ahufinger et al. Phys. Rev. A **61**, 053814 (2000).
18. F. Silva et al. Europhys. Lett. **51**, 286 (2000).
19. F. Silva et al. Laser Phys. **9**, 858 (1999).
20. F. Silva et al. Opt. Commun. **114**, 519 (1995).
21. R. Corbalán et al. Opt. Commun. **133**, 225 (1997).
22. See, e.g., S. Stenholm and W. E. Lamb, Jr., Phys. Rev. **181**, 618 (1969).
23. B. J. Feldman and M. S. Feld, Phys. Rev. A**1**, 1375 (1970).
24. See, e.g., B. J. Feldman and M. S. Feld, Phys. Rev. A**5**, 899 (1972).
25. R. Vilaseca et al. Appl. Phys. B **34**, 73 (1984).
26. C. Cohen-Tannoudji et al. *Atom-Photon Interactions* (John Wiley & Sons, 1992).
27. L. Roso et al., Appl. Phys. B **31**, 115 (1983).
28. J. Mompart, and R. Corbalán, Optics Comm. **156**, 133 (1998).
29. P. B. Sellin et al. Phys. Rev. A **54**, 2402 (1996).
30. J. Mompart et al. Opt. Commun. **147**, 299 (1998)
31. J. Mompart and R. Corbalán, in "3rd Iberoamerican Optics Meeting", Ed. A. M. Guzmán, SPIE, Vol. 3572, 33 (1999)
32. J. Mompart et al. Laser Phys. **9**, 844 (1999).

Frequency Up-Conversion to the Vacuum Ultra-Violet in Coherently Prepared Media

J.P. Marangos, I. Kuçukkara, and M. Anscombe

Laser Optics and Spectroscopy Group
Blackett Laboratory
Imperial College of Science, Technology and Medicine, London, SW7 2BW

Abstract. In this contribution we discuss recent work that has demonstrated that electromagnetically induced transparency (EIT) can greatly increase the intensity of coherent vacuum ultra-violet (VUV) radiation (at wavelengths shorter than 130nm) generated in resonant four-wave mixing schemes. We will start by introducing the basic concept of EIT enhanced resonant four-wave mixing, comparing it briefly with other coherently enhanced non-linear frequency up-conversion schemes. Experiments proving this concept in atomic hydrogen in the mid-1990's will be summarised. Our own recent work on EIT enhanced four-wave mixing in Kr will be presented and recent results demonstrating high conversion efficiencies into the VUV, and the limits placed on this process at high density-length products, will be discussed.

1 Introduction

An important objective of laser and non-linear optics has been to extend the range of useful sources of coherent radiation into the long (infra-red) and short (ultra-violet and vacuum ultraviolet) wavelength ranges. This work is motivated by the continuing impact of coherent radiation throughout the sciences in providing new tools to study and control physical, chemical and biological processes. Producing bright sources of coherent radiation into the VUV and XUV has proved one of the greatest challenges. Extensive work on X-ray/XUV lasers has been carried out [1], as has work using non-linear optical frequency up-conversion techniques to produce short wavelength coherent radiation [2] [3]. X-ray and XUV lasers face formidable challenges due to the requirement for optics effective at these short wavelengths required to fabricate the laser resonator and because of the difficulties of obtaining inversion on short wavelength transitions. Non-linear optical frequency up-conversion offers, in contrast, some important adavantages since the coherence, spatial and temporal properties of the generated short wavelength field are directly tied to the properties of the visible laser systems from which they are driven. Also many such schemes are inherently tunable.

All frequency up-conversion schemes to the VUV/XUV have, however, been found to be limited to very modest energy conversion efficiencies from the laser fields that drive them. Typically conversion efficiencies are restricted to 10^{-6}. The reason for this is that the only media transparent to radiation below 150nm

M. Orszag, J.C. Retamal (Eds.): LNP 575, pp. 195–211, 2001.

are atomic and molecular gases. In contrast at longer wavelengths optical crystals can be used for frequency mixing, [4] these have; (a) high non-linearities, (b) transparency to the applied and generated fields, (c) the property of birefringence permits almost perfect phase-matching to be achieved. As a consequence SHG and SFM in crystal can often be carried out with conversion efficiencies approaching unity. Vapours and gases are transparent to non-resonant radiation but the available non-linearites are small. On resonance, at finite density, the non-linearity will be high but the medium is then very opaque. Moreover there are no simple techniques to give near perfect phase-matching. Attempts to improve phase-matching by introducing a buffer gas have met with some success with conversion efficiencies of 10^{-3} being reported [5], but this is still far lower than we would wish.

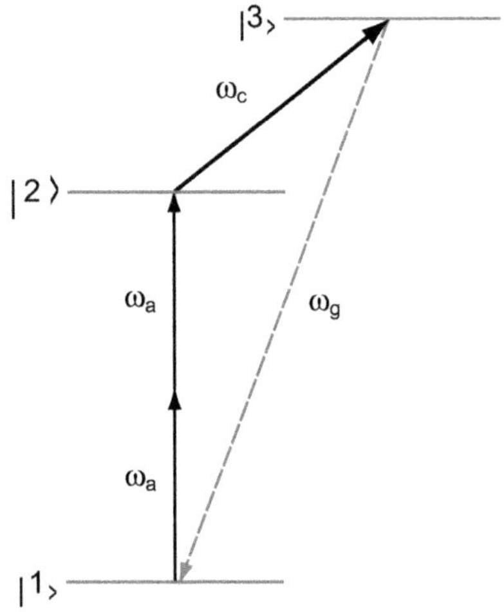

Fig. 1. Electromagnetically induced transparency scheme

A potential solution to these difficulties was proposed by Harris in 1990 [6]. Here a scheme with resonant four-wave mixing was presented that results in a large non-linear susceptibility(Figure 1). In this proposal one of the applied fields was single-photon resonant with an atomic transition between a pair of excited states of the atom. One of the excited state ($|3\rangle$) is dipole coupled to the ground-state ($|1\rangle$), and it is on this transition that the generated field is produced. The other excited state ($|2\rangle$) was not connected by a dipole allowed transition with the ground-state and instead was coupled by a two-photon transition to $|1\rangle$ from a second applied laser field. Under these conditions the first laser field (the coupling field) creates electromagnetically induced transparency, i.e. the "dressed" linear susceptibility vanishes due to destructive interference between the amplitudes associated with the various possible absorption pathways [7]. As a result of this quantum mechanical interference the absorption and refractive index acquire essentially their vacuum values. In contrast in the dressed system the amplitudes leading to the "dressed" non-linear susceptibility undergo constructive interference. As a consequence the non-linear susceptibility remains resonantly enhanced whilst the linear susceptibility vanishes, as can be seen in figure 2 (which was calculated using the method described in 3). Large non-linearity accompanying vanishing absorption (transparency) of course match conditions (a) and (b) that lead to high conversion efficiencies in optical crystals. Moreover the dispersion (controlling the refractive index) also vanishes at resonance, this leads to perfect phase-matching (i.e. zero wave-vector mismatch between the fields) in the limit of a simple three-level system. As a result of these features a large enhancement of the conversion efficiency in this type of scheme was predicted.

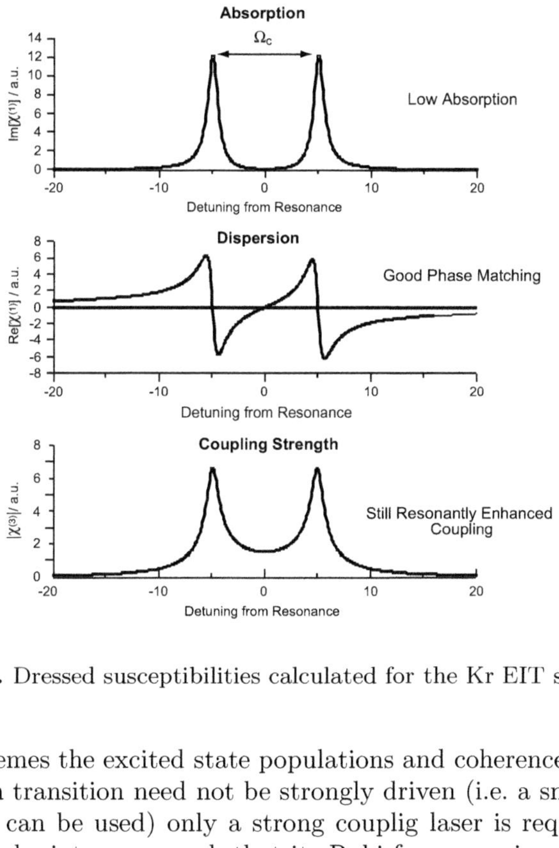

Fig. 2. Dressed susceptibilities calculated for the Kr EIT scheme

In these schemes the excited state populations and coherences remain small. The two-photon transition need not be strongly driven (i.e. a small two-photon Rabi frequency can be used) only a strong couplig laser is required. The coupling laser must be intense enough that its Rabi frequency is comparable to or exceeds the inhomogenous widths in the system (i.e. Doppler width), a laser intensity of above $1\mathrm{MW}/cm^2$ is required for a typical transition. This is trivially achieved even for unfocused pulsed lasers but does present a serious limit to CW lasers unless a specific Doppler free configuration is employed. The latter is not normally suitable for a frequency up-conversion scheme, however recent experiments report significant progress in CW frequency up-conversion using

EIT and likewise a number of other possibilities e,g, laser cooled atoms [8] and standing wave fields [9] (this volume) have been proposed. Ideally two transform limited single mode laser pulse should be used for driving the wave-mixing. This is essential for the coupling laser field, a multi-mode field will cause a serious dephasing effect on the coherence that leads to interference. In contrast, although it is adavantageous for the field driving the two-photon transition to be single mode (in order to achieve optimal temporal and spectal overlap with the EIT hole induced by the dressing laser), this is not essential since this field does not drive the coherence responsible for interference.

Before moving on to talk in more detail about this experimental work that has been carried out on this topic we should first introduce another related concept (again originated by Harris) [10] that has been demonstrated to give near unity conversion efficiencies into the far-UV [11]. Again an appropriate atomic vapour is first coherently prepared. This time two strong driving fields, both single-mode, are applied close to Raman resonace with a Lamdba system and create a coherence between the groundstate $|1\rangle$ and an excited metastable state $|2\rangle$ that is large in magnitude (0.5) but with a negative sign, i.e corresponding to a dark state of the kind involved in coherent population trapping. The strong, near maximal coherence, established between ground and metastable state, then plays the role of a local oscillator with which a third laser field mixes. Essentially the third laser field can mix with the large coherence to generate a new field at the sum- or difference of the two frequencies (i.e. that of the coherence and that of the third field). The strength of the non-linear coupling to the new field is governed by the single photon detuning of the third field from any atomic resonance and so is of the same magnitude as the linear terms. The large size of the coupling means that (unlike usual non-linear optics) the efficient conversion of energy into the new field can be achieved over a distance less than the coherence length, thereby allowing near unity frequency conversion efficiencies from the third field to the new field. This is a very promising route that may become widely used for short wavelength generation. Recently a similar technique was used in a gas of D_2 molecules to generate a very broad spectrum of Raman side-bands (from 195nm to 2940nm) [12]. At the time of writing, however, maximal coherence techniques have not been used to generate wavelengths shorter than 186nm [13][14]. A restriction is placed by the requirements on the atomic system. These schemes require an atom with a metastable state at a high energy, and that can be excited into a coherent superposition with the groundstate via a Raman excitation. The highest energy photon obtainable via frequency mixing is limited by the energy of this metastable state plus the energy of a photon from the third field. To go to shorter wavelengths it will be necessary to develop techniques that excite maximal coherences with states at high energy. We will confine the remainder of our discussions to the EIT schemes producing coherent VUV below 130nm.

This contribution is organised as follows. In section 2 we will briefly review earlier related work on EIT especially the work on atomic hydrogen by Hakuta and co-workers. Following this in section 3 we will outline a theoretical approach for understanding and modelling EIT enhanced four-wave mixing schemes. In section 4 we will describe our own experimental work and present the principle results. Finally we will conclude in section 5 with a brief look at the prospects for future developments.

2 Review of Previous Work on EIT Enhanced Non-linear Mixing

The proposal to utilise EIT in improving the efficiency of non-linear frequency mixing was first made in 1990 by Harris and co-workers [6]. Prior to this other workers had considered related laser induced coherence and interference effects that might be used to enhance the conversion efficiency non-linear mixing schemes. Tewari and Agarwal [15] recognised that an additional control field applied to a resonant mixing scheme could be used to alter the refractive index of the medium so as to improve the phase-matching for the frequency mixing. Harris's proposal however recognised that EIT could be used both to create conditions of near perfect phase matching whilst the non-linear susceptibility was resonantly enhanced and the medium was made optically thin.

A demonstration of a closely related process was made very shortly after by Hakuta, Marmet and Stoicheff [16], who recognised that an experiment that showed strong second-harmonic generation in hydrogen when a DC electric field was applied could be interpreted in terms of EIT. In these experiments (and other subsequent work mentioned below) atomic hydrogen was created from H_2 gas in a dc glow discharge and diffused into the interaction region via a 0.5mm diameter nozzle. The hydrogen density was relatively low ($2 \times 10^{13} cm^{-3}$). A dc-field up to 13 kV/cm was applied in the interaction region. This had the role of causing an Stark splitting between the 2s and 2p states that are strongly mixed by the interaction. The result is an EIT type situation with a finite second-order non-linearity (normally zero in an atomic system) that permits three-wave mixing with a UV field applied at 243nm frequency doubled to 123.6nm (the Ly-alpha transition). A much higher conversion efficiency (up to 2%) was predicted in the limit of a large density x length sample [17].

The first demonstrations of EIT enhanced four-wave mixing were made in 1993. The group in Stanford showed how, in a four-wave mixing scheme in Pb, the conversion efficiency increased by a factor of 60 in a step-like manner when the coupling laser Rabi frequency overcame the Doppler broadening of the medium [18]. In this scheme radiation at 283nm was generated. The coupling laser field in this experiment was derived from an injection seeded Nd:YAG (at 1064nm) and this was $6cm^{-1}$ from resonance. As a consequence of this detuning only the effect of the dressing laser field on phase-matching was important and not the opacity modification.

The first experiment to show EIT in a resonant scheme, where both the EIT effect on opacity and phase-matching were important, was reported by Zhang, Hakuta and Stoicheff [19]. They employed a four-wave mixing scheme in Hydrogen (equivalent to figure 1), where the EIT was created on the 3p-1s transition at 103nm in a Lambda scheme by the application of a field at 656nm on the 2s-3p transition. A field at 103nm is generated by the four-wave mixing process that was enhanced by this EIT. In these experiments, and subsequent work by the same authors conversion efficiencies up to 2×10^{-4} were reported [20]. To achieve this a microwave discharge was used this gave an atomic density as high as $2 \times 10^{14} cm^{-3}$. Given that path lengths of several centimeters were available density x length (NL) products of up to $3.5 \times 10^{15} cm^{-2}$ were available. These experiments serve to illustrate the importance of a large value for the density length product NL, with the VUV yield increasing by a factor of 250 when the value of NL was increased from 2×10^{14} to $3 \times 10^{15} cm^{-2}$. At even higher NL products larger conversion efficiency were found ($> 10^{-3}$ being found [23]. At the higher densities the influence of undissociated molecules also increased [21]. Investigations by the same workers have shown EIT enhancement of mixing of resonant schemes involving higher levels (4p up to 8p). In these experiments shorter wavelengths down to 92.6nm were generated [22].

A limit to the conversion in an EIT enhanced four-wave mixing scheme is set by the Doppler width. A large Doppler width dictates that a comparable Rabi frequency is required to create good transparency, this in turn leads to a reduction in the non-linear susceptibility (see figure 2). This is one of the most important factors limiting the conversion efficiency in the H scheme (Hakuta private communication). A combination of the low mass of the H atom and the elevated temperatures required in the dissociation process leads to especially large Doppler widths (> 10 GHz). To overcome such a severe limit we have undertaken related investigations in Kr atomic gas where the relevant Doppler width is only 1 GHz. Moreover achieving high values of NL was straightfoward in this scheme since Kr occurs naturally as a monoatomic gas.

The frequency mixing scheme utilised in these experiments can be seen in the insert in figure 3. For simplicity only the three energy levels that are resonantly coupled have been shown. The groundstate ($|1\rangle$) in Kr is the $4p^6 1_0^S$ state whilst the two upper states in this scheme are $4p^5 5p[0, 1/2]$ ($|2\rangle$) and the $4p^5 5s[1, 1/2]$ ($|3\rangle$). The latter state is coupled to the ground state by the strong resonance transition at 123.6 nm and has a dipole allowed coupling to state $|2\rangle$. State $|2\rangle$ is not dipole coupled to the ground state (but can be excited via a two-photon process) and so these three levels form a Ladder system. It should be remembered that the remaining atomic energy levels also play a role especially in determining the dispersion of the medium. A feature of the Kr scheme was that it was a sum-difference mixing scheme in contrast to the original sum-frequency mixing of Harris (figure 1). This means the couplings leading to EIT are in a Ladder rather than a Lambda configuration, as a consequence the EIT is not perfect in our scheme due to the finite decay rate of $|2\rangle$. Since this decay rate is significantly smaller than that from $|3\rangle$, however, there is still a large effect on transparency and phase matching from the coupling field. This still results in a large efficiency in the frequency mixing process as we will see. Before discussing in detail our experiments we will introduce the theoretical model employed to interpret our results.

3 Theoretical Treatment of EIT Enhanced Four-Wave Mixing in Kr

Our treatment was developed along similar lines to that of Harris et al [6] but used the density matrix formulation to compute the susceptibilities, and included the decay rates, dipole moments and other parameters specific to the Kr scheme [24]. The interaction Hamiltonian V of the atom with the electromagnetic fields is:

$$V = \hbar\Omega_a e^{-i2\omega_a t}\,|2\rangle\langle 1| + \hbar\Omega_c e^{-i\omega_c t}\,|2\rangle\langle 3| + \hbar\Omega_g e^{-i\omega_g t}\,|3\rangle\langle 1| + c.c \tag{1}$$

where the coupling strengths are given by the three Rabi frequencies Ω_a, Ω_c and Ω_g. These are respectively; the effective two-photon Rabi frequency (in practice finite, but of smaller magnitude than the radiative decays), the Rabi frequency characterising the strong resonant coupling field (this will typically be factors of 10 and 10^3 larger than the Doppler and radiative widths respectively) and the Rabi frequency of the coupling field (this is assumed small in our analysis to derive the susceptibilities, but in the limit of high conversion efficiency this assumption may cease to be valid).

The evolution of each of the density matrix elements is described by the Liouville equation:

$$\hbar\frac{\partial}{\partial t}\rho_{nm}(t) = -i\sum_k H_{nk}(t)\rho_{km}(t) + i\sum_k \rho_{nk}(t)H_{km}(t) + \Lambda_{nm} \tag{2}$$

where the decay rates Λ_{mn} are incorporated into the equation for each density matrix element.

This system of equations was solved under the assumption of steady state conditions. The rotating wave approximation was made and the terms of the equation were transformed into a rotating frame such that all the frequencies are eliminated with the exception of the detunings:

$$\begin{aligned}\Delta_a &= \omega_{21} - 2\omega_a \\ \Delta_c &= \omega_{23} - \omega_c\omega_g\end{aligned}$$

The assumption that the two-photon coupling was always weak could be relaxed in our treatment. This leads to the prediction of a number of novel features including gain at the frequency corresponding to the Autler-Townes satellites of the coupling field [24]. However in the situation relevant to our experiments this coupling was in fact always small relative to the coupling field. In this situation the assumption that $\rho_{11} >> \rho_{22}, \rho_{33}$ could be made. The phenomenological dampings include the spontaneous decay rates appropriate to these states of Kr.

For the results given later however we ignored the laser dephasing, collisional dephasing and the photoionisation. Monochromatic laser fields were assumed in this calculation since the inclusion of laser phase diffusion is unnecessary for the case of the transform limited coupling field employed in this experiment, and would in fact lead to innaccurate results when employed in a steady-state model. The densities employed in this work were always smaller than $10^{17} cm^{-3}$, and it is only above this density that collisional effects become large compared to the radiative decays. Likewise the photoionisation rates, for the intensities used in the experiments discussed below, are less than the spontaneous decay rates.

The set of differential equations was solved in the steady state limit by a straighfoward matrix inversion of what reduces to an 8x8 matrix [24]. The coherence ρ_{13} that is obtained from the matrix inversion is critical in computing the susceptibilities (linear and non-linear) that determine the response of the medium at the frequency of the generated field. These are obtained from the expression for the macroscopic polarisation:

$$P(\omega_g) = 2N\mu_{13}\rho_{13} \tag{3}$$

The matrix element contains, to all orders, the terms relevant to the calculating the dressed susceptibilities of the system i.e. $\chi_D^1(\omega_g, \omega_g)$ and $\chi_D^3(\omega_g; 2\omega_a, \omega_c)$ These are extracted from the expression for the density matrix and the equation for the macroscopic polarisation. The form of these susceptibilities as a function of detuning of the "probe" field applied at the generated field frequency are essentially as those plotted in figure 2.

Random Doppler shifts due to the Maxwellian velocity distribution of the atoms lead to a corresponding distribution in the detunings for the atomic ensemble. The response of the medium, as characterized by the susceptibilities, must therefore include the Doppler effect by performing a weighted sum over these detunings. The weighting is determined by the Gaussian form of the Maxwellian velocity distribution. From this the effective value of the susceptibilities are obtained, and these can be used to calculate the generated field characteristics.

To simplify the calculation of the generated field the applied fields were assumed to be in the plane wave limit. This was appropriate for the weak focussing conditions employed in the experiments. The resulting propagation equation was derived from Maxwells equations and had the form:

$$\begin{aligned} \frac{\partial}{\partial z} A_g &= i\frac{3\omega_g}{4c}\chi^{(3)}(-\omega_g, 2\omega_p, -\omega_c)A_p^2 A_c e^{-i\Delta k_g z} \\ &- \frac{\omega_g}{2c} Im[\chi^{(1)}]A_g + i\frac{\omega_g}{2c} Re[\chi^{(1)}]A_g. \end{aligned} \tag{4}$$

where A_j's are the electric field amplitudes and Δk_g is the wave-vector mismatch for the processes (i.e. $\Delta k_g = k_g + k_c - 2k_a$). The wave-vectors are computed from the refractive index obtained from $Re[\chi^{(1)}]$ plus the contribution to the real part of the atomic susceptibility from all of the other levels not included in the three level scheme. This background susceptibility was calculated from the available oscillator strength data for Kr using the Sellmier formula.

The equations were solved analytically for a homogeneous medium with a well defined length and this operation lead to a general analytical equation for the generated intensity. The inclusion of the residual susceptibility in the calculation of phase-matching is, as we shall see below, very important for the correct calculation of the generated field for higher values of NL and also for understanding some of the limits to the conversion efficiency.

4 Experimental Investgation of EIT Enhanced Four-Wave Mixing in Kr

4.1 Experimental system and results

There are a number of factors that determined our choice of Kr in the investigation of high conversion efficiency EIT enhanced frequency conversion in the VUV. Krypton is a naturally occurring atomic gas, this immediately simplifies the gas handling. Densities greater than $10^{18} cm^{-3}$ can be used without significant dimer and cluster formation. Therefore we could increase density x length product (NL) to large values, although for densities $> 10^{17} cm^{-3}$ we would expect collisional dephasing to begin to have a significant influence on the coherence. Kr has an atomic weight of 83.8 (average of 4 naturally occuring isotopes) and so the Doppler width of the VUV transitions was only 1GHz (compared to $>$ 10GHz in the hydrogen experiments). The resonance transitions in Kr are also in the VUV (i.e. 124nm for the scheme investigated).

The experimental layout and procedures have been described in detail previously [25] [26] . Here we will only outline the apparatus and highlight the most important features of this equipment and methodology. The experimental layout is shown schematically in Figure 3. The coupling field at 759 nm was produced by an optical parametric oscillator (OPO) pumped by a single-mode, frequency-doubled Nd:YAG laser (Continuum Mirage 800 and Powerlite 7010). The OPO output was then amplified in titanium- sapphire to generate pulse energies up to 10mJ. This coupling field was single mode and near transform limited (<250 MHz, 4 ns); kept to within 1 GHz of the resonance by reference to laser induced fluorescence in a krypton hollow cathode lamp. The coupling laser intensity was varied over a range of values using a wave-plate/polariser combination. The ultraviolet field was provided by a XeCl excimer-pumped dye laser, frequency doubled to produce a pulse of length 16ns, energy 150mJ and 2 GHz bandwidth [25]. Exact two-photon resonance was determined by measurement of photo-ionisation in a separate krypton gas cell. The relative jitter in arrival time of the laser pulses was minimised using computer controlled active feedback to around 2.5 ns, short compared to the 16ns pulse length of the UV field.

The UV and IR beams were directed into either a gas jet or a 10 mm long cell containing krypton gas. The gas jet provided an interaction length of only a few

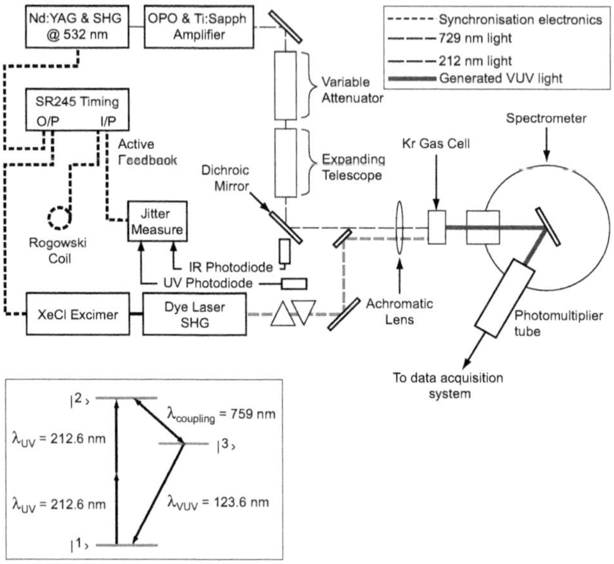

Fig. 3. Kr EIT experimental layout with the atomic scheme (inset)

millimetres with densities estimated to be $< 10^{16} cm^{-3}$, so the maximum NL values obtained in this case were around $10^{15} cm^{-2}$. For later studies the 10mm cell was used which provided higher density and NL values [26]. The density-length product (NL) could therfore be varied form 0 - $2.510^{17} cm^{-2}$. The UV laser field was focussed into the interaction region by a 330 mm focal length lens giving a confocal parameter of 100 mm. At the interaction region the coupling beam diameter was significantly larger, providing uniform IR intensity throughout the focussed region of the UV beam. The coupling beam size changed very little over the length of the cell. These beam sizes were measured with a CCD array placed at the interaction plane. The coupling beam was measured to be 2.3 ± 0.1 mm (fwhm) in diameter at the interaction region, and the UV beam 0.10 ± 0.02 mm (fwhm) in diameter at the interaction region.

Measurements of the variation of the VUV pulse intensity as a function of the coupling laser intensity and density-length product were made using a solar blind photomultiplier tube (PMT). This provided a linear detection to enable the measurement of the relative, but not absolute, VUV intensities. Absolute intensity measurements were also made using a calibrated VUV photodiode located behind the monochromator exit plane. This had an aluminium oxide photocathode and was issued by NIST as a transfer standard calibrated in 1987. The sensitivity of this photodiode at 123.6nm was extrapolated to be 7.5 ± 110^{-3} from the quantum efficiency curves. Further details of the measurement of absolute VUV yield can be found in [26].

The data acquisition system comprised three gated integrators (Stanford Research Systems SR250) feeding output voltages proportional to integrated signals to a computer-controlled analogue to digital converter (SR245). This system permitted simultaneous integrated acquisition of the generated VUV signals, time-delays between pulses and the coupling laser energy. Cell pressures did not require integration and were connected directly to the analogue to digital converter. The simultaneous record of laser pulse energies, cell pressure and the VUV signal enabled the study of correlation between these parameters. It was also necessary to monitor the variation in time delay (2ns) between the two laser pulses that arose from the jitter between the Nd.YAG and Excimer lasers, as well as the laser pulse energies.

The generated VUV signal was measured as a function of coupling laser fluence and gas density-length product. With krypton pressure (NL product) kept constant, at a range of discrete values, the coupling laser energy (fluence) was slowly and smoothly varied between 0 and 4 mJ per pulse (Fluence 0 - 0.1 Jcm^{-2})using the polariser-waveplate combination. The data shows good qualitative agreement with the results of a calculation using the method outlined above. For high coupling laser fluences ($> 0.05Jcm^{-2}$) the VUV yields are far larger in the high NL regime compared to when the density-length product is small ($10^{15}cm^{-2}$). The VUV yield also reaches it's maximum at much lower coupling pulse fluences for low NL values. For example at an NL of $0.3 \times 10^{16}cm^{-2}$ the VUV yield reaches its maximum value at a fluence of $0.02Jcm^{-2}$. Under the high NL conditions ($> 410^{16}cm^{-2}$) it should be noted that the coupling laser fluence required to provide a Rabi frequency larger than the Doppler width (just $10^{-5}Jcm^{-2}$) is much less than that needed to satisfy the coupling laser preparation energy condition outlined by Harris and Luo [27] (for which a fluence of about $10^{-2}Jcm^{-2}$ is required).

For the second set of experimental data the coupling laser fluence was held constant and NL value of the krypton gas medium was slowly varied through a range of pressures using a needle valve feeding the gas into the 10-mm cell. 1000 discrete data points were taken as the pressure was varied, over the period of the data acquisition, between 0 and 10 mbar. Figure 4 shows that for all the coupling laser fluences there is a steep initial rise in VUV intensity with density length product. The peak in the VUV intensity occurs at a higher NL values as the coupling laser fluence was increased. There exists, for all coupling fluences, an NL value at which the VUV yield achieves a maximum value. For larger values of NL the yield falls significantly.

The detection system employed to obtain figures 4 could provide an accurate measurement of the relative VUV intensity. Measurements of the absolute VUV yield were however made for the density conditions ($NL = 510^{16}cm^{-2}$) that gave the largest VUV intensity. These showed that the conversion efficiency achieved (referenced to the energy effectively available for frequency mixing in the laser beams) was 5×10^{-3}, although it was likely to be a factor of two larger since this

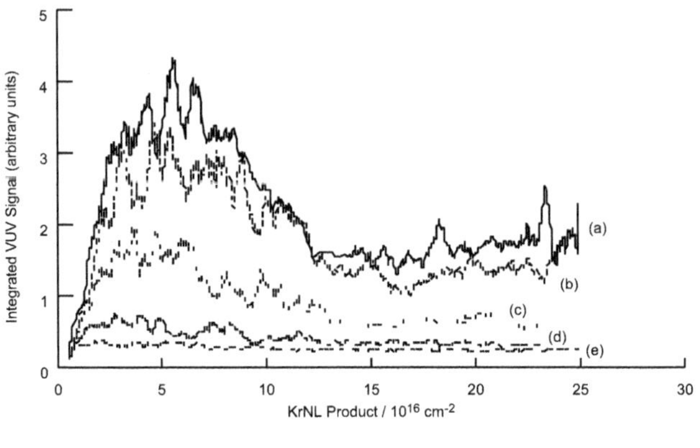

Fig. 4. VUV intensity (arb.units) versus NL

figure did not include unquantifiable losses in the monochromator and detection systems [26].

4.2 Discussion

The measured conversion efficiency of almost 1% for this mixing scheme provides a confirmation of EIT enhancement in this four-wave mixing scheme. This value of conversion efficiency was found for an optimal NL value of $5 \times 10^{16} cm^{-2}$. When the NL value was increased from 1.2×10^{15} to $5 \times 10^{16} cm^{-2}$ the yield of the frequency mixing scheme was measured to increase by a factor of 5×10^{3}. The scaling of the VUV yield with density length product was therefore close to the NL^2 behaviour expected under conditions of complete transparency and perfect phase matching. In the absence of the coupling field the absorption and coherence lengths for the generated fields would only have been of the order of a few tens of microns. Therefore we deduce that the coupling field has rendered the entire sample length transparent and properly phase matched for the resonant radiation. There is, however, a limit to the scaling with NL set not by the transparency of the sample (which is almost complete) but by the residual wave-vector mismatch of the medium (due to the levels outside of the 3 level EIT scheme). This limit is now discussed.

The results shown in Figure 5 were obtained from a calculation using the method described in section 3. These differ from the experimental data in Figure 4 in several respects. The numerical calculations show no evidence that the intensity of the VUV generated should falls off as NL is increased for values of the coupling laser fluences $< 0.026 Jcm^{-2}$. Such a fall off is clearly seen in the experimental data in Figure 4. This is because in the simple theoretical steady-state model there is no consideration of the preparation energy required to render the medium transparent. In the experimental data as the density-length (NL) product rises the generated intensity of VUV radiation initially increases. A peak

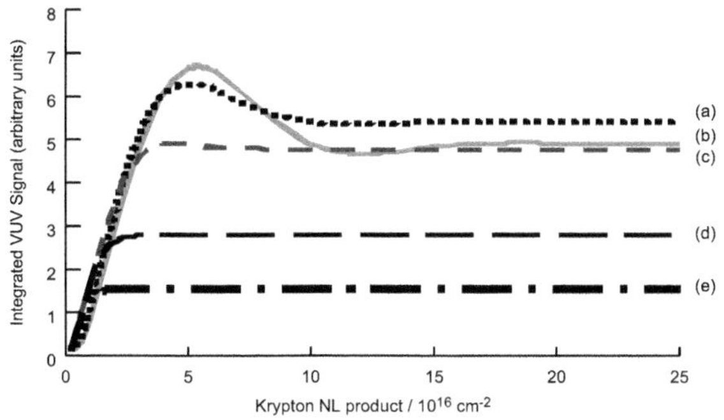

Fig. 5. VUV intensity versus NL calculated with model

in intensity is reached at $7.5 \times 10^{16} cm^{-2}$ for the higher intensity coupling laser fluences ($0.072 Jcm^{-2}$ and $0.048 Jcm^{-2}$ per pulse) and at slightly lower pressures for lower coupling laser intensities. For the lower coupling laser fluences (0.027, 0.012 and 0.006 Jcm^{-2}) the VUV signal falls away at higher NL products as the coupling laser energy was insufficient to prepare the entire length of the medium. The initial parts of the medium will be rendered transparent by the coupling field, but the number of photons necessary to induce transparency along the entire length of the medium is not reached. For example the coupling pulse of fluence $0.006 Jcm^{-2}$ the coupling pulse is only capable of preparing transparency up to an NL product of $5 \times 10^{16} cm^{-2}$ and for greater NL full EIT will not be achieved at this fluence.

In the experimental and theoretical data at higher coupling pulse fluences the intensity of the generated VUV radiation is seen to fall from the peak intensity at NL $7.5 \times 10^{16} cm^{-2}$ to a minimum at NL $1.5 \times 10^{17} cm^{-2}$, and then begin to rise again. In this case the fluence was sufficient to meet the preparation energy condition for all NL values investigated. These results match closely the numerical data as the effect of preparation energy is minimal for these higher fluences. The experimental and theoretical results both show the appearance of modulations at higher coupling pulse intensities (0.072 and $0.048 Jcm^{-2}$). For high coupling fluences we have an essentially completely transparent medium due to electromagnetically induced transparency and with phase matching conditions that are perfect in so far as the 3-levels involved in the EIT scheme are included. The susceptibilities are then dominated by residual contributions from other levels. In this regime we see periodic intensity fluctuation with NL product due to changes in the phase matching conditions, the residual phase matching contribution is directly proportional to the density-length product of the sample. This periodicity can been used to measure the residual contributions to phase matching. From the position of the first minimum of the experimental graph we can calculate a Δk_{resid} of $4.8 \times 10^{-21} Nm^{-1}$ which agrees, within experimental error, to the value calculated using the Sellmeier formula for the residual levels which is $4.83 \times 10^{-21} m^{-1}$ [29].

A more complete picture of the effects of the variation of the density-length product of the medium and the coupling laser pulse fluence can be seen in the three dimensional plot Figure 6, calculated using the numerical model outlined above. It is informative to consider three regions of Figure 6; these are (a) relatively low coupling laser fluences ($< 0.1 Jcm^{-2}$), below the fluence of the peak in VUV generation; (b) the region where the coupling laser fluence is high ($> 0.2 Jcm^{-2}$) where oscillatory behaviour is seen as NL varies and (c) the region where the density-length product is lower than the first peak at $0.5 \times 10^{16} cm^{-2}$. As the Rabi frequency of the coupling field (proportional to the square root of the fluence of the coupling pulse) is increased from zero the medium becomes increasingly transparent at the generated wavelength. This at first results in an increase in the intensity of the VUV generated. When the medium is rendered completely transparent, however, further increase in the coupling fluence causes the generated intensity to fall as the magnitude of the non-linear part of the susceptibility governing the mixing process begins to decrease. Region (a) is thus dominated by the absorption of the generated field due to the lack of complete transparency of the medium. As the coupling fluence is increased the medium will be rendered more transparent and the size of the generated VUV field grows rapidly. Region (b) is a region dominated by the residual phase matching. The medium is in the perfectly transparent regime, with minimal absorption and a dispersion determined only by the levels outside of the 3-level system. Region (c) was the regime investigated in the previous studies [26] where the VUV yield was found to still be increasing strongly with NL (approximately as NL^2).

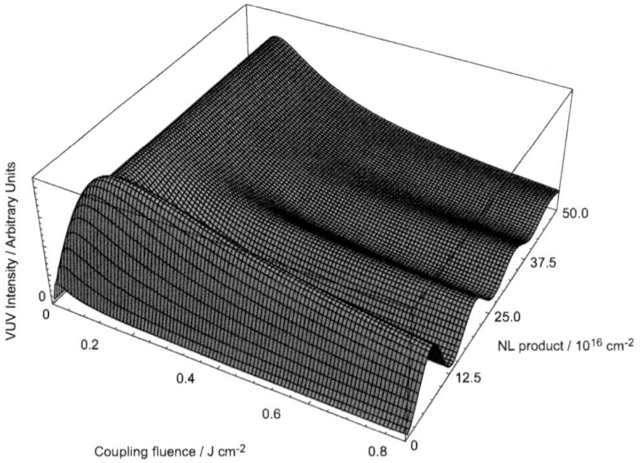

Fig. 6. VUV intensity versus NL and coupling field fluence calculated from model

The optimum value of NL was essentially determined by the residual phase matching effects of the medium. Maxima in the generated field will occur at distances that are an odd multiple of the coherence length in the medium. This translates to a similar periodicity of the generated field as the pressure of a fixed length cell is varied. As the pressure is increased, such that the length of the cell equals the coherence length of the medium (where the coherence length is dominated by Δk_{resid}), the generated field will have slipped in phase by $\pi/2$ radians relative to the driving polarisation. If the NL product is increased further then the non-linear polarisation will take power from the generated wave and return it to the applied fields. For a general value of NL the VUV field will only be generated in the coherence length, or fraction thereof. The VUV yield at the optimum NL value will scale approximately as NL^2. If Δk_{resid} were reduced then the optimum value of NL would increase and thus the VUV yield would increase. This increases the range of NL over which the behaviour outlined for region (c) is valid.

For unbuffered Kr with the inherent finite value of Δk_{resid} the maximum generated VUV intensity will occur for an NL of $5 \times 10^{16} cm^{-2}$. This corresponds to the conditions for which we measure a conversion efficiency of near 1%. If the residual phas-mismatch could be reduced it would be possible to arrange the value of NL for maximum conversion to be larger and therefore for the absolute yield to be increased by a comensurate factor. This could for instance be achieved by introduction of a negatively dispersive buffer gas at the appropriate density. A suitable choice for the 123.6nm wavelength would be Hg vapour.

5 Further Developments and Conclusion

Earlier experiments showed [30] how an autoionising transition can be made transparent by EIT. We have undertaken an experiment, still in progress, to test how this may be used to enhance wave-mixing to still shorter wavelengths. We are studying XUV generation in Kr via a sum-frequency mixing schemes involving auto-ionising resonances at 117 716cm^{-1} . The auto- ionising state selected lies in the ns'[1/2] series and was chosen due to a relatively narrow auto-ionising width (17s' width $< 1.4\,\mathrm{cm}^{-1}$) [31]. A strong field couples this state ($|3\rangle$) to the $4p^5 5p[0, 1/2]$ ($|2\rangle$) to form a Lambda EIT system. With two-photon excitation at 212nm to ($|2\rangle$) as in the VUV mixing experimentys fields at 88nm are generated. Presently the effect on XUV yield of the coupling laser between ($|2\rangle$)and ($|2\rangle$)is being investigated

Energy conversion efficiencies of 0.01 have been found for the EIT enhanced frequency up-conversion in Kr discussed in this contribution. The wave-vector mismatch from other levels outside of the 3-level EIT scheme was found to be the principle limiting factor to the production of still larger yields of coherent VUV radiation. By addition of a buffer gas with negative dispersion at 124nm (Hg vapour for example) this limit could in principle be overcome. In that case still larger NL values could be utilised in the four-wave mixing process, resulting in larger conversion efficiency. Further improvements are anticipated if identical transform limited lasers are employed for the fields driving both the coupling and the two-photon transitions. In this case there will be improved matching of both the frequency spectrum and temporal profiles of the lasers. Moreover a two-photon field of higher power will increase the VUV yield further. Together these improvements, along with a careful choice of focus geometry to best utilise the laser powers without causing excessive photoionisation, are anticipated to lead to conversions efficiencies from the UV laser energy of between 1 and 5 %. A 2mJ pulse of 212nm radiation would then generate a 20-100 μJ pulse of VUV with peak powers of 10-50kW. These powers of transform limited coherent VUV open up the possibility of various techniques of non-linear VUV spectroscopy.

References

1. "X-ray Lasers 1998", Institute of Physics Conference Series, **159** (1999).
2. J.F. Reintjes, *Nonlinear Optical Parametric Processes in Liquids and Gases*, (Quantum electronics-principles and applications), Academic Press (1984).
3. C.R.Vidal, in *Tunable Lasers*, L.F.Mollenauer and J.C.White, eds., vol **59** of Springer-Verlag Topics in Applied Physics,(Springer-Verlag, Berlin, 1987), (Chapter 3).
4. Boyd, *Non-linear Optics*,(Academic Press, Boston;London) (1992).
5. J.P.Marangos, N.Shen, H.Ma, M.H.R.Hutchinson and J.P.Connerade , J.Opt. Soc. Am B **7**, 1254, (1990).
6. S.E. Harris, J.E. Field, A. Imamoglu, Phys.Rev.Lett. **64**, 1107 (1990).
7. J.P.Marangos, Journal of Modern Optics, **45**, 471 (1998).
8. S.A.Hopkins, E.Usadi, H.X.Chen and A.V.Durrant, Optics.Commun. **134**, 185 (1997).
9. F.Silva, J.Mompart, V.Ahufinger and R.Corbalan, Europhys.Lett. **51**, 286 (2000).
10. S.E.Harris, G.Y.Yin, M.Jain, H.Xia and A.J.Merriam, Philos.Trans. R.Soc.London A **355**, 2291 (1997).
11. M.Jain, H.Xia,G.Y.Yin, A.J.Merriam, and S.E.Harris, Phys.Rev.Lett. **77**, 4326 (1996).
12. A.V.Sokolov, D.R.Walker, D.D.Yavuz, G.Y.Yin, and S.E.Harris, PRL **85**, 562 (2000).
13. A.J.Merriam, S.J.Sharpe, H.Xia, D.Manuszak, G.Y.Yin and S.E.Harris, Opt. Lett. **24**, 625 (1999).
14. A.J.Merriam, S.J.Sharpe, M.Shverdin, M.Manuszak, G.Y.Yin and S.E.Harris, PRL, **84**, 5308 (2000).
15. S.P.Tewari and G.S.Agarawal, PRL **56**, 1811 (1986).
16. K. Hakuta, L. Marmet and B.P. Stoicheff, Phys. Rev. Lett. **66**, 596 (1991).
17. K.Hakuta et al, in Laser Spectroscopy X, ed. M.Duclois, E.Giaciobino and G.Camy), 301 (1991).
18. M. Jain, G.Y. Gin, J.E.Field and S.E. Harris Opt. Lett **18**, 998, (1993)
19. G.Z. Zhang, K. Hakuta, and B.P. Stoicheff, Phys.Rev.Lett. **71**, 3009 (1993).
20. D.Z.Zhang, M.Katsuragawa, K.Hakuta, R.I.Thompson and B.P.Stoicheff, Phys.Rev.A, **52**, 1584 (1995).
21. M.Katsuragawa, G.Z.Zhang, and K.Hakuta, Opt. Communication, **129** 212 (1996).
22. G.Z.Zhang, D.W.Tokaryk, B.P.Stoicheff and K.Hakuta, **56** 813 (1997).
23. K.Hakuta, private communication.
24. J.C.Petch, C.H.Keitel, P.L.Knight and J.P.Marangos, Phys.Rev A, **53** 543 (1996).
25. C.Dorman and J.P.Marangos, Phys.Rev A, **58**, 4121 (1998).

26. C.Dorman, I.Kucukkara and J.P.Marangos, Phys.Rev A, **61** 013802-1 (2000).
27. S.E.Harris and Z.-F.Luo, Phys.Rev A **52** R928 (1995).
28. C.Dorman, I.Kucukkara and J.P.Marangos, Opt.Communication, **180**, 263 (2000).
29. A.Bideau-Mehu, Y.Guern, R.Abjean, A.Johannin-Gills, J.Quantum Spectrosc. Radiat. Transfer **25** 395 (1981).
30. K.J.Boller, A.Imamoglu, and S.E.Harris, PRL **66**, 2593 (1991).
31. K.Ito, K.Maeda, Y.Morioka and T.Namoka, Appl.Optics **28**, 1813 (1989)

Optical Lattice Dynamics and Scattering Processes Resulting from Dipole-Dipole Interaction

A. Guzmán[1,2] and J. Zapata[1,3,4]

[1] Depto. de Física, Universidad Nacional de Colombia, Bogotá, Colombia
[2] Georgia Tech Lorraine, 2-3 Rue Marconi, F-57070 Metz, France
[3] Depto. de Física, Universidad del Atlántico, Barranquilla, Colombia
[4] Depto. de Física, Universidad del Norte, Barranquilla, Colombia

Abstract. The hopping of atoms between lattice sites and matter-wave diffraction, both caused by the quantum dipole-dipole interaction in *antiferromagnetic* optical lattices, are described here by means of two alternative and complementary physical pictures: Wannier and Bloch representations. From dipole-dipole matrix elements in the Wannier representation we infer favored hopping between sublattice sites and alignment of confined atoms on *strings* parallel to the quantization axis. In the Bloch representation, differential cross-sections for scattering in one-dimensional optical lattices result in spin-polarized diffraction patterns. Scattering processes without spin-flip give rise to atomic Bragg diffraction, whereas, because of total angular momentum conservation, changes in the total magnetic quantum number lead to suppression of forward and Bragg scattering.

1 Introduction

During the last decade, it has been well established that atoms can be cooled and trapped into discrete quantum states of optical lattices by means of Sisyphus cooling [1,2,3,4], and experiments with Bose-Einstein-Condensates [5] have demonstrated the coherent properties of matter waves.

Because of their relative simplicity, optical lattices have opened a new field for the study of a wide variety of phenomena that combine concepts from atomic, optical, and condensed-matter physics, and of many effects usually associated with electrons in a crystalline solid [6,7,8,9,10,11]. Moreover, optical lattices offer the advantage over real crystals that atoms are bound by optical potentials easily controllable by laser parameters. This fact has led to recent proposals to confine atoms in optical lattices in the possible realization of *qubits* for quantum computing. Optical lattices have also been suggested as means to achieve coherent manipulation of atomic wave packets [12] and to build up quantum logic gates [13,14,15]. For these purposes it is crucial to have detailed knowledge and control of the interactions between confined atoms and their environment, which might be used for external manipulation of the system or that could cause decoherence of a many-body quantum state.

M. Orszag, J.C. Retamal (Eds.): LNP 575, pp. 212–226, 2001.

Much theoretical work has been devoted to the study of laser cooling dynamics in optical lattices [16,17,18]. The overall effects of the dipole-dipole interaction on optical confinement have been discussed recently within a mean-field framework [19], and effects on the movement of atoms inside atomic resonators have been observed [20]. Dipole-dipole interaction is fairly long-ranged, and the δ-function approach (Fermi contact potential) commonly used to describe atomic interactions might oversimplify its properties. The consideration of its quantum character [21], its anisotropy, and its spin dependence adds additional richness to atomic scattering processes that, if controlled, might be used to generate entangled states of bound atoms [22].

Optical lattices constitute a unique scenario for the theoretical study of atom-atom interactions, amenable to a relatively simple and fully quantum mechanical description. Well-developed techniques from condensed matter physics can be applied straightforwardly to describe the bound and quasi-free atomic quantum states within the lattice, and to consider the quantum nature of the interatomic dipole-dipole potential. We evaluate here differential cross-sections for dipole-dipole atomic scattering. The paper is organized as follows: Section II reviews the quantum mechanical description of non-interacting atoms in optical lattices. We introduce two alternative and complementary pictures, the Bloch and Wannier representations, and discuss their properties and major adequacy for specific scattering problems. In Section III we express the effective dipole-dipole potential between atoms in the electronic ground state [21,23] within the framework of second quantization, and give insight into its origin and quantum character. In section IV we use a Wannier representation to describe hopping caused by dipole-dipole interaction and summarize results given in [24]. In section V Bloch eigenfunctions are used to determine differential cross-sections for spin-polarized matter-matter diffraction in one-dimensional optical lattices. Finally section VI is a summary and outlook that we consider to give some valuable information for both quantum informatics and condensed matter physicists.

2 Atomic States in Optical Lattices

Three-dimensional optical lattices are generated through effective optical potentials that arise from light-matter interaction; they can have a wide variety of Bravais structures depending on the geometry of the laser beam arrangement. We specialize here to a laser frequency largely detuned from a $J_g = 1/2 \rightarrow J_e = 3/2$ off-resonant atomic transition, and to a four-beam geometry [25], sketched in Fig. 1(a), that results from splitting each of the two counterpropagating beams of a 1D *lin* $\perp$ *lin* configuration. The y-linearly-polarized beam is split into two beams of equal amplitudes that propagate in the xOz plane with the Oz-axis as bisector. Similarly the x-polarized beam is split into two beams propagating in the yOz plane. Different lattice structures can be generated by varying the angle θ between the beams and the Oz-axis. For $\cos\theta = 1/\sqrt{5}$ this geometry results in a *fcc* lattice with a lattice constant $\Lambda = \pi\sqrt{5}/k_L$ (Fig. 1(b)), where k_L is the

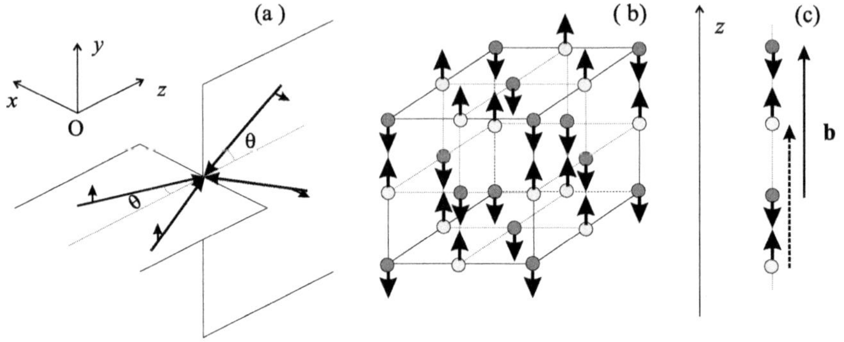

Fig. 1. (**a**) Four-beam geometry used to generate 3D optical lattices. Schematic representation of the *antiferromagnetic* structure of (**b**) a 3D optical lattice ($\cos\theta = 1/\sqrt{5}$) and (**c**) a 1D lattice, showing a primitive translation vector $\boldsymbol{b}$.

laser wavenumber. A 1D-lattice corresponds to the particular case $\theta = 0$ (Fig. 1(c)).

The circular polarized components of the field giving rise to the lattice are

$$E_{\pm}(\boldsymbol{r}) = \frac{\sqrt{2}}{2} E_0 \sum_{\alpha=1}^{4} (\mp 1)^{\alpha} \exp\left(\imath \boldsymbol{K}_{\alpha} \cdot \boldsymbol{r}\right), \tag{1}$$

where E_0 is their amplitude, and the $\boldsymbol{K}_{\alpha}$ are the wavevectors of the cooling beams.

In optical lattice experiments the detuning δ between the laser and the atomic transition frequencies is usually large enough that the excited states can be adiabatically eliminated; the laser-atom interaction can then be described through an effective optical potential $U_M(z)$ that becomes diagonal in the basis of the ground-state magnetic sublevels $|g, M = \pm 1/2\rangle$. Ignoring heating effects caused by optical pumping and atomic interactions, the lattice actually consists of two independent sublattices built up from the optical potentials, one associated with $M = 1/2$, the other with $M = -1/2$. In the following we use indexes $M = \pm$ instead of $\pm 1/2$ when no ambiguity is likely to arise. General expressions for the potentials have been given in [25] that correspond to simple expressions for the 1D optical potential [26]:

$$U_{\pm}(z) = \langle g, pm|\hat{U}|g, \pm\rangle = (U_0/2)(2 \mp \cos 2k_L z), \tag{2}$$

where $U_0 = (2/3)\hbar\delta s_0$ is the potential depth. The saturation parameter,

$$s_0 = \frac{\Omega^2/2}{\delta^2 + \Gamma^2/4}, \tag{3}$$

depends on δ, the Rabi frequency Ω, and the spontaneous decay rate of the atomic transition Γ. The atomic Hamiltonian $\hat{H}$ can be written as the sum of two Hamiltonians $\hat{H}_M$ associated with noninteracting atoms that have different angular momentum components along the quantization z-axis :

$$\hat{H} = \sum_{M=\pm} \hat{H}_M |M\rangle\langle M| = \sum_{M=\pm} \left[\hat{\boldsymbol{P}}^2/2m + \hat{U}_M(\boldsymbol{r})\right] |M\rangle\langle M|, \tag{4}$$

where m, $\hat{P}$, $\boldsymbol{r}$ are respectively the atomic mass, and the center-of-mass momentum and position operators.

The minima of the *total* optical potential define the lattice sites:

$$\boldsymbol{R}_\eta = \frac{\Lambda}{2}\left(N_1\boldsymbol{e}_x + N_2\boldsymbol{e}_y + N_3\boldsymbol{e}_z\right), \tag{5}$$

where $\boldsymbol{e}_x, \boldsymbol{e}_y, \boldsymbol{e}_z$ are unit vectors along the coordinate axes. The site η is specified by the integers $N_i, i = 1,2,3$. Sublattice sites $\boldsymbol{R}_\eta^M$ are sorted by the parity of $p = N_1 + N_2 + N_3$. If the origin is placed at a minimum of U_-, odd (even) values of p sort out the $M = +\ (-)$ sublattice sites. Thus *fcc* as well as 1D lattices, can be thought of as having an *antiferromagnetic* structure, as suggested in Fig. 1(b,c).

Differences between pairs of beam wavevectors give the primitive translation vectors of the reciprocal lattice, which reduce to one for a 1D lattice, namely $\boldsymbol{b} = 2k_L\boldsymbol{e}_z$. Optical potentials for the two sublattices are related by

$$U_-(\boldsymbol{r}) = U_+(\boldsymbol{r} + \boldsymbol{R}_\eta^+). \tag{6}$$

with the result that $\hat{H}_-$ and $\hat{H}_+$ have identical energy band spectra $E_n(\boldsymbol{q})$, given by the eigenvalue equation

$$\hat{H}_M \phi_{n,\boldsymbol{q}}^M(\boldsymbol{r}) = E_n(\boldsymbol{q})\phi_{n,\boldsymbol{q}}^M(\boldsymbol{r}), \tag{7}$$

and their eigenstates are related by a spatial translation analogous to (6). The function $\phi_{n,\boldsymbol{q}}^M(\boldsymbol{r})$ is a Bloch eigenfunction with quasi-momentum $\hbar\boldsymbol{q}$ and fixed magnetic quantum number M. It can be conveniently expressed as

$$\phi_{n,\boldsymbol{q}}^M(\boldsymbol{r}) = \exp\left[\imath\boldsymbol{q}\cdot\boldsymbol{r}\right]\Xi_{n,\boldsymbol{q}}^M(\boldsymbol{r}), \tag{8}$$

where the functions $\Xi_{n,\boldsymbol{q}}^M(\boldsymbol{r})$ have the spatial lattice periodicity.

The eigenvalue problem for a 1D lattice can be cast in the form of a well-known Mathieu eigenvalue problem [27,28], whose energy band structure in the first Brillouin Zone (1BZ) [29] is displayed in Fig. 2(a) for reference. The functions $\Xi_{n,q_z}^M(z)$ are easily obtained by numerical evaluation [28] of the coefficients in the Fourier series

$$\Xi_{n,q_z}^M(z) = \frac{1}{\sqrt{2\pi}} \sum_{s=-\infty}^{s=+\infty} c_s^{n,q_z,M} \exp\left[\imath sbz\right]. \tag{9}$$

The optical potential $U_-(z)$, in units of the recoil energy $E_R = \hbar^2 k_L^2/2m$, and some of its Bloch eigenfunctions are shown in Fig. 2(b).

An alternative physical picture, better suited to describe confined atoms, is given by a "localized" basis set, namely, the Wannier functions, which are defined in terms of Bloch functions by

$$\psi_n\left(\boldsymbol{r} - \boldsymbol{R}_\eta^M\right) = \left[\frac{\mathcal{V}}{(2\pi)^3}\right]^{1/2} \int_{1BZ} d\boldsymbol{q} \, \exp\left[-\imath \boldsymbol{q} \cdot \boldsymbol{R}_\eta^M\right] \phi_{n,\boldsymbol{q},M}(\boldsymbol{r}), \tag{10}$$

where $\mathcal{V}$ is the volume of the unit cell, and where the integral runs over the quasimomenta $\boldsymbol{q}$ in the first Brillouin zone (1BZ). Wannier functions can be thought of as a Fourier transform of Bloch functions corresponding to a single energy band, the result being a wavefunction that is peaked at the site $\boldsymbol{R}_\eta^M$. They constitute a set of orthogonal functions, i.e.,

$$\int d\boldsymbol{r} \, \psi_n^*\left(\boldsymbol{r} - \boldsymbol{R}_\eta^M\right) \psi_{n'}\left(\boldsymbol{r} - \boldsymbol{R}_{\eta'}^{M'}\right) = \delta_{nn'}\delta_{MM'}\delta_{\eta\eta'}, \tag{11}$$

which can be used to describe long-lived and well-localized impurity states in condensed matter. They are not, however, eigenfunctions of (7), and the Hamiltonian in the Wannier representation includes coupling between sites within a sublattice, leading to intrasublattice hopping. Nevertheless, in the tight-binding approximation, Wannier functions decay rapidly with distance from the lattice site, and the hopping probability, which depends on the overlap between functions at different sublattice sites, is negligible. We can thus safely consider tight-bound atoms as being confined within a lattice site for times shorter than the characteristic time for intralattice hopping. Additionally, for low-lying bands, the *cos*^{2}-shape of the optical potential can be approximated by a parabolla and Wannier functions by harmonic oscillator eigenfunctions, as schematically shown in Fig. 2(c).

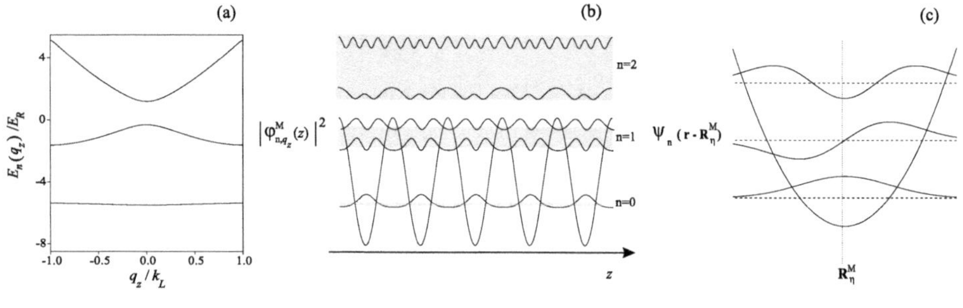

Fig. 2. (**a**) Energy band structure $E_n(q_z)/E_R$ of an 1D optical lattice as a function of the quasi-momentum q_z in the 1BZ. (**b**) Optical potential $U_-(z)$ and some Bloch functions at the edges of the bands $n = 0, 1, 2$, for $U_0 = -8E_R$. (**c**) Schematic representation of a Wannier function in the tight-binding limit.

For compactness, in the following we use the composite indexes $\nu = \{n, \boldsymbol{q}, M\}$, and $\sigma = \{n, \eta, M\}$ to label Bloch and Wannier functions respectively, whenever their use does not lead to ambiguities.

Either Bloch or Wannier functions can be used as a basis set for second quantization whereby the Schrödinger field operator is expanded in terms of bosonic atomic annihilation operators either on Bloch ($\hat{b}_\nu$) or Wannier ($\hat{w}_\sigma$) states:

$$\Psi^M(\boldsymbol{r}) = \sum_n \int d\boldsymbol{q}\ \phi_\nu(\boldsymbol{r})\hat{b}_\nu = \sum_{n,\eta} \psi_\sigma(\boldsymbol{r})\hat{w}_\sigma. \quad (12)$$

Annihilation operators in both representations annihilate an atom with quantum magnetic number M in band n; however in Bloch representation the atom annihilated has well defined quasimomentum $\boldsymbol{q}$, whereas $\hat{w}_{n,\eta,M}$ annihilates an atom at the site $\boldsymbol{R}_\eta^M$. Bloch and Wannier annilitation operators are related by

$$\hat{w}_{n,\eta,M} = \left[\frac{\mathcal{V}}{(2\pi)^3}\right]^{1/2} \int_{1BZ} d\boldsymbol{q}\ \exp\left[\imath\boldsymbol{q}\cdot\boldsymbol{R}_\eta^M\right]\hat{b}_\nu, \quad (13)$$

with the inverse transformation given by

$$\hat{b}_\nu = \left[\frac{\mathcal{V}}{(2\pi)^3}\right]^{1/2} \sum_\eta \exp\left[-\imath\boldsymbol{q}\cdot\boldsymbol{R}_\eta^M\right]\hat{w}_{n,\eta,M}. \quad (14)$$

For bosonic atoms, as considered here, both Bloch and Wannier annihilation and creation operators satisfy bosonic commutation relationships.

3 The Dipole-Dipole Interaction in an Optical Lattice

In the previous section we summarized properties of optical lattices for noninteracting atoms. But atoms in the optical lattice carry an electric dipole moment induced by the light field to which they are subjected and, hence, at least at low densities, they interact primarily through the near-resonant dipole-dipole interaction. For tight-bound atoms the dipole-dipole interaction might cause intersublattice hopping, i.e, hopping between sites of different sublattices, which are the nearest neighbors in an *antiferromagnetic* structure. For quasi-free atoms, scattering processes from bound atoms might occur. We study both, hopping and scattering processes, using the more appropriate representation for each case, namely, the Wannier representation for hopping of tight-bound atoms and the Bloch representation for scattering of quasi-free atoms. Since second-quantization expressions are quite similar in Bloch and Wannier representations, we write all following expressions in the Bloch representation. The correspondent expressions in the Wannier representation are obtained by replacing the Bloch operators and wavefunctions with their analog in the Wannier representation.

In the large detuning approximation, and after adiabatic elimination of the excited electronic states, two-body dipole-dipole interactions between ground-state atoms are described by the second-quantization operator

$$\hat{V}_{dd} = \frac{1}{2} \sum_{\nu_1 \nu_2 \nu_1' \nu_2'} \hat{b}^\dagger_{\nu_1'} \hat{b}^\dagger_{\nu_2'} \hat{b}_{\nu_2} \hat{b}_{\nu_1} V_{\nu_1 \nu_2}^{\nu_1' \nu_2'} + H.c., \tag{15}$$

where

$$V_{\nu_1 \nu_2}^{\nu_1' \nu_2'} = \int d\boldsymbol{r}_1 \int d\boldsymbol{r}_2 \phi^*_{\nu_1'}(\boldsymbol{r}_1) \phi^*_{\nu_2'}(\boldsymbol{r}_2) V_{M_1 M_2}^{M_1' M_2'}(\boldsymbol{r}_1, \boldsymbol{r}_2) \phi_{\nu_1}(\boldsymbol{r}_1) \phi_{\nu_2}(\boldsymbol{r}_2). \tag{16}$$

Explicit expressions for the effective dipole-dipole interaction operator $V_{M_1 M_2}^{M_1' M_2'}(\boldsymbol{r}_1, \boldsymbol{r}_2)$ are given in [23]. However, it is useful to backtrack from these results and to consider $V_{M_1 M_2}^{M_1' M_2'}(\boldsymbol{r}_1, \boldsymbol{r}_2)$ in its integral form prior to carrying out the integration over the wavevector $\boldsymbol{k}$-space of the continuum of vacuum modes of the electromagnetic field [24]:

$$V_{M_1 M_2}^{M_1' M_2'}(\boldsymbol{r}_1, \boldsymbol{r}_2) = \frac{V_0}{k_L^3} \sum_{\alpha,\beta} \mathcal{P} \int \frac{k d\boldsymbol{k}}{k - k_L} \times$$

$$\left[\mathcal{A}_{\alpha\beta}^{M_1 M_2, M_1' M_2'}(\Omega_{\boldsymbol{k}}) \chi_{\alpha\beta}(\boldsymbol{r}_1, \boldsymbol{r}_2) + \mathcal{B}_{\alpha\beta}^{M_1 M_2, M_1' M_2'}(\Omega_{\boldsymbol{k}}) \chi_{\alpha\beta}(\boldsymbol{r}_2, \boldsymbol{r}_1) \right], \tag{17}$$

where $\mathcal{P}$ stands for the principal part of the integral and $V_0 = \hbar\Gamma(\Omega^2/8\delta^2)$. The functions $\mathcal{A}_{\alpha\beta}^{M_1 M_2, M_1' M_2'}(\Omega_{\boldsymbol{k}})$, $\mathcal{B}_{\alpha\beta}^{M_1 M_2, M_1' M_2'}(\Omega_{\boldsymbol{k}})$ are linear combinations of spherical harmonics [24], and $\Omega_{\boldsymbol{k}}$ is the solid angle in the $\boldsymbol{k}$-space .

Note that the dependency on $\mathbf{r}_1$ and $\mathbf{r}_2$ in Eq. (17) is only through the momentum transfer operators

$$\chi_{\alpha\beta}(\mathbf{r}_i, \mathbf{r}_j) = \exp\left[\imath\left(\mathbf{Q}_\alpha \cdot \mathbf{r}_i - \mathbf{Q}_\beta \cdot \mathbf{r}_j\right)\right], \quad i, j = 1, 2, \tag{18}$$

where $\boldsymbol{Q}_{\alpha,\beta} = \boldsymbol{k} - \boldsymbol{K}_{\alpha,\beta}$, which are not symmetric under exchange of particles. The dipole-dipole interaction in the adiabatic approximation, where both atoms remain in their electronic ground state, involves four photons, as sketched in Fig. (3). Atom 1 absorbs a photon from a laser and emits one with wavevector $\boldsymbol{k}$ into the vacuum field. The latter is reabsorbed by Atom 2, which emits a photon into a laser mode and remains in the ground state.

The matrix elements of $\chi_{\alpha\beta}(\boldsymbol{r}_i, \boldsymbol{r}_j)$ are conveniently evaluated by carrying out the spatial integration of (16) prior to the $\boldsymbol{k}$-space integration that appears in (17). One finds, for example,

$$\langle \nu_1', \nu_2' | \chi_{\alpha\beta}(\boldsymbol{r}_1, \boldsymbol{r}_2) | \nu_1, \nu_2 \rangle = \mathcal{S}_{\nu_1, \nu_1'}(\boldsymbol{Q}_\alpha) \mathcal{S}_{\nu_2, \nu_2'}(-\boldsymbol{Q}_\beta), \tag{19}$$

where

$$\mathcal{S}_{\nu,\nu'}(\boldsymbol{Q}) = \int d\boldsymbol{r} \phi^*_{\nu'}(\boldsymbol{r}) e^{\imath \boldsymbol{Q}\cdot\boldsymbol{r}} \phi_\nu(\boldsymbol{r}) \tag{20}$$

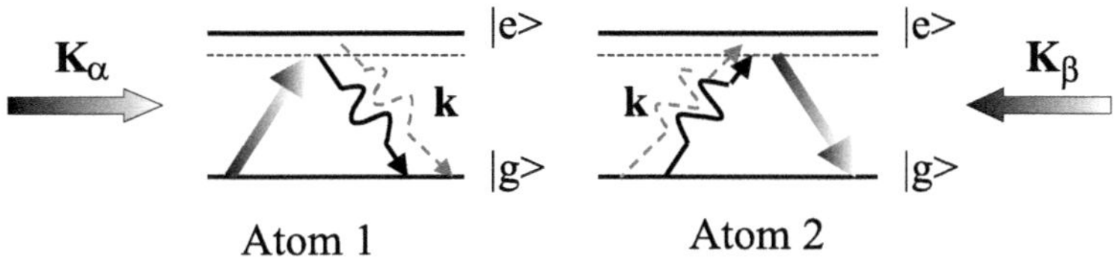

Fig. 3. Schematic representation of the dipole-dipole interaction in the adiabatic approximation.

is a matrix element of the momentum translation operator $\exp[\imath \boldsymbol{Q} \cdot \boldsymbol{r}]$ between two single-atom states, i.e, the individual momentum change suffered for one of the collisional partners. Both integrals in (19) depend through $\boldsymbol{Q}$ on the wave vector $\boldsymbol{k}$ of the electromagnetic vacuum, integration over whose space still remains.

4 Hopping Within the Wannier Representation

In this section we highlight the main results obtained in [24] for the effects of dipole-dipole interaction between bound atoms. Probabilities for hopping between lattice sites caused by the dipole-dipole interaction are, to first order, proportional to the squared absolute value of the matrix element (16) in the Wannier representation. We work within the tight-binding approximation for atoms in low-lying eigenstates and approximate the Wannier functions by 3D harmonic oscillator eigenfunctions, thereby obtaining a simple analytical result for the integrals (20):

$$\mathcal{S}_{\sigma\sigma'}(\boldsymbol{Q}) = \exp[\imath \boldsymbol{Q} \cdot \boldsymbol{R}_{\sigma\sigma'}] \exp\left[-|P_{\sigma\sigma'}(\boldsymbol{Q})|^2\right] \mathcal{F}_{\sigma\sigma'}(\boldsymbol{P}_{\sigma\sigma'}(\boldsymbol{Q})), \qquad (21)$$

where $\boldsymbol{R}_{\sigma\sigma'} = (\boldsymbol{R}_\sigma + \boldsymbol{R}_{\sigma'})/2$ is the center-of-mass coordinate of the lattice sites. Expressions for $\mathcal{F}_{\sigma\sigma'}$ and $\boldsymbol{P}_{\sigma\sigma'}(\boldsymbol{Q})$, and several plots showing the dependence of the matrix elements (21) on the individual atomic momentum transfer $\boldsymbol{Q}$, are given elsewhere [24]. For hopping resonant processes, where the atoms remain in the same energy state but change site, the matrix elements $\mathcal{S}_{\sigma,\sigma'}(\boldsymbol{Q})$ are maximum when the individual atomic momentum is conserved, i.e., $Q \to 0$. In that case, $P_{\sigma\sigma'}(0)$ denotes a scaled distance between sites [24]. The integration over $\boldsymbol{k}$-space yields simple expressions that involve only the locations of the sites and the overlap between the corresponding Wannier states,

$$\hat{V}_{dd} = \frac{1}{2} \sum_{\sigma,\sigma',\sigma'',\sigma'''} \hat{w}^\dagger_{\sigma'}, \hat{w}^\dagger_{\sigma'}, \hat{w}_{\sigma''}, \hat{w}_{\sigma'''} V_{\sigma'''\sigma'',\sigma\sigma'} \left(\boldsymbol{R}_{\eta,\eta'''}, \boldsymbol{R}_{\eta',\eta''}\right) \times$$
$$\exp\left[-P^2_{\eta,\eta'''}\right] \exp\left[-P^2_{\eta',\eta''}\right] \mathcal{F}_{\sigma,\sigma'''}\left(P^2_{\eta,\eta'''}\right) \mathcal{F}_{\sigma',\sigma''}\left(P^2_{\eta',\eta''}\right). \qquad (22)$$

Depending upon whether the final state is in the same sublattice or not, these processes may or may not involve a spin-flip in the electronic state of the atoms.

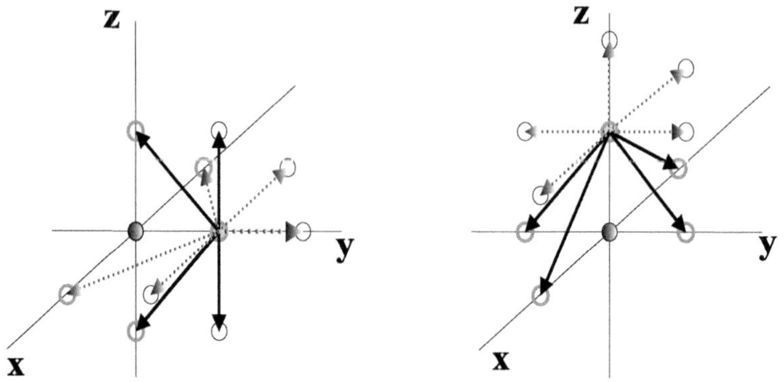

Fig. 4. Schematic representation of allowed (solid arrows) and forbidden (dashed arrows) near-neighbor hopping caused by dipole-dipole interaction. The site at the origin corresponds to the $M = -1/2$ sublattice, and the hopping atom belongs initially to the other sublattice.

Dominant terms in (22) are those that do not involve the transport of atoms on the lattice and that give the dipole-dipole contribution to the total lattice energy. Because of the Gaussian factors appearing in (22), the next lower-order terms correspond to transport processes where only one of the atoms moves to a nearest neighboring site, the other remaining in place. Allowed transport processes are indicated with solid arrows in Fig. 4, while the dashed arrows show forbidden processes.

In an *antiferromagnetic* structure, nearest neighbors belong to different sublattices, and the gaussian overlap factor in (22) favors hopping between sublattices. If atoms lie initially in the xy-plane, favored hopping occurs in the z direction, as shown at the left side of Fig. 4. For atoms initially aligned along the z direction, intralattice hopping is the only weakly allowed process. These facts taken together would seem to favor the alignment of atoms on *strings* along the z direction, where the dipole-dipole interaction energy between nearest neighbors is zero. However numerical simulation of the lattice dynamics induced by the anisotropy of the dipole-dipole interaction remains still to be performed.

5 Atom-Atom Diffraction in 1D Optical Lattices

The Bloch representation is better suited to describing diffraction by atomic lattices, since the atomic probability density is periodically distributed over the entire lattice, in close resemblance to gratings. We will determine here the differential cross-sections for spin-polarized atom-atom scattering from which diffraction patterns may be obtained. The asymptotic states and scattering geometry are illustrated schematically in Fig. 5. The projectile is initially in a quasi-free Bloch state that is approximated here by a plane wave with well-defined momentum $\hbar \boldsymbol{q}_1$ along the x axis. Target atoms in the lattice are in a bound Bloch

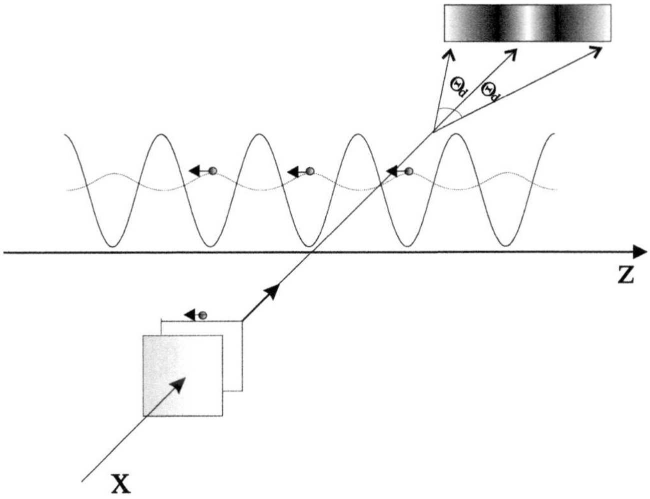

Fig. 5. Schematic representation of the scattering geometry. The optical potential and the bound Bloch function of the target atoms are represented above the z-axis.

state, whose probability density resembles a matter grating. We consider elastic processes where the bound atoms remain in the initial energy band, even if experiencing change in momentum and/or spin-flip. Through the dipole-dipole interaction, the incident beam is scattered into new quasi-free Bloch states with momentum $\hbar \boldsymbol{q}_1'$ that give rise to spin-dependent diffraction patterns. Diffractive effects due to the finite spatial width of the laser field are not included in the analysis.

Since the effective dipole-dipole potential is not central, partial wave scattering analysis into waves of well-defined angular momentum, specifically s-wave scattering, is not appropriate. Therefore, as a plausible starting point, we compute differential scattering cross sections within the first-order Born approximation, which is valid when the kinetic energy of the projectile is large in comparison to the binding energy of the target. To emphasize spin-dependence scattering channels are distinguished here by the initial M_i and final M_f values of the total magnetic number $M_1 + M_2$. Differential cross sections for multichannel scattering processes in the first Born approximation are proportional to the squared modulus of the interaction matrix element evaluated between the asymptotic states that define the channel:

$$\sigma^{M_i,M_f}_{\nu_1\nu_2\nu_1'\nu_2'}(\Theta_d, \Phi_d) = (2\pi)^4 \frac{m^2}{\hbar^4} \frac{|\boldsymbol{q}_1'|}{|\boldsymbol{q}_1|} |\langle \nu_1'\nu_2'|\hat{V}_{dd}|\nu_1\nu_2\rangle|^2, \tag{23}$$

where Θ_d and Φ_d specify the angular deflection of $\boldsymbol{q}_1'$ with respect to the direction of incidence, i.e., that of $\boldsymbol{q}_1$. If projectile and target belong initially to the $M = -1/2$ sublattice, $\sigma^{-1,-1}$ is the differential cross-section for scattering without spin-flip, $\sigma^{-1,0}$ corresponds to processes where the outgoing particle has spin opposite to that of the projectile, and $\sigma^{-1,1}$ to those where both particles change their spin orientation. Calculation of the matrix elements of the dipole-dipole operator in the Bloch representation is relatively straightforward, since the matrix elements of the translation momentum operator have a simple form. As an example, we show results for the direct process, where the projectile is also the scattered particle, and when Bloch states ν_1 and ν_1' are approximated by plane waves. The matrix element for the momentum translation operators (20) become

$$\begin{aligned} \mathcal{S}_{\nu_1,\nu_1'}(\mathbf{Q}_\alpha) &= \delta\left(\boldsymbol{k} - \Delta\boldsymbol{q}_1 - \boldsymbol{K}_\alpha\right), \\ \mathcal{S}_{\nu_2,\nu_2'}(-\mathbf{Q}_\beta) &= \sum_{s',s} c_{s'}^{\nu_2'} c_s^{\nu_2} \delta\left(\boldsymbol{k} + \Delta\boldsymbol{q}_2 - \boldsymbol{K}_\beta + [s'-s]\boldsymbol{b}\right), \end{aligned} \tag{24}$$

where $\Delta\boldsymbol{q}_i = \boldsymbol{q}_i' - \boldsymbol{q}_i$, $i = 1, 2$. The Dirac δ-functions clearly express the momentum transfer between the atoms, the vacuum field, the laser fields, and the reciprocal lattice.

A typical matrix element of the momentum transfer operator (19) involves the product of two Dirac δ-functions. The dependence on $\boldsymbol{k}$ can be then removed from one while the other is used in the remaining integral over the $\boldsymbol{k}$-space. Integration over the solid angle fixes a value for $\Omega_{\boldsymbol{k}}$ in (19). Away from the pole, the contributions from the upper and lower half $\boldsymbol{k}$-spaces cancel one another. The only contribution from the integral over k is that at the pole $k = k_L$, i.e., when the laser and vacuum photons have the same energy, imposing also the condition $|\boldsymbol{\Delta}_{\alpha,\beta}^{\pm}| = |\boldsymbol{K}_{\alpha,\beta} \pm \Delta\boldsymbol{q}_1| = k_L$. A pictorial representation is given in Fig. 6, where the spheres represent all directions of the $\boldsymbol{k}$ vector of magnitud k_L, as well as the allowed values of $\boldsymbol{\Delta}_\alpha^+, \alpha = 1, 2$. The plane wave approximation leads to $|\Delta q_{1z}| \leq 2k_L$ for direct processes and restricts the diffraction pattern to first order, or, equivalently, to elastic processes, where the projectile remains in the same energy band. A full Bloch wave treatment that allows for higher diffraction orders is being currently conducted.

The final integration over the continuum of modes of the electromagnetic field gives

$$\begin{aligned} \langle \nu_1'; \nu_2' | \hat{V}_{dd} | \nu_1; \nu_2 \rangle &= \frac{V_0}{k_L^3} \delta\left(\Delta\boldsymbol{q}_{1_\perp} + \Delta\boldsymbol{q}_{2_\perp}\right) \sum_{\alpha\beta=1}^{2} \sum_s c_s^{\nu_2} c_{s-\ell_{\alpha\beta}-\ell_z}^{\nu_2'} \times \\ &\left[\mathcal{A}_{\alpha\beta}^{M_1 M_2, M_2' M_1'}\left(\Omega_{\boldsymbol{\Delta}_\alpha^+}\right) + \mathcal{B}_{\alpha\beta}^{M_1 M_2, M_2' M_1'}\left(\Omega_{\boldsymbol{\Delta}_\beta^-}\right)\right], \end{aligned} \tag{25}$$

where $\ell_{\alpha\beta}$ takes the values $0, \pm 1$ and is determined by $K_\alpha - K_\beta = 2\ell_{\alpha\beta} k_L$. Irrespective of the channel, $\Delta q_{1z} + \Delta q_{2z} = \ell_z b$, ℓ_z being an integer, with the implication that the overall momentum conservation law for scattering processes in 1D optical lattices reads

$$\Delta\boldsymbol{q}_1 + \Delta\boldsymbol{q}_2 = \ell_z \boldsymbol{b}. \tag{26}$$

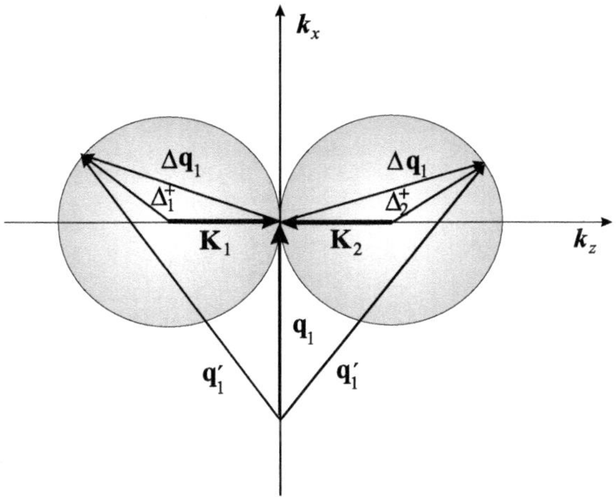

Fig. 6. Schematic representation of the scattering of a quasifree atom by a bound atom. The projectile has initial momentum $\hbar \boldsymbol{q}_1$, and emerges with final momentum $\hbar \boldsymbol{q}_1'$, the direction and magnitud of which are subject to the condition $|\boldsymbol{\Delta}_{1,2}^+| = k_L$.

Expressions analogous to (11) are found for exchange processes that result from the symmetrization of the atomic wave function, as well as for elastic processes that involve spin-flip of one or both collision partners. The method applied here gives analytical expressions in terms of the coefficients that appear in (9), which are obtained numerically.

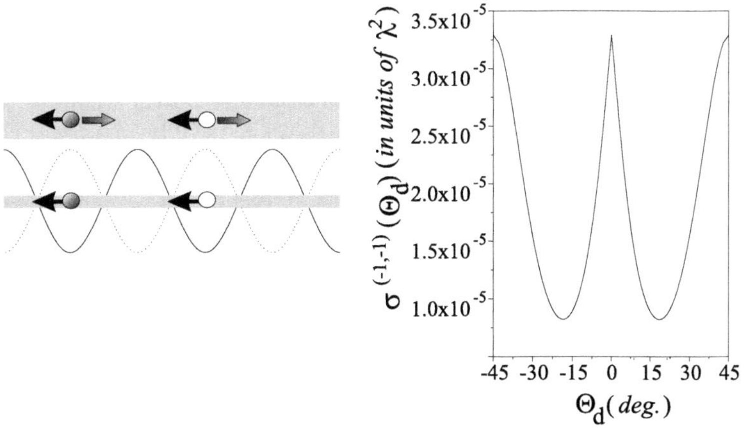

Fig. 7. Differential cross-section for scattering without spin-flip at incidence along the x-axis. Θ_d is the angle between the momenta of the scattered and incident atoms, and $\Phi_d = 0$. At the left side appears a schematic sketch of the process: full (empty) balls represent initial (final) states of bound and quasi-free atoms.

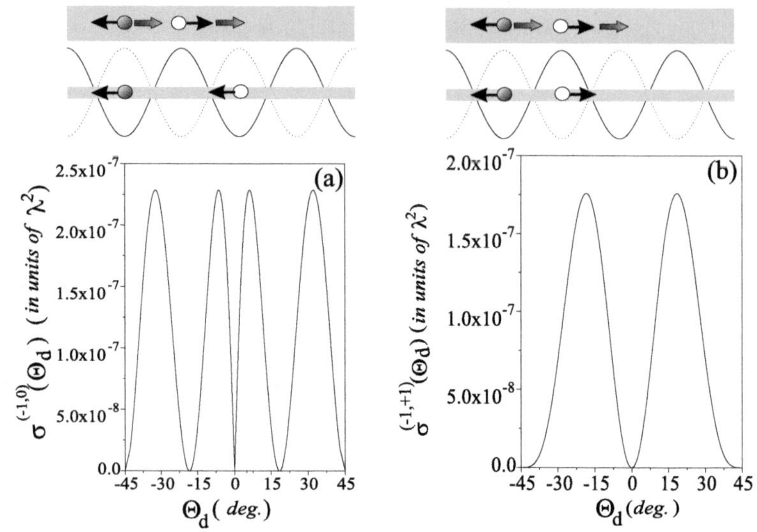

Fig. 8. Differential cross-section for dipole-dipole scattering involving (**a**) spin-flip of the projectile, (**b**) spin-flip of both interacting atoms. Schematic sketches for each process appear above the plots of the correspondent cross-sections. Scattering geometry and other conventions are as specified above in Fig. 7.

Differential cross-sections were calculated for a shallow optical potential with $U_0 = -2.1E_R$, where the band $n = 1$ is already a conduction band with energy $E_1(q_z = 0) = 3.97E_R$, and the band $n = 0$ is bound with $E_0(q_z = 0) = -0.13E_R$. The initial momentum of the projectile corresponds to incidence perpendicular to the quantization axis. The result for elastic scattering without spin-flip, $\sigma^{-1,-1}$, is shown in Fig. 7 as a function of the longitudinal momentum transfer as well as of the scattering angle Θ_d for $\Phi_d = 0$. Dependence on the azimuth scattering angle Φ_d can also be obtained from expression (11). The maximum at $\Theta_d = 0$ corresponds to scattering in the forward direction, whereas the maxima at $\Theta_d = \pm 45°$ are due to Bragg scattering that results when the momentum of the projectile is changed by the primitive vector of the reciprocal lattice, $\pm\boldsymbol{b}$.

Cross sections for scattering processes involving spin-flip are smaller, since they require atomic hopping between the two coexisting sublattices. Whenever the process leads to a change of the total magnetic quantum number, both forward and Bragg scattering are suppressed. This behavior is observed in the cross-section $\sigma_{el}^{-1,0}(\Theta_d)$, Fig. 8(a), for a process where only the spin of the scattered atom is flipped, and also in the function $\sigma^{-1,+1}(\Theta_d)$, shown in Fig. 8(b), which corresponds to a scattering process where both atoms undergo spin-flip. Both cross-sections clearly lack forward and Bragg scattering maxima. The spatial translation factor that relates eigenfunctions belonging to different sublattices implies a change in orbital angular momentum that compensates the spin-flip in order to conserve the total angular momentum.

6 Summary and Conclusions

The quantum character of the two-body dipole-dipole interaction taking place as a momentum transfer through the vacuum modes of the electromagnetic field has been emphasized. This character is important when the quantum nature of the translation motion of the atoms becomes significant. In optical lattices the translational eigenstates of non-interacting atoms can be represented, as in condensed matter physics, either by Bloch eigenfunctions or by localized Wannier functions, the latter being better suited to the description of atoms in the tight-binding approximation. The Wannier representation allowed us to find analytical expressions for the matrix elements of the effective dipole-dipole operator responsible for intersublattice hopping and to conclude that the dipole-dipole interaction in *fcc* lattices should favor atom alignment in strings along the quantization axis. The analytical approach used here is completely general and can be applied to different Bravais lattices. We have also studied scattering processes to first order between quasifree and bound atoms in 1D optical lattices, and obtained differential cross-sections for spin-polarized scattering that lead to diffraction patterns where Bragg scattering is suppressed whenever the total magnetic number is not conserved, as could be expected from the requirement of total angular momentum conservation. Our results might be relevant for models that involve bound atoms in optical lattices as a possible realization of quantum computers, or that involve atom-atom interactions as a means of entanglement control.

The authors acknowledge financial support from COLCIENCIAS and greatfully thank P. Meystre, J. Charris, W. Rhodes, and P. Jessen for helpful discussions.

References

1. J. Dalibard and C. Cohen-Tannoudji: Opt. Soc. Am. B **6**, 2023 (1989).
2. P. J. Ungar, D. S. Weiss, E. Riis, and S. Chu: Opt. Soc. Am. B **6**, 2058 (1989).
3. G. Grynberg, B. Lounis, P. Verkerk, J.-Y. Courtois, and C. Salomon: Phys. Rev. Lett. **70**, 2249 (1993).
4. P. S. Jessen, C. Gerz, P. D. Lett, W. D. Phillips, S. L. Rolston, R. J. C. Spreeuw, and C. I. Westbrook: Phy. Rev. Lett. **69**, 49 (1992).
5. M.R. Andrews, C.G. Townsend, H.-J. Miesner, D.S. Durfee, D.M. Kurn, and W. Ketterle: Science **275**, 637 (1997) M.-O. Mewes, M.R. Andrews, D.M. Kurn, D.S. Durfee, C.G. Townsend, and W. Ketterle: Phys. Rev. Lett. **78**, 582 (1997).
6. M. Ben Dahan, E. Peik, J. Reichel, Y. Castin and C. Salomon: Phy. Rev. Lett. **76**, 4508 (1996).
7. Q. Niu, X.-G. Zhao, G. A. Georgakis, and M. G. Raizen: Phys. Rev Lett. **76**, 4504 (1996).
8. S. R. Wilkinson, C. F. Bharucha, K. W. Madison, Q. Niu, and M. G. Raizen: Phys. Rev. Lett. **76**, 4512 (1996).
9. M. Raizen, C. Salomon, and Q. Niu: Phys. Today **50**, 30 (1997).
10. C. F. Bharucha, K. W. Madison, P. R. Morrow, S. R. Wilkinson, B. Sundaram, and M. G. Raizen: Phys. Rev. A **55**, R857 (1997).

11. Q. Niu and M. G. Raizen: Phys. Rev. Lett. **80**, 3491 (1998).
12. I. H. Deutsch and P. S. Jessen; Phys. Rev. A **57**, 1972 (1998).
13. G. K. Brennen, C. M. Caves, P. S. Jessen, and I. H. Deutsch: Phys. Rev. Lett. **82**, 1060 (1999).
14. D. Jaksch, H.-J. Briegel, J. I. Cirac, C. W. Gardiner, and P. Zoller: Phys. Rev. Lett. **82**, 1975 (1999).
15. A. Hemmerich: Phys. Rev. A **60**, 943 (1999).
16. K. Berg-Sørensen, Y. Castin, E. Bonderup and K. Mølmer: J. Phys. B **25**, 4195 (1992) K. Berg-Sørensen, Y. Castin, K. Mølmer, and J. Dalibard: Europhys. Lett. **22**, 663 (1993).
17. S. Marksteiner, K. Ellinger, and P. Zoller: Phys. Rev A **53**, 3409, (1996).
18. W. Greenwood, P. Pax, and P. Meystre: Phys. Rev. A **56**, 2109 (1997).
19. C. Menotti and H. Ritsch: Appl. Phys. B **69**, 311 (1999).
20. P. Münstermann, T. Fischer, P. Maunz, P.W.H. Pinkse, and G. Rempe: Phys. Rev. Lett. **84**, 4068 (2000).
21. G. Lenz and P. Meystre: Phys. Rev. A **48**, 3365 (1993).
22. H.-J. Briegel, T. Calarco, D. Jaksch, J.I. Cirac, and P. Zoller: J. Mod. Opt. **47**, 415 (2000) T. Calarco, E. A. Hinds, D. Jaksch, J. Schmiedmayer, J.I. Cirac, and P. Zoller: Phys. Rev. A **61**, 022304 (2000).
23. E. V.Goldstein, P. Pax and P. Meystre: Phys. Rev. A **53**, 2604 (1996).
24. A. M. Guzmán and P. Meystre: Phys. Rev. A **57**, 1139 (1998).
25. K. J. Petsas, A. B. Coates, and G. Grynberg: Phys. Rev. A. **50**, 5173 (1994).
26. Y. Castin, J. Dalibard, and C. Cohen-Tannoudji: *L*ight Induced Kinetic Effects on Atoms, ed. by L. Moi, S. Gozzini, C. Gabbanini, E. Arimondo, and F. Strumina (ETS Editrice, Pisa, 1992).
27. Y. Castin and J. Dalibard: Europhys. Lett. **14**, 761 (1991).
28. M. Abramowitz and I. A. Stegun: *H*andbook of Mathematical Functions, (Dover, New York, 1970).
29. N. W. Ashcroft and N. D. Mermin: *S*olid State Physics (Saunders College, Philadelphia, 1976).
30. C. J. Joachain: *Q*uantum Collision Theory, (North-Holland, Amsterdam, 1975)
31. P. J. Martin, B. G. Oldaker, A. H. Miklich, and D. E. Pritchard: Phys. Rev. Lett. **60**, 515 (1988).

Part IV

Quantum Optics and Applications

Time Delay and Tunneling

H.M. Nussenzweig

Instituto de Fisica da UFRJ.C.P. 68528, Rio de Janeiro, R.J., Brazil

Abstract. A brief critical survey is given of the main approaches to the tunneling time problem: when a particle has tunneled through a one-dimensional barrier, how long did it take to cross it? A new approach, based on Faddeev's formulation of one-dimensional scattering and on the evaluation of the average dwell time for an arbitrary wave packet, indicates that the problem is ill-posed. Furthermore, observations of apparent superluminal propagation do not violate causality.

1 Introduction

Few problems have given rise to as much controversy as that of the existence and proper evaluation of the time taken by a particle in tunneling through a potential barrier. Review articles [1] and conference proceedings [2] have been devoted to this question. Here we present a brief survey of various proposals, including our own recent contribution [3] leading to the conclusion that the tunneling time problem is an ill-posed one. Contrary to claims that have sometimes been made, no superluminal causality-violating effects are found.

2 The Eisenbud-Wigner Time Delay

In nonrelativistic quantum scattering, consider an s-wave incoming wave packet,

$$r\psi_{in} \approx \int_0^\infty \mid A(E) \mid \exp\left\{ i \left[-\frac{p}{\hbar} r - \frac{E}{\hbar} t + \delta(E) \right] \right\} dE, \ (r \to \infty) \tag{1}$$

where $\mid A(E) \mid$ is peaked at $E = E_0$. As is done in the derivation of group velocity [4] one usually identifies the time of passage of its "center" at a given r with the *stationary phase point*

$$t = t_0 - \frac{r}{\nu}, \ \nu = \frac{dE}{dp} \mid_{E_0}, \ t_0 = \hbar \frac{d\delta}{dE} \mid_{E_0} . \tag{2}$$

The corresponding outgoing wave packet,

$$r\psi_{out} \approx \int_0^\infty \mid A(E) \mid \exp\left\{ i \left[\frac{p}{\hbar} r - \frac{E}{\hbar} t + 2\eta(E) + \delta(E) \right] \right\} dE, \ (r \to \infty) \tag{3}$$

M. Orszag, J.C. Retamal (Eds.): LNP 575, pp. 229–243, 2001.

where $S(E) = \exp[2i\eta(E)]$ is the S-function(η is the phase shift), centered at:

$$t = t_0 + \frac{r}{\nu} + \Delta t \tag{4}$$

where

$$\Delta t = 2\hbar\frac{d\eta}{dE} = -i\hbar S^{-1}(E)\frac{dS}{dE}, \tag{5}$$

(evaluated at E_0) is the *Eisenbud-Wigner time delay*[5].

For scattering by a hard core of radius a, for which $S(E) = \exp(-2ika)$ where $k \equiv p/\hbar = m\nu/\hbar$, this leads to a time advance, $\Delta t = -2a/\nu$ ($\nu = p/m = \hbar k/m$), arising from the surface reflection, whereas, for a sharp isolated resonance at E_r, with (for E near E_r)

$$S(E) \approx \frac{E - E_r - \frac{1}{2}i\Gamma}{E - E_r + \frac{1}{2}i\Gamma}, \tag{6}$$

we get

$$\Delta t = \frac{\hbar\Gamma}{(E-E)^2 + \frac{1}{4}\Gamma^2} \sim \frac{\hbar}{\Gamma} \; for \; E \sim E_r, \tag{7}$$

a delay of the order of the *resonance lifetime.*

3 Tunneling Time as Group Delay

Now apply the same idea to the tunneling of a one-dimensional wave packet through a *rectangular potential barrier* (e. g., a semiconductor layer) of height V_0 and width a. The stationary solution, respectively in the regions (left, within, right), is:

$$\exp(ikx) + R\exp(-ikx), c\exp(-\varkappa x) + d\exp(\varkappa x), T\exp(ikx) \tag{8}$$

where $T = \tau\exp(i\varphi)$ is the complex transmission amplitude and $\varkappa$ the penetration depth. The group velocity approach leads to the *barrier transmission group delay*

$$\Delta t_g = \hbar\frac{d\varphi}{dE}, \tag{9}$$

interpreted as a tunneling time delay, meant to represent the difference between the time spent within the barrier (variously called *barrier interaction time*, *transversal time* and *tunneling time*) and the time spent in the same region for free particles(no barrier). For opaque barrier ($\tau \ll 1$),

$$\Delta t_g = \frac{2m}{\hbar k\varkappa} = \frac{2}{\nu\varkappa}, \tag{10}$$

is the flight time through two penetration depths, independent of the barrier width a. Therefore, a thick enough barrier would apparently lead to *superluminal* propagation.[6]

3.1 Critique of tunneling time as group delay

The idea that one can always extrapolate from comparing the asymptotic behavior of a transmitted wave packet "center" with the corresponding asymptotic "center" of the incident one, to draw inferences on what happens within the interaction region, is unwarranted. Besides the shape distortion due to interference with the reflected wave packet (reflection is strong for an opaque barrier), the barrier behaves like a dispersive medium, reshaping also the transmitted component. Because of these effects, a transmitted peak need not be causally related with an incident peak. Fast components are favored in transmission, so a transmitted peak can even *precede* the arrival of the incident one at the barrier [7].

In absorbing or amplifying media, reshaping can also lead to apparent superluminality. In inverted (nonlinear) media, one can even have *superluminal solitons*: the weak leading edge of a pulse unleashes the energy already stored in the medium, building up the advanced peak, like a fuse setting up an explosion. It has been shown [8] that these self-transparent solitons are rigorously causal: the domain of dependence and the domain of influence of the solution, at a given spacetime point, are respectively contained within the backward and the forward light cones. The apparent superluminality is an optical illusion, as in the well-known rotating searchlight effect: the intersection of its beam with a fixed wall may appear to move arbitrarily fast, but the succession of light spots along the wall does not represent causal propagation.

4 The Larmor Times

Imagine introducing a *small* perturbing uniform magnetic field $\mathbf{B}$, of magnitude B_0 and direction z, within the barrier, with the incident electron spin polarized along the direction x of propagation. Inside the barrier, the spin component in the (xy) plane, precessing at the Larmor frequency ω_L, plays the role of a clock. The total precession angle $\theta_y$defines the Baz-Rybachenko [9] measure of the "tunneling time", the *in-plane Larmor time* $\mathbf{T}_y$, given by

$$\theta_y = \omega_L \mathbf{T}_y. \tag{11}$$

However, Büttiker [10] pointed out that an incident x-polarized spin is a superposition of $\uparrow$ and $\downarrow$ in z, which, by magnetic splitting, see different effective barriers. This tends to align the spin with $\mathbf{B}$, leading to an equivalent out-of-plane z rotation (much larger than the y one, for an opaque barrier)

$$\theta_z = \omega_L \mathbf{T}_z. \tag{12}$$

defining the *out-of-plane Larmor time* $\mathbf{T}_z$.

The full 3-dimensional rotation is associated with the *Büttiker-Landauer time* [11]

$$\mathbf{T}_x = \sqrt{(\mathbf{T}_y)^2 + (\mathbf{T}_z)^2} \equiv \mathbf{T}_{B+L}. \tag{13}$$

This is interpreted as the *barrier interaction (tunneling) time* for particles that get transmitted. A similar analysis for reflected particles leads to corresponding *reflection times.*

For a rectangular barrier of height V_0 and width a, the following results are found:

$$\mathbf{T}_y = -\hbar \frac{\partial \varphi}{\partial V_0}, \mathbf{T}_z = -\hbar \frac{\partial}{\partial V_0} \ln \tau, \tag{14}$$

where $T = \tau \exp(i\varphi)$ is the complex transmission amplitude defined in (8). For an opaque barrier ($\tau \ll 1$) ,$\mathbf{T}_z$ is dominant and is approximately equal to the "bounce time" $a/ \mid v \mid$, where $\mid v \mid \equiv \hbar\varkappa/m$ is the magnitude of the (imaginary) *intrabarrier* "velocity"

5 Stationary Dwell Time

A new definition of delay time was given by Smith [12]. For a three dimensional potential scattering, let $\mathbf{T}_{d,R}$ be the time spent by the particle within a distance R from the scattering center. This is called the *dwell time* within this region. Let $\mathbf{T}_{d0,R}$ be the corresponding dwell time in the absence of the scatterer. The *dwell time delay* due to scattering is defined as

$$\Delta\mathbf{T}_d = \mathbf{T}_{d,R} - \mathbf{T}_{d0,R} \ (R \to \infty). \tag{15}$$

Smith treated only *stationary* scattering, defining $\mathbf{T}_{d,R}$ by

$$\mathbf{T}_{d,R} \equiv \frac{\textit{total probability that } r \leq R}{\textit{flux of particles through } \partial R}, \tag{16}$$

where ∂R is the boundary of the sphere, $\mid r \mid = R$. This measures a *relative phase space weight.* For s waves, it is found that

$$\mathbf{T}_{d,R} = 2\hbar \frac{d\eta_0}{dE} + \frac{2R}{\nu} - \frac{1}{k\nu} \sin\left[2(kR + \eta_0)\right], \tag{17}$$

so that

$$\mathbf{T}_{d0,R} = \frac{2R}{\nu} - \frac{1}{k\nu} \sin\left[2(kR)\right] \tag{18}$$

is the *free* dwell time.

To get rid of the oscillatory terms, one averages over R:

$$\langle .. \rangle_R \equiv \frac{1}{R} \int_R^{2R} \langle .. \rangle dR, \textit{ so that } \langle \mathbf{T}_{d0,R} \rangle_R = \frac{2R}{\nu}. \tag{19}$$

Thus, one finds, in agreement with the Eisenbud-Wigner (group) delay,

$$\langle \Delta\mathbf{T}_d \rangle \equiv \lim_{R \to \infty} \left[\langle \mathbf{T}_{d,R} \rangle_R - \langle \mathbf{T}_{d0,R} \rangle_R \right] = 2\hbar \frac{d\eta_0}{dE}. \tag{20}$$

5.1 Remarks

1. The oscillatory term is related to the uncertainty principle [13]: in the low-energy limit $kR \ll 1$, $\mathbf{T}_{d0,R} \ll 2R/\nu$: a particle cannot be localized within a domain much smaller than its de Broglie wavelength.

2. For a rectangular barrier in one dimension, $\mathbf{T}_d$ is equal to the in-plane Larmor time $\mathbf{T}_y$.

3. $\mathbf{T}_d$ tends to be dismissed as a measure of tunneling time, since it does not distinguish between transmitted and reflected particles. It becomes very small for an opaque barrier, when most particles are reflected.

6 Other Approaches to Tunneling Time

6.1 Modulation of the barrier or of the incident wave

Introduce a small oscillatory modulation of the barrier height [14] or of the incident wave amplitude [15], by superposing energies E and $E + \Delta E$. Look for a temporal modulation period $\hbar/\Delta E$ such that the transmitted wave begins to depart from an adiabatic following of the incident profile. This is attributed to the effect of the barrier, yielding a "barrier interaction time delay" $\mid \Delta\mathbf{T}_E \mid$, where

$$\Delta\mathbf{T}_E = -i\hbar\frac{d}{dE}\left[\ln \mathbf{T}(E)\right], \tag{21}$$

and $\mathbf{T}(E)$ is the complex transmission amplitude. Note that $\mathrm{Re}\mathbf{T}(E) = \Delta t_g$ is the transmission group delay (9).

6.2 Conditional dwell time

Restricting attention only to particles that get transmitted (or reflected) amounts to choosing a subensemble, *pre*selected (by taking particle incidence from the left) - and *post*selected (by restriction to transmitted or reflected particles). The "tunneling time" would be a property of the system during the time interval between the initial and final situations.

Aharonov, Albert and Vaidman [16,17] treated such *conditional* (weak) measurements. The following is a simplified (nonrigorous!) version of their argument: *assume* that, given an initial state $\mid i\rangle$ and a final state $\mid f\rangle$, the conditional probability $P(A_i \mid f)$ for the particle to be found in an eigenstate A_i, given that it is subsequently found in $\mid f\rangle$, is

$$P(A_i \mid f) = \frac{\langle \mathrm{Proj}(f)\mathrm{Proj}(A_i)\rangle}{\langle \mathrm{Proj}(f)\rangle}, \text{where } \mathrm{Proj}(f) \equiv \mid f\rangle\langle f \mid . \tag{22}$$

Note that the product of projectors need not be Hermitian! The weak value of A is then given by

$$\langle A \rangle_{fi} \equiv \frac{\langle f \mid A \mid i \rangle}{\langle f \mid i \rangle} \tag{23}$$

This is generally *complex*, because of the non-hermiticity.

Weak values of observables have many paradoxical properties: the spin component of a spin 1/2 particle in a given direction may be arbitrarily large, superluminal velocities are found, etc.

Steinberg [18] used weak measurements to evaluate the *conditional dwell time* $\mathbf{T}_{d,c}$, defined as the value of the projector onto the region within the barrier, for particles that finally get transmitted.

The result is a complex combination of the in-plane and out-of-plane Larmor times :

$$\mathbf{T}_{d,c} = \mathbf{T}_y - i\mathbf{T}_z. \tag{24}$$

By considering the weak value in terms of a von Neumann-style measurement, he also found that $\mathbf{T}_y$ is a property of the tunneling process, regardless of how it is observed, while $\mathbf{T}_z$ has the character of a *susceptibility*, describing the sensitivity of the tunneling probability to small perturbations.

6.3 Path integrals

Approaches based on Feynman's path integral formulation of quantum mechanics were taken by Sokolovski and Baskin [19] and by Fertig [20]. One begins by defining, for a *classical path* $x(t)$, *the time spent under the barrier* by

$$t_{cl}\left[x(t)\right] = \int_0^\infty dt' \theta_B \left[x(t')\right], \tag{25}$$

where $\theta_B = 1$ under the barrier, $= 0$ otherwise. Then, the "*Feynman*" *tunneling time* $\mathbf{T}_F$ is *defined* as the functional average of $t_{cl}\left[x(t)\right]$ over paths that start far to the left and end far to the right, with the weight function $\exp\{iS\left[x(t)/\hbar\right]\}$, where S is the action associated with the path $x(t)$. The result is, again, the same complex "tunneling time" (24):

$$\mathbf{T}_F = \mathbf{T}_{d,c} = \mathbf{T}_y - i\mathbf{T}_z. \tag{26}$$

6.4 Critique of the "Feynman" approach

The weighting by $\exp\{iS\left[x(t)/\hbar\right]\}$is as little justified as trying to define an "effective position" by

$$x_{eff} \equiv \frac{\int x\psi(x)dx}{\int \psi(x)dx}. \tag{27}$$

One is replacing probability *density* by probability *amplitude*. Further discussion [21] leads to the conclusion that a probability distribution of tunneling times cannot be defined (is "unspeakable").

Another nonconventional approach is based on Bohm trajectories [22]. Since such trajectories do not cross, one is led to a very peculiar picture of the transmission process [11].

7 Average Wave Packet Dwell Time

In three-dimensional Schrödinger scattering, by probability conservation, the total probability of finding the particle within a distance R of the scattering center, at time t, is

$$P(R,t) = \int_{-\infty}^{t} \Phi(R,t')dt'. \tag{28}$$

where $\Phi(R,t)$ is the inward probability flux through $\mid r \mid \equiv R$ at time t, and the incoming *wave packet* is normalized to 1 incident particle for $t \to \infty$.

The *average dwell time* of the wave packet within $\mid r \mid \leq R$ is

$$\langle \mathbf{T}_d(R) \rangle = \int_{-\infty}^{\infty} P(R,t)dt. \tag{29}$$

The *average dwell time delay* is given by

$$\langle \Delta\mathbf{T}_d \rangle = \langle \mathbf{T}_d(R) \rangle - \langle \mathbf{T}_{d,0}(R) \rangle, \tag{30}$$

where R is large enough, so that the interaction is negligible at $r = R$ (asymptotic domain). The last term of (30) is the free-particle dwell time.

For an incoming partial wave l, one finds [13]

$$\langle \Delta\mathbf{T}_{d,l} \rangle = 2\hbar \langle \frac{d\eta_l}{dE} \rangle_{in}, \tag{31}$$

where $\langle .. \rangle_{in}$ is the *expectation value over the energy spectrum of the incident wave packet*. The result holds for an arbitrary wave packet, includes shape effects and gives a reasonable account of the average time delay in all cases.

This treatment was extended to a *plane incident wave train* [23], yielding a similar expression for the average nonrelativistic *angular dwell time delay* in a given observation direction θ. The result is a spectral average, over the incident wave train, of the expression $2\hbar\frac{\partial}{\partial E}[\arg f(E,\theta)]$, obtained (through stationary phase arguments) by Froissart, Goldberger and Watson [24], where f is the total scattering amplitude.

An extension to a relativistic case was also made [25], for scattering of an *electromagnetic wave train*, with similar results (but the partial wave result is 1/2 of the Eisenbud-Wigner one!)

We finally turn to our new approach to the tunneling time problem [3], based upon computing the average wave packet dwell time for *one-dimensional scattering*. We begin by reconsidering one-dimensional quantum scattering theory.

8 One-Dimensional Quantum Scattering Theory

We assume that the potential $V(x)$ is real and that $| V(x) |, | xV(x) |$ are integrable. The following are independent scattering solutions:

$$\begin{aligned} \psi_{\rightarrow}(k,x) &\approx \exp(ikx) + R_{-}(k)\exp(-ikx), \quad & x \to -\infty \\ &\approx T(k)\exp(ikx), & x \to \infty \end{aligned} \tag{32}$$

$$\begin{aligned} \psi_{\leftarrow}(k,x) &\approx T(k)\exp(-ikx), & x \to -\infty \\ &\approx \exp(-ikx) + R_{+}(k)\exp(ikx), \quad & x \to \infty \end{aligned} \tag{33}$$

The equality of $T(k)$ follows from time reversal invariance.

The S-matrix is given by [26]

$$\mathbf{S}(k) = \begin{bmatrix} T(k) & R_{+}(k) \\ R_{-}(k) & T(k) \end{bmatrix} \tag{34}$$

For real k, it satisfies the symmetry relations $S_{ij}(-k) = [S_{ij}(k)]^*$ and the unitarity relations

$$| R_{-}(k) |^2 + | T(k) |^2 = | R_{+}(k) |^2 + | T(k) |^2 = 1$$

$$T(k)R_{-}^{*}(k) + R_{+}(k)T^{*}(k) = 0 \tag{35}$$

These allow the parametrization:

$$R_{-}(k) \equiv \rho(k)\exp\left[i\psi(k)\right], \; T(k) \equiv \tau(k)\exp\left[i\varphi(k)\right]$$

$$R_{+}(k) \equiv -\rho(k)\exp\left[i\left\{2\varphi(k) - \psi(k)\right\}\right], \tag{36}$$

where

$$\left[\rho(k)\right]^2 + \left[\tau(k)\right]^2 = 1. \tag{37}$$

If $V(x) = V(-x)$ (parity), one has $R_{+} = R_{-}$, so that

$$\psi = \varphi - \frac{\pi}{2} \; (V \text{ symmetric}) \tag{38}$$

A symmetric V is the analogue of a central potential in three dimensions.

8.1 The time delay matrix

We define it (as is done in three-dimensional scattering) by

$$\mathbf{T} = -i\hbar \mathbf{S}^{-1} \frac{d\mathbf{S}}{dE} = \mathbf{T}^{\dagger}. \tag{39}$$

The eigenvalues of $\mathbf{T}$ are found to be

$$(\Delta t)_{\pm} = \hbar\varphi' \pm \hbar \sqrt{\rho^2(\psi' - \varphi')^2 + (\frac{\tau}{\rho})^2 \left[(\ln\tau)'\right]^2}, \tag{40}$$

where $f' \equiv df/dE$. By (9), $\hbar\varphi' = \Delta t_g$ is the "transmission group delay".Similarly, $\hbar\psi'$ is a reflection group delay and [cf. (14)] $\hbar(\ln\tau)'$ is reminiscent of the out-of-plane Larmor time $\mathbf{T}_z$.

In particular, for a *symmetric* potential, $V(-x) = V(x)$,

$$(\Delta t)^{sym}_{\pm} = \hbar\varphi' \pm \hbar(\frac{\tau}{\rho})(\ln\tau)' \tag{41}$$

What is the physical interpretation of these eigenvalues?

8.2 New basis functions

Instead of $\psi_{\rightarrow}(k,x)$ and $\psi_{\leftarrow}(k,x)$ (*unilateral* incidence), related to the *transfer matrix,* we take *symmetric* and *antisymmetric* bilateral incidence, appropriate to the *S-matrix*:

$$\begin{aligned} \psi_s(k,x) &= \psi_{\rightarrow}(k,x) + \psi_{\leftarrow}(k,x), \\ \psi_a(k,x) &= \psi_{\rightarrow}(k,x) - \psi_{\leftarrow}(k,x). \end{aligned} \tag{42}$$

It follows that

$$\begin{aligned} \psi_{s,a}(k,x) &\approx \exp(ikx) + [R_-(k) \pm T(k)]\exp(-ikx), \quad & x \to -\infty, \\ &\approx \pm\exp(-ikx) + [\pm R_+(k) + T(k)]\exp(ikx), \quad & x \to +\infty, \end{aligned} \tag{43}$$

$\psi_s(k,x)$ is a one-dimensional analogue of an s wave, and $\psi_a(k,x)$ of a p wave. The *most general one-dimensional wave packet* is a superposition of these solutions:

$$\psi(x,t) = \int_0^{\infty} [A_s(E)\psi_s(k,x) + A_a(E)\psi_a(k,x)] \exp(\frac{-iEt}{\hbar})dE, \tag{44}$$

$$E \equiv \frac{(\hbar k)^2}{2m} = \frac{1}{2}m\nu^2.$$

In terms of the momentum eigenfunction components $a(k)$, one has

$$A_{s,a}E \equiv (\frac{m}{2\hbar^2})\,[a(k) \pm a(-k)]\,/\mid k \mid$$

Therefore, one needs *bilateral* incidence to represent even a free ("unidirectional") Gaussian wave packet, as well as (by analyticity) any wave packet initially confined by a shutter.

9 The Average One-Dimensional Wave Packet Dwell Time

We normalize (44) to represent one *incident particle*

$$\lim_{X\to\infty}\int_{-\infty}^{\infty}\left[j_{in}(-X,t)-j_{in}(X,t)\right]dt=1, \tag{45}$$

where $j_{in}(X,t)$ is the *incident* probability current. This leads to

$$4\pi\hbar\int_{0}^{\infty}\left[\mid A_s(E)\mid^2+\mid A_a(E)\mid^2\right]\nu dE=1 \tag{46}$$

With this normalization,

$$P(\mid x\mid< X,t)=1-\int_{|x|>X}\mid\psi(x,t)\mid^2 dx \tag{47}$$

is the probability to find the particle inside $(-X,X)$ at time t, and

$$\langle\mathbf{T}_d(X)\rangle=\int_{-\infty}^{\infty}P(\mid x\mid< X,t)dt, \tag{48}$$

is the *average dwell time* for the wave packet within this region.

The associated *average dwell time delay* in the scattering process is

$$\langle\Delta\mathbf{T}_d\rangle=\lim_{X\to\infty}\int_{-\infty}^{\infty}dt\int_{|x|>X}\left[\mid\psi_0(x,t)\mid^2-\mid\psi(x,t)\mid^2\right]dx \tag{49}$$

where $\psi_0(x,t)$ is the corresponding free-particle wave packet. Note that (51) can be computed entirely from the asymptotic behavior in the scattering region (large enough X). The *average free-particle dwell time* in $\mid x\mid< X$ is found to be

$$\langle\mathbf{T}_{d,0}(X)\rangle=4\pi\hbar\int_{0}^{\infty}\{\left[\mid A_s(E)\mid^2+\mid A_a(E)\mid^2\right]\frac{2X}{\nu}$$
$$+\left[\mid A_s(E)\mid^2-\mid A_a(E)\mid^2\right]\frac{\sin(2kX)}{k\nu}\}\nu dE \tag{50}$$

which is the *spectral average* of the classical free time of flight $2X/\nu$ over the energy spectrum of the incident wave packet, plus additional oscillatory terms (again connected with the uncertainty relation). Such terms are always of order $\langle\frac{\lambda_{DB}}{\nu}\rangle$ and can be eliminated by averaging in X over several de Broglie wavelengths λ_{DB}. One can now compute the average *interacting* dwell time. We give the results [3] for a symmetric potential.

9.1 Average one-dimensional dwell time for a symmetric potential

For $V(-x) = V(x)$, one finds

$$\langle \Delta \mathbf{T}_d \rangle^{sym} = 4\pi\hbar \int_0^\infty \{[\mid A_s(E) \mid^2 (\Delta t)_+^{sym} + \mid A_a(E) \mid^2 (\Delta t)_-^{sym}]$$

$$- \left[\mid A_s(E) \mid^2 + \mid A_a(E) \mid^2\right] \rho \frac{\lambda_{DB}}{\nu} \cos(\Delta\varphi)$$

$$+ \left[\mid A_s(E) \mid^2 - \mid A_a(E) \mid^2\right] \left[\tau \sin(\Delta\varphi) - \sin(kX)\right] \frac{\lambda_{DB}}{\nu}\} \nu dE \qquad (51)$$

where $\Delta\varphi \equiv kX + \varphi$, and $(\Delta t)_\pm^{sym}$ are the eigenvalues (41) of the time delay matrix. This yields their *physical interpretation:* $(\Delta t)_+^{sym} [(\Delta t)_-^{sym}]$ is associated with symmetric [antisymmetric] bilateral incidence.

Unilateral incidence is a particular case, with $A_s(E) = \pm A_a(E) \equiv \frac{A(E)}{2}$(+ for $\rightarrow$, - for $\leftarrow$),

$$\langle \Delta \mathbf{T}_d \rangle_{\rightarrow}^{sym} = 2\pi\hbar \int_0^\infty \mid A(E) \mid^2 \left[\hbar\varphi' - \rho \frac{\lambda_{DB}}{\nu} \cos(\Delta\varphi)\right] \nu dE \qquad (52)$$
$$= \langle \Delta \mathbf{T}_d \rangle_{\leftarrow}^{sym},$$

which (apart from the oscillatory term) is the spectral average of the transmission group delay (9). This result generalizes an identity obtained by Hauge et al. [27].

The completely general case (nonsymmetric potential and arbitrary incidence) leads to somewhat more complicated formulas.

The *average dwell time*, for symmetric V, satisfies

$$\frac{1}{2}(\langle \mathbf{T}_d \rangle_s + \langle \mathbf{T}_d \rangle_a)^{sym} = \mathrm{Re}\langle \mathbf{T}_X^E - \frac{\lambda_{DB}}{\nu} R_X(E) \rangle = \langle \mathbf{T}_d \rangle_{\rightarrow}^{sym} = \langle \mathbf{T}_d \rangle_{\leftarrow}^{sym},$$

$$\frac{1}{2}(\langle \mathbf{T}_d \rangle_s - \langle \mathbf{T}_d \rangle_a)^{sym} = - \mathrm{Im}\langle \frac{\tau}{\rho} (\mathbf{T}_X^E - \frac{\lambda_{DB}}{\nu} R_X(E)) \rangle \qquad (53)$$

where $R_X(E) \equiv \mid R(E) \mid \exp(i\Delta\varphi(E))$, and $\mid \mathbf{T}_X^E \mid$ is the *Büttiker-Landauer barrier interaction time* (for modulated incident amplitude).

10 Rectangular Potential

The transmission ($\mathbf{T}$) and reflection ($\mathbf{R}$) amplitudes for a *potential step* of height V_0, with $E > V_0$, are $\mathbf{R} = (N-1)/(N+1)$, $\mathbf{T} = 2\sqrt{N}/(N+1)$,where $N = \sqrt{1 - V_0/E}$ is the "*refractive index*". For a rectangular barrier of width a, define $\delta \equiv 4Nka$ (double traversal phase) and $\chi \equiv 2(N-1)ka$ (single traversal phase shift). Then, the barrier transmission and reflection amplitudes, for energies above the barrier, are

$$T = \frac{\mathbf{T}^2 \exp(i\chi)}{1 - \mathbf{R}^2 \exp(i\delta)}, R = \frac{2i\mathbf{R} \exp(i\chi)}{1 - \mathbf{R}^2 \exp(i\delta)} \sin(2Nka). \tag{54}$$

Since

$$\frac{1}{1 - \mathbf{R}^2 \exp(i\delta)} = 1 + \mathbf{R}^2 \exp(i\delta) + \left[\mathbf{R}^2 \exp(i\delta)\right]^2 + ... \tag{55}$$

(54) has an immediate interpretation, in terms of an infinite series of multiple internal reflections and transmissions (Fabry-Perot interferometer). Thus, *transmission and reflection are inextricably entangled*: this is at the root of the problems in attempting to define a "tunneling time".

Around a *transmission resonance*, $Nka = m\pi (m = 1, 2, 3..)$, one finds

$$\langle \mathbf{T}_d(a) \rangle_{\rightarrow} \approx \langle \frac{a}{\nu_1} + \frac{\mathbf{R}^2}{\mathbf{T}^2} \frac{2a}{\nu_1} \rangle, \tag{56}$$

where $\nu_1 = N\nu$ is the internal velocity over the barrier. For *a tall barrier* ($N \ll 1$) and a near-barrier-top resonance, the dwell time is $\gg \langle a/\nu_1 \rangle$ (the classical time of flight), because of the large average number $\approx (\mathbf{R}^2/\mathbf{T}^2)$ of oscillations before emergence.

10.1 Average dwell time in tunneling

For tunneling, $E < V_0$, $N \rightarrow i \mid N \mid$. We find

$$\langle \mathbf{T}_d(a) \rangle_{\rightarrow} = \langle \mathbf{T}_y \rangle, \tag{57}$$

$$\langle \mathbf{T}_d(a) \rangle_{s,a} = \langle \mathbf{T}_y \pm \frac{\tau}{\rho} \mathbf{T}_z \rangle, \tag{58}$$

where $\mathbf{T}_y$ is Büttiker's *in-plane Larmor time*, $\mathbf{T}_z$ is Büttiker's *out-of-plane Larmor time*, and $\tau \ll 1$ for an opaque barrier.

This leads to a reinterpretation of the Larmor time results:

$$\langle \mathbf{T}_y \rangle = \frac{1}{2}(\langle \mathbf{T}_d \rangle_s + \langle \mathbf{T}_d \rangle_a),$$

$$\langle \frac{\tau}{\rho} \mathbf{T}_z \rangle = \frac{1}{2}(\langle \mathbf{T}_d \rangle_s - \langle \mathbf{T}_d \rangle_a). \tag{59}$$

Remarks

(i) Büttiker's out-of-plane Larmor "reflection time" is

$$\mathbf{T}_{zR} = -(\frac{\tau}{\rho})^2 \mathbf{T}_z.$$

(ii) While $\langle \mathbf{T}_y \rangle$ is indeed an average barrier dwell time, the differential character of $\langle \mathbf{T}_z \rangle$ shows that it does behave like a *susceptibility,* associated with the difference between symmetric and antisymmetric excitation, which affects the interference between transmission and reflection.

(iii) The oscillatory terms transform $\partial/\partial E \to -\partial/\partial V_0$ [cf.(14)].

11 Main Problems with Previous Treatments

We sum up the above critical discussions and add some comments:

1. *Group (phase) delay*: the extrapolation from the asymptotic to the near-barrier region, as well as the idea of following the wave packet "center", are unjustified, because of reshaping and self-interference effects.

2. Larmor times: as pointed out by Steinberg, T_z measures a susceptibility to small perturbations, dependent on the observation process, rather than an intrinsic property of tunneling. Also: for a barrier of arbitrary shape (not rectangular), the *rate* of the spin clock is not uniform, varying along the barrier.

3. *Weak measurement*: because of ensemble postselection, it tends to amplify the effect of rare fluctuations (the results are very imprecise), leading, for instance, to apparent superluminality. Also: it predicts *negative* values for the average time spent by the reflected particles on the right side of the barrier [28]!

4. *Feynman (and related) path summations*:: As already follows from the Fabry-Perot picture (infinite number of internal reflections), one cannot separate "to-be-transmitted" and "to-be-reflected" portions of paths. Interference (in time!) between paths that spend different amounts of time within the barrier does not allow one to define a "probability distribution of tunneling times" [21].

Conclusion:

"Tunneling time" is an ill-defined concept.

However, as was previously found for nonrelativistic three-dimensional quantum scattering [13,23], as well as for electromagnetic scattering [25], *average wave packet dwell time is well defined.*

12 Ten Good Features of the Average Dwell Time

1. It is defined entirely within the principles and standard interpretation of quantum theory: no "exotic" reinterpretations are employed.

2. It is positive definite (this is van Kampen's [29] "causality condition" for nonrelativistic quantum scattering).

3. It holds for arbitrary wave packets, and thereby accounts for shape effects.

4. Combined with Faddeev's formulation, it extends the S-matrix theory of time delay to one-dimensional scattering.

5. It leads to a "normal" physical interpretation of the Larmor times, in terms of the *eigenvalues of the time delay matrix.*

6. It leads to reasonable results both for tunneling and for above-barrier energies (in particular, at transmission resonances).

7. It does *not* distinguish between reflected and transmitted particles. This has been taken as a defect, but it is actually a virtue, since transmission and reflection are inextricably intertwined.

8. It is, in principle, *measurable.* As was first suggested by Pippard, the average dwell time within the barrier can, in principle, be measured, by making the barrier weakly absorbing [30]. Absorption - like the dwell time - does not distinguish between reflected and transmitted portions of a wave packet.

9. The average dwell time also has a well-defined physical meaning in terms of its connection with *statistical mechanics*: contributions to the partition function, from each phase space domain, must be weighted by the average time which the system spends in that domain.

10. The time delay matrix has a close relationship with the *density of states* [31]:

$$\rho(E) - \rho^{(0)}(E) = \frac{1}{2\pi i} Tr \frac{d}{dE} \log \mathbf{S} = \frac{1}{2\pi i} Tr(\mathbf{S}^{-1} \frac{d\mathbf{S}}{dE}) \tag{60}$$

where $\rho^{(0)}(E)$ is the free-particle density of states.

Acknowledgments

It is a pleasure to thank Carlos Alberto Aragão de Carvalho for valuable discussions. This work was partially supported by Conselho Nacional de Desenvolvimento Científico e Tecnológico.

References

1. E. H. Hauge and J. A. Støvneng: Rev. Mod. Phys. **61,** 917 (1989); R. Landauer and Th. Martin, Rev. Mod. Phys. 66, 217 (1994); R. Y. Chiao and A. M. Steinberg, in: Progress in Optics, ed. by E. Wolf (Elsevier, Amsterdam, 1997), vol. 37, p. 345.
2. Tunneling and its Implications, ed. by D. Mugnai, A. Ranfagni and L. S. Schulman (World Scientific, Singapore, 1997).
3. H. M. Nussenzveig: Phys. Rev. A 62, 042107-1 (2000).

4. H. Lamb : Hydrodynamics, 6th edn.(Cambridge University Press, Cambridge 1953), p. 395.
5. Wigner, E. P., 1955 Phys. Rev. 98, 145 .
6. T. E.Hartman: J. Appl. Phys. 33, 3427 (1962).
7. R. Landauer and Th. Martin: Solid State Commun. 84, 115 (1992).
8. R. C. T. da Costa: J. Math. Phys. 11, 2799 (1970).
9. A. I. Baz, Yad. Fiz. 4, 252 (1967) [Sov. J. Nucl. Phys. 4, 229 (1967)]; V. F. Rybachenko, ibid. 5, 895 (1967) [ibid. 5, 635 (1967)].
10. M. Büttiker: Phys. Rev. B 27, 6178 (1983).
11. R. Landauer and Th. Martin, Rev. Mod. Phys. 66, 217 (1994).
12. F. T. Smith: Phys. Rev. 118, 349 (1960).
13. H. M. Nussenzveig: Phys. Rev. 177, 1848 (1969).
14. M. Büttiker and R. Landauer: Phys. Rev. Lett. 49, 1739 (1982).
15. M. Büttiker and R. Landauer: Phys. Scr. 32, 429 (1985).
16. Y. Aharonov, D. Z. Albert and L. Vaidman: Phys. Rev. Lett. 60, 1351 (1988).
17. Y. Aharonov and L. Vaidman: Phys. Rev. A 14, 11 (1990).
18. A. M. Steinberg: Phys. Rev. Lett. 74, 2405 (1995).
19. D. Sokolovski and L. M. Baskin: Phys. Rev. A 36, 4604 (1987).
20. H. A. Fertig: Phys. Rev. Lett. 65, 2321 (1990).
21. N. Yamada: Phys. Rev. Lett. 83, 3350 (1999).
22. C. R. Leavens and G. C. Aers, in: Scanning Tunneling Microscopy III, eds. R. Wiesendanger and H. J. Güntherodt (Springer, New York, 1993), p. 105.
23. H. M. Nussenzveig: Phys. Rev. D 6, 1534 (1972).
24. M. Froissart, M. L. Goldberger and K. M. Watson: Phys. Rev. 131, (1963).
25. H. M. Nussenzveig: Phys. Rev. A 55, 1012 (1997).
26. L. D. Faddeev: Trudy Mat. Inst. Steklov 73, 314 (1964) [Am. Math. Soc. Transl. 2, 139 (1964)]; K. Chadan and P. C. Sabatier: Inverse Problems in Quantum Scattering Theory, 2nd. ed. (Springer, New York, 1989).
27. E. H. Hauge, J. P. Falck and T. A. Fjeldly: Phys. Rev. B 36, 4203 (1987).
28. G. Iannaconne, in: Tunneling and its Implications, ed. by D. Mugnai, A. Ranfagni and L. S. Schulman (World Scientific, Singapore, 1997), p. 292.
29. N. G. van Kampen: Phys. Rev. 91, 1267 (1953).
30. Y. Z. Huang and C. M. Wang: J. Phys. Condens. Matter 3, 5915 (1991).
31. J. Friedel: Philos. Mag. 43, 153 (1952). R. Balian and C. Bloch: Ann. Phys. (N.Y.) 85, 514 (1974).

Giant Intensity-Intensity Correlations and Quantum Interference in a Driven Three-Level Atom

S. Swain[1] and Z. Ficek[2]

[1] School of Mathematics and Physics, The Queen's University of Belfast, Belfast BT7 1NN, The United Kingdom
[2] Department of Physics, The University of Queensland, Brisbane, QLD 4072, Australia

Abstract. We analyze the role of quantum interference in a three-level Vee system using symmetric and antisymmetric superposition states. For the laser coupling to both transitions, we show that both transitions may exhibit extended periods of darkness, and that huge values for the intensity-intensity correlation function are possible. For the laser driven on just one transition, using the same technique we show that the gain of a probe beam, recently predicted by Menon and Agarwal, is due to an unexpected amplification on a completely-inverted, nondecaying transition.

1 Introduction

Quantum interference is a very old technique which has recently returned to prominence because of its utility in manipulating spontaneous emission and other radiative properties of atomic systems. The suppression of spontaneous emission and the control of radiative processes are topics of considerable current importance in spectroscopy, optical communication and laser physics. Spontaneous emission, which leads to noise and decoherence, arises from the coupling of a system to the environment and imposes limits on precision spectroscopic measurements, population inversions and quantum information processing. Different schemes have been proposed to suppress the spontaneous process, including frequency dependent reservoirs [1], optical waveguides [2] and squeezed vacuum fields [3].

One of the most interesting approaches is based on quantum interference, and involves the modification of spontaneous emission through the vacuum induced coherences between two atomic transitions which occurs only if the dipole moments of the two transitions are not perpendicular. The spontaneous emission from one transition modifies the spontaneous emission of the other. Quantum interference lies at the heart of many important radiative effects such as quenching of spontaneous emission [4,5,6], high-contrast resonances [7,8], electromagnetically induced transparency [9], amplification without population inversion [15], ultra-slow group velocities of light [11], photon superbunching [12], phase dependent population inversions and phase control of spontaneous emission [14],

M. Orszag, J.C. Retamal (Eds.): LNP 575, pp. 244–261, 2001.

and the intensity control of very narrow spectral lines [15]. The essential feature of these phenomena is the generation of superposition states which can be used to modify the interaction with the environment—for example, by adjusting the polarizations of the atomic dipole moments, or the amplitudes and phases of the external driving fields. With a suitable choice of parameters, the superposition states can be decoupled from the environment and the population can be trapped in these states without decaying to the lower levels. Such states are called "dark states".

Here, we are concerned with some effects of quantum interference in a three-level V-type atom consisting of two excited levels coupled to a singlet ground level by electric dipole interactions. The atom is driven by a single-mode laser which couples to both atomic transitions. The basis of our treatment is a unitary transform to superposition states of closely-spaced levels which diagonalizes the spontaneous emission for certain values of the parameters. We also consider the situation where the laser couples to only one of the transitions. This case can be treated by the same approach, except that the superposition states now involve dressed states.

Many interesting effects have been predicted for the singly-coupled system which would not appear if the field couples to both atomic transitions. For example, Paspalakis and Knight [16] have shown that spontaneous emission from a three-level V-system can be controlled via the phase difference of two lasers each coupled to only one of the two transitions. Gong *et al.* [17] have found unexpected population inversions, and Paspalakis *et al.* [18] have shown that this system can exhibit gain without population inversion,

electromagnetically-induced transparency and enhancement of the index of refraction. Kocharovskaya *et al.* [19] have found drastic modifications in the relaxation processes of the system when one of the dressed states of a driven transition crosses a nearby unperturbed atomic level. Menon and Agarwal [20] calculated the Autler-Townes spectrum of a weak probe beam monitoring a three-level V-type atom driven by a strong laser field, and showed that the spectrum can exhibit gain instead of the usual absorption. We analyze this situation here, and demonstrate that the essential process is the interesting effect of amplification on a dark transition that is not accompanied by spontaneous emission [21].

Where the atom couples to both transitions, we are concerned with the effects of quantum interference on the intensity-intensity correlations in a three-level atom. These correlations have been investigated under a variety of conditions [13,22,23,24,25,26]. Here we study a coherently driven V-type atom, and find two major results. Firstly, in the presence of quantum interference there exist extended *simultaneous* periods of darkness in the fluorescence from the two atomic transitions, even for equal decay rates of the excited levels. This is in contrast to the dark periods previously predicted [27,28] for a V-type atom with uncorrelated transitions and significantly different decay rates, where the atom "prefers" to stay in the transition with the larger decay rate and there is a small probability of finding the system in the other transition. We show that the atom occupies superposition states and emits a stream of photons exhibiting strong correlations. Secondly, we show that under appropriate conditions, the maximum value of the normalized, second-order field correlation function can be made huge (values of the order of hundreds or thousands) under conditions of quantum interference, whereas the corresponding maximum in the absence of quantum interference is 'normal' (values of the order of unity). We are not aware of such large values being previously reported in the literature for single atoms, but indefinitely large correlation functions for a pair of two-level atoms were reported by Wiegand [29]. There the effect arises because in the three dimensional problem with two atoms, there are positions where the field vanishes.

2 The Three-Level Model: Both Transitions Excited

The atom consists of two closely-spaced excited levels $|1\rangle$ and $|2\rangle$ and the ground level $|0\rangle$. The transition between the excited levels is forbidden in the electric dipole approximation. The excited states are separated in frequency by $\omega_{21} = \omega_2 - \omega_1$, where ω_k is the transition frequency from $|k\rangle$ to $|0\rangle$, $k = 1, 2$. The atom is driven by a strong laser field of frequency ω_L which may couple to both transitions.

In the frame rotating with the laser frequency the master equation is of the form

$$\dot{\rho} = -i[H, \rho] + \mathcal{L}\rho, \tag{1}$$

where the Hamiltonian is

$$H = (\Delta - \omega_{12}) A_{11} + \Delta A_{22} + [(\Omega_1 A_{10} + \Omega_2 A_{20}) + \text{H.c.}], \tag{2}$$

and the damping term is

$$\begin{aligned}\mathcal{L}\rho = &\frac{\gamma_1}{2}(2A_{01}\rho A_{10} - A_{11}\rho - \rho A_{11}) + \frac{\gamma_2}{2}(2A_{02}\rho A_{20} - A_{22}\rho - \rho A_{22}) \\ &+\frac{\gamma_{12}}{2}(2A_{01}\rho A_{20} - A_{21}\rho - \rho A_{21} + 2A_{02}\rho A_{10} - A_{12}\rho - \rho A_{12}),\end{aligned} \tag{3}$$

with units such that $\hbar = 1$. In (2), $\Delta = \omega_2 - \omega_L$ is the detuning between the frequency ω_2 of the $|0\rangle \rightarrow |2\rangle$ transition and the driving laser frequency, $2\Omega_k$ is the Rabi frequency of the kth transition, and $A_{ij} \equiv |i\rangle\langle j|$ is a transition operator. The spontaneous decay rate for the $|k\rangle \rightarrow |0\rangle$ transition is γ_k and γ_{12} is the vacuum induced cross-damping rate which arises from quantum interference between the transitions $|1\rangle \rightarrow |0\rangle$ and $|2\rangle \rightarrow |0\rangle$. It satisfies

$$\gamma_{12} = p\sqrt{\gamma_1\gamma_2}, \tag{4}$$

where $p \equiv \cos\theta$ and θ is the angle between the atomic dipole moments. The most interesting features arise for $p \simeq 1$, that is, for almost parallel dipole moments.

The properties of this system can be understood by transforming to new states which diagonalize the damping term in certain circumstances. We introduce the symmetric and antisymmetric superposition states of the closely-spaced excited levels:

$$|s\rangle = \frac{1}{\sqrt{\gamma_1+\gamma_2}}\left(\sqrt{\gamma_1}\,|1\rangle + \sqrt{\gamma_2}\,|2\rangle\right), \tag{5}$$

$$|a\rangle = \frac{1}{\sqrt{\gamma_1+\gamma_2}}\left(\sqrt{\gamma_2}\,|1\rangle - \sqrt{\gamma_1}\,|2\rangle\right). \tag{6}$$

In terms of these new states, the damping Liouvillean may be written

$$\mathcal{L}\rho = \frac{\gamma_s}{2}\left(2A_{0s}\rho A_{s0} - A_{ss}\rho - \rho A_{ss}\right) + \frac{\gamma_a}{2}\left(2A_{0a}\rho A_{a0} - A_{aa}\rho - \rho A_{aa}\right)$$
$$+\frac{\gamma_{as}}{2}\left(2A_{0s}\rho A_{a0} - A_{as}\rho - \rho A_{as} + 2A_{0a}\rho A_{s0} - A_{sa}\rho - \rho A_{sa}\right) \tag{7}$$

where

$$\begin{aligned}
\gamma_s &= \left(\gamma_1^2+\gamma_2^2+2p\gamma_1\gamma_2\right)/\left(\gamma_1+\gamma_2\right),\\
\gamma_a &= 2\gamma_1\gamma_2\left(1-p\right)/\left(\gamma_1+\gamma_2\right),\\
\gamma_{as} &= \sqrt{\gamma_1\gamma_2}\left(\gamma_1-\gamma_2\right)\left(1-p\right)/\left(\gamma_1+\gamma_2\right).
\end{aligned} \tag{8}$$

We see that $\gamma_{as} = 0$ when $\gamma_1 = \gamma_2$. The cross-relaxation terms, which are difficult to interpret, then vanish, and the decay from the symmetric and antisymmetric states is purely diagonal. However, when $p = 1$, both γ_a and γ_{as} are zero, and so the antisymmetric state becomes a *trapping state*. (Population can decay *into* the state, but not *out of* it.)

For simplicity, we restrict attention henceforth to the case $\gamma_1 = \gamma_2 = \gamma$. Then the Hamiltonian takes the form

$$H = \left(\Delta - \frac{1}{2}\omega_{12}\right)\left(A_{ss}+A_{aa}\right) - \frac{1}{2}\omega_{12}\left(A_{sa}+A_{as}\right)$$
$$+\left[\frac{(\Omega_1+\Omega_2)}{\sqrt{2}}A_{s0} + \frac{(\Omega_1-\Omega_2)}{\sqrt{2}}A_{a0} + \text{H.c.}\right]. \tag{9}$$

2.1 Second-order correlation functions

Assuming detection at a fixed field point $\boldsymbol{r}$, the normalized second-order two-time correlation function (intensity-intensity correlation) is defined as:

$$g^{(2)}\left(t;t+\tau\right) = \frac{G^{(2)}\left(t;t+\tau\right)}{G^{(1)}\left(t\right)G^{(1)}\left(t+\tau\right)}, \tag{10}$$

where the first- and second-order correlation functions, appearing in (10), can be expressed in terms of the positive and negative frequency parts of the electric field operator as

$$G^{(1)}(t) = \left\langle \boldsymbol{E}^{(+)}(t)\,\boldsymbol{E}^{(-)}(t)\right\rangle, \tag{11}$$

$$G^{(2)}(t;\tau) = \left\langle \boldsymbol{E}^{(+)}(t)\,\boldsymbol{E}^{(+)}(t+\tau)\,\boldsymbol{E}^{(-)}(t+\tau)\,\boldsymbol{E}^{(-)}(t)\right\rangle. \tag{12}$$

In the far-field zone, $r \gg c/\omega_0$, and $t > r/c$, the positive frequency part of the electric field operator is given by

$$\boldsymbol{E}^{(+)}(t) = \boldsymbol{E}_0^{(+)}(\boldsymbol{r},t) - \sum_{i=1}^{2} \frac{\boldsymbol{r}\times(\boldsymbol{r}\times\boldsymbol{\mu}_i)}{c^2 r^3}\omega_i^2 A_{i0}(t-r/c),$$

where $\boldsymbol{\mu}_k$ is the dipole moment of the kth transition. Since the field is initially in the vacuum state, the vacuum part $\boldsymbol{E}_0^{(+)}(\boldsymbol{r},t)$ does not contribute to the expectation values of the normally-ordered field operators and then we obtain the following expression for the correlation functions

$$G^{(1)}(t) = K\sum_{i,j=1}^{2} \gamma_{ij}\left\langle A_{i0}(t)\,A_{0j}(t)\right\rangle, \tag{13}$$

$$G^{(2)}(t,\tau) = K^2 \sum_{i,j,k,l=1}^{2} \gamma_{il}\gamma_{jk}\left\langle A_{i0}(t)\,A_{j0}(t+\tau)\,A_{0k}(t+\tau)\,A_{0l}(t)\right\rangle, \tag{14}$$

where K is a constant, and the γ_{ij} are the spontaneous decay rates introduced in (3) with $\gamma_{kk} \equiv \gamma_k$ and $\gamma_{12} = \gamma_{21}$.

In terms of the symmetric and antisymmetric superposition states, the correlation functions (13) and (14) can be written as

$$G^{(1)}(t) = K\left[\gamma_s\left\langle A_{s0}(t)\,A_{0s}(t)\right\rangle + \gamma_a\left\langle A_{a0}(t)\,A_{0a}(t)\right\rangle\right] \tag{15}$$

and

$$G^{(2)}(t,\tau) = K^2\left[\gamma_s\left\langle A_{s0}(t)\,U(t+\tau)\,A_{0s}(t)\right\rangle + \gamma_a\left\langle A_{a0}(t)\,U(t+\tau)\,A_{0a}(t)\right\rangle\right] \tag{16}$$

where from (8) $\gamma_s = \gamma(1+p)$ and $\gamma_a = \gamma(1-p)$, and

$$U(t+\tau) = \gamma_s A_{s0}(t+\tau)\,A_{0s}(t+\tau) + \gamma_a A_{a0}(t+\tau)\,A_{0a}(t+\tau). \tag{17}$$

For parallel dipole moments ($p = 1$) only the transition $|s\rangle \to |0\rangle$ contributes to the fluorescence intensity and the second-order correlation function, indicating that in this case the system reduces effectively to a two-level system. However, correlations between the emitted photons can be significantly different from those one would expect for a two-level system.

We here consider the second-order correlation function only for the case of a strong driving field, since the effects of quantum interference are most pronounced in this case. The weak driving field is discussed in [12]. We consider two experimental situations. The first is where the detector distinguishes between photons emitted from the two transitions (distinguishable photons), and the second where it does not (indistinguishable photons).

2.2 Distinguishable photons

If the photons emitted from the excited states to the ground state are distinguishable, e.g. by having significantly different polarizations or frequencies, then the following normalized second-order correlation functions of the steady-state fluorescence intensity can be defined [13]:

$$g_{ij}^{(2)}(\tau) = \lim_{t\to\infty} g^{(2)}(t;t+\tau) = \frac{\langle A_{i0}A_{j0}(\tau)A_{0j}(\tau)A_{0i}\rangle}{\langle A_{i0}A_{0i}\rangle}, \quad i,j=1,2,$$

where the angular brackets denote steady-state averages. $g_{ij}^{(2)}(\tau)$ is related to the joint probability that if a photon from the transition $|0\rangle \to |i\rangle$ is detected at time t, a photon is detected at time $t+\tau$ from the transition $|0\rangle \to |j\rangle$. In particular, we consider the following correlation functions

$$g_{11}^{(2)}(\tau) = g_{21}^{(2)}(\tau) = \langle A_{10}A_{10}(\tau)A_{01}(\tau)A_{01}\rangle / P_1 \tag{18}$$

$$g_{22}^{(2)}(\tau) = g_{12}^{(2)}(\tau) = \langle A_{20}A_{20}(\tau)A_{02}(\tau)A_{02}\rangle / P_2. \tag{19}$$

where the angular brackets denote the steady-state average and $P_k = \langle A_{k0}A_{0k}\rangle = \rho_{kk}(t\to\infty)$. The behavior of the correlation functions depends strongly on the value of ω_{21}. We find that for degenerate transitions, $\omega_{12}=0$, $g_{11}^{(2)}(\tau)$ and $g_{22}^{(2)}(\tau)$ oscillate with the Rabi frequency of the driving field and there is no qualitative difference between them for different values of p. The shape of the oscillations resembles that for a two-level atom [30,31], and indicates that the atomic dipole moments oscillate independently.

For the non-degenerate case, shown in Fig. 1 for $\omega_{12}=5\gamma$ and $\Delta=\omega_{12}/2$, the behavior of the correlation functions is qualitatively different to the case where $\omega_{12}=0$. For $p=0.99$, the values of $g_{11}^{(2)}(\tau)$ and $g_{22}^{(2)}(\tau)$ remain below unity for all times, which shows that for any τ the probability of emission of two photons from levels $|1\rangle$ or $|2\rangle$ is very small. We can interpret this as extended *simultaneous* periods of darkness in the fluorescence from the two atomic transitions: after detection of a photon at time $\tau=0$, detection of another photon at time $\tau>0$, emitted from levels $|1\rangle$ or $|2\rangle$, is very unlikely. The simultaneous periods of darkness appear only for correlated transitions ($p\simeq 1$). Dark periods of fluorescence have been predicted before [27,28] in a three-level atom with $p=0$ and significantly different transition rates γ_1 and γ_2, but they appeared on only one of the two atomic transitions.

As we later show, oscillations occur at the Rabi frequency 2Ω as well as at Ω.

2.3 Indistinguishable photons

We consider here the situation where the detector does not distinguish between the photons emitted from the two atomic transitions. The detector responds to the total field so that the correlation functions are given by (15) and (16). It is convenient to consider separately the following correlation functions, defined analogously to (18) and (19)

$$g_{ss}^{(2)}(\tau) = \langle A_{s0} A_{s0}(\tau) A_{0s}(\tau) A_{0s} \rangle / P_s, \tag{20}$$

$$g_{aa}^{(2)}(\tau) = \langle A_{a0} A_{a0}(\tau) A_{0a}(\tau) A_{0a} \rangle / P_a. \tag{21}$$

For the case $p \approx 1$, the case of greatest interest, by far the major contribution to the correlation function (16) comes from $g_{ss}^{(2)}(\tau)$, as the contribution from the asymmetric state, which is proportional to $(1-p)$, is negligible.

In Fig. 2, we plot the correlation functions (20) and (21) for $\Omega_1 = \Omega_2 = \Omega = 5\gamma, \omega_{12} = 5\gamma$ and $\Delta = \omega_{12}/2$. Again, the solid line is for $p = 0.99$ and the dashed line for $p = 0$. For the $p = 0.99$ case, there are very strong correlations of photons on the $|s\rangle \to |0\rangle$ transition, whereas the photons are strongly anticorrelated on the $|a\rangle \to |0\rangle$ transition. The function $g_{ss}^{(2)}(\tau)$ oscillates with the frequency $2\sqrt{2}\Omega$, which is the Rabi frequency in the symmetric basis, and attains its maximum value at time $\tau = (2\sqrt{2}\Omega)^{-1}\pi$. The decay rate is very small, and it takes a time in excess of $300\pi/\gamma_1$ before $g_{ss}^{(2)}$ gets close to unity. The function $g_{aa}^{(2)}(\tau)$ oscillates with frequency $\sqrt{2}\Omega$, and for $p \simeq 1$ is less than unity for all times, whereas for $p = 0$, it has a maximum value of around 2.8.

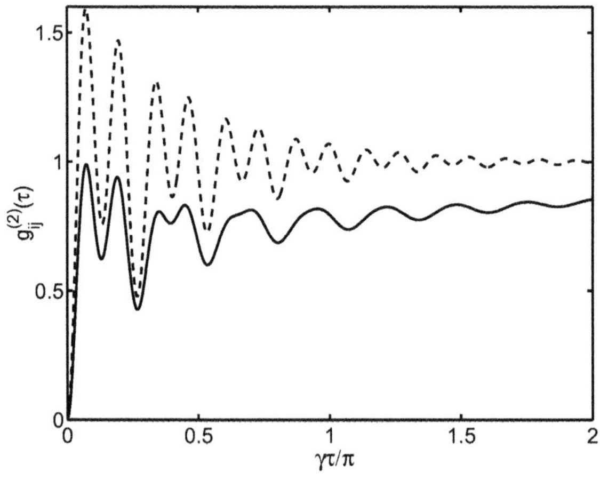

Fig. 1. Second-order correlation functions $g_{ij}^{(2)}(\tau), i, j = 1, 2$ for the case of distinguishable photons, and $\omega_{12} = 5\gamma, \Delta = \omega_{12}/2$ and $\Omega = 5\gamma$. We plot graphs for different β: $\beta = 0.99$ (solid line), and $\beta = 0$ (dashed line).

We should emphasize that for $p \simeq 1$, very large values of $g_{ss}^{(2)}(\tau)$ are possible, whereas for $p = 0$ its maximum value is of the order of unity. We have chosen the value $\Omega = 5\gamma_1$ in Fig. 2 so that the curves for $g_{ss}^{(2)}(\tau)$ for $p = 0.99$ and $p = 0$ are visible on the same graph: the maximum value of $g_{ss}^{(2)}(\tau)$ for $p \simeq 1$ is then about 22.5. However, much larger values are possible: for example, if we reduce the value of Ω to 0.5γ, leaving other parameters unchanged, then the maximum value increases to almost 1500!

As we reduce the value of ω_{12}, the difference between the $p = 0$ and $p = 0.99$ graphs for $g_{ss}^{(2)}(\tau)$ becomes less pronounced. For example, for $\omega_{12} = 0.1\gamma$, the correlation functions for $p = 0$ and $p = 0.99$ oscillate in a similar fashion with $g_{ss}^{(2)}(\tau) < 2$ for all times τ.

The dominant frequency in $g_{ss}^{(2)}(\tau)$ is the Rabi frequency $2\sqrt{2}\Omega$, whereas in $g_{aa}^{(2)}(\tau)$ it is $\sqrt{2}\Omega$, as we discuss later.

2.4 Interpretation of the results

The effect of quantum interference on the second-order correlation functions depends very sensitively on the splitting ω_{12} of the excited levels. For degenerate or nearly degenerate excited levels ($\omega_{12} \simeq 0$) the photon emissions are similar to those of a two-level atom, independent of quantum interference. For large splittings, the correlation functions $g_{ij}^{(2)}(\tau)$, $i, j = 1, 2$ and $g_{aa}^{(2)}(\tau)$ are smaller

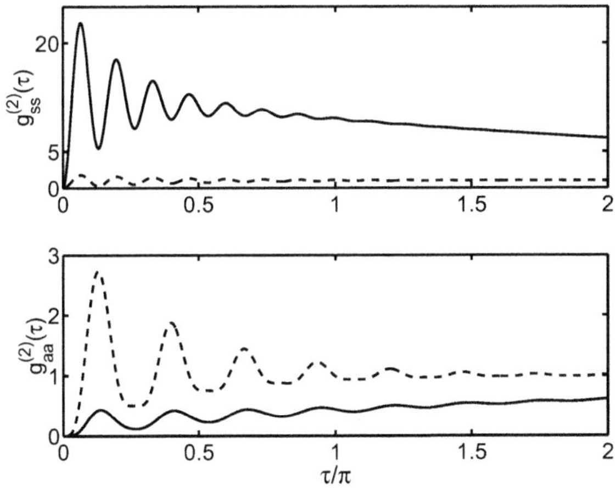

Fig. 2. Second-order correlation functions for the case of indistinguishable photons for $\omega_{12} = 5\gamma, \Delta = \omega_{12}/2$ and $\Omega = 5\gamma$. In the upper plot we present $g_{ss}^{(2)}(\tau)$ and in the lower $g_{aa}^{(2)}(\tau)$. The solid line is for $\beta = 0.99$ and the dashed line for $\beta = 0$. If we change β and Δ to $\beta = 1$ and $\Delta = 0.4\omega_{12}/2$, the graphs are almost identical.

than unity for all times τ, while $g_{ss}^{(2)}(\tau)$ exhibits strong correlations ($g_{ss}^{(2)}(\tau) \gg 2$) for $\tau \approx (2\sqrt{2}\Omega)^{-1}\pi$ which decay at a very low rate.

We can explain these features by considering the equations for the coherences obtained from the master equation (9) . They are

$$\begin{aligned}\dot{\rho}_{10} &= \left(-\frac{1}{2}\gamma + i\left(\Delta - \omega_{12}\right)\right)\rho_{10} - \frac{1}{2}\gamma_{12}\rho_{20} + i\Omega\left(\rho_{11} - \rho_{00}\right)\\ \dot{\rho}_{20} &= \left(-\frac{1}{2}\gamma + i\Delta\right)\rho_{20} - \frac{1}{2}\gamma_{12}\rho_{10} + i\Omega\left(\rho_{22} - \rho_{00}\right). \end{aligned} \tag{22}$$

For $\omega_{12} = 0$ these equations are symmetric between the states $|1\rangle$ and $|2\rangle$ and the coherences ρ_{10} and ρ_{20} oscillate in phase with frequency Δ. They are directly coupled by the cross-damping term γ_{12}, but for $\Omega \gg \gamma_i, \gamma_{12}$ the Rabi oscillations dominate over the spontaneous exchange of photons, resulting in independent oscillations of the atomic dipole moments.

When $\omega_{12} \neq 0$ and $\Delta = \omega_{12}/2$ however, the coherences oscillate with opposite phases indicating an exchange of photons between the states $|1\rangle$ and $|2\rangle$ which prevents photons being emitted from these levels. The coherences oscillate at frequencies $\pm\omega_{12}/2$ which accounts for the modulation of the Rabi oscillations seen in Fig. 1.

The situation is better understood by working in the symmetric/antisymmetric basis (5) and (6). Setting $\Delta = \omega_{12}/2$ and $\Omega = \Omega_1 = \Omega_2$ for simplicity, the master equation (9) using (7) and (9) yields the equations:

$$\dot{\rho}_{ss} = -\frac{1}{2}\gamma\left(1+p\right)\rho_{ss} - \frac{1}{2}i\omega_{12}\left(\rho_{sa} - \rho_{as}\right) - i\sqrt{2}\Omega\left(\rho_{s0} - \rho_{0s}\right), \tag{23}$$

$$\dot{\rho}_{aa} = -\frac{1}{2}\gamma\left(1-p\right)\rho_{aa} + \frac{1}{2}i\omega_{12}\left(\rho_{sa} - \rho_{as}\right). \tag{24}$$

We see that the laser field couples only to the symmetric state and both states decay independently to the ground state with different decay rates. For $\omega_{12} \neq 0$ the antisymmetric state is coupled to the symmetric state. Since the decay rate of the antisymmetric state, $\gamma(1-p)$, is very small for $p \approx 1$, the population stays in this state for a long time. The transfer of the population to a slowly decaying state leaves the symmetric state almost unpopulated even if the driving field is strong. However, if $\omega_{12} = 0$ the antisymmetric state is decoupled from the symmetric state and $\rho_{aa}(t)$ is zero if its initial value is zero: the system reduces effectively to a two-level atom.

We illustrate these features in Fig. 3, where we plot the steady-state populations ρ_{ss}, ρ_{aa} and ρ_{00} as a function of Δ for $\Omega = 5\gamma_1, \omega_{12} = 5\gamma_1$ and $p = 0.99$. It is clear that the symmetric state is almost unpopulated for $\Delta = \omega_{12}/2$, indicating that in the presence of quantum interference,

the driving field does not saturate the transition $|0\rangle \rightarrow |s\rangle$, even for very large Rabi frequencies. The small population in the state $|s\rangle$ increases the probability of the atom being returned to this state from the ground state by the driving field. Consequently, $g_{ss}^{(2)}(\tau)$ attains a very large value at time $\tau = (2\sqrt{2}\Omega)^{-1}\pi$ corresponding to half of the Rabi cycle between $|0\rangle$ and $|s\rangle$. However, $g_{aa}^{(2)}(\tau)$ attains a maximum at $\tau \approx (\sqrt{2}\Omega)^{-1}\pi$, i.e. at the Rabi period. This is because the driving laser takes the population from $|0\rangle$ to $|s\rangle$ in a time equal to half of the Rabi period, and then the population is transferred to $|a\rangle$ in time equal to that that the population will stay in $|s\rangle$ i.e. in time equal to half of the Rabi period. Therefore, the total time of transferring the population from $|0\rangle$ to $|a\rangle$ is equal to the Rabi period. We also note that the value of $\rho_{ss} = P_s$ is small when $\Delta = \omega_{12}/2$, and this will act to further increase the maximum values of $g_{ss}^{(2)}(\tau)$.

3 Single-Transition Excitation

Here we consider the same model as studied in the previous section except that now the laser couples to only the $|0\rangle \rightarrow |1\rangle$ transition, and we no longer assume $\gamma_1 = \gamma_2$. The observable we wish to calculate is the absorption rate of a weak probe beam coupled to the $|0\rangle \longleftrightarrow |2\rangle$ transition. Here, it is convenient to use a quantum description for the Hamiltonian, which takes the form

$$H = \omega_1 A_{11} + \omega_2 A_{22} + \omega_L a_L^\dagger a_L + \frac{1}{2}\left(g_1 A_{10} a_L + \text{H.c.}\right), \tag{25}$$

where $a_L^\dagger$ is the creation operator of the driving field, and $g_1 = (2\omega_1/\hbar\epsilon_0 V)^{1/2}$ is the coupling constant of the laser field to the $|0\rangle \rightarrow |1\rangle$ transition. We first

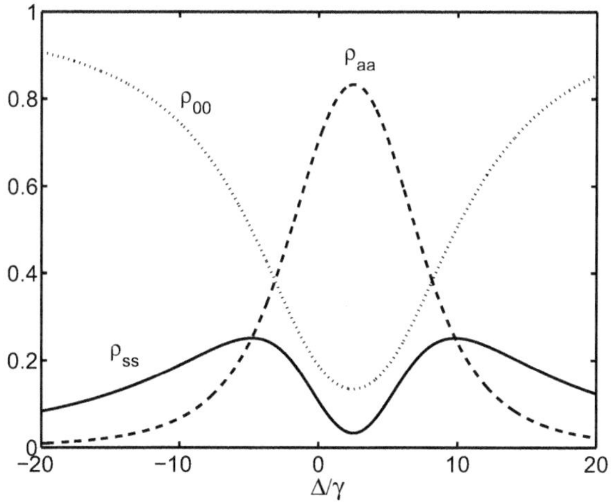

Fig. 3. The steady-state populations ρ_{ss}, ρ_{aa} and ρ_{00} as a function of Δ/γ for $\Omega = 5\gamma, \omega_{12} = 5\gamma$ and $\beta = 0.99$.

diagonalize the Hamiltonian (25) using the dressed states:

$$
\begin{aligned}
|+,N\rangle &= \frac{1}{\sqrt{2}}\left(|0,N\rangle + |1,N-1\rangle\right), \\
|-,N\rangle &= \frac{1}{\sqrt{2}}\left(|0,N\rangle - |1,N-1\rangle\right), \\
\left|\tilde{2},N\right\rangle &= |2,N-1\rangle,
\end{aligned}
\tag{26}
$$

with energies

$$
\begin{aligned}
E_{N,\pm} &= N\omega_1 \pm \Omega, \\
E_{N,2} &= N\omega_1 + \Delta,
\end{aligned}
\tag{27}
$$

where N is the photon number for the laser field and $2\Omega = g_1\sqrt{\langle N\rangle}$ is the Rabi frequency of the driving field.

The dressed states (26) form an infinite ladder of nondegenerate three-state manifolds, with adjacent manifolds separated by ω_1. Inside a manifold, the states $|+,N\rangle$ and $|-,N\rangle$ are separated by 2Ω, while the state $\left|\tilde{2},N\right\rangle$ is separated from $|+,N\rangle$ by $\Delta - \Omega$. Clearly, the detuning $\Delta = \Omega$ is significant, as then the latter two states become degenerate.

The interaction of the atom with the vacuum field allows spontaneous emission to occur from states $|i,N+1\rangle$ to states $|j,N\rangle$, where $i,j = \pm,\tilde{2}$. These transitions occur with probabilities

$$
\Gamma_{i,j} = \frac{\gamma_n}{|\boldsymbol{\mu}_{n0}|^2}\left|\langle i,N+1|\,\boldsymbol{\mu}\,|j,N\rangle\right|^2, \quad n = 1,2,
\tag{28}
$$

given by the expressions

$$
\begin{aligned}
\Gamma_{\pm,\pm} &= \Gamma_{\pm,\mp} = \frac{1}{4}\gamma_1, \\
\Gamma_{\tilde{2},\tilde{2}} &= \Gamma_{\pm,\tilde{2}} = 0, \\
\Gamma_{\tilde{2},\pm} &= \frac{1}{2}\gamma_2,
\end{aligned}
\tag{29}
$$

from which we find that the total spontaneous emission rates from $|i,N\rangle$ are

$$
\begin{aligned}
\Gamma_{\pm} &= \sum_{i=\pm,\tilde{2}} \Gamma_{\pm,i} = \frac{1}{2}\gamma_1, \\
\Gamma_{\tilde{2}} &= \sum_{i=\pm,\tilde{2}} \Gamma_{\tilde{2},i} = \gamma_2.
\end{aligned}
\tag{30}
$$

It may be seen from (9) that the state $\left|\tilde{2},N\right\rangle$ in a given manifold is not coupled to the states of the manifold above. Thus, this state cannot be populated by spontaneous emission. It is easily verified from the master equation that the steady-state population of $|\tilde{2}, N\rangle$ is zero. However, the state $|\tilde{2}, N\rangle$ may be connected to the dressed states of the manifold below through the dipole moment μ_{20} so that this state can be populated by the absorption of a weak probe beam coupled to $|\pm, N\rangle \to |\tilde{2}, N\rangle$ transitions. The absorption occurs at the two frequencies $\omega_{\pm,2} = \pm\Omega$ with probabilities $\Gamma_{\tilde{2},\pm} = \frac{1}{2}\gamma_2$, giving the familiar Autler-Townes absorption doublet [33].

3.1 Superposition dressed states

The absorption properties on the $|\pm, N\rangle \longrightarrow |\tilde{2}, N\rangle$ transitions are not affected by the presence of quantum interference between the atomic transitions, unless $\Delta \simeq \Omega$, when $|+, N\rangle$ and $|\tilde{2}, N\rangle$ become degenerate. This may be seen from the equation for $\rho_{+,\tilde{2}} = \langle +, N| \rho |\tilde{2}, N\rangle$:

$$\dot{\rho}_{+,\tilde{2}} = -\left[\frac{1}{2}\left(\Gamma_+ + \gamma_2\right) - i\left(\Delta - \Omega\right)\right]\rho_{+,\tilde{2}} - \frac{1}{2\sqrt{2}}\gamma_{12}\left(\rho_{\tilde{2},\tilde{2}} + \rho_{+,+}\right), \tag{31}$$

where $\rho_{\tilde{2},\tilde{2}}$ and $\rho_{+,+}$ are the populations of the states $|\tilde{2}, N\rangle$ and $|+, N\rangle$ respectively. The coherence oscillates with frequency $(\Delta - \Omega)$ and averages to zero over a long period of time. However, for $\Delta = \Omega$ the coherence does not oscillate and attains a non-zero value in the steady-state due to the coupling to the populations through the cross-damping term γ_{12}. Analogous to (5) and (6), we introduce the following superposition of the two closely-spaced dressed states

$$\begin{aligned} |s, N\rangle &= \alpha\,|+, N\rangle + \beta\,|\tilde{2}, N\rangle\,, \\ |a, N\rangle &= \beta\,|+, N\rangle - \alpha\,|\tilde{2}, N\rangle\,, \end{aligned} \tag{32}$$

where

$$\alpha = \frac{\sqrt{\Gamma_+}}{\sqrt{\gamma_2 + \Gamma_+}}, \quad \beta = \frac{\sqrt{\gamma_2}}{\sqrt{\gamma_2 + \Gamma_+}}. \tag{33}$$

The states (32) diagonalize spontaneous emission between the dressed states of two neighboring manifolds.

It is straightforward to evaluate the spontaneous emission rates from the states $|k, N+1\rangle$ to states $|l, N\rangle$, where $k, l = a, s, -$, using (9), (8) and (32). The expressions may be found in [21]. Here we just remark that Γ_{as}, Γ_{a-} and Γ_{aa} all contain the factor $(1-p)$, and so vanish for $\theta = 0$.

Figure 4 shows the superposition dressed states with the allowed spontaneous emission rates, for $\Delta = \Omega$. The dashed lines show those transitions, Γ_{as}, Γ_{a-} and Γ_{aa}, which can be canceled by quantum interference, leading to trapping of the population in the state $|a, N+1\rangle$. To find the steady-state population of the antisymmetric state, we take diagonal matrix elements of the master equation with respect to $|a, N+1\rangle$, make the secular approximation [34], valid for $\Omega \gg \gamma_1, \gamma_2$, in which we ignore coupling of the populations to the coherences, and thus find the "reduced" steady-state population $P_a = \sum_N \langle a, N+1 | \rho | a, N+1 \rangle$ to be

$$P_a = \frac{1}{1 + \frac{u}{w}(1 - p)}, \tag{34}$$

where

$$\begin{aligned} u &= 2\left(\Gamma_{sa} + \Gamma_{s-} + \Gamma_{-s}\right) + \frac{\gamma_1 \left(\Gamma_s + \frac{1}{4}\gamma_1\right)}{\left(\gamma_2 + \frac{1}{2}\gamma_1\right)}, \\ w &= \left(\Gamma_{sa} + \Gamma_{s-}\right) + \frac{1}{2}\gamma_1 \Gamma_{sa}/\gamma_2 . \end{aligned} \tag{35}$$

We see that $P_a = 1$ when $p = 1$.

Despite the fact that the spontaneous emission rates of the transitions from $|a, N+1\rangle$ to the states of the manifold below depend strongly on θ (through p), a probe beam can couple to these transitions with a rate that is *independent of* θ. To show this, we assume that the probe couples with the atom exclusively via the transition dipole moment $\boldsymbol{\mu}_{20}$. In this case, the interaction is given by

$$H_p = \frac{1}{2}\boldsymbol{\mu}_{20} \cdot \boldsymbol{E}_p \left(A_{20} e^{-i\omega_p t} + H.c.\right) . \tag{36}$$

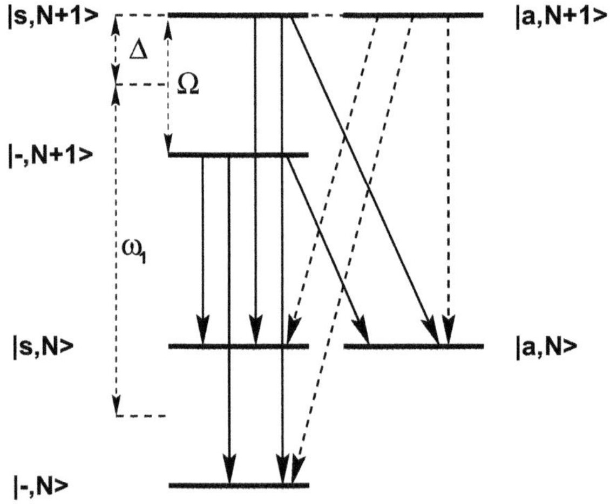

Fig. 4. Energy level diagram of the superposition dressed states for $\Delta = \Omega$. The solid lines indicate spontaneous transitions which occur independent of quantum interference, whereas the dashed lines indicate transitions which can be canceled by quantum interference.

In terms of the superposition states the interaction (36) can be written as

$$H_p = \frac{1}{2\sqrt{2}}\boldsymbol{\mu}_{20} \cdot \boldsymbol{E}_p \sum_N \{\alpha\beta\,|s,N+1\rangle\langle s,N| + \beta^2\,|s,N+1\rangle\langle a,N| - \alpha\beta\,|a,N+1\rangle\langle a,N| - \alpha^2\,|a,N+1\rangle\langle s,N| + \beta\,|s,N+1\rangle\langle -,N| - \alpha\,|a,N+1\rangle\langle -,N|\}\,e^{-i\omega_p t} + \text{H.c.} \qquad (37)$$

We see immediately that the probe field couples to all transitions between the manifolds $N+1$ and N. Since α and β are independent of p, the coupling is independent of θ. Thus for $\theta \approx 0$ the transition rates *from* the state $|a,N+1\rangle$ can be canceled without affecting the coupling strength of the probe beam *to* these transitions.

For $\Omega \gg \gamma_1, \gamma_2$, we can treat the transitions between the $N+1$ and N manifolds as separate two-level systems [34,9,36], allowing us to write the total absorption rate $\bar{W}_{a,-}$ on the $|a,N+1\rangle - |-,N\rangle$ transition as

$$\bar{W}_{a,-} = \frac{\pi\alpha^2}{2\hbar}\,|\boldsymbol{\mu}_{20} \cdot \boldsymbol{E}_p|^2\,(P_{-,N} - P_{a,N+1})\,. \qquad (38)$$

Since the $|a,N+1\rangle \to |-,N\rangle$ transition can be inverted through the action of quantum interference, we anticipate gain at the frequency $\omega_1 + 2\Omega$. The coupling strength of the probe field to the transition, however, remains constant, independent of the population difference between the states. Thus, the amplification is controlled solely by the population difference and attains its maximum value when the population is trapped in the state $|a,N+1\rangle$.

When the probe couples to both atomic transitions, the interaction Hamiltonian is given by

$$H_p = \frac{1}{2}\left([(\boldsymbol{\mu}_{20} \cdot \boldsymbol{E}_p)\,A_{20} + (\boldsymbol{\mu}_{10} \cdot \boldsymbol{E}_p)\,A_{10}]\,e^{-i\omega_p t} + \text{H.c.}\right). \qquad (39)$$

Assuming the simple case where the polarization of the probe field is chosen to satisfy

$$\frac{\boldsymbol{\mu}_{10} \cdot \boldsymbol{E}_p}{|\boldsymbol{\mu}_{10}|\,|\boldsymbol{E}_p|} = \frac{\boldsymbol{\mu}_{20} \cdot \boldsymbol{E}_p}{|\boldsymbol{\mu}_{20}|\,|\boldsymbol{E}_p|} = \cos\frac{\theta}{2}, \qquad (40)$$

the Hamiltonian (39) can be written in the basis of the superposition states as

$$H_p = \frac{1}{4}\,|\boldsymbol{\mu}_{10}|\,|\boldsymbol{E}_p|\cos\frac{\theta}{2}\sum_N \Big\{|s,N+1\rangle\langle s,N| + \frac{\alpha}{\beta}\,|s,N+1\rangle\langle a,N| + \frac{1}{\beta}\,|s,N+1\rangle\langle -,N| - \alpha\,|-,N+1\rangle\langle s,N| + \beta\,|-,N+1\rangle\langle a,N| - |-,N+1\rangle\langle -,N|\}\,e^{-i\omega_p t} + \text{H.c.} \qquad (41)$$

It is clear from (41) that the probe does not couple the transitions from $|a,N+1\rangle$ to the states below. Thus, in order to observe the amplification on dark transitions it is important that the probe couples selectively to one of the atomic transitions.

In Fig. 5, we plot $W(\omega_p) = \mathrm{Re}\rho_{02}^{(+1)}/\Omega_p$, the absorption rate of a probe beam coupled to the $|2\rangle - |0\rangle$ transition [20,9,37], where Ω_p is the Rabi frequency of the probe beam, and $\rho_{02}^{(+1)}$ is the stationary component of the coherence ρ_{02} oscillating with the probe detuning $\delta = \omega_p - \omega_2$. The probe experiences 33% gain at $\delta = 2\Omega$, which appears on the dark transition $|a, N+1\rangle \to |-, N\rangle$. Apart from the gain at $\delta = 2\Omega$, the probe can experience a gain at $\delta = 0$. As predicted by Menon and Agarwal [20] it happens for $\gamma_2 < \frac{1}{2}\gamma_1$. Again, we can explain the gain as arising from an amplification of the probe on a dark transition. It is seen from Fig. 4 that at $\delta = 0$ the probe couples to three transitions. The transition $|a, N+1\rangle \to |a, N\rangle$ is transparent for the probe because it occurs between two states of the same population. Therefore, the absorptive/emissive properties result from the coupling of the probe to the $|s, N+1\rangle \to |a, N\rangle$ and $|a, N+1\rangle \to |s, N\rangle$ transitions. For $\theta \approx 0$ almost all the population is trapped in the antisymmetric state, and then the probe is strongly absorbed on the $|s, N+1\rangle \to |a, N\rangle$ transition, but is amplified on the $|a, N+1\rangle \to |s, N\rangle$ transition. Because Γ_{as} is proportional to $(1-p)$ and vanishes for $\theta = 0$, the latter transition is dark. Since the absolute values of the population differences are the same for both transitions, the absorptive/emissive properties depend solely on the relation between the transition rates. Using (37), we find that the

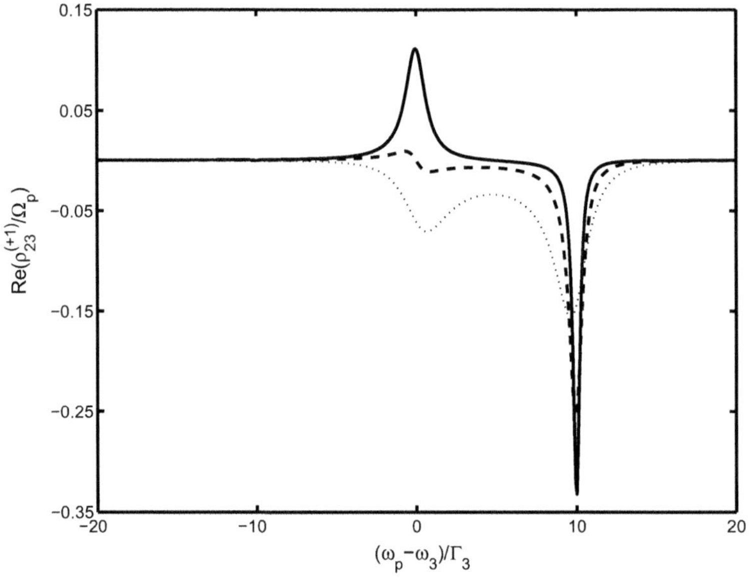

Fig. 5. The absorption rate $\rho_{02}^{(+1)}/\Omega_p$ as a function of the probe detuning $\delta/\gamma_2 = (\omega_p - \omega_2)/\gamma_2$ for $p = 0.999, 2\Omega = 10\gamma_2$, $\Delta = 5\gamma_2$, and different γ_1: $\gamma_1 = \gamma_2$ (solid line), $\gamma_1 = 2\gamma_2$ (dashed line), $\gamma_1 = 5\gamma_2$ (dotted line). The driving laser and probe fields couple to the $|0\rangle - |1\rangle$ and $|0\rangle - |2\rangle$ transitions, respectively.

net magnitude of the absorption rate on the transitions at $\delta = 0$ is given by

$$\bar{W}_0 = \frac{\pi}{2} \left| \boldsymbol{\mu}_{eg} \cdot \boldsymbol{E}_p \right|^2 \frac{\left(\gamma_2 - \frac{1}{2}\gamma_1\right)}{\left(\frac{1}{2}\gamma_1 + \gamma_2\right)} \left| \left(P_{a,N+1} - P_{s,N}\right) \right| . \tag{42}$$

Thus, for $\gamma_2 > \frac{1}{2}\gamma_1$ the probe is absorbed, whereas for $\gamma_2 < \frac{1}{2}\gamma_1$ stimulated emission on the dark $|a, N+1\rangle \to |s, N\rangle$ transition outweighs absorption on the $|s, N+1\rangle \to |a, N\rangle$ transition resulting in an amplification of the probe beam.

4 Conclusions

We have examined the effect of quantum interference on a V-type three-level atom driven by a single-mode laser field, using a unitary transform to symmetric and antisymmetric superposition states. When the field couples to both transitions, we have calculated various correlation functions $g_{ij}^{(2)}(\tau)$. When a strong driving field is tuned to the middle of the two excited levels the correlations of photons emitted from the atomic transitions exhibit anticorrelations which persist for all times. The atom emits strongly correlated photons from a symmetric superposition of the excited levels. The correlations result from a coherent transfer of populations to the antisymmetric state which leaves the symmetric state unpopulated even for very strong driving fields. We have shown that very large values of the intensity-intensity correlation function are possible.

When the laser couples to only one transition, we have shown that the gain features in the Autler-Townes spectrum of a driven three-level V-system, recently found by Menon and Agarwal [20], are an interesting example of gain on dark transitions with a completely inverted population. We have found that the gain features depend upon whether the probe beam couples to one or both atomic transitions. When the probe field couples only to the undriven transition, its coupling strength to the superposition states remains constant, independent of the angle θ between the dipole moments. This property allows amplification of the probe field on dark transitions between the completely populated trapping state and unpopulated lower levels. The gain features are absent when the probe couples to both atomic transitions.

Acknowledgments

This research was supported by the United Kingdom Engineering and Physical Sciences Research Council. We acknowledge helpful conversations with Dr. Peng Zhou.

References

1. P.W. Milonni and P.L. Knight, Opt. Commun. **9**, 119 (1973); R. G. Hulet, E. S. Hilfer, and D. Kleppner, Phys. Rev. Lett. **55**, 2137 (1985); M. Lewenstein and T.W. Mossberg, Phys. Rev. A **37**, 2048 (1988); C. H. Keitel, P. L. Knight, L. M. Narducci, and M. O. Scully, Optics Commun. **118**, 143 (1995); S. John and T. Quang, Phys. Rev. A **50**, 1764 (1994); S. Bay, P. Lambropoulos, and K. Molmer, Phys. Rev. Lett. **79**, 5238 (1997).
2. D. Kleppner, Phys. Rev. Lett. **47**, 233 (1981); V. Frerichs, D. Meschede, and A. Schenzle, Optics Commun. **117**, 325 (1995).
3. For a recent review see B.J. Dalton, Z. Ficek, and S. Swain, J. Mod. Opt. **46**, 379 (1999).
4. P. Zhou and S. Swain, Phys. Rev. Lett. **77**, 3995 (1996); Phys. Rev. A**56**, 3011 (1997).
5. S.-Y. Zhu and M. O. Scully, Phys. Rev. Lett. **76**, 388 (1996).
6. H. Lee, P. Polynkin, M. O. Scully, and S. Y. Zhu, Phys. Rev. A **55**, 4454 (1997); F.-L. Li and S.-Y. Zhu, Phys. Rev. A **59**, 2330 (1999).
7. P. Zhou and S. Swain, Phys. Rev. Lett. **77**, 3995 (1996); Phys. Rev. A **56**, 3011 (1997).
8. M.D. Lukin, S.F. Yelin, M. Fleischhauer, and M.O. Scully, Phys. Rev. A **60**, 3225 (1999).
9. K. J. Boller, A. Imamoglu, and S. E. Harris, Phys. Rev. Lett. **66**, 2593 (1991); K. Hakuta, L. Marmet, and B. Stoicheff, Phys. Rev. Lett. **66**, 596 (1991); J. C. Petch, C. H. Keitel, P. L. Knight, and J. P. Marangos, Phys. Rev. A**53**, 543 (1996).
10. S. E. Harris, Phys. Rev. Lett. **62**, 1033 (1989); M. O. Scully, S.-Y. Zhu, and A. Gavrielides, Phys. Rev. Lett. **62**, 2813 (1989); G. S. Agarwal, Phys. Rev. A **44**, R28 (1991); C. H. Keitel, O. Kocharovskaya, L. M. Narducci, M. O. Scully, S.-Y. Zhu, and H. M. Doss, Phys. Rev. A **48**, 3196 (1993); J. Kitching and L. Hollberg, Phys. Rev. A **59**, 4685 (1999).
11. L.V. Hau, S.E. Harris, Z. Dutton, and C.H. Behroozi, Nature **397**, 594 (1999); M. Kash, V. Sautenkov, A. Zibrov, L. Hollberg, G. Welch, M. Lukin, Y. Rostovtsev, E. Fry, and M.O. Scully, Phys. Rev. Lett. **82**, 5229 (1999); M.D. Lukin and A. Imamoglu, Phys. Rev. Lett. **84**, 1419 (2000).
12. S. Swain, P. Zhou, and Z. Ficek, Phys. Rev. A **61**, 0343410 (2000).
13. H. Huang, S.-Y. Zhu, M. S. Zubairy, and M. O. Scully, Phys. Rev. A **53**, 1834 (1996).
14. A. K. Patnaik and G. S. Agarwal, J. Mod. Opt. **45**, 2131 (1998); E. Paspalakis, C. H. Keitel, and P. L. Knight, Phys. Rev. A**58**, 4868 (1998); S. Menon and G. S. Agarwal, Phys. Rev. A**57**, 4014 (1998); S.-Q. Gong, E. Paspalakis, and P. L. Knight, J. Mod. Opt. **45**, 2433 (1998).
15. C. H. Keitel, Phys. Rev. Lett. **83**, 1307 (1999).
16. E. Paspalakis and P.L. Knight, Phys. Rev. Lett. **81**, 293 (1998).
17. S.-Q. Gong, E. Paspalakis, and P.L. Knight, J. Mod. Opt. **45**, 2433 (1998); S.-Q. Gong, Y. Li, S.-D. Du, and Z.-Z. Xu, Phys. Lett. A **259**, 43 (1999).
18. E. Paspalakis, S.-Q. Gong, and P.L. Knight, Opt. Commun. **152**, 293 (1998).
19. O. Kocharovskaya, Y.V. Radeonychev, P. Mandel, and M.O. Scully, Phys. Rev. A **60**, 3091 (1999).
20. S. Menon and G.S. Agarwal, Phys. Rev. A **61**, 013807 (2000).
21. Z. Ficek, S. Swain and Uzma Akram, J. Phys. B — to be published.

22. L. M. Narducci, M. O. Scully, G.-L. Oppo, P. Ru and J. R. Tredicci, Phys. Rev. A **42**,1630 (1990).
23. D. J. Gauthier, Yifu Zhu and T. W. Mossberg, Phys. Rev. Lett. **66**, 2460 (1991).
24. B. N. Jagatap, Q. V. Lawande, and S. V. Lawande, Phys. Rev. A **43**, 535 (1991).
25. A. S. Manka, E. J. D'Angelo, L. M. Narducci and M. O. Scully, Phys. Rev. A **47**, 4236 (1993).
26. G. C. Hegerfeldt and M. B. Plenio, Phys. Rev. A **47**, 2186 (1993).
27. R. J. Cook and H. J. Kimble, Phys. Rev. Lett. **54**, 1023 (1985).
28. D. T. Pegg, R. Loudon, and P. L. Knight, Phys. Rev. A **33**, 4085 (1986).
29. M. Wiegand, J. Phys. **B 16,** 1133 (1983).
30. H. J. Kimble and L. Mandel, Phys. Rev. A **13**, 2123 (1976).
31. M. Daganais and L. Mandel, Phys. Rev. A **18**, 2217 (1978).
32. M. Lax, Phys. Rev. **172**, 350 (1968).
33. S.H. Autler and C.H. Townes, Phys. Rev. **100**, 703 (1955).
34. C. Cohen-Tannoudji and S. Reynaud, J. Phys. B **10**, 345 (1977); C. Cohen-Tannoudji, J. Dupont-Roc, and G. Grynberg, *Atom-Photon Interactions*, (Wiley, New York, 1992).
35. B.R. Mollow, Phys. Rev. A **5**, 1522, 2217 (1972).
36. G.A. Wilson, K.K. Meduri, P.B. Sellin, and T.W. Mossberg, Phys. Rev. A **50**, 3394 (1994).
37. P. Meystre and M. Sargent III, *Elements of Quantum Optics*, (Springer, Berlin, 1990), p. 230.

A Cavity QED Test of Quantum Mechanics

Z. Ficek[1] and S. Swain[2]

[1] Department of Physics, The University of Queensland, Brisbane, QLD 4072, Australia
[2] School of Mathematics and Physics, The Queen's University of Belfast, Belfast BT7 1NN, The United Kingdom

Abstract. We establish a connection between a harmonic oscillator and a strongly driven two-level atom coupled to a quantized cavity mode. Employing a weak probe which connects the ground state to an auxiliary level, the Autler-Townes spectrum is calculated. There are two distinct regions of the atomic dynamics depending on the relative magnitude of the Rabi frequencies of the cavity field, Ω_c, and the driving field, Ω. For $\Omega_c < \Omega$ and moderate cavity coupling strengths, the spectral peaks are composed of multiplets. A quantized dressed-atom approach shows that the oscillations in the spectral components arise from the oscillations of the population distribution in the dressed states. This distribution is an analog of the harmonic oscillator probability distribution function, and should be experimentally observable. For $\Omega_c \geq \Omega$ there is no Autler-Townes splitting and the spectrum is composed of a single peak located at the frequency of the probe transition. We show that this effect results from the collapse of the atom to the ground state.

1 Introduction

The fundamental model of the atom-field interaction is the Jaynes-Cummings model [1] consisting of an excited two-level atom strongly coupled to a single mode of the radiation field. The model has been extensively studied and many interesting quantum effects have been predicted and observed, among these the most well known are the collapse and revival of the Rabi frequencies [2], subnatural linewidths [3], multiple fluorescence spectra [4,5], and nonclassical photon statistics [6]. These features result from the presence of multiple exchange of photons between the radiating atom and the cavity mode and occur when the coupling strengths between the atom and the cavity mode are larger than the damping rates of the system.

The Jaynes-Cummings model has been extended to include spontaneous emission, cavity damping and external driving fields. Two different configurations of atom driving have been analyzed. In the first case the external field drives the cavity mode [6,7], and in the second the driving field couples to the atom through an auxiliary field, different to the cavity mode [5,8]. The cases of strong and weak atom-cavity couplings have been considered. In the case of the atom driven through an auxiliary mode and weak atom-cavity coupling the system behaves formally the same as in free space, but with significantly modified spontaneous emission rates. For instance, the fluorescence spectrum of a

M. Orszag, J.C. Retamal (Eds.): LNP 575, pp. 262–278, 2001.

strongly driven atom is a triplet, as in free space [9], but with widely differing linewidths. Depending on the detuning of the cavity mode from the atomic resonance, the central or even all three spectral components can be significantly narrowed [10]. For strong atom-cavity coupling, each Mollow triplet component is composed of a multiplet, whose detailed structure depends on the atom-cavity coupling strength, the cavity and spontaneous emission decay rates, and the photon-number distribution of the cavity field [4,5]. Moreover, in the case of the lossless cavity and exact resonance of the cavity and the driving fields to the atomic transition frequency, the atom can remain in its ground state resulting in the disappearance of the atomic resonance fluorescence [8].

Recent studies of the Jaynes-Cummings model have been devoted to observing those signatures of the discrete nature of field quanta in the atom-cavity interaction that are sensitive to the presence of single quanta in the cavity mode. With the successful experiments in the laser cooling and trapping of a single atom within a single mode of a microscopic cavity [11], it is now possible to test the theoretical predictions of quantum physics [12], and the cavity quantum electrodynamics of the strong interaction of atoms with single quanta of the radiation field. The most recent experiments detected quantum Rabi oscillations [13], and produced Fock states of the radiation field [14] and a quantum phase gate [15]. However, the basic signature of a discrete small number of photons in the cavity mode is the dependence of the energy spectrum of the Jaynes-Cummings model on the number of photons n. The energy spectrum is composed of a single ground ($n = 0$) level, and a ladder of doublets separated by $\hbar\omega_0$. The intradoublet splitting is equal to $\hbar g\sqrt{n}$, where ω_0 is the resonance frequency and g is the atom-cavity coupling constant. The splitting of the lowest energy doublet ($n = 1$), called the vacuum Rabi splitting, has been observed experimentally [16], and a photon correlation spectroscopy technique involving a weak multichromatic field has been proposed to measure the unequal splitting of the second and third excited state doublets [17]. Since the splittings are unequal, the absorption resonances of the multichromatic field will be proportional to $(\sqrt{n} - \sqrt{n-1})$, where $n = 1, 2, \ldots$. These resonances vanish in the classical limit of $n \gg 1$.

The one-dimensional harmonic oscillator of mass m and frequency ω_0, a system of fundamental importance in quantum physics, has a simpler energy spectrum—a ladder of equally-spaced levels separated by $\hbar\omega_0$. It is well-known [17] that the energy eigenfunctions of this system are given by

$$\phi_n(\xi) = \left(\sqrt{2\pi}2^n n!\right)^{-\frac{1}{2}} H_n(\xi)\, e^{-\frac{1}{2}\xi^2}, \tag{1}$$

where $\xi = (m\omega/\hbar)^{1/2} x$ is the dimensionless coordinate, and $H_n(\xi)$ is a Hermite polynomial. The probability distribution, $|\phi_n(\xi)|^2$, shows $n+1$ oscillations which reveal the discrete energy spectrum of the harmonic oscillator. In the classical limit of a large number of quanta ($n \gg 1$) the change in the number of oscillation with n is unnoticable. In spite of its importance, the probability distribution of the harmonic oscillator has never been observed.

Another elementary system which is sensitive to the presence of single quanta is a particle confined in a box [17]. This system is qualitatively similar to a harmonic oscillator whose position operator exhibits a discrete spectrum such that the probability of finding the particle at the dimensionless position ξ inside the box depends on ξ and the number of quanta n. The probability is maximal in the vicinity of the classical turning points of the harmonic oscillator eigenfunction $\phi_n(\xi)$, i.e., for $\xi_n \sim \pm 2\sqrt{n+\frac{1}{2}}$, or for energies $|\xi_n|\hbar\omega_0$. For $|\xi| > |\xi_n|$ the probability goes rapidly to zero, while for $\xi < |\xi_n|$ the probability is nonzero and oscillates with ξ. The oscillations in the probability density of an electron confined in a quantum dot have recently been observed [19].

In this paper, we draw an analogy between the harmonic oscillator and a strongly driven two-level atom interacting with a quantized cavity mode. We investigate the Autler-Townes spectrum of a three-level atom where one of the transitions is coupled to a single cavity mode and is driven, through an auxiliary mode, by a strong coherent laser field. The cavity and the laser field frequencies are assumed to be equal and resonant with the atomic transition frequency. We show that the Autler-Townes spectrum can exhibit oscillations which mirror the probability distribution, $|\phi_n(\xi)|^2$ of the harmonic oscillator. The mere presence of oscillations is a signature of the quantum nature of the cavity field.

2 The Eigenstructure of the Driven Two-Level Atom in a Cavity

We consider a two-level atom with ground state $|2\rangle$ and excited state $|1\rangle$. The transition $|2\rangle - |1\rangle$ is coupled to a cavity mode and driven through an auxiliary mode by a classical laser field. For simplicity, we assume that the cavity mode and the driving field are both exactly resonant with the atomic transition frequency ω_{12}. Both the cavity and driving laser fields are quantized, and we wish to find the dressed states of the system. The Hamiltonian is

$$H = H_0 + H_{int}, \tag{2}$$

where

$$H_0 = \hbar\omega_{12}\left(a_c^\dagger a_c + a_L^\dagger a_L + S_1^+ S_1^-\right) \tag{3}$$

is the unperturbed Hamiltonian, and

$$H_{int} \equiv V + V_L = \hbar g\left(S_1^+ a_c + a_c^\dagger S_1^-\right) + \hbar g_L\left(S_1^+ a_L + a_L^\dagger S_1^-\right) \tag{4}$$

is the interaction between the cavity and driving fields, where a_L is the annihilation operator of the laser mode, a_c that of the cavity mode, $S_1^+ = |1\rangle\langle 2|$ and $S_1^- = |2\rangle\langle 1|$ are the atomic dipole raising and lowering operators of the $|1\rangle - |2\rangle$ transition, g is the atom-cavity coupling constant, and g_L is the atom-laser field coupling constant.

The eigenstates of H_0 are the product states $|i, N_L, n\rangle = |i\rangle \otimes |N_L\rangle \otimes |n\rangle$, where $|i\rangle$ is an atomic state ($i = 1, 2$), N_L is the number of photons in the laser mode, and n is the number of photons in the cavity mode. The states $|i, N_L, n\rangle$ form an infinite ladder of highly degenerate sets of states of energies $\hbar N\omega_{12}$, where $N = N_L + n$ is the total number of photons in the field modes. We first diagonalize the Hamiltonian consisting of the unperturbed Hamiltonian plus the interaction with the laser field, $H_1 \equiv H_0 + V_L$. This gives rise to the singly-dressed states, which have the form

$$H_1 |N_L\pm, n\rangle = \hbar \left(N\omega_{12} \pm \frac{1}{2}\Omega \right) |N_L\pm, n\rangle \tag{5}$$

where $\Omega = 2g_L\sqrt{\langle N_L\rangle}$, $|N_L\pm, n\rangle \equiv |N_L, \pm\rangle\, |n\rangle$ and the $|N_L, \pm\rangle$ are the conventional two-level atom dressed states [20]. Finally, including the weak interaction V, we obtain doublet continua $E_{N,d_i,\lambda}$ and eigenstates $|N, d_i, \lambda\rangle$ which satisfy the eigenvalue equation

$$H |N, d_i, \lambda\rangle = E_{N,d_i,\lambda} |N, d_i, \lambda\rangle\,, \quad (i = 1, 2) \tag{6}$$

where the label d_i indicates dressed states,

$$E_{N,d_i,\lambda} = N\hbar\omega_{12} + (-1)^i \left(\frac{1}{2}\Omega + \lambda g \right), \tag{7}$$

$$|N, d_i, \lambda\rangle = \sum_{n=1}^{\infty} \phi_n \left((-1)^i \frac{\lambda}{\sqrt{2}} \right) |i, N, n\rangle\,, \tag{8}$$

with

$$\phi_n(x) = \left(\sqrt{2\pi} 2^n n! \right)^{-\frac{1}{2}} H_n(x)\, e^{-\frac{1}{2}x^2}, \tag{9}$$

and λ is an arbitrary real number ($-\infty < \lambda < \infty$).

The quantity $\phi_n(x)$ of (9) is just the harmonic oscillator eigenfunction described in the Introduction, with $|\phi_n(x)|^2$ being the corresponding probability distribution function. We have thus established an intimate connection between the driven two-level atom in a good cavity, and the harmonic oscillator. This connection is further reinforced by noting that the matrix elements of the cavity interaction V in terms of the singly-dressed eigenstates of H_1 are

$$\langle N_L\pm, n| V |N_L\pm, m\rangle = \pm\frac{1}{2}\hbar g \left(\sqrt{n+1}\delta_{n+1,m} + \sqrt{n}\delta_{n-1,m} \right). \tag{10}$$

We have assumed negligible interaction between the + and - manifolds, which is true when the laser Rabi frequency exceeds the cavity Rabi frequency. This matrix is identical to that which represents the position operator of the harmonic oscillator in the basis of the energy eigenstates [17]. In fact, it is shown in [21] that the eigenstates of the position operator of the harmonic oscillator in the basis of its energy eigenstates are just the functions (9).

From (7) and (8) we can easily predict the energy spectrum of the system. Since λ is an arbitrary real number, analogous to the coordinate of the one-dimensional harmonic oscillator, discrete energy levels of the system *do not exist.* In this case the energy spectrum of the system is composed of an infinite ladder of doublet continua with interdoublet separation $\hbar\omega_{12}$ and intradoublet splitting Ω.

A major purpose here to suggest ways of observing $|\phi_n(x)|^2$, the probability distribution function of the harmonic oscillator [17]. One way of achieving this is to couple the ground state to a third level which may be probed by a second, weak, tunable laser. This is considered in the next section, where we make our model more realistic by introducing damping and allowing for detuning between the atomic transition and the driving laser frequency.

3 Master Equation of the System

We now consider a three-level atom with ground state $|2\rangle$ and two excited states $|1\rangle$ and $|3\rangle$. As before, the transition $|2\rangle - |1\rangle$ is coupled to a quantized cavity mode and driven through an auxiliary mode by a resonant laser field, but now the latter field is classical. The transition $|2\rangle - |3\rangle$, which is not coupled to the cavity mode, is probed by a weak laser field that monitors the cavity effects on the coherently driven transition. We assume that the dipole moments of the transitions $\boldsymbol{\mu}_{12}$ and $\boldsymbol{\mu}_{32}$ are perpendicular to each other that no correlation exists between the transitions. In the interaction picture, the master equation of the system is

$$\frac{d\rho}{dt} = -\frac{i}{\hbar}[H,\rho] + L_c\rho + L_a\rho, \tag{11}$$

where

$$\begin{aligned} H = {} & \hbar\Delta_a S_1^+ S_1^- + \hbar\Delta_p S_3^+ S_3^- + \hbar g\left(S_1^+ a_c + a_c^\dagger S_1^-\right) \\ & + \frac{\hbar}{2}\Omega\left(S_1^+ + S_1^-\right) + \frac{\hbar}{2}\Omega_p\left(S_3^+ + S_3^-\right) \end{aligned} \tag{12}$$

is the Hamiltonian of the probed and driven atom-cavity system in the frame rotating with the laser frequency ω_L,

$$L_c\rho = \frac{1}{2}\kappa\left(2a_c\rho a_c^\dagger - a_c^\dagger a_c\rho - \rho a_c^\dagger a_c\right) \tag{13}$$

is the damping operator of the cavity field,

$$\begin{aligned} L_a\rho = {} & \frac{1}{2}\Gamma_1\left(2S_1^-\rho S_1^+ - S_1^+ S_1^-\rho - \rho S_1^+ S_1^-\right) \\ & + \frac{1}{2}\Gamma_3\left(2S_3^-\rho S_3^+ - S_3^+ S_3^-\rho - \rho S_3^+ S_3^-\right) \end{aligned} \tag{14}$$

is the damping operator for the atom. Here, $\Delta_a = \omega_{12} - \omega_L$ is the detuning of the driving field from the $|2\rangle - |1\rangle$ transition, $\Delta_p = \omega_{32} - \omega_p$ is the detuning of the probe field from the $|2\rangle - |3\rangle$ transition, $S_3^+ = |3\rangle\langle 2|$ and $S_i^- = |2\rangle\langle 3|$ are the atomic dipole raising and lowering operators of the $|3\rangle - |2\rangle$ transition, Ω and Ω_p are the Rabi frequencies of the driving and probe fields respectively, and κ, Γ_{1i} and Γ_3 are the cavity and spontaneous emission damping rates. We assume the strong field limit, that is, the Rabi frequency of the driving field is so strong that

$$\Omega \gg \kappa, \Gamma_1, \Gamma_3. \tag{15}$$

Since the $|2\rangle - |1\rangle$ transition is strongly driven by the laser field, it is convenient to work in the basis of the semiclassical dressed states of the transition [20]

$$\begin{aligned} |d_1\rangle &= \cos\phi\,|2\rangle + \sin\phi\,|1\rangle \\ |d_2\rangle &= \sin\phi\,|2\rangle - \cos\phi\,|1\rangle\,, \end{aligned} \tag{16}$$

where $\cos^2\phi = \frac{1}{2} + \Delta_a/2\Omega$. In the dressed-atom approach we express the atomic operators $S_i^{\pm}$ in terms of dressed-state operators $R_{ij} = |d_i\rangle\,\langle d_j|$ as

$$\begin{aligned} S_1^+ &= \frac{1}{2}\sin 2\phi R_0 - \sin^2\phi R_{12} + \cos^2\phi R_{21}, \\ S_1^- &= \frac{1}{2}\sin 2\phi R_0 - \sin^2\phi R_{21} + \cos^2\phi R_{12}, \\ S_3^+ &= \frac{1}{\sqrt{2}}\left(\sin\phi R_{32} + \cos\phi R_{31}\right), \end{aligned} \tag{17}$$

where $R_0 = R_{22} - R_{11}$. We substitute (17) into the master equation (11) and make the unitary transformation

$$\tilde{\rho}(t) = e^{-i\tilde{H}_L t/\hbar}\rho e^{i\tilde{H}_L t/\hbar}, \tag{18}$$

where

$$\tilde{H}_L = \hbar\left(\Delta_c a_c^\dagger a_c + \frac{1}{2}\Omega R_0\right). \tag{19}$$

Then, ignoring terms oscillating rapidly at the frequencies Ω and 2Ω, we find that the master equation reduces to

$$\begin{aligned} \frac{d}{dt}\tilde{\rho}(t) &= \frac{1}{2}g\sin 2\phi\left[a_c^\dagger R_0 - R_0 a_c, \tilde{\rho}(t)\right] - \frac{i}{\hbar}\left[\tilde{H}_p(t), \tilde{\rho}(t)\right] + L_c\tilde{\rho}(t) \\ &-\frac{1}{8}\Gamma_1\sin^2 2\phi\left(R_0^2\tilde{\rho}(t) + \tilde{\rho}(t)R_0^2 - 2R_0\tilde{\rho}(t)R_0\right) \\ &-\frac{1}{2}\cos^4\phi\Gamma_1\left(R_{11}\tilde{\rho}(t) + \tilde{\rho}(t)R_{11} - 2R_{21}\tilde{\rho}(t)R_{12}\right) \\ &-\frac{1}{2}\sin^4\phi\Gamma_1\left(R_{22}\tilde{\rho}(t) + \tilde{\rho}(t)R_{22} - 2R_{12}\tilde{\rho}(t)R_{21}\right) \\ &-\frac{1}{4}\Gamma_3\cos^2\phi\left(R_{33}\tilde{\rho}(t) + \tilde{\rho}(t)R_{33} - 2R_{13}\tilde{\rho}(t)R_{31}\right) \\ &-\frac{1}{4}\Gamma_3\sin^2\phi\left(R_{33}\tilde{\rho}(t) + \tilde{\rho}(t)R_{33} - 2R_{23}\tilde{\rho}(t)R_{32}\right). \end{aligned} \tag{20}$$

The master equation (20) is diagonal with respect to the driving field, but is nondiagonal with respect to the cavity field. Therefore the secular approximation used in the derivation of (20) is valid only when the Rabi frequency $\Omega_c = 2g\sqrt{\langle n\rangle}$ of the cavity field is much smaller than the Rabi frequency of the driving field. In other words, the cavity field produces oscillations which are much slower than those produced by the driving field. When $\Omega_c \approx \Omega$ the atomic dynamics can change dramatically. We will address this point in Sec. 5.

4 The Autler-Townes Absorption Spectrum

The Autler-Townes absorption spectrum is obtained by applying a weak probe field which couples the driven system to an undriven level. The population of the undriven level is measured as a function of the probe field frequency, or equivalently, one can measure the coherence of the probed transition.

4.1 Population of the undriven level

We calculate the population of the state $|3\rangle$, which is not coupled to the driving and cavity fields, as a function of the probe field detuning Δ_p. As we show below, the major features of this quantity are determined by the populations of the dressed states. From the master equation (20) and assuming $\Delta_a = 0$, we find that the equations of motion for the population of the state $|3\rangle$, and for the coherences between state $|3\rangle$ and the dressed states $|d_1\rangle$ and $|d_2\rangle$, are given by

$$
\begin{aligned}
\tilde{\rho}_{33} &= -\Gamma_3\tilde{\rho}_{33} + \frac{\Omega_p}{2\sqrt{2}}\left(\tilde{\rho}_{3d_1} + \tilde{\rho}_{3d_2} + \tilde{\rho}_{d_13} + \tilde{\rho}_{d_23}\right) + L_c\tilde{\rho}_{33}, \\
\tilde{\rho}_{3d_1} &= -\left[\frac{1}{2}\left(\Gamma_3 + \frac{1}{2}\Gamma_1\right) + i\left(\Delta_p - \frac{1}{2}\Omega\right)\right]\tilde{\rho}_{3d_1} - \frac{1}{2}g\tilde{\rho}_{3d_1}\left(a_c - a_c^\dagger\right) \\
&\quad + \frac{\Omega_p}{2\sqrt{2}}\left(\tilde{\rho}_{d_1d_1} - \tilde{\rho}_{33}\right) + L_c\tilde{\rho}_{3d_1}, \\
\tilde{\rho}_{3d_2} &= -\left[\frac{1}{2}\left(\Gamma_3 + \frac{1}{2}\Gamma_1\right) + i\left(\Delta_p + \frac{1}{2}\Omega\right)\right]\tilde{\rho}_{3d_2} - \frac{1}{2}g\tilde{\rho}_{3d_2}\left(a_c - a_c^\dagger\right) \\
&\quad + \frac{\Omega_p}{2\sqrt{2}}\left(\tilde{\rho}_{d_2d_2} - \tilde{\rho}_{33}\right) + L_c\tilde{\rho}_{3d_2}.
\end{aligned} \tag{21}
$$

Equations (21) are still operators with respect to the cavity field. Since the probe field is very weak, the population $\tilde{\rho}_{33}$ is much smaller than the populations $\tilde{\rho}_{d_2d_2}$ and $\tilde{\rho}_{d_1d_1}$, and thus we can ignore the effect of $\tilde{\rho}_{33}$ on the evolution of the coherences $\tilde{\rho}_{3d_1}$ and $\tilde{\rho}_{3d_2}$.

First, we consider the population of the state $|3\rangle$ in the absence of the cavity. In this case the density matrix elements (21) become c-numbers and it is easy to find that the steady-state population $\tilde{\rho}_{33}$ is given by

$$\tilde{\rho}_{33} = \frac{\Omega_p^2}{8\Gamma_3}\left\{\frac{\Gamma_A\tilde{\rho}_{d_1d_1}}{\frac{1}{4}\Gamma_A^2+\left(\Delta_p-\frac{1}{2}\Omega\right)^2}+\frac{\Gamma_A\tilde{\rho}_{d_2d_2}}{\frac{1}{4}\Gamma_A^2+\left(\Delta_p+\frac{1}{2}\Omega\right)^2}\right\}, \tag{22}$$

where $\Gamma_A = \frac{1}{2}\left(\Gamma_3 + \frac{1}{2}\Gamma_1\right)$.

The spectrum is composed of two lines of equal linewidths, Γ_A, located at frequencies shifted from the $|3\rangle - |2\rangle$ resonance by $\pm\frac{1}{2}\Omega$. The intensities of the lines depend on the population distribution between the dressed states of the driven transition.

Including the cavity, and transforming the density matrix elements into the photon-number representation, we find that the coherence $X_{n,m} = \langle n|\tilde{\rho}_{3d_1}|m\rangle$ satisfies the recurrence relation

$$\begin{aligned} &A_{n,m}X_{n,m} + \frac{1}{2}g\sqrt{m}X_{n,m-1} - \frac{1}{2}g\sqrt{m+1}X_{n,m+1} \\ &+\kappa\sqrt{(n+1)(m+1)}X_{n+1,m+1} = \frac{\Omega_p}{2\sqrt{2}}\tilde{\rho}_{n,m}^{(d_1d_1)}, \end{aligned} \tag{23}$$

where

$$A_{n,m} = \frac{1}{2}\left(\Gamma_3 + \frac{1}{2}\Gamma_1\right) + \frac{1}{2}\kappa(n+m) + i\left(\Delta_p - \frac{1}{2}\Omega\right), \tag{24}$$

and $\tilde{\rho}_{n,m}^{(d_1d_1)} = \langle n|\tilde{\rho}_{d_1d_1}|m\rangle$. The recurrence relation for the coherence $Y_{n,m} = \langle n|\tilde{\rho}_{3d_2}|m\rangle$ has the same form as (23) with $\Omega \to -\Omega$ and $d_1 \to d_2$.

In general, the formula (23) is a two-dimensional recurrence relation and reduces to an one-dimensional recurrence relation in the limit of vanishing cavity damping $\kappa \to 0$. The coherences $X_{n,m}$ depend on the populations $\tilde{\rho}_{n,m}^{(d_id_i)}$ of the cavity modified dressed states and are also coupled to the "non-diagonal" elements $X_{n,m\pm1}$ and $X_{n+1,m+1}$. In the limit of vanishing κ and to lowest order in the coupling g, where we ignore the coupling of the diagonal elements to the off-diagonal elements, the coherence $X_{n,m}$ reduces to its free-space value,

$$X_{n,m} = \frac{\Omega_p}{2\sqrt{2}}\frac{\tilde{\rho}_{n,m}^{(d_1d_1)}}{A_{n,m}}. \tag{25}$$

In the next approximation, we include the coupling of $X_{n,m}$ to $X_{n,m\pm1}$, and we find that near the resonance $\Delta_p \approx \frac{1}{2}\Omega$, and for weak cavity damping, the coherence is given by

$$X_{n,m} = \frac{\Omega_p}{2\sqrt{2}}\frac{\tilde{\rho}_{n,m}^{(d_1d_1)}}{\Gamma_A/2 + 2g^2n/\Gamma_A + i\left(\Delta_p - \frac{1}{2}\Omega\right)}. \tag{26}$$

Thus, the effect of the off-diagonal terms in the recurrence relation (23) is to produce a power broadening of the Autler-Townes lines. Therefore, similarly to the case of free space, the major factor determining the shape and intensities of the Autler-Townes lines is the population distribution between the dressed states of the driven transition. These quantities are calculated in the next section.

4.2 Population of the dressed states

From the master equation (20), we find that the equations of motion for the populations of the dressed states $\tilde{\rho}_{d_i d_i} \equiv \rho_{ii}$ are of the form

$$\begin{aligned}\dot{\rho}_{11} &= \frac{g}{2}\left[\left(a_c - a_c^\dagger\right)\rho_{11} - \rho_{11}\left(a_c - a_c^\dagger\right)\right] - \frac{\Gamma_1}{4}\left(\rho_{11} - \rho_{22}\right) + L_c\rho_{11},\\ \dot{\rho}_{22} &= -\frac{g}{2}\left[\left(a_c - a_c^\dagger\right)\rho_{22} - \rho_{22}\left(a_c - a_c^\dagger\right)\right] - \frac{\Gamma_1}{4}\left(\rho_{22} - \rho_{11}\right) + L_c\rho_{22}. \end{aligned} \quad (27)$$

It is interesting to note that the equations of motion for the populations are decoupled from the equations of motion for the coherences ρ_{12} and ρ_{21}.

It is convenient to introduce the following combinations of the populations of the dressed states

$$\begin{aligned} u &= \rho_{11} + \rho_{22},\\ w &= \rho_{22} - \rho_{11}, \end{aligned} \quad (28)$$

which satisfy the following equations of motion

$$\dot{u} = -\frac{g}{2}\left[\left(a_c - a_c^\dagger\right)w - w\left(a_c - a_c^\dagger\right)\right] + L_c u, \quad (29)$$

$$\dot{w} = -\frac{g}{2}\left[\left(a_c - a_c^\dagger\right)u - u\left(a_c - a_c^\dagger\right)\right] - \frac{\Gamma_1}{2}w + L_c w. \quad (30)$$

Note that $u = \mathrm{Tr}_{atom}(\rho)$ is the reduced density operator of the cavity field. Moreover, the combination u, which gives the cavity photon distribution, does not depend on the atomic evolution.

In the photon-number representation, equations of motion for $u_{n,m} = \langle n|u|m\rangle$ and $w_{n,m} = \langle n|w|m\rangle$ are given by

$$\begin{aligned}\dot{u}_{n,m} &= -\frac{g}{2}\left(\sqrt{n+1}w_{n+1,m} - \sqrt{n}w_{n-1,m} - \sqrt{m}w_{n,m-1} + \sqrt{m+1}w_{n,m+1}\right)\\ &\quad +\kappa\left(\sqrt{(n+1)(m+1)}u_{n+1,m+1} - \frac{1}{2}(n+m)u_{n,m}\right),\\ \dot{w}_{n,m} &= -\frac{g}{2}\left(\sqrt{n+1}u_{n+1,m} - \sqrt{n}u_{n-1,m} - \sqrt{m}u_{n,m-1} + \sqrt{m+1}u_{n,m+1}\right)\\ &\quad +\kappa\left(\sqrt{(n+1)(m+1)}w_{n+1,m+1} - \frac{1}{2}(n+m)w_{n,m}\right)\\ &\quad -\frac{\Gamma_1}{2}w_{n,m}. \end{aligned} \quad (31)$$

It is seen from (31) that the diagonal elements $u_{n,m}(w_{n,m})$ are coupled to the off-diagonal elements $w_{n\pm1,m}(u_{n\pm1,m})$ and $w_{n,m\pm1}(u_{n,m\pm1})$. By setting the left-hand side of (31) equal to zero we obtain the steady-state solutions of these equations. A straightforward manipulation of (31) leads to the following six-term two-dimensional recurrence relation

$$\begin{aligned} &G_{n,m}Z^{(n,m)} + D_{n+1,m+1}Z^{(n+1,m+1)} + B_{0,m+1}Z^{(n,m+1)} \\ &+B_{n+1,0}Z^{(n+1,m)} - B_{0,m}Z^{(n,m-1)} - B_{n,0}Z^{(n-1,m)} = 0, \end{aligned} \tag{32}$$

where

$$\begin{aligned} Z^{(n,n)} &= \begin{pmatrix} u^{(n,m)} \\ w^{(n,m)} \end{pmatrix}, G_{n,m} = \begin{pmatrix} \kappa n & 0 \\ 0 & \kappa n + \frac{1}{2}\Gamma_1 \end{pmatrix}, \\ B_{n,m} &= \begin{pmatrix} 0 & g\sqrt{n+m} \\ g\sqrt{n+m} & 0 \end{pmatrix}, D_{n,m} = \begin{pmatrix} -\kappa\sqrt{nm} & 0 \\ 0 & -\kappa\sqrt{nm} \end{pmatrix}, \end{aligned} \tag{33}$$

with $\sum_{n,m} u_{n,m} = 1$. Note that the degeneracy in the cavity and driving field frequencies leads to a doubling of the dimension of the dynamics of the system, in the sense that the diagonal elements $u_{n,n}(w_{n,n})$ are not decoupled from the off-diagonal elements $u_{n,n+m}(w_{n,n+m})$. In order to calculate the Autler-Townes absorption spectrum, we include a weak probe field coupled to the $|3\rangle - |2\rangle$ transition and calculate the steady-state coherence ρ_{32} from the recurrence relation (23). We then find the population ρ_{33} as a function of Δ_p.

5 Autler-Townes Spectra

We investigate the modifications in the dynamics of the driven atomic transition produced by the cavity in the strong field limit of $\Omega \gg \Gamma_1, \Gamma_3, \kappa$. We consider separately two cases: In the first case we calculate the Autler-Townes spectrum in a truncated basis of a small and fixed number of photons in the cavity mode. It has recently been shown that it is experimentally possible to achieve this [22]. This is relevant to the situation in which the cavity is initially prepared in a state with a definite number of photons present. In the second case, we calculate the spectrum numerically in the basis of a large and undetermined number of photons.

5.1 Fixed number of photons

In Fig. 1, we show the steady-state population ρ_{33} as a function of the detuning Δ_p for $g \ll \kappa, \Gamma_1$. The spectrum, known as the Autler-Townes doublet [23], consists of two lines of equal linewidths $\frac{1}{2}\left(\Gamma_3 + \frac{1}{2}\Gamma_1\right)$ and separated by the Rabi frequency Ω of the driving field. As the coupling constant g increases the lines broaden and for $g \gg \kappa, \Gamma_1$, split into multiplets. Figure 2 shows the spectrum for two different values of n and g such that $2g\sqrt{n} = \text{const}$. Here, we see that the number of peaks in the structures increases with n and the oscillations vanish for a large n. The case of $n \gg 1$ corresponds to the Autler-Townes spectrum of a two-level transition driven by two lasers of the same frequency [24].

The multipeak-structure of the Autler-Townes spectrum could suggest that in the presence of the cavity the dressed states $|d_1\rangle$ and $|d_2\rangle$ split into discrete multiplets. However, the analysis of Section 2 shows that this is not the case in the present system. We showed that the energy spectrum of the system is composed of an infinite ladder of doublet continua with interdoublet separation $\hbar\omega_{12}$ and interdoublet splitting Ω. Since the λ of (7) is an arbitrary real number, discrete energy levels of the system *do not* exist.

Thus, a weak laser field coupled to the $|3\rangle - |2\rangle$ transition will probe the steady-state population distribution in the *continua* induced by the cavity and driving laser fields. In order to find the steady-state population of the dressed states (8) we project the master equation (11) onto $|N, d_i, \lambda\rangle$ on the right and $\langle N, d_i, \lambda|$ on the left. We make the secular approximation [20], valid for $\Omega \gg \Gamma_1, \Gamma_3$, in which we ignore the coupling of the populations to the coherences, and find the "reduced" steady-state populations $P_{d_i} = \sum_N \langle N, d_i, \lambda| \rho |N, d_i, \lambda\rangle$ to be

$$P_{d_1} = P_{d_2} = \frac{1}{2}\left|\phi_n\left(\frac{\lambda}{\sqrt{2}}\right)\right|^2 . \tag{34}$$

The population is distributed equally between the continua of the dressed states and is spread across these states with the weight function $\left|\phi_n\left(\lambda/\sqrt{2}\right)\right|^2$. Clearly the oscillations seen in Figs. 1 and 2 result from the oscillatory distribution of the population inside the continuum of the dressed states. The oscillations

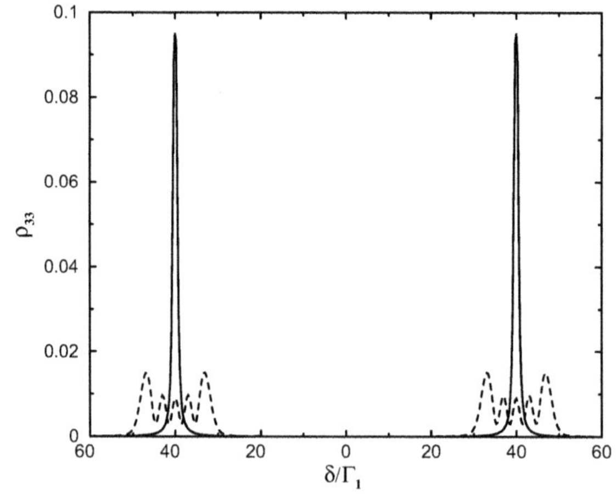

Fig. 1. The Autler-Townes spectrum for $\Omega = 80\Gamma_1, \kappa = \Gamma_3 = 0.1\Gamma_1$ and different atom-cavity couplings: $g = 0.01\Gamma_1$ (solid line), $g = 1.5\Gamma_1$ (dashed line).

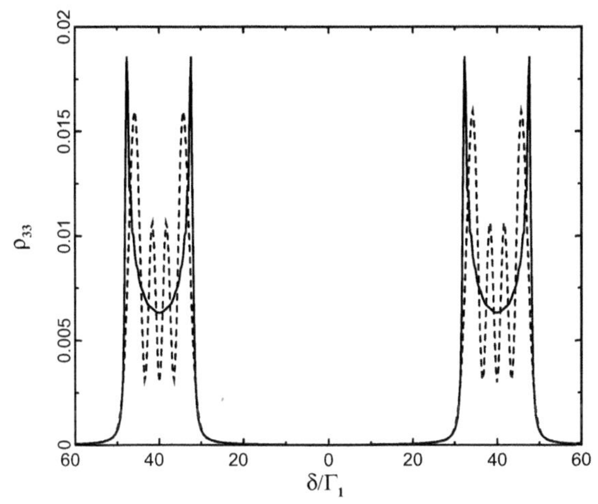

Fig. 2. The Autler-Townes spectrum for $\Omega = 80\Gamma_1$ and the other parameters chosen such that $g\sqrt{n}=$ const: $g = 0.666\Gamma_1, n = 36$ (solid line), $g = 2\Gamma_1, n = 4$ (dashed line).

provide an evidence of field quantization in the cavity and their observation would provide a direct measurement of the probability distribution function of the harmonic oscillator.

5.2 Numerical results

Figure 3 shows the Autler-Townes spectrum for $\Omega = 100\Gamma_1, \kappa = 0.1\Gamma_1, \Gamma_3 = 0.1\Gamma_1$ and different g. This spectrum has been obtained by averaging the population over a truncated basis of seventy number states. One sees that for small g the spectrum is the familiar Autler-Townes doublet. For moderate g the spectral lines split into multiplets with the overall width of the features equal to the Rabi frequency of the cavity field, $\Omega_c = 2g\sqrt{\langle n \rangle}$, where $\langle n \rangle$ is the average number of photons in the cavity mode. The width of the multiple structures increases linearly with g indicating that the average number of photons in the cavity mode is constant, independent of g. Surprisingly, for large g the splitting disappears and the Autler-Townes multiple doublet reduces to a single peak with linewidth approximately equal to Γ_3, located at the atomic transition frequency ω_{32}.

In order to get a physical insight into the cancellation of the Autler-Townes splitting, we plot in Fig. 4 the quantity $\mathrm{Tr}(\rho^2)$ which provides information about the number of states of the system and their purity. We find $\mathrm{Tr}(\rho^2)$ by a direct numerical integration of the master equation (11). In this approach we use the Fock state representation for the cavity field and write the density matrix ρ as a vector composed of the density-matrix elements. We solve this system of linear, ordinary, differential equations using the symbolic representational power of the quantum optics toolbox for matlab [25].

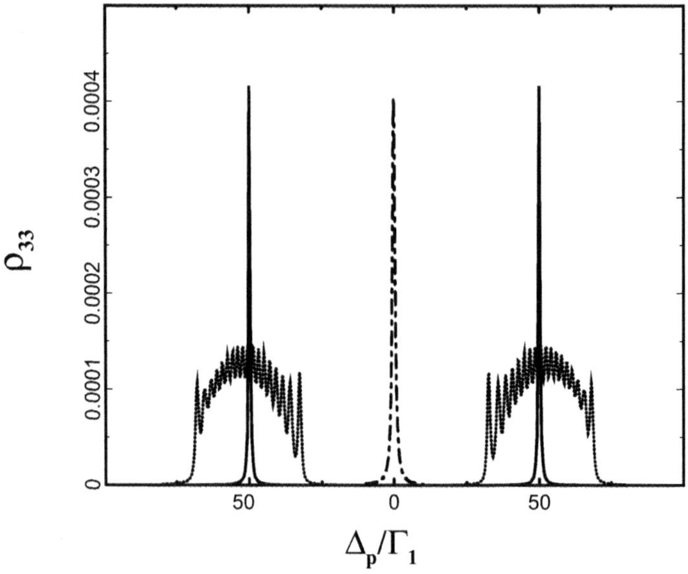

Fig. 3. The Autler-Townes spectrum for $\Omega = 100\Gamma_1, \kappa = \Gamma_3 = 0.1\Gamma_1$ and different g: $g = 0.01\Gamma_1$ (solid line), $g = 3\Gamma_1$ (multiple peaks), $g = 20\Gamma_1$ (dashed-dotted line).

For $g = 0$, $\mathrm{Tr}(\rho^2) = \frac{1}{2}$ indicating that the system is in a mixed state of the two dressed states $|d_1\rangle$ and $|d_2\rangle$. When $g \neq 0$ and κ is small, the state of the system becomes more mixed with $\mathrm{Tr}(\rho^2) \approx 0$. This indicates that the system is in a mixed state involving a very large number of states. This is easy to understand if one refers to (7) which shows that the energy spectrum of the system is composed of two continua each containing an infinite number of states. As g increases the purity remains constant and at $g \approx 7.5\Gamma_1$ changes rapidly from the maximally-mixed state $\mathrm{Tr}(\rho^2) = 0$ to a maximally pure state $\mathrm{Tr}(\rho^2) = 1$. The threshold value corresponds to $\Omega_c = \Omega$. For $g > 7.5\Gamma_1$ the system remains in the pure state independent of g. The threshold value of g depends on κ and moves towards larger g as κ increases. In Fig. 5, we plot the populations of the atomic bare states $|1\rangle$ and $|2\rangle$ for the same parameters as the $\kappa = 0.1\Gamma_1$ case of Fig. 4. We see that for $g < 7.5\Gamma_1$ the atomic states are equally populated and the population does not changes with g. When $g \approx 7.5\Gamma_1$ the atom collapses into its ground state and remains in this state independent of g.

Alsing *et al.* [8] have predicted that a two-level atom located inside a lossless cavity can remain in its ground state even if is continuously driven by a coherent laser field. They have explained this effect as arising from the destructive interference between the driving and the cavity fields which cancels the effective driving of the atom. Here, we present an alternative explanation which involves linear

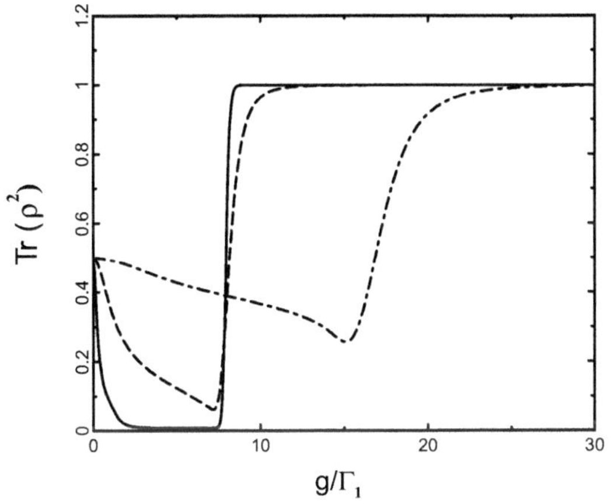

Fig. 4. $\mathrm{Tr}(\rho^2)$ as a function of g for $\Gamma_3 = 0.1\Gamma_1$, $\Omega = 100\Gamma_1$ and different κ: $\kappa = 0.1\Gamma_1$ (solid line), $\kappa = 1.0\Gamma_1$ (dashed line), $\kappa = 5\Gamma_1$ (dashed-dotted line).

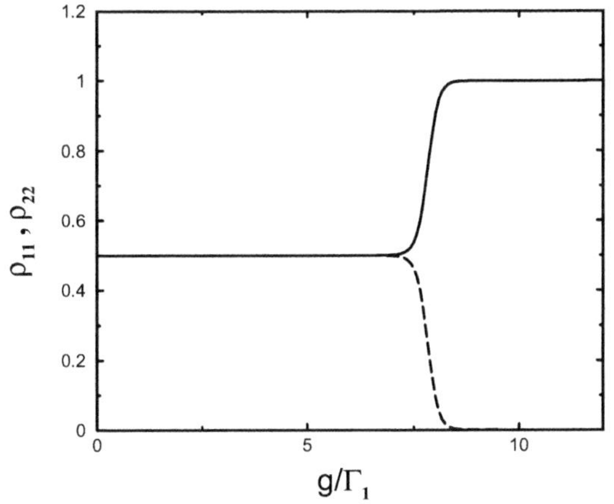

Fig. 5. Populations of the atomic bare states as a function of g for the same parameters as the $\kappa = 0.1\Gamma_1$ case of Fig. 4: ρ_{22} (solid line), ρ_{11} (dashed line).

superpositions of two degenerate dressed states of the system, and accounts for the fact that the degeneracy appears only when $\Omega_c \geq \Omega$.

According to the dressed states analysis, the continua are separated by $\Omega - \Omega_c$ and they start to overlap when $\Omega_c = \Omega$. In the case of non-overlapping continua the nonvanishing dipole matrix elements are only between two neighboring man-

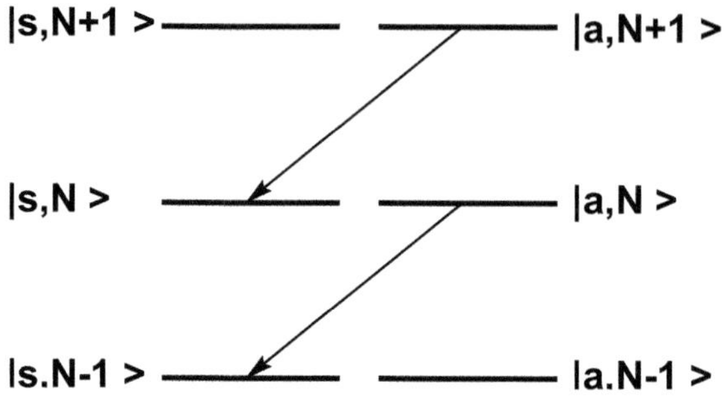

Fig. 6. Superpositions states of the overlapping continua. The solid arrows indicate the allowed spontaneous transitions.

ifolds and are given by

$$\begin{aligned}
\langle N, d_1, \lambda | \boldsymbol{\mu} | N-1, d_2, \lambda' \rangle &= \frac{1}{2}\boldsymbol{\mu}_{12}\delta(\lambda+\lambda'), \\
\langle N, d_2, \lambda | \boldsymbol{\mu} | N-1, d_1, \lambda' \rangle &= -\frac{1}{2}\boldsymbol{\mu}_{12}\delta(\lambda+\lambda'), \\
\langle N, d_1, \lambda | \boldsymbol{\mu} | N-1, d_1, \lambda' \rangle &= \frac{1}{2}\boldsymbol{\mu}_{12}\delta(\lambda-\lambda'), \\
\langle N, d_2, \lambda | \boldsymbol{\mu} | N-1, d_2, \lambda' \rangle &= -\frac{1}{2}\boldsymbol{\mu}_{12}\delta(\lambda-\lambda'). \qquad (35)
\end{aligned}$$

Consider two extremal states of the continua $|N, d_1, \lambda_{max}\rangle$ and $|N, d_2, \lambda_{min}\rangle$ which overlap for $\Omega_c = \Omega$. The degeneracy in the energy of the two overlapping states leads to the following linear superposition states

$$\begin{aligned}
|s, N\rangle &= \frac{1}{\sqrt{2}}\left(|N, d_1, \lambda_{max}\rangle + |N, d_2, \lambda_{min}\rangle\right), \\
|a, N\rangle &= \frac{1}{\sqrt{2}}\left(|N, d_1, \lambda_{max}\rangle - |N, d_2, \lambda_{min}\rangle\right). \qquad (36)
\end{aligned}$$

It is easily verified from (35) and (36) that the *only* nonzero transition moments are between the states $|a, N+p\rangle$ and $|s, N+p-1\rangle$

$$\langle N+p, a | \boldsymbol{\mu} | s, N+p-1\rangle = \frac{1}{4}\boldsymbol{\mu}_{12}, \qquad (37)$$

where $p = 0, 1, 2, \ldots$. In Fig. 6, we plot the superposition states and the allowed spontaneous transitions. It is clear from Fig. 6 that the population flows from the antisymmetric state to the symmetric state of the manifold below but cannot escape from the symmetric state, resulting in the trapping of the population in that state.

Using (36) it is easy to show that the symmetric state can be written as

$$|s, N\rangle = \sum_n \phi_n \left(\frac{\lambda}{\sqrt{2}}\right) |2, N-n, n\rangle \,. \tag{38}$$

Thus, the trapping state of the system involves only the ground state of the atom.

6 Summary

We have demonstrated a relationship between the harmonic oscillator and a two-level atom coupled to a cavity mode and driven by a resonant laser field. The properties of the latter system have been investigated by connecting the ground state to a third level, not coupled to the cavity, by a weak probe field, and calculating the Autler-Townes spectrum. In the strong-coupling limit the components of the Autler-Townes doublet are composed of multiplets, whose detailed structure depends on the atom-cavity coupling and the cavity and spontaneous emission damping rates. We have shown that the multiplets do not correspond to any discrete energy levels of the system, but result from the oscillatory distribution of the population inside the continuum of the dressed states. The significance of the oscillations is twofold: (1) they are a signature of the quantum nature of the cavity field, and (2), they are a direct analog of the well-known oscillations of the probability distribution function of the harmonic oscillator. They should be measurable for this system. The probability distribution function of the harmonic oscillator has not yet been measured, and it is obviously of great interest to observe this fundamental property. We have also shown that there is a threshold value for the Rabi frequency of the cavity field at which the Autler-Townes splitting disappears and the atom collapses to the ground state, and interpreted it as a population trapping effect.

Acknowledgments

This research was supported by the United Kingdom Engineering and Physical Sciences Research Council.

References

1. E.T. Jaynes and F.W. Cummings, Proc. IEEE **51**, 89 (1963).
2. J.H. Eberly, N.B. Narozhny, and J.I. Sanchez-Mondragon, Phys. Rev. Lett. **44**, 1323 (1980); P.L. Knight and P.M. Radmore, Phys. Lett. **90A**, 342 (1982); G. Rempe, H. Walther, and N. Klein, Phys. Rev. Lett. **58**, 353 (1987).
3. H.J. Carmichael, R.J. Brecha, M.G. Raizen, H.J. Kimble, and P.R. Rice, Phys. Rev. A **40**, 5516 (1989).

4. J.I. Cirac, H. Ritsch, and P. Zoller, Phys. Rev. A **44**, 4541 (1991).
5. T. Quang and H. Freedhoff, Phys. Rev. A **47**, 2285 (1993); H. Freedhoff and T. Quang, J. Opt. Soc. Am. B **10**, 1337 (1993); Phys. Rev. Lett. **72**, 474 (1994).
6. R.J. Brecha, P.R. Rice and M. Xiao, Phys. Rev. A **59**, 2392 (1999); J.P. Clemens and P.R. Rice, Phys. Rev. A **61**, 063810 (2000).
7. C.M. Savage, Phys. Rev. Lett. **60**, 1828 (1988); M. Lindberg and C.M. Savage, Phys. Rev. A **38**, 5182 (1988); C.M. Savage, Phys. Rev. Lett. **63**, 1376 (1989); P. Alsing and H.J. Carmichael, Quantum Opt. **3**, 13 (1991); P.M. Alsing, D.-S. Guo, and H.J. Carmichael, Phys. Rev. A **45**, 5135 (1992); H. Nha, Y.-T. Chough, and K. An, Phys. Rev. A **62**, 021801(R) (2000).
8. P.M. Alsing, D.A. Cardimona, and H.J. Carmichael, Phys. Rev. A **45**, 1793 (1992).
9. B.R. Mollow, Phys. Rev. **188**, 1969 (1969).
10. M. Lewenstein, T.W. Mossberg, and R.J. Glauber, Phys. Rev. Lett. **59**, 775 (1987); M. Lewenstein and T.W. Mossberg, Phys. Rev. A **37**, 2048 (1988); P. Zhou and S. Swain, Phys. Rev. A **58**, 1515 (1998).
11. J. Ye, D.W. Vernooy, and H.J. Kimble, Phys. Rev. Lett. **83**, 4987 (1999); P. W. H. Pinkse, T. Fischer, P. Maunz, and G. Rempe, Nature **404**, 365 (2000); A.C. Doherty, T.W. Lynn, C.J. Hood, and H.J. Kimble, Phys. Rev. A **63**, 013401 (2001).
12. A. Whitaker, Prog. Quant. Elect. **24**, 1 (2000).
13. M. Brune, F. Schmidt-Kaler, A. Maali, J. Dreyer, E. Hagley, J.M. Raimond, and S. Haroche, Phys. Rev. Lett. **76**, 1800 (1996).
14. B.T.H. Varcoe, S. Brattke, B.G. Englert, and H. Walther, Laser Phys. **10**, 1 (2000); B.T.H. Varcoe, S. Brattke, M. Weldinger, and H. Walther, Nature **403**, 743 (2000).
15. Q.A. Turchette, Phys. Rev. Lett. **75**, 4710 (1995); A. Rauschenbeutel, G. Nogues, S. Osnaghi, P. Brune, J.M. Raimond, and S. Haroche, Phys. Rev. Lett. **83**, 5166 (1999).
16. J.I. Sanchez-Mondragon, N.B. Narozhny, and J.H. Eberly, Phys. Rev. Lett. **51**, 550 (1983); Y. Kaluzny, P. Goy, M. Gross, J.N. Raimond, and S. Haroche, Phys. Rev. Lett. **51**, 1175 (1983); M.G. Raizen, R.J. Thompson, R.J. Brecha, H.J. Kimble, and H.J. Carmichael, Phys. Rev. Lett. **63**, 240 (1989); Y. Zhu, D.J. Gauthier, S.E. Morin, Q. Wu, H.J. Carmichael, and T.W. Mossberg, Phys. Rev. Lett. **64**, 2499 (1990).
17. H.J. Carmichael, P. Kochan, and B.C. Sanders, Phys. Rev. Lett. **77**, 631 (1996); B.C. Sanders, H.J. Carmichael, and B.F. Wielinga, Phys. Rev. A **55**, 1358 (1997); L. Horvath, B.C. Sanders, and B.F. Wielinga, J. Opt. B: Quantum Semiclass. Opt. **1**, 446 (1999).
18. See e.g., A.S. Davydov, *Quantum Mechanics* (Pergamon Press, Oxford, 1965), p. 116.
19. E. E. Vdovin, A. Levin, A. Patane, L. Eaves, P. C. Main, Y. N. Khanin, Y. V. Dubrovskii, M. Henini, and G. Hill, Science **290**, 122 (2000).
20. C. Cohen-Tannoudji and S. Reynaud, J. Phys. B **10**, 345 (1977); C. Cohen-Tannoudji, J. Dupont-Roc, and G. Grynberg, *Atom-Photon Interactions*, (Wiley, New York, 1992).
21. A. Böhm, *Quantum Mechanics*, (Springer, Berlin, 1979).
22. S. Brattke, B-G. Englert, B. T. H. Varcoe and H. Walther, J. mod. Opt. **47**, 2857 (2000).
23. S.H. Autler and C.H. Townes, Phys. Rev. **100**, 703 (1955).
24. H.S. Freedhoff and Z. Ficek, Phys. Rev. A **55**, 1234 (1997); A.D. Greentree, C. Wei, S.A. Holmstrom, J.P.D. Martin, N.B. Manson, K.R. Catchpole, and C. Savage, J. Opt. B: Quantum Semiclass. Opt. **1**, 240 (1999)
25. S.M. Tan, J. Opt. B: Quantum Semiclass. Opt. **1**, 424 (1999).

The Method of Quantum Jumps and Quantum White Noise

W. von Waldenfels

Institut für Angewandte Mathematik der Universität Heidelberg,
Im Neuenheimer Feld 294, 69120 Heidelberg, Germany

Abstract. The density matrix $\varrho(t)$ of a small quantum system in a heatbath can be given by the master equation in its Lindblad form. Numerically it is often more advantageous to calculate the density matrix by a Monte Carlo method, the method of quantum jumps, simulating the influence of the heatbath on the gas by a classical Markov process. The mathematical background is investigated. Physically more appropriate seems to be to put the small system into a quantum heatbath and to approximate the quantum heatbath by quantum white noise. The resulting quantum stochastic differential equation yields again Lindblad's equation and some information more. Two examples are discussed , the two level atom in a heatbath of oscillators and the oscillator in an atomic heatbath.

1 Introduction

The density matrix of a small quantum system S in a heatbath is given by the master equation in Lindblad's form

$$\dot{\varrho} = \mathcal{L}(\varrho) \tag{1}$$

with

$$\mathcal{L}(\varrho) = -\mathrm{i}[H_S, \varrho] - 1/2 \sum_{m=0}^{M} (C_m^\dagger C_m \varrho - 2C_m \varrho C_m^\dagger + \varrho C_m^\dagger C_m). \tag{2}$$

We show in section 2, that this equation can be solved in the the following way. Introduce

$$\begin{aligned} &\Phi_0 = \mathrm{e}^{-\mathrm{i}Ht} \\ &\Phi_n(t_1, m_1; \cdots; t_n, m_n)(t) \\ &= \mathrm{e}^{-\mathrm{i}H(t-t_n)} C_{m_n} \mathrm{e}^{-\mathrm{i}H(t_n - t_{n-1})} C_{m_{n-1}} \cdots \mathrm{e}^{-\mathrm{i}H(t_2-t_1)} C_{m_1} \mathrm{e}^{-\mathrm{i}H(t_1)}. \end{aligned} \tag{3}$$

with

$$H = H_S - \mathrm{i}/2 \sum_m C_m^\dagger C_m \tag{4}$$

then

$$\varrho(t) = \Phi_0(t)\, \varrho\, \Phi_0(t)^\dagger + \sum_{n=1}^{\infty} \sum_{m_1, \cdots, m_n} \int_{0<t_1<\cdots<t_n<t} \mathrm{d}t_1 \cdots \mathrm{d}t_n$$

$$\Phi_n(t_1, m_1; \cdots; t_n, m_n)(t) \varrho(0) \Phi_n(t_1, m_1; \cdots; t_n, m_n)(t)^\dagger. \tag{5}$$

M. Orszag, J.C. Retamal (Eds.): LNP 575, pp. 279–293, 2001.

This can be interpreted as if the system moves in a deterministic way under the influence of the non-hermitian operator H and undergoes some jumps due to multiplication with C_m. These jumps are called quantum jumps. In concrete calculations the Hilbert space $\mathcal{H}$ of the quantum system S is approximated by a finite dimensional space, say with n dimensions. Then the density matrix has n^2 and the Lindblad operator has n^4 entries. So if e.g. $n = 100$ we encounter enormous difficulties. A way out are socalled stochastic Schroedinger equations, where one can use Monte Carlo methods. In section 3 we will show, how in the form of the stochastic Schroedinger equation using quantum jumps the density matrix can be presented as

$$\varrho(t) = \mathbf{E}\,|\Psi(t)\rangle\langle\Psi(t)|, \tag{6}$$

where $\Psi(t)$ is a classical Markov process on the unit sphere $\mathcal{B}_S$ of $\mathcal{H}_S$ and $\mathbf{E}$ is the stochastic average. Assuming for simplicity, that

$$\varrho(0) = |\psi_0\rangle\langle\psi_0| \tag{7}$$

the paths of the Markov process are given by

$$\mathrm{un}(\Phi_k(t_1, m_1; \cdots ; t_k, m_k)(t)\psi_0), \tag{8}$$

assuming,that there are k quantum jumps between 0 and t.Here we defined for a vector ϕ in a Hilbertspace the unitvector in the same direction by

$$\mathrm{un}(\phi) = \phi/\|\phi\|. \tag{9}$$

The probability that for fixed k and the m_j the jumping times are between t_j and $t_j + \mathrm{d}t_j$ is

$$\|\Phi_k(t_1, m_1; \cdots ; t_k, m_k)(t)\psi_0\|^2 \mathrm{d}t_1 \cdots \mathrm{d}t_k. \tag{10}$$

¿From there one concludes immediately (5).
Introducing a Markov process means coupling the system S to a classical heat bath and to approximate the interaction by a Markovian interaction, i.e. without memory to the past.There is an other way, and it seems physically more intuitive. The system S is coupled to a quantum heat bath and then the quantum heat bath is approximated by quantum white noise. One arrives at a quantum stochastic differential equation. The solution is a unitary operator on the tensorspace $\mathcal{H}_S \otimes \mathcal{H}_B$, where B stands for white noise. Then $\varrho(t)$ can be calculated by

$$\mathrm{Tr} X \varrho(t) = \mathrm{Tr}((X \otimes 1)U(t)(\varrho(0) \otimes |0\rangle\langle 0|)U(t)^\dagger), \tag{11}$$

where we can calculate in an explicite way

$$U(t)|0\rangle = \Phi_0|0\rangle + \sum_{n=1}^{\infty}(-\mathrm{i})^n \int_{0<t_1\cdots<t_n<t} \sum_{m_1,\cdots,m_n}$$
$$\Phi_n(t_1, m_1; \cdots ; t_n, m_n)(t) a^\dagger_{m_n}(t_n) \cdots a^\dagger_{m_1}(t_1)|0\rangle \mathrm{d}t_1 \cdots \mathrm{d}t_n. \tag{12}$$

Remark, that here again the functions Φ appear.
After these considerations in section 4 we study two examples. In section 5 we discuss the Wigner Weisskopf atom, a two level atom in the radiation field, which is considered as a heat bath of oscillators. In section 6 we present the slight modification of a model due to M.Orszag et al. [4],[5]. A standing wave between two mirrors is exposed to a beam of three level atoms. This is considered as a harmonic oscillator in an atomic heatbath. In both cases we obtain the approximate quantum stochastic differential equation.

2 The Master Equation

Consider a Quantum system in a heatbath. Its density matrix at time t obeys in some approximation the socalled master equation. The master equation in Lindblad form is

$$\dot{\varrho} = \mathcal{L}(\varrho)$$

with

$$\mathcal{L} = -\mathrm{i}[H_S, \varrho] - 1/2 \sum_m (C_m^\dagger C_m \varrho - 2C_m \varrho C_m^\dagger + \varrho C_m^\dagger C_m).$$

where H_S is the unperturbed Hamiltonian of the system and the C_m are operators in $\mathcal{H}_S$, the Hilbert space of the system. Denote

$$\mathcal{L}_0(\varrho) = -\mathrm{i}[H_S, \varrho] - 1/2 \sum_m (C_m^\dagger C_m \varrho + \varrho C_m^\dagger C_m) = -\mathrm{i}H\varrho + \mathrm{i}\varrho H$$

with H given by (4) and and

$$J_m(\varrho) = C_m \varrho C_m^\dagger.$$

We obtain

$$\varrho(t) = \mathrm{e}^{\mathcal{L}t}(\varrho)$$

and the Dyson expansion

$$\mathrm{e}^{\mathcal{L}t} = \mathrm{e}^{\mathcal{L}_0 t} + \int_0^t \mathrm{d}t_1 \sum_m \mathrm{e}^{(t-t_1)\mathcal{L}_0} J_m \mathrm{e}^{t_1 \mathcal{L}_0} + \cdots$$

$$+ \int_{0<t_1<\cdots<t_n} \mathrm{d}t_1 \cdots \mathrm{d}t_n \sum_{m_1,\cdots,m_n} \mathrm{e}^{(t-t_n)\mathcal{L}_0} J_{m_n} \mathrm{e}^{(t_n)-t_{n-1}\mathcal{L}_0} \mathrm{e}^{(t_2-t_1)\mathcal{L}_0} J_{m_1} \mathrm{e}^{t_1 \mathcal{L}_0}.$$

From there one establishs (5) imediately.

3 The Quantum Jump Method

In the quantum jump method the usual Schroedinger equation will be replaced by a classical Markov process $\Psi(t)$ on the unitball $\mathcal{B}_S$ of the Hilbert space $\mathcal{H}_S$

of the quantum system S. We will show,that

$$\varrho(t) = \mathbf{E}\,|\Psi(t)\rangle\langle\Psi(t)|,$$

where $\mathbf{E}$ denotes the stochastic average.
We recall the notion of a Markov process. A Markov process is a special kind of a stochastic process. A stochastic process is a random function usually on an interval of the real line. The space, where the random function takes its values, is called the statespace.The different realizations of the random function are called sample paths. . The stochastic average means averaging over all sample paths .In our case $\Psi(t)$ is defined for $t \geq 0$
If $X(t)$ is a stochastic process and t_0 is a fixed time point, then the future,i.e. the values of $X(t)$ for $t \geq t_0$, depends generally on the present and on the past, i.e. on the values for $t \leq t_0$. If the future depends only on present, i.e. on $X(t_0)$, then the stochastic process is called a Markov process.A Markov process is given by its initial probability distribution and by its transition probabilities.
In order to clarify these notions and because we shall need these processes later on, we want at first to discuss Markov processes with a finite state space $\mathcal{M} = \{1, \cdots, M\}$.A probability distribution on $\mathcal{M}$ is a family of numbers $\{p_m \geq 0; m = 1, \cdots, M\}$, such that $\sum_m p_m = 1$. A transition probability is a $M \times M$ matrix P_{mn} with entries ≥ 0, such that $\sum_m P_{mn} = 1$ for all $n = 1, \cdots, M$. A Markov $X(t)$ process on the positive real axis with state space $\mathcal{M}$ is given by the initial probability π and by a family $P(t,s), s \leq t$ of transition probabilities.So the probability that $X(0) = m$ is π_m and the conditional probability, that $X(t) = m$ given $X(s) = n$ is $P_{mn}(t,s)$. One has

$$\begin{aligned}&\mathbf{P}\{X(0) = m_0, X(t_1) = m_1, \cdots, X(t_k) = m_k\}\\&= \pi_{m_0} P_{m_1 m_0}(t_1, t_0) \cdots P_{m_k m_{k-1}}(t_k, t_{k-1}).\end{aligned}$$

The Markov process is called stationary,if the transition probabilities depend only on the time difference, i.e.$P(t,s) = P(t-s)$. One has

$$P(0) = 1$$

$$P(t+s) = P(t)P(s).$$

Hence

$$P(t) = \mathrm{e}^{At},$$

where A is a $M \times M$ matrix with the properties

$$A_{mn} \geq 0 \text{ for } m \neq n$$

and

$$\sum_m A_{mn} = 0.$$

Hence

$$A_n = \sum_{m:m \neq n} A_{mn} = -A_{nn} \geq 0.$$

The matrix A together with the probability distribution π determine the Markov process in an unique way.The infinitesimal properties of the process are very intuitive : Given $X(t) = n$, we have

$$X(t + \mathrm{d}t) = \begin{cases} n \text{ with probability } 1 - A_n \mathrm{d}t \\ m \neq n \text{ with probability } A_{mn} \mathrm{d}t \end{cases}$$

The sample paths of the process are jump functions. Let us calculate the probability $q(t, s)$, that $X(u) = n$ for $t < u < t + s$. By the Markov property

$$q(t, s + \mathrm{d}s) = (1 - A_n \mathrm{d}s) q(t, s),$$

hence

$$q(t, s) = q_n(s) = \mathrm{e}^{-A_n s}.$$

The probability, that$X(u) = n$ for $t < u < t + s$ and that the process jumps between $t + s$ and $t + s + \mathrm{d}s$ from n to $m \neq n$ is again by the Markov property

$$A_{mn} \mathrm{d}s \, \mathrm{e}^{-A_n s}.$$

A stepfunction beween 0 and t with k jumps is given by its values $m_0, \cdots, m_k$ and its jumping times $t_1, \cdots, t_k$. The probability, that the process X is a stepfunction with k jumps and values $m_0, \cdots, m_k$ and jumping times between t_j and $t_j + \mathrm{d}t_j$ is given by

$$\begin{aligned} &\exp(-A_{m_k}(t - t_k)) \, A_{m_k, m_{k-1}} \exp(-A_{m_{k-1}}(t_k - t_{k-1})) \cdots \\ &A_{m_1, m_0} \exp(-A_{m_0 t_1}) \mathrm{d}t_1 \cdots \mathrm{d}t_k. \end{aligned}$$

In this way we defined a probability law on the set of step functions.
We go now back to our Markov process $\Psi(t)$ on $\mathcal{B}_s$. In finite times the process makes finitely many jumps, called quantum jumps, and moves in a deterministic way beween the jumps. The infinitesimal law of the Markov process $\Psi(t)$ is determined by the following rules: Given that $\Psi(t) = \psi$ we have

$$\Psi(t + \mathrm{d}t) = \begin{cases} \mathrm{un}(\mathrm{e}^{-\mathrm{i}H\mathrm{d}t}\psi) \text{ with probability } \|\mathrm{e}^{-\mathrm{i}H\mathrm{d}t)}\psi\|^2 = 1 - \sum_m \|C_m\|^2 \mathrm{d}t \\ \mathrm{un}(C_m \psi) \text{ with probability } \|C_m \psi\|^2 \mathrm{d}t \end{cases}.$$

So the vector moves on in a deterministic way or it jumps by multiplication with C_m with probability$\|C_m \psi\|^2 \mathrm{d}t$ for $m = 1, \cdots, M$.
We want to calculate the probability $q(t, s; \psi)$, that there is no quantum jump between t and $t + s$ given that $\Psi(t) = \psi$ .Then

$$\Psi(t + s) = \mathrm{un}(\mathrm{e}^{-\mathrm{i}Hs}\psi) = \psi(s).$$

Using the Markov property

$$q(t, s + \mathrm{d}s; \psi) = q(t + s, \mathrm{d}s; \psi(s)) q(t, s; \psi) = \|\mathrm{e}^{-\mathrm{i}H\mathrm{d}s}\psi(s)\|^2 q(t, s; \psi).$$

Hence

$$(\mathrm{d}/\mathrm{d}s)q(t,s;\psi) = -\sum \|C_m\psi(s)\|^2\, q(t,s;\psi) = (\mathrm{d}/\mathrm{d}s)\|\mathrm{e}^{-\mathrm{i}Hs}\psi\|^2.$$

As for $s = 0$ both funcions coincide, we obtain

$$q(s,t;\psi) = \|\mathrm{e}^{-\mathrm{i}Hs}\psi\|^2.$$

The probability, that there is no quantum jump between t and $t+s$ and that there is a quantum jump between $t+s$ and $t+s+\mathrm{d}s$ due to C_m given that $\Psi(t) = \psi$ is

$$q(t,s;\psi)\,\|C_m\psi(s)\|^2\,\mathrm{d}s = \|C_m\mathrm{e}^{-\mathrm{i}Hs}\psi\|^2\mathrm{d}s$$

We have to fix the initial distribution of our Markov process. Given an initial density matrix $\varrho(0)$ there are usually many probability distributions π on $\mathcal{B}_S$ such that

$$\mathbf{E}_\pi|\psi >< \psi| = \varrho(0).$$

We could choose any one of them. In order to facilitate the exposure we prefer a fixed vector ψ_0 as initial condition, cf.(7).
A sample path between 0 and t is given by the number of jumps $k = 0,1,2,\cdots$ and the jumping times $t_1,\cdots,t_k$ and the jumping factors $C_{m_1},\cdots,C_{m_k}$. We will denote it by $\psi(t_1,m_1;\cdots;t_k,m_k)$. Recalling (3) the value of the path ψ at time t is

$$\begin{aligned}
&\psi(t_1,m_1;\cdots;t_k,m_k)\\
&= \mathrm{un}(\mathrm{e}^{-\mathrm{i}H(t-t_k)}\,C_{m_k}\,\mathrm{e}^{-\mathrm{i}H(t_k-t_{k-1})}C_{m_{k-1}}\cdots\mathrm{e}^{-\mathrm{i}H(t_2-t_1)}\,C_{m_1}\,\mathrm{e}^{-\mathrm{i}H(t_1)}\psi_0)\\
&= \mathrm{un}(\Phi_k(t_1,m_1;\cdots;t_k,m_k)(t)\psi_0)
\end{aligned}$$

The probability that for fixed k and the m_j the jumping times are between t_j and $t_j + \mathrm{d}t_j$ is

$$\|\Phi_k(t_1,m_1;\cdots;t_k,m_k)(t)\psi_0\|^2\mathrm{d}t_1\cdots\mathrm{d}t_k$$

Let S be a functional on the set of sample functions from 0 to t.

$$\begin{aligned}
\mathbf{E}S[\Psi] = S[\psi_0]\|\Phi_0(t)\psi_0\|^2 + \sum_{k=1}^{\infty}\int_{0<t_1<\cdots<t_k}\sum_{m_1,\cdots,m_k}\\
S[\psi(t_1,m_1;\cdots;t_k,m_k)]\|\Phi_k(t_1,m_1;\cdots;t_k,m_k)(t)\psi_0\|^2\mathrm{d}t_1\cdots\mathrm{d}t_k.
\end{aligned}$$

Apply this formula to

$$S[\Psi] = |\Psi(t) >< \Psi(t)|$$

and obtain immediately

$$\mathbf{E}|\Psi(t) >< \Psi(t)| = \mathrm{e}^{\mathcal{L}t}(|\psi_0 >< \psi_0|) = \varrho(t).$$

4 Quantum White Noise Integrals

Bosonic quantum white noise is a one dimensional Bose quantum field defined on the usual Fock space, i.e. at 0 temperature. It is given by a family of annihilation and creation operators $a^\dagger(t)$ and $a(t)$, where t runs through all real numbers. The nontrivial commutation relations are

$$[a(t), a^\dagger(s)] = \delta(t-s).$$

The Hilbert space of the white noise is spanned by integrals of the form

$$\int f(t_1, \cdots, t_n) a^\dagger(t_1) \cdots a^\dagger(t_n) \mathrm{d}t_1 \cdots \mathrm{d}t_n |0\rangle$$

Typical elements of the Fock space are the coherent or exponential vectors

$$|e(f)\rangle = \exp(\int f(t) a^\dagger(t) \mathrm{d}t)|0\rangle.$$

In the previous section we considered a quantum system connected with a cassical heatbath. The influence of the heatbath was approximated in such a way that the quantum system performed a Markov process. We treat in this section a quantum heatbath and approximate it by white noise. In the remaining two sections we shall study two more concrete examples. Here we state only the quantum stochastic differential equation and its solution by a white noise integral.

We introduce M independent white noises $a_m(t), a_m^\dagger(t)$ with the nontrivial commutation relations

$$[a_m(t), a_n^\dagger(s)] = \delta_{mn}\delta(t-s).$$

Recalling the definitions of the introduction we formulate the stochastic differential equation

$$\frac{\mathrm{d}U(t)}{\mathrm{d}t} = (-\mathrm{i}H_S - \mathrm{i}\sum_m (C_m^\dagger a_m(t) + C_m a_m^\dagger(t)) - 1/2 \sum_m C_m^\dagger C_m) U(t). \tag{13}$$

Introducing H (cf.(4))and

$$V(t) = \sum_m (C_m^\dagger a_m(t) + C_m a_m^\dagger(t))$$

we write

$$\frac{\mathrm{d}U(t)}{\mathrm{d}t} = (-\mathrm{i}(H + V(t))U(t).$$

The Dyson expansion yields

$$U(t) = \mathrm{e}^{-\mathrm{i}Ht} + \sum_{n=1}^{\infty} (-\mathrm{i})^n \int_{0<t_1 \cdots <t_n<t}$$

$$\mathrm{e}^{-\mathrm{i}H(t-t_n)} V(t_n) \mathrm{e}^{-\mathrm{i}H(t_n - t_{n-1})} V(t_{n-1}) \cdots V(t_1) \mathrm{e}^{-\mathrm{i}Ht_1} \mathrm{d}t_1 \cdots \mathrm{d}t_n. \tag{14}$$

The operator $U(t)$ lives on the tensorproduct $\mathcal{H}_S \otimes \mathcal{H}_B$ where $\mathcal{H}_B$ is the tensorproduct of all M white noise Fockspaces. By$|0\rangle$ we denote the vacuum of $\mathcal{H}_B$ and by $U(t)|0\rangle$ the partial application to the vacuum, the result being a tensor product of an operator on $\mathcal{H}_S$ and a white noise vector. As

$$V(t)|0\rangle = \sum_m C_m a_m^\dagger(t)|0\rangle$$

we obtain

$$U(t)|0\rangle = \Phi_0|0\rangle + \sum_{n=1}^{\infty}(-\mathrm{i})^n \int_{0<t_1\cdots<t_n<t} \sum_{m_1,\cdots,m_n}$$
$$\Phi_n(t_1,m_1;\cdots;t_n,m_n)(t)a_{m_n}^\dagger(t_n)\cdots a_{m_1}^\dagger(t_1)|0\rangle \mathrm{d}t_1\cdots\mathrm{d}t_n. \tag{15}$$

If X is an operator on the small quantum system and if $\varrho(0)$ is its initial density matrix we claim that

$$\mathrm{Tr}X\varrho(t) = \mathrm{Tr}((X\otimes 1)U(t)(\varrho(0)\otimes|0\rangle\langle 0|)U(t)^\dagger) = \mathrm{Tr}(X\exp(\mathcal{L}t(\varrho(0)). \tag{16}$$

This can be easily checked by using the equality

$$\langle 0|a_{m_1}(s_1)\cdots a_{m_k}(s_k)a_{n_l}^\dagger(t_l)\cdots a_{n_1}^\dagger(t_1)|0\rangle =$$
$$\delta_{kl}\delta_{m_1,n_1}\cdots\delta_{m_k,n_k}\delta(s_1-t_1)\cdots\delta(s_k-t_k)$$

The equality(16) can be shown as well by using the quantum stochastic differential equation and the quantum stochastic calculus. It is wellknown, that the operator $U(t)$ is unitary on the product space $\mathcal{H}_S\otimes\mathcal{H}_B$. We refer to the literature [3],[6],[8].

5 The Two Level Atom

In this section we want to discuss the Wigner Weisskopf atom. We consider a two level atom with transition frequency ω_0 in a heatbath of oscillators labelled by λ with frequencies $\omega_0 + \omega_\lambda$. Assuming rotating wave approximation the time development operator in interaction representation is given by

$$\mathrm{d}U(t)/\mathrm{d}t = -\mathrm{i}H(t)U(t)\ ;\ U(0) = 1$$

with

$$H(t) = \sum_\lambda \left(g_\lambda a_\lambda \mathrm{e}^{-\mathrm{i}\omega_\lambda t}\sigma_+ + \overline{g_\lambda} a_\lambda^\dagger \mathrm{e}^{+\mathrm{i}\omega_\lambda t}\sigma_-\right)$$

Here $a_\lambda, a_\lambda^\dagger$ are the annihilation and creation operators of the independent bath oscillators with the usual commutation relations. We denote the two eigenstates of the atom by $|\pm\rangle$ and $\sigma_+ = |+\rangle\langle -|$ and $\sigma_+ = |+\rangle\langle -|$. Following e.g. H.Hakens's [2] sugestion we interprete

$$F(t) = \sum g_\lambda a_\lambda \mathrm{e}^{-\mathrm{i}\omega_\lambda t}$$

as a quantum coloured noise and write

$$H(t) = F(t)\sigma_+ + F(t)^\dagger\sigma_-.$$

The vacuum correlation function is

$$k(t-s) = \langle 0|F(t)F(s)^\dagger|0\rangle = \sum |g_\lambda|^2 \mathrm{e}^{-\mathrm{i}\omega_\lambda(t-s)}$$

and the non trivial commutator is

$$k(t-s)1 = [F(t), F(s)^\dagger].$$

We want now to make the transition from colored noise to white noise,i.e. we want to approximate the correlation function by a multiple of a δ-function. At first we assume a continuum of oscillators. We write

$$F(t) = \int g(\omega)a_\omega \mathrm{d}\omega, \tag{17}$$

where $g(\omega)$ vanishes outside a frequency interval $\Delta\omega \ll \omega_0$. The correlation function becomes

$$k(t-s) = \int |g(\omega)|^2 \mathrm{e}^{-\mathrm{i}\omega(t-s)} \mathrm{d}\,\omega. \tag{18}$$

So if g becomes a constant, the correlation function becomes the multiple of a δ-function. We want to discuss this convergence more carefully. The factor of the δ-function has to be

$$\gamma = \int k(t)\mathrm{d}t = 2\pi|g(0)|^2.$$

This is the chacteristic frequency of the problem , and, as it will be seen later, the decay rate of the atom.The function $k(t)$ vanishes outside an interval of the length $\Delta t = 1/\Delta\omega$. So if

$$\gamma\,\Delta t = \gamma/\Delta\omega \ll 1,$$

we can safely replace

$$k(t) = \gamma\delta(t).$$

Without problem we may extend the range of the ω from $(-\omega_0, \infty)$ to $(-\infty, +\infty)$ and introduce

$$a(t) = \frac{1}{\sqrt{2\pi}} \int_{-\infty}^{+\infty} a_\omega \mathrm{d}\omega$$

As $a(t)$ and $a^\dagger(t)$ have the same commutation relations and live on the same Hilbert space as the operators introduced in the last section, they form a representation of quantum white noise as well .Assuming that the function $g(\omega)$ becomes in the limit the constant g and, without loss of generality, that g is positive,equation(17) becomes in the limit

$$F(t) = \sqrt{\gamma}a(t).$$

We generalize a little bit and write

$$\mathrm{d}U(t)/\mathrm{d}t = (AF(t) + BF^\dagger(t))\,U(t), \tag{19}$$

calculate $U(t+\Delta t)$ to the second order and obtain

$$U(t+\Delta t) = (1 + \int_{t<t_1<t+\Delta t} (AF(t_1) + BF^\dagger(t_1))\mathrm{d}t_1$$
$$+ \int_{t<t_1<t_2<t+\Delta t} (AF(t_2) + BF^\dagger(t_2))(AF(t_1) + BF^\dagger(t_1))\mathrm{d}t_1\mathrm{d}t_2 + \cdots)\,U(t).$$

As

$$a(t)|e(f)\rangle = f(t)|e(f)\rangle,$$

we obtain by sandwiching between two exponential vectors

$$\langle e(g)| \int_{t<t_1<t_2<t+\Delta t} F^\dagger(t_2)F(t_1)\mathrm{d}t_1\mathrm{d}t_2|e(f)\rangle$$
$$= \int_{t<t_1<t_2<t+\Delta t} \overline{g(t_2)}f(t_1)\mathrm{d}t_1\mathrm{d}t_2\,\langle e(g)|e(f)\rangle = O(\Delta t^2).$$

In a similar way one can show, that the FF and the $F^\dagger F^\dagger$ terms are of order $O(\Delta t^2)$.But for the $FF^\dagger$ term one obtains

$$\int_{t<t_1<t_2<t+\Delta t} F(t_2)F^\dagger(t_1)\mathrm{d}t_1\mathrm{d}t_2$$
$$= \int_{t<t_1<t_2<t+\Delta t} F^\dagger(t_1)F(t_2)\mathrm{d}t_1\mathrm{d}t_2 + \int_{t<t_1<t_2<t+\Delta t} k(t_2-t_1)\mathrm{d}t_1\mathrm{d}t_2$$
$$= \int_{t<t_1<t_2<t+\Delta t} k(t_2-t_1)\mathrm{d}t_1\mathrm{d}t_2 + O(\Delta t^2).$$

Because of the singularity of the δ-function it is not possible to perform the limit under the integral right away. We obtain for the integral over k before taking the limit(cf.(18))

$$\int_{0<t_1<t_2<\Delta t} \mathrm{d}t_1\mathrm{d}t_2 \int |g(\omega)|^2 \mathrm{e}^{-\mathrm{i}\omega(t_2-t_1)}\mathrm{d}\omega$$
$$= (1/2)\int_{0<t_1,t_2<\Delta t} \mathrm{d}t_1\mathrm{d}t_2 \int |g(\omega)|^2 \cos\omega(t_2-t_1)\mathrm{d}\omega,$$

if we neglect by simplicity the imaginary sinus term, which physically would give rise to an energy shift.Under the last integral we can perform the limit to a constant without any difficulty and arrive at

$$(1/2)\int_{0<t_1,t_2<\Delta t} \mathrm{d}t_1\mathrm{d}t_2\gamma\delta(t_2-t_1) = \gamma\Delta t/2.$$

So finally (19) becomes the quantum stochastic differential equation

$$\mathrm{d}U(t)/\mathrm{d}t = (A\sqrt{\gamma}a(t) + B\sqrt{\gamma}a\dagger(t) + (1/2)AB\gamma)\,U(t),$$

These very heuristic onsiderations can be made perfectly rigorous [9].
The quantum stochastic differential equation for the Wigner Weisskopf atom becomes

$$\mathrm{d}U(t)/\mathrm{d}t = (-\mathrm{i}\sqrt{\gamma}a(t)\sigma_+ - \mathrm{i}\sqrt{\gamma}a^\dagger(t)\sigma_+ - (1/2)\gamma\pi_+)\,U(t),$$

with $\pi_\pm = |\pm\rangle\langle\pm|$. In the expansion (15) only two terms appear

$$U(t)|0\rangle = (\mathrm{e}^{-\gamma\pi_+ t/2} - \mathrm{i}\sqrt{\gamma}\int_0^t \mathrm{e}^{-\gamma\pi_+(t-t_1)/2}\,\sigma_- a^\dagger(t_1)\mathrm{e}^{-\gamma\pi_+ t_1/2}\mathrm{d}t_1)|0\rangle.$$

We have

$$U(t)|0,-\rangle = |0,-\rangle.$$

So, if the atom is at time 0 in the lower state and no photon around , then the atom stays in the lower state all the time and no photon is emitted.

$$U(t)|0,+\rangle = \mathrm{e}^{-\gamma t/2}|0,+\rangle - \mathrm{i}\sqrt{\gamma}\int_0^t a^\dagger(t_1)\mathrm{e}^{-\gamma t_1/2}|0,-\rangle\mathrm{d}t_1.$$

If the atom is in the upper state at time 0 and no photon around it stays in the upper state up to time t with probability amplitude $\mathrm{e}^{-\gamma t/2}$ or its emitts a photon with formal time label $t_1, 0 < t_1 < t$. So the decay rate of the upper state is γ and the label t of the photon can be interpreted as emission time.

$$U(\infty)|0,+\rangle = -\mathrm{i}\sqrt{\gamma}\int_0^\infty a^\dagger(t_1)\mathrm{e}^{-\gamma t_1/2}|0,-\rangle\mathrm{d}t_1.$$

Going back to the frequency representation

$$a^\dagger(t) = (1/\sqrt{2\pi})\int \mathrm{e}^{-\mathrm{i}\omega t}a^\dagger_\omega\mathrm{d}\omega$$

we obtain

$$U(\infty)|0,+\rangle = -\mathrm{i}\sqrt{\gamma}/\sqrt{2\pi}\int \frac{\mathrm{d}\omega}{\gamma/2 + \mathrm{i}\omega}a^\dagger_\omega|0,-\rangle.$$

The probability density for a photon of frequency ω is

$$\frac{\gamma}{2\pi}\frac{1}{\gamma^2/4 + \omega^2}$$

the usual Lorentz profile.
There is associated a classical Markov process on the set $\{+,-\}$. If

$$\varrho(0) = \begin{pmatrix} p_+ & 0 \\ 0 & p_- \end{pmatrix},$$

then for

$$p_{\pm}(t) = \mathrm{Tr}(\pi_{\pm}\varrho(t)) = \mathrm{Tr}((1 \otimes \pi_{\pm})U(t)\,(|0\rangle\langle 0| \otimes \varrho(0))U(t)^{\dagger})$$

we obtain

$$\begin{pmatrix} p_+(t) \\ p_-(t) \end{pmatrix} = \begin{pmatrix} \mathrm{e}^{-\gamma t} & 0 \\ 1-\mathrm{e}^{-\gamma t} & 1 \end{pmatrix} \begin{pmatrix} p_+ \\ p_- \end{pmatrix} = \mathrm{e}^{At} \begin{pmatrix} p_+ \\ p_- \end{pmatrix}$$

with

$$A = \begin{pmatrix} -\gamma & 0 \\ \gamma & 0 \end{pmatrix}.$$

We have a very simple example of a Markov process. If it starts in $-$, it stays there all the time. If it starts in $+$, it jumps to $-$ in an exponentially distributed random time.
Let us now assume, that at time 0 the atom is in the lower state and a photon of time label t_0 is present then we have to consider the first three terms of the expansion (14)

$$U(t)a^{\dagger}(t_0)|0,-\rangle = a^{\dagger}(t_0)|0,-\rangle$$
$$-\mathrm{i}\sqrt{\gamma}\theta(t-t_0)\mathrm{e}^{-\gamma(t-t_0)/2}|0,+\rangle - \gamma\theta(t-t_0)\int_{t_0}^{t} \mathrm{d}t_1 \mathrm{e}^{-\gamma(t_1-t_0)/2} a^{\dagger}(t_1)|0,-\rangle.$$

So the photon passes without any interaction at all or the photon is absorbed at time t_0 and either reimitted or not reimitted.

$$U(\infty)a^{\dagger}(t_0)|0,-\rangle = a^{\dagger}(t_0)|0,-\rangle - \gamma\theta(t-t_0)\int_{t_0}^{\infty} \mathrm{d}t_1 \mathrm{e}^{-\gamma(t_1-t_0)/2} a^{\dagger}(t_1)|0,-\rangle.$$

So eventually the absorbed photon will be reemitted.

6 The Oscillator in an Atomic Heat Bath

Consider the master equation for a harmonic oscillator

$$\dot{\varrho} = \mathcal{L}(\varrho) = -(\gamma_+/2)(bb^{\dagger}\varrho - 2b^{\dagger}\varrho b + \varrho bb^{\dagger}) - (\gamma_-/2)(b^{\dagger}b\varrho - 2b\varrho b^{\dagger} + \varrho b^{\dagger}b),$$

where b and $b^{\dagger}$ are the usual oscillator operators. If $|n\rangle$ is the n-th oscillator eigenvector, then one obtains for

$$p_n(t) = \langle n|\varrho(t)|n\rangle$$

the equation

$$\dot{p_n} = -(\gamma_+(n+1) + \gamma_- n)p_n + \gamma_+(n+1)p_{n-1} + \gamma_- n p_{n+1}.$$

This is the equation of a classical stationary Markov process $X(t)$ on the set $\{0,1,2,\cdots\}$, so $p_n(t) = \mathbf{P}\{X(t) = n\}$. The transition probabilities can be written in the following way, given $X(t) = n$ then

$$X(t + \mathrm{d}t) = \begin{cases} n \text{ with probability } 1 - (\gamma_+(n+1) + \gamma_- n)\mathrm{d}t \\ n+1 \text{ with probability } \gamma_+(n+1)\mathrm{d}t \\ n \text{ with probability } \gamma_- n dt \end{cases}$$

So $X(t)$ may only jump by ± 1 . If $\gamma_+ = 0$, then for $t \to \infty$ the process converges to 0, if $\gamma_- = 0$ then the process diverges exponentially fast. The probabilities are stationary for

$$p_n = \mathrm{const}(\gamma_+/\gamma_-)^n.$$

This implies that $\gamma_+/\gamma_- < 1$.
The related quantum stochastic differential equation is (cf.(13))

$$\begin{aligned} &\mathrm{d}U/\mathrm{d}t = \\ &\left(-\mathrm{i}\sqrt{\gamma_+}(ba_+(t) + b^\dagger a_+(t)^\dagger) - \mathrm{i}\sqrt{\gamma_-}(b^\dagger a_-(t) + ba_-(t)^\dagger) - K/2\right) U(t) \end{aligned} \quad (20)$$

with

$$K = \gamma_+ bb^\dagger + \gamma_- b^\dagger b$$

As in the previous section the heatbath could be realized by a system of oscillators. There is however a new and very intuitive model, by M. Orszag et al.[4],[5], which we present in a slightly modified version.
We assume a standing wave of frequency ω_0 between two mirrors and a beam of three level atoms passing through the beam, perpendicular to the direction of wave propagation. The atoms have the energy levels 0 and $\pm\omega_0$. The eigenstates of the n-th atom are $|-1,n\rangle, |0,n\rangle, |1,n\rangle$. We assume that there are N atoms and that each atom interacts only during a time τ with the oscillator, we observe during the time from 0 to $T = N\tau$. During each interaction time the interaction is constant. Denote

$$E^n_{ij} = |i,n\rangle\langle j,n|.$$

The unperturbed Hamiltonian is

$$H_0 = \omega_0(b^\dagger b + \sum (E^n_{1,1} - E^n_{-1,-1})).$$

The interaction Hamiltonian is

$$H_I(t) = H^n = g_-(E^n_{10} b + E^n_{01} b^\dagger) + g_+(E^n_{0,-1} b + E^n_{-1,0} b^\dagger)$$

for $(n-1)\tau < t < n\tau$. As H_0 commutes with H_I we may neglect the trivial time dependence due to H_0 and calculate only the time development due to H_I. The time development operator from 0 to T is $V = W^N \cdots W^1$ with

$$W^n = \mathrm{e}^{-\mathrm{i}\tau H^n}.$$

The atomic bath, i.e. the Hilbertspace of the N atoms is spanned by the vectors of the form $|\zeta_1, \cdots, \zeta_N\rangle$ with $\zeta_n = -1, 0, 1$.
We fix T, go with N to ∞, so $\tau = T/N \to 0$ and assume $g_\pm = \sqrt{\gamma_\pm}/\sqrt{\tau}$. We assume furthermore, and that is very essential, that all N incoming atoms are in state 0.Then we claim, that for N going to ∞

$$V|0, \cdots, 0\rangle \approx U(T)|0\rangle, \tag{21}$$

where $U(T)$ is the solution of the quantum stochastic differential equation (20)and $|0\rangle$ is the white noise vacuum.
We calculate the scalar product

$$\langle \zeta_1 \cdots \zeta_N | V | 0, \cdots, 0\rangle = \langle \zeta_N | W | 0\rangle \cdots \langle \zeta_1 | W | 0\rangle,$$

where we erazed the upper index $n = 1, \cdots, N$, because the right hand side brackets are independent of it. We calculate

$$\begin{aligned}\langle 0|W|0\rangle &= 1 - \tau K + O(\tau^{3/2})\\ \langle 1|W|0\rangle &= -\mathrm{i}\sqrt{\tau}\sqrt{\gamma_-} b + O(\tau^{3/2})\\ \langle -1|W|0\rangle &= -\mathrm{i}\sqrt{\tau}\sqrt{\gamma_+} b^\dagger + O(\tau^{3/2}).\end{aligned}$$

We have

$$\langle 0, \cdots, 0|V|0, \cdots, 0\rangle = \langle 0|W|0\rangle^N \to \mathrm{e}^{-KT} = \langle 0|U(T)|0\rangle.$$

So for the case, that there is no quantum jump between 0 and T, the two expressions coincide in the limit. In order to study the general case, imbed the atomic bath into the white noise bath. We can write in a unique way

$$|\zeta_1, \cdots, \zeta_N\rangle = E^{j_1}_{\varepsilon_1, 0} \cdots E^{j_n}_{\varepsilon_n, 0} |0, \cdots, 0\rangle$$

with$j_1 < \cdots < j_n$ and $\varepsilon = \pm 1$. We define

$$\eta|\zeta_1, \cdots, \zeta_N\rangle = \tau^{-n/2} a^\dagger_{\varepsilon_1}(\Delta_{j_1}) \cdots a^\dagger_{\varepsilon_n}(\Delta_{j_n})|0\rangle$$

with $\Delta_j = ((j-1)\tau, j\tau)$ and

$$a^\dagger_\varepsilon(\Delta_j) = \int_{(j-1)\tau}^{j\tau} a^\dagger_\varepsilon(t)\mathrm{d}t$$

We claim, that for $N \to \infty$

$$\eta(V|0, \cdots, 0\rangle \to U(T)|0\rangle. \tag{22}$$

We do not want to give the general proof. One gets a good insight considering only the case of only one quantum jump.In white noise space the vectors including only one creation are of the form

$$|\phi\rangle = (\int f_+(t) a^\dagger_+(t)\mathrm{d}t + \int f_-(t) a^\dagger_-(t)\mathrm{d}t)|0\rangle.$$

We have

$$\langle\phi|U(T)|0\rangle = -\mathrm{i}\int_0^T \mathrm{e}^{-K(T-t)}(\sqrt{\gamma_+}f_+(t) + \sqrt{\gamma_-}f_-(t))\mathrm{e}^{-Kt}\mathrm{d}t \tag{23}$$

On the other hand

$$\begin{aligned}
&\langle\phi|\eta(V|0,\cdots,0\rangle) \\
&= (1/\sqrt{\tau})\sum_{j=1}^{N}\sum_{\varepsilon=\pm1}\langle\phi|a_\varepsilon^\dagger(\Delta_j)|0\rangle\langle 0,\cdots,0|E_{0,\varepsilon}^j V|0,\cdots,0\rangle \\
&= \sum_j\sum_\varepsilon(1/\sqrt{\tau})\int_{\Delta_j} f_\varepsilon(t)\mathrm{d}t\langle 0|W|0\rangle^{N-j}\langle\varepsilon|W|0\rangle\langle 0|W|0\rangle^{j-1} \\
&\approx (-\mathrm{i})\sum_j\int_{\Delta_j}\mathrm{d}t\mathrm{e}^{-K(N-j)\tau}(\sqrt{\gamma_+}f_+(t) + \sqrt{\gamma_-}f_-(t))\mathrm{e}^{(j-1)K\tau}.
\end{aligned}$$

Clearly this last expression converges to (23).

References

1. C.W.Gardiner *Quantum Noise*(Springer, Berlin,Heidelberg,1991.)
2. H.Haken, *Laser Theory*,(Springer, Berlin,Heidelberg,1970)
3. :P.A.Meyer *Quantum Probability for Probabilists* Lecture Notes in Mathematics **1538**(Springer, Berlin,Heidelberg 1993)
4. M.Orszag:*Quantum Optics*(Springer, Berlin,Heidelberg 2000)
5. T.B.Kist,M.Orszag,T.Brun,L.Davidovich: J.Optics B **1** 251 (1999).
6. K.R.Parthasarathy:*An Introduction to Quantum Stochastic Calculus*(Birkhaeuser, Basel,Boston,Berlin 1992).
7. W.von Waldenfels:Spontaneous Light Emission described by Quantum Stochastic Differential Equation. Lecture Notes in Mathematics **1136**, p.515-534 (Springer, Berlin,Heidelberg 1985)
8. W. von Waldenfels:'Continuos Maassen Kernels and the Inverse Oscillator. Sem.de Probabilités XXX, Springer Lecture Notes im Mathematics 1626 (1996).
9. W. von Waldenfels:'An Example 0f the singular coupling limit'. Anestoc 1998. Birkhaeuser 2000.

Quantum Orbits in Intense-Laser Atom Physics

R. Kopold and W. Becker

Max-Born-Institut, Max-Born-Str. 2a, 12489 Berlin, Germany

Abstract. Atoms interacting with intense laser fields exhibit effects of extreme multiphoton order such as above-threshold ionization. These effects are modeled in terms of one electron initially bound by a zero-range potential, employing the strong-field approximation. The evaluation of the quantum-mechanical probability amplitude for ionization leads to the concept of quantum orbits. For a laser field with linear and, in particular, with elliptical polarization, the calculated spectra and angular distributions of the ionized electron are analyzed in terms of the properties of the quantum orbits and their constructive and destructive interference.

1 Introduction

The irradiation of atoms by intense laser fields has revealed a number of unexpected and fascinating effects. For example, the atom may emit extremely high harmonics of the laser frequency, up to the order of about 300 (high-harmonic generation), or an electron generated by ionization of a single atom may be detected with an energy corresponding to hundreds of laser photons (above-threshold ionization). High-harmonic generation (HHG) is becoming a versatile table-top source of coherent radiation extending well into the water window that may shortly be able to deliver pulses of subfemtosecond duration. High-order above-threshold ionization (HATI) of a single atom is a subtle effect providing a unique mixture of almost classical features and genuinely quantum-mechanical effects. For reviews, see [1,2,3,4].

The theory of HHG and HATI appears to constitute a formidable challenge, even if one restricts oneself to considering only one active electron: the minimum number N of photons that must be absorbed is so large that even lowest-order perturbation theory is completely out of reach to any analytical approach. In addition, for the laser field strengths of recent experiments, lowest-order perturbation theory or indeed perturbation theory of any finite order is no longer sufficient. Numerical simulations, based on the solution of the time-dependent Schrödinger equation, are very demanding owing to the required large extent of the temporal and the spatial grid and the smallness of the effects of interest; see, however, Ref. [5]. In particular, the excursion amplitude of a free electron, that quivers along the direction of the laser field, is a large multiple of the Bohr radius, and this excursion must be covered by the spatial grid. These problems are compounded for a laser field of elliptical polarization, since in such a field the quivering motion of the electron takes place in a plane rather than in one dimension as it does for linear polarization.

M. Orszag, J.C. Retamal (Eds.): LNP 575, pp. 294–309, 2001.

It is illuminating to envisage the quantities that set the scale for the motion of a free electron in a laser field with frequency ω, electric-field amplitude $E = \omega A$, and intensity I, assuming the field is not strong enough to cause relativistic motion. In a laser field described by the vector potential $\boldsymbol{A}(t)$, an electron's velocity is $m\boldsymbol{v}(t) = \boldsymbol{p} - e\boldsymbol{A}(t)$. We assume a gauge such that the time average of the vector potential over one cycle T vanishes, $\langle \boldsymbol{A}(t) \rangle_T = 0$. Then, on top of its drift motion with momentum $\boldsymbol{p}$ (its conserved canonical momentum), the electron performs a wiggling motion with the excursion amplitude

$$\alpha\,[\text{atomic units}] = \frac{eA}{m\omega} = 3.95 \times 10^{-6} \sqrt{I[\text{W/cm}^2]}\,\omega^{-2}[\text{eV}]. \tag{1}$$

The associated kinetic energy, the so-called ponderomotive energy, is

$$U_P\,[\text{eV}] \equiv \left\langle \frac{e^2\boldsymbol{A}^2}{2m} \right\rangle_T = 1.43 \times 10^{-13} I[\text{W/cm}^2]\,\omega^{-2}[\text{eV}]. \tag{2}$$

For a titanium-sapphire laser with $\hbar\omega = 1.5\,\text{eV}$ and $I = 5 \times 10^{14}\,\text{W/cm}^2$ we have $\alpha = 39$ Bohr radii and $U_P = 31.8\,\text{eV}$. These numbers suggest an approximation that is appropriate for this situation: during most of its motion, the electron is far away from its parent ion and its average kinetic energy is much higher than the atomic binding energy. The strong-field approximation (SFA) [6,7,8] then neglects the atomic binding potential for an electron in the continuum. For such a strong field, the electron enters the continuum via tunneling which, to exponential accuracy, just depends on its binding energy in the absence of the laser field. Hence, the SFA also neglects the effect of the laser field on an electron bound in the atomic ground state. The SFA provides expressions for the transition amplitudes that are amenable to a largely analytical evaluation, typically up to one final quadrature that is left for numerical evaluation.

A semiclassical approximation of the ionization amplitude leads to the concept of "quantum orbits" [8,9]. For fields of not too low intensity, the quality of this approximation is excellent. More importantly, the quantum orbits yield an intuitively appealing physical picture of the ionization process which explains many of its detailed properties in terms of simple concepts, such as constructive and destructive interference. These quantum orbits are the object of this paper. We will restrict ourselves to HATI, but HHG can be treated along precisely the same lines.

Figure 1 shows electron spectra of above-threshold ionization. Electrons leave the atom by tunneling. Thereafter, electrons may or may not re-encounter the ion. If they do not, their energies are restricted to values below about $2U_P$ [10]. Higher energies can only be reached if the electron rescatters off the ion at just the right times [11,12]. The probability of this to happen is low, owing to the smallness of the electron-ion cross section, but on the logarithmic scale of Fig. 1 such events stand out and form the so-called "plateau" [13] with its cutoff at $10U_P$. The two afore-mentioned cutoffs, $2U_P$ and $10U_P$, are particularly well developed for high intensity; the best examples are ATI spectra in helium at $10^{15}\,\text{W/cm}^2$ [14].

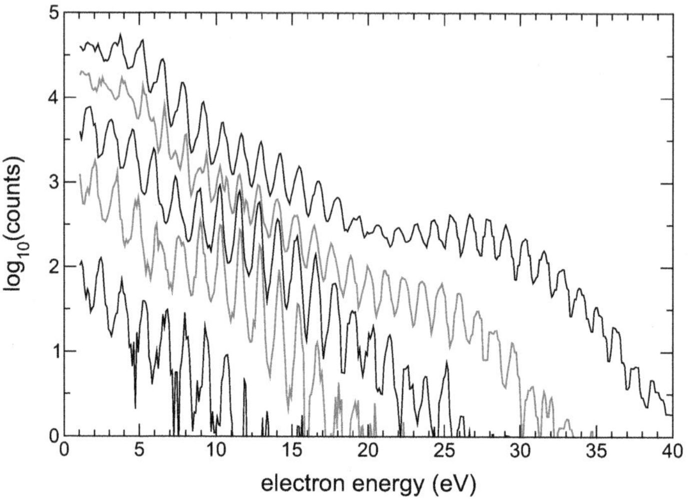

Fig. 1. Experimental ATI electron spectra in xenon for emission in the direction of the linearly polarized laser field (G. G. Paulus et al., unpublished). The laser intensity increases from bottom to top from about 3 to $5 \times 10^{13}\,\mathrm{W/cm^2}$, the highest intensity corresponding to $U_P = 3.2\,\mathrm{eV}$. The frequency is 800 nm, the pulse length about 40fs. With increasing intensity, a plateau develops.

The organization of this article is as follows. In the second Section, making use of the SFA we derive an expression for the ATI transition amplitude which is the basis of all developments that follow. We introduce the concept of the "quantum orbits", expand the amplitude in terms of them, and discuss their relation to Newtonian orbits. In the third Section, we apply this concept to HATI caused, in particular, by an elliptically polarized laser field. We present spectra and angular distributions and their expansions in terms of quantum orbits. In the concluding Section, we briefly comment on several experiments that were able to test the concept of quantum orbits.

2 The S Matrix for Ionization

2.1 General formalism

The S-matrix element for laser-induced ionization from the ground state $|\psi_0\rangle$ of a binding potential $V(\boldsymbol{r})$ into a scattering state $|\psi_{\boldsymbol{p}}\rangle$ of the same potential is

$$M_{\boldsymbol{p}} = \lim_{t\to\infty, t'\to-\infty} \langle \psi_{\boldsymbol{p}}(t)|U(t,t')|\psi_0(t')\rangle, \tag{3}$$

where $U(t,t')$ denotes the time-evolution operator of the entire system taking into account both the binding potential and the interaction $H_I(t)$ with the laser field. It satisfies the integral equation

$$U(t,t') = U_0(t,t') - \mathrm{i}\int_{t'}^{t} \mathrm{d}t'' U(t,t'')H_I(t'')U_0(t'',t'). \tag{4}$$

Here $U_0(t,t')$ is the time-evolution operator in the presence of just the binding potential, so that $U_0(t,t')|\psi_0(t')\rangle = |\psi_0(t)\rangle$. Owing to the two limits, the form (3) is not very useful. If we make use of the integral equation (4) in the matrix element (3) we obtain, in view of the orthogonality of the ground state and the scattering state,

$$M_{\boldsymbol{p}} = -\mathrm{i} \lim_{t\to\infty} \int_{-\infty}^{t} \mathrm{d}t' \langle \psi_{\boldsymbol{p}}(t)|U(t,t')H_I(t')|\psi_0(t')\rangle. \tag{5}$$

The time-evolution operator also satisfies the integral equation

$$U(t,t') = U^{(\mathrm{Vv})}(t,t') - \mathrm{i}\int_{t'}^{t} \mathrm{d}t'' U^{(\mathrm{Vv})}(t,t'')VU(t'',t'), \tag{6}$$

where

$$\langle \boldsymbol{r}|U^{(\mathrm{Vv})}(t,t')|\boldsymbol{r}'\rangle = \int \mathrm{d}^3\boldsymbol{k}\psi_{\boldsymbol{k}}^{(\mathrm{Vv})}(\boldsymbol{r}t)\psi_{\boldsymbol{k}}^{(\mathrm{Vv})*}(\boldsymbol{r}'t') \tag{7}$$

is the time-evolution operator of an otherwise free electron in the presence of a laser field, the so-called Volkov propagator, see, e.g. [15]. There is a compact expression for this quantity, but here we will only require its expansion (7) in terms of the wave functions of an electron in the presence of the laser field, specified by the vector potential $\boldsymbol{A}(t)$. In the length gauge, the Volkov wave functions have the form

$$\psi_{\boldsymbol{p}}^{(\mathrm{Vv})}(\boldsymbol{r}t) = (2\pi)^{-3/2}\mathrm{e}^{\mathrm{i}[\boldsymbol{p}-e\boldsymbol{A}(t)]\cdot\boldsymbol{r}} \exp\left(-\frac{\mathrm{i}}{2m}\int^{t} \mathrm{d}\tau[\boldsymbol{p}-e\boldsymbol{A}(\tau)]^2\right). \tag{8}$$

Next, we insert the integral equation (6) into the matrix element (5). This results in two terms,

$$\begin{aligned} M_{\boldsymbol{p}} = &- \mathrm{i} \lim_{t\to\infty} \int_{-\infty}^{t} \mathrm{d}t' \langle \psi_{\boldsymbol{p}}(t)|U^{(\mathrm{Vv})}(t,t')[H_I(t')|\psi_0(t')\rangle \\ &- \mathrm{i}\int_{-\infty}^{t'} \mathrm{d}t'' VU(t',t'')H_I(t'')|\psi_0(t'')\rangle\Bigg]. \end{aligned} \tag{9}$$

This expression is still exact. We will now approximate the scattering state $\langle\psi_{\boldsymbol{p}}|$ by a plane-wave state so that $\langle\psi_{\boldsymbol{p}}|U^{(\mathrm{Vv})}(t,t')$ becomes a Volkov state with the complex conjugate of the wave function (8). Thereby, the remaining limit in the matrix element (9) is carried out. In the second term, we will replace the exact time-evolution operator $U(t',t'')$ by the Volkov operator $U^{(\mathrm{Vv})}(t',t'')$. After these approximations, the first term on the right-hand side of Eq. (9) is the so-called

KFR (Keldysh-Faisal-Reiss) matrix element [6,7]. It describes ionization such that the electron once freed only feels the laser field. The second term allows for exactly one additional interaction with the binding potential. Hence this term describes "rescattering" processes where the ionized electron propagates in the laser field and returns (at the time t') to within the range of the binding potential and interacts with it before it finally moves away towards the detector. These approximations constitute the afore-mentioned strong-field approximation (SFA). They are the better justified, the larger the quivering amplitude of a free electron in the laser field is in comparison to the range of the binding potential. Hence, the quality of the SFA improves with increasing intensity.

With the help of an integration by parts, the two terms in Eq. (9) can be combined into one [16],

$$M_{\boldsymbol{p}} = -\int_{-\infty}^{\infty} \mathrm{d}t \int_{-\infty}^{t} \mathrm{d}t' \langle \psi_{\boldsymbol{p}}^{(\mathrm{Vv})}(t)|VU^{(\mathrm{Vv})}(t,t')V|\psi_0(t')\rangle, \qquad (10)$$

which is the expression we will further analyze[1].

The S-matrix element (10) simplifies enormously if we choose a zero-range potential for the atomic binding potential so that the spatial integrations become trivial. Inserting the expansion (7) of the Volkov time-evolution operator we arrive at the representation

$$M_{\boldsymbol{p}} \sim \int_{-\infty}^{\infty} dt \int_{0}^{\infty} d(t-t') \int d^3k \exp\left[\mathrm{i}S_{\boldsymbol{p}}(t,t',\boldsymbol{k})\right]. \qquad (11)$$

The action in the exponent consists of three parts

$$\begin{aligned} S_{\boldsymbol{p}}(t,t',\boldsymbol{k}) &= -\frac{1}{2m}\int_t^{\infty} \mathrm{d}\tau [\boldsymbol{p} - e\boldsymbol{A}(\tau)]^2 \\ &\quad - \frac{1}{2m}\int_{t'}^{t} \mathrm{d}\tau [\boldsymbol{k} - e\boldsymbol{A}(\tau)]^2 - \int_{-\infty}^{t'} \mathrm{d}\tau E_0, \end{aligned} \qquad (12)$$

which recall the strong-field approximation made in the derivation: at time t', the electron tunnels out of the binding potential $V(\boldsymbol{r})$ [cf. Eq. (10)], and at the later time $t > t'$ it rescatters off the same potential. For times earlier than t' when the electron is still bound, the laser field is neglected. For later times, the binding potential is ignored except at the instant t of rescattering.

[1] Similar expressions were suggested and discussed by various authors. An expansion in powers of the binding potential was written down in [7] while caution with such expansions is requested by [17]. The "atomic-antenna" picture [18] also should be mentioned in this context. Recent references include [19] where Coulomb effects are partly taken into account, [20] which presents results in partly analytical form that are applicable in the deep tunneling limit, and [21] in the context of a many-electron formalism.

2.2 Approximation by quantum orbits

The matrix element (11) will be the starting point of all subsequent developments. It can be quickly evaluated numerically [16]: the Volkov propagator (7) has a compact closed form, one of the two remaining temporal integrations yields the delta function of energy conservation so that just one integration is left for numerical computation. We will refer to the corresponding results as "exact". The point of this paper is, however, the comparison of this "exact" calculation to a saddle-point evaluation whose description follows.

To this end, we determine those values of the ionization time t', the rescattering time t, and the canonical momentum $\boldsymbol{k}$ that render the action (12) stationary. They are given by the solutions of the three conditions [9]

$$[\boldsymbol{k} - e\boldsymbol{A}(t')]^2 = 2mE_0, \tag{13}$$

$$[\boldsymbol{k} - e\boldsymbol{A}(t)]^2 = [\boldsymbol{p} - e\boldsymbol{A}(t)]^2, \tag{14}$$

$$(t - t')\boldsymbol{k} = \int_{t'}^{t} \mathrm{d}\tau e\boldsymbol{A}(\tau), \tag{15}$$

respectively. Their physical interpretation is, in turn, "energy conservation" in the process of tunneling, elastic scattering when the electron returns to the parent ion, and the condition that it return. Since $E_0 < 0$, the condition of "energy conservation" at the time of tunneling cannot be satisfied for any real time t' and, therefore, all solutions $[t'_s, t_s(\mathrm{Re}\, t_s > \mathrm{Re}\, t'_s), \boldsymbol{k}_s]$ with $s = 1, 2, \ldots$ of the saddle-point equations (13)-(15) are complex.

In terms of these solutions the matrix element can be written as

$$M_{\boldsymbol{p}} \sim \sum_s \left(\frac{(2\pi \mathrm{i}\hbar)^5}{\det(\partial^2 S_{\boldsymbol{p}} / \partial q_j^{(s)} \partial q_k^{(s)})_{j,k=1,\ldots 5}} \right)^{1/2} \exp\left[\mathrm{i} S_{\boldsymbol{p}}(t_s, t'_s, \boldsymbol{k}_s)\right], \tag{16}$$

where $q_i^{(s)}$ $(i = 1, \ldots 5)$ combines the five variables t_s, t'_s and $\boldsymbol{k}_s$. The sum must be extended over an appropriate subset of the solutions of the saddle-point solutions which is identified from consistency considerations [22].

Any solution to the saddle-point equations (13)-(15) defines an electronic trajectory

$$m\boldsymbol{x}(t) = (t - t'_s)\boldsymbol{k}_s - \int_{t'_s}^{t} \mathrm{d}\tau e\boldsymbol{A}(\tau) \tag{17}$$

before, and

$$m\boldsymbol{x}(t) = (t - t_s)\boldsymbol{p} - \int_{t_s}^{t} \mathrm{d}\tau e\boldsymbol{A}(\tau) \tag{18}$$

after rescattering. Such a complex trajectory we call a quantum orbit. Its real part as a function of real t with $t > \mathrm{Re}\, t'_s$ is interpreted as the orbit of the electron in space after it appeared in the continuum at the time $\mathrm{Re}\, t'_s$. We will see below that the saddle-point equations yield solutions t_s for the *rescattering* time whose imaginary part is very small, in contrast to the imaginary part of the *ionization* time that can be substantial.

The saddle-point equations have infinitely many solutions. However, just a handful, those corresponding to travel times $t_s - t'_s$ up to a few periods T of the laser field, are of any significance; exactly how many, depends on the electron's momentum $\boldsymbol{p}$. The conceptual simplification achieved by the saddle-point evaluation is immense: in the form (11), the S-matrix element integrates up contributions from all start times, all return times, and all momenta of the orbit in between. In contrast, in the form (16), it is approximated by the contributions of just a few quantum orbits characterized by the saddle-point solutions $[t'_s, t_s(\mathrm{Re}\, t_s > \mathrm{Re}\, t'_s), \boldsymbol{k}_s]$. These solutions are complex which reflects the initial condition of tunneling.

2.3 Quantum orbits and the simple-man model

If we let the binding energy E_0 equal to zero then, for linear polarization, the saddle-point equation (13) says that the electron starts its orbit with zero velocity at the real time t' (provided that the final momentum $\boldsymbol{p}$ is classically accessible under this condition). The additional parameters t and $\boldsymbol{k}$ then come out real as well. The resulting equations are those of the so-called simple-man model [11,12]. Solutions typically come in pairs which can be ordered by their travel times $\tau_s \equiv t_s - t'_s$, such that the two solutions of the first pair have travel times $T/2 \leq \tau \leq T$, those of the second have $T \leq \tau \leq 3T/2$, and so on [23,24]. This is explained in some detail in Fig. 2, see also Ref. [22]. For increasing electron energy, the two solutions of a pair approach each other and finally merge. The respective energy defines the cutoff energy of this pair. It is the maximal kinetic energy that an electron having started with zero velocity classically can gain. For energies higher than that, this pair is no longer a solution. The number of pairs of solutions depends on the electron energy.

Let us now allow for a nonvanishing binding energy. If the electron energy is classically accessible (that is, below the cutoff energy), then the afore-mentioned solutions $(t_s, t'_s, \boldsymbol{k}_s)$ acquire small imaginary parts while their real parts do not change by much. After their cutoffs, however, they no longer disappear from the set of solutions; rather, when the electron energy increases beyond the cutoff, their real parts remain essentially frozen while their imaginary parts quickly increase. Irrespective of the electron energy, the orbits [Eqs. (17,18)] now are complex. Below their respective cutoffs, their real parts are still quite close to the simple-man orbits.

In the classically accessible regime, the return times t_s (in contrast to the ionization times t'_s) come out to be almost real, which reflects the fact that the electron at the detector has only little recollection of its quantum-mechanical origin by tunneling. The solution of the saddle-point equations (13)-(15) can be

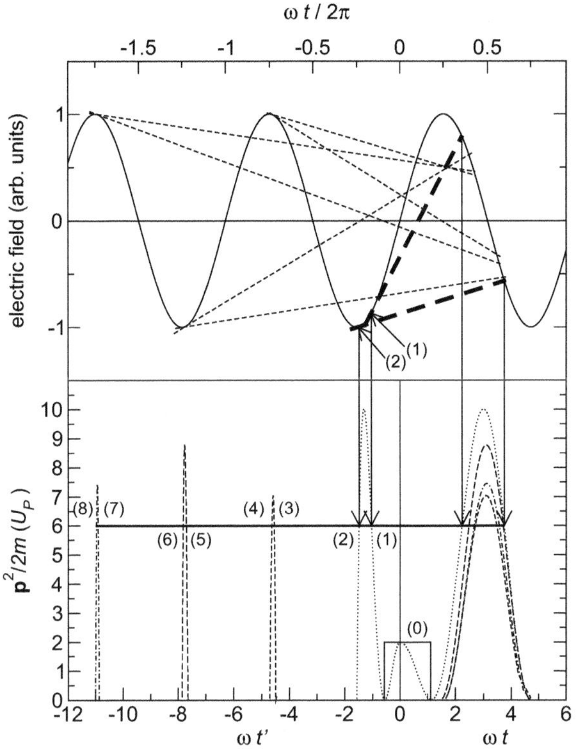

Fig. 2. Illustration of the simple-man model for high-order ATI and linear polarization. The upper panel depicts the function $F(t) = \int^t dt A(t)$ which for a sinusoidal monochromatic field is proportional to the electric field. In terms of this function, the return condition (15) can be written as $F(t) = F(t') + (t - t')\mathrm{d}F(t')/\mathrm{d}t'$, that is, in order to find the start time t' for given return time t, one has to draw a tangent from $F(t)$ to the curve $F(t')$ for $t' < t$. The various dashed lines are such tangents. The lower panel summarizes the results of a numerical determination of the start time t' and the return time t, based on Eqs. (13)-(15) for zero binding energy E_0 and for specified electron energy E_p and emission in the direction of the field. In the figure the case $E_p = 6U_p$ is considered, indicated by the horizontal line in the lower panel. Its intersection with the rightmost set of curves identifies pairs of return times, one pair for each type of curves (dotted, long-dashed, dashed-dotted, short-dashed). Each type of curve is once recovered in the left-hand part of the lower panel where its intersection with the horizontal line at $6U_p$ gives the associated start times. The numbers in parentheses refer to the numbers of individual trajectories used in the subsequent figures and in the text. The tangent construction mentioned above is exemplified in the upper panel. It shows how the start times are determined once the return times are known. Consider, for example, the dotted curve: its intersections with the horizontal line identify return times of $\omega t \approx 2.3$ and 3.75, marked by vertical lines with arrows. Then, in the upper panel, from the points $F(t)$ the tangents must be drawn to $F(t')$ for $t' < t$ which determine the associated start times t'; in the example, these two tangents are represented by thick dashed lines, and the start times are again marked by vertical lines with arrows. Similar tangent lines are also drawn for the earlier start times, corresponding to the long-dashed, dashed-dotted, and the short-dashed curves.

significantly simplified if the return time is taken as real [25]. This entails a small imaginary part of the final energy which is ignored.

For elliptical polarization, the vector $\boldsymbol{k}$ is no longer restricted to the direction of the field and, consequently, Eq. (13) no longer implies that the initial velocity is zero for $E_0 = 0$. We can merely conclude that it is a complex null vector [26].

3 Results

3.1 Spectra for linear polarization

Figure 3 presents a HATI spectrum calculated from the first seven pairs of quantum orbits. The first three pairs, however, already produce a spectrum that can hardly be told apart from the exact calculation. The figure also shows the contributions of the various pairs of quantum orbits. Each pair displays a well defined cutoff (for the first four pairs, the cutoffs can be read off from Fig. 2) and a pronounced interference pattern below the cutoff. The contributions of all pairs add up to produce the highly irregular total spectrum. Just below the cutoff at $10U_P$, however, the total spectrum is dominated by the pair (1,2) with the shortest travel times and, therefore, regular.

3.2 Spectra for elliptical polarization

Figure 4 shows a spectrum for the same parameters [26], except that the polarization is now elliptical. One notices immediately that the plateau characteristic of HATI for linear polarization has turned into a staircase. While for linear polarization the contributions of the various pairs of orbits yield a complicated pattern which does not exhibit clear imprints of individual pairs, here we can attribute each step of the staircase to one particular pair of orbits. The corresponding orbits are reproduced in the figure. The orbits of the pair (5,6) are quite complicated already: they pass the ion twice before they rescatter. Yet these orbits are responsible for the middle part of the spectrum. The fact that the shortest orbits 1 and 2 quantitatively make the smallest contribution may be surprising, since generally wave-packet spreading favors short orbits [27,28]. It is also in marked contrast to the case of linear polarization. The reason is the following: in order that the orbits for elliptical polarization return to the ion, they have to have a nonzero drift-momentum component k_y in the direction of the small component of the field ellipse. The action (12) assigns a small weight to such orbits. The shortest orbits 1 and 2 require a particularly large such component [26]. Figure 4 also shows that the cutoff of the spectrum recedes for increasing ellipticity; for the case of the figure ($\xi = 0.5$), it is at 8.6 U_P in place of $10\,U_P$ for $\xi = 0$.

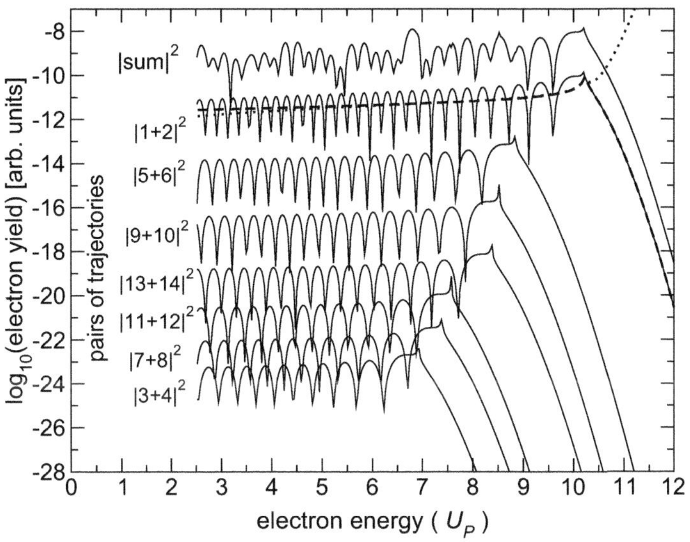

Fig. 3. High-order ATI spectrum in the direction of the field, approximated according to Eq. (16) by the first 7 pairs of quantum trajectories, for a laser intensity of 10^{15} W/cm^2, $\hbar\omega = 0.0584$a.u., and a binding energy of $E_0 = -0.9$ a.u. simulating neon. The electron energy is specified in multiples of the ponderomotive energy U_P. For energies exceeding $3U_P$, the uppermost curve is virtually identical with the exact calculation from Eq. (10). The lower curves, which are displaced for visual convenience, depict the contributions from the pairs $(2n-1, 2n)$ of quantum trajectories. They are ordered from top to bottom by the decrease of their cutoffs which oscillate with increasing travel time about a final value of about $8U_p$. The pair (3,4) is the one with the next-to-shortest travel times, and the lowest cutoff. The dotted and dashed line near the contribution of the first pair (1,2) are the contributions of $|1|^2|$ and $|2|^2$ separately. After the cutoff, one of them increases exponentially and has to be dropped from the sum (16) while the other one describes the spectrum virtually exactly. The first three pairs already give an excellent fit of the exact results, particularly for the upper part of the plateau.

3.3 Angular distributions for elliptical polarization

In all of the figures so far, electrons were counted in the direction of the large component of the laser field. For linear polarization, the classical simple-man model implies that this is the direction where the electron can reach the highest energy after rescattering. For elliptical polarization, it is not immediately clear in which direction the highest energies will be emitted, but as long as the ellipticity is moderate, one expects it to be near the large axis. For circular polarization, of course, all directions in the plane of polarization are equivalent.

For the parameters of Fig. 4, angular distributions (AD's) in the plane of polarization are shown in Fig. 5. Owing to the inversion symmetry of the elliptically polarized laser field, we can restrict ourselves to angles between $\phi = 0°$ and

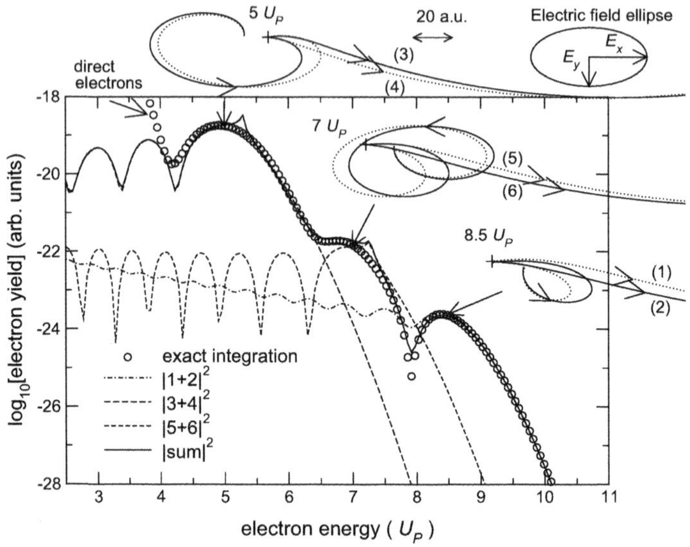

Fig. 4. ATI spectrum in the direction of the large component of the elliptically polarized driving laser field $\boldsymbol{E}(t) = (E/\sqrt{1+\xi^2})(\hat{\boldsymbol{x}} \sin \omega t - \xi \hat{\boldsymbol{y}} \cos \omega t)$ for ξ=0.5 (see the field ellipse in the upper right corner of the figure) and electron energies between 2.5 and 10.5 eV. The other parameters are the same that underlie Fig. 3. The open circles give the yields of the individual ATI peaks calculated from the exact integral (10) for the zero-range potential. The other curves represent the contributions to the quantum-path approximation (16) of the shortest trajectories 1 and 2 (dot-dashed line), 3 and 4 (long-dashed line), and 5 and 6 (short-dashed line), as well as the sum of all six (solid line). Please, note that some of these curves partly or entirely overlap. The orbits that are responsible for each part of the spectrum, viz. 1 and 2, 3 and 4, and 5 and 6, are presented near the margins of the figure. In each case, the position of the ion is marked by a cross. The orbits do not depart from there, but rather from a point several atomic units away from it. This is the point where the electron tunnels into the continuum. The electron travels the orbits in the direction of the arrows.

$\phi = 180°$ with respect to the large axis as the yields satisfy $R(\phi) = R(\phi + \pi)$. The figure shows that beyond the cutoff at the energy $E = 8.59\,U_P$ the yields quickly decrease and the AD is maximal at $\phi = -5°$. At energies below the cutoff, interferences begin to develop. The pattern with the highest contrast appears at $E = 7.92 U_P$, and the destructive interference is strongest at $\phi = 5°$. The corresponding maxima are the analogs for elliptical polarization of the so-called "anomalous angular distributions" first observed several years ago [29]. They are a manifestation of rainbow scattering. For lower energies, the number of maxima and minima increases and finally, for $E \sim 5\,U_P$, a staircase similar to Fig. 4 has developed, with three steps for negative and two for positive angles.

Figure 6 illustrates for some selected energies how the AD's of Fig. 5 are built up by the quantum orbits. We consider the contributions of the first six pairs. The AD for the highest energy of $8.59 U_P$ is entirely generated by the pair (1,2)

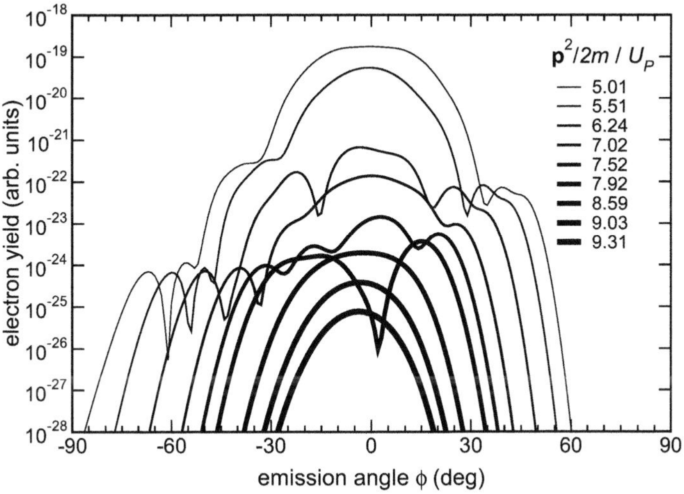

Fig. 5. ATI angular distributions for $\xi = 0.5$ and various electron energies specified by their ponderomotive potentials and the parameters of Fig. 4. The angle ϕ is with respect to the large axis of the polarization ellipse.

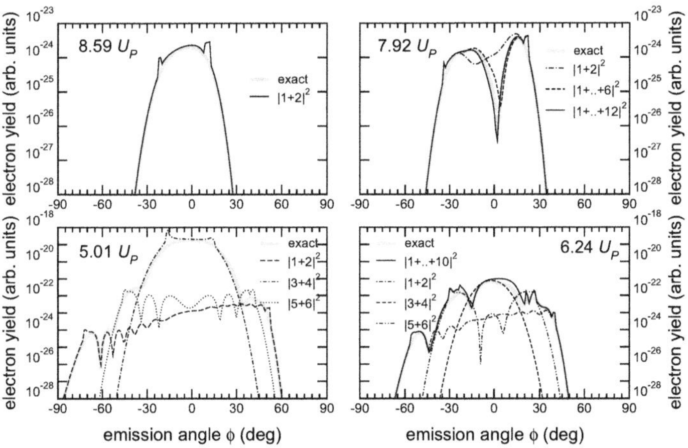

Fig. 6. Approximation of some selected angular distributions of Fig. 5 in terms of quantum orbits.

with the shortest travel times. However, already for the slightly lower energy of $7.92U_P$ the massive dip near 0° requires a coordinated participation of many more orbits. Note that the contributions of the pair (1,2) not yet exhibit this dip. For lower energies, the staircase structure begins to develop, and we have a situation much like in Fig. 4: in the latter case, each part of the energy spectrum is generated by one particular pair of orbits while here pairs of orbits form the steps of the staircase for constant energy.

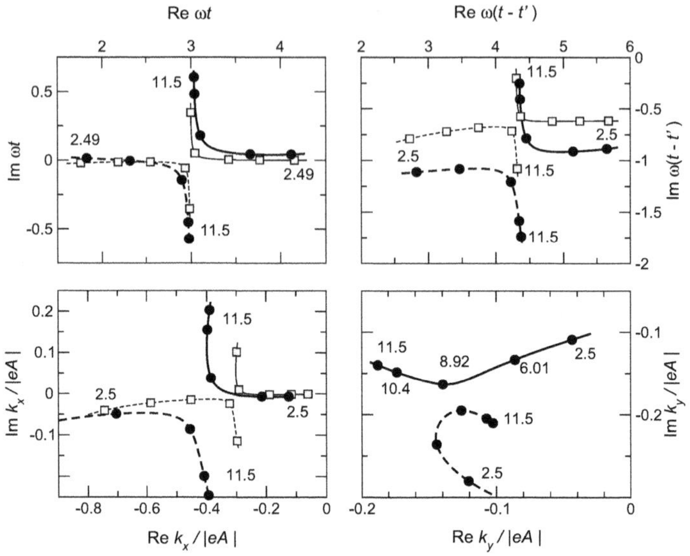

Fig. 7. Saddle points $(t_s, t'_s, \boldsymbol{k}_s)$ for the orbits $(s = 1, 2)$ having the shortest two travel times. The figure shows a comparison of elliptical polarization ($\xi = 0.5$, filled circles) and linear polarization (open squares). The values of the other parameters are those of Fig. 4 ($eA = 2.04\,\text{a.u.}$). The symbols identify electron energies of 11.5, 10.4, 8.92, 6.01, and 2.49, all in multiples of U_P. With the scaling of k_i given on the ordinate, the saddle points only depend on the Keldysh parameter $\gamma = \sqrt{|E_0|/2U_P}$.

Finally, in Fig. 7 we display, as a function of the electron energy and for emission in the direction of the large component of the field, the parameters t, $t - t'$, k_x, and k_y that define the first two orbits. These are the orbits that generate the last step of the staircase in Fig. 4. Several features are worth emphasizing: In contrast to the imaginary parts of the travel times, the imaginary parts of the return times are very small (for energies below the cutoff), albeit somewhat larger than for linear polarization. The imaginary parts of the drift momentum, and in particular of the component k_y, are quite large. They are much larger than those of the next two orbits 3 and 4 (not shown). As mentioned above, this is why the orbits 1 and 2 quantitatively make a small contribution.

We have, in this Section, applied the zero-range-potential based quantum orbits to the analysis of ATI spectra and angular distributions generated by elliptical polarization. We should not fail to add a remark of caution: it is well known that the SFA does not usually work in this case for electrons of low energy [30,31]. Here, however, our interest lay in the electrons of the plateau, for which the model is better justified. In the final Section we will review a comparison to some pertinent experiments.

4 Comparison to Experimental Data and Conclusions

Throughout this paper, we tacitly took for granted that intense-laser atom experiments take place in the realm of a single electron interacting with the field. This assumption, far from being evident at first glance, has recently been justified, by a comparison of numerical one-electron simulations to experiments, to an accuracy that exceeded all expectations [32]. In this paper, we considered the additional drastic approximation of the effective electronic binding potential, which in addition to its Coulomb tail includes various soft-core contributions, by the zero-range potential that underlies all explicit results displayed in this article. The results that follow from this model have been compared with experimental data in various cases. First, it should be mentioned that this model is largely equivalent to the Lewenstein model [8,2] of high-harmonic generation which has very successfully explained a variety of data and been used as a guide to the design of experiments. In above-threshold ionization, the interferences that are the most obvious footprint of quantum orbits have been observed in elliptically polarized laser fields for the direct electrons (those that do not undergoing rescattering) in the dependence of the spectra on the ellipticity [33]. In the same context, interferences between direct and rescattered electrons were recently identified in the angular distributions [34]. Additional support for the zero-range model derives from recent ATI experiments [35,36] that established the existence of resonant-like behavior in ATI spectra: ATI peaks rise and fall by up to one order of magnitude while the laser intensity changes by only a few percent. This behavior has been explained by multiphoton resonance with ponderomotively upshifted Rydberg states [37]. The effect can be explained [38,39] with comparable accuracy in terms of the zero-range model where it can be traced to a constructive interference of many quantum orbits that happens at channel closings. This is quite surprising since the zero-range potential does not support any excited states let alone a Rydberg series. For a comparison of zero-range spectra to spectra calculated by a solution of the three-dimensional Schrödinger equation for hydrogen [40], see Ref. [38].

In principle, the concept of quantum orbits can be generalized to arbitrary (non-zero-range) binding potentials, but owing to the twofold role of the binding potential in this formalism – in the act of rescattering and in the propagation in the continuum – this cannot be done in an unambiguous fashion. Further work is desirable to understand the origin of the surprising success of the zero-range model and its quantum orbits in the description of high-intensity laser-atom phenomena.

Acknowledgments

We enjoyed useful discussions with M. Kleber, D. B. Milošević, and G. G. Paulus. We are indebted to G. G. Paulus for the permission to present unpublished data of his. This work was supported in part by Deutsche Forschungsgemeinschaft.

References

1. L. F. DiMauro, P. Agostini: Adv. At. Mol. Opt. Phys. **35**, 79 (1995)
2. P. Salières, A. L'Huillier, Ph. Antoine, M. Lewenstein: Adv. At. Mol. Opt. Phys. **41**, 83 (1999)
3. C. Joachain, M. Dörr, N. Kylstra: Adv. At. Mol. Opt. Phys. **42**, 225 (2000)
4. Th. Brabec, F. Krausz: Rev. Mod. Phys. **72**, 545 (2000)
5. H. G. Muller: Laser Phys. **9**, 138 (1999)
6. L. V. Keldysh: Zh. Eksp. Teor. Fiz. **47**, 1945 (1964) [Sov. Phys. JETP **20**, 1307]; A. M. Perelomov, V. S. Popov, M. V. Terent'ev: Zh. Eksp. Teor. Fiz. **51**, 309 (1966) [Sov. Phys. JETP **24**, 207 (1967)]; F. H. M. Faisal: J. Phys. B **6**, L89 (1973)
7. H. R. Reiss: Phys. Rev. A **22**, 1786 (1980)
8. M. Lewenstein, Ph. Balcou, M. Yu. Ivanov, A. L'Huillier, P. B. Corkum: Phys. Rev. A **49**, 2117 (1994)
9. M. Lewenstein, K. C. Kulander, K. J. Schafer, P. Bucksbaum: Phys. Rev. A **51**, 1495 (1995)
10. H. B. van Linden van den Heuvell, H. G. Muller: in *Multiphoton Processes*, ed. by S. J. Smith, P. L. Knight (Cambridge University Press, 1988), pp. 25-34
11. K. C. Kulander, K. J. Schafer, J. L. Krause: in *Super-Intense-Laser Atom Physics*, Vol. 316 of NATO Advanced Study Institute, Series B: Physics, ed. by B. Piraux, A. L'Huillier, K. Rzążewski (Plenum, New York, 1993), p. 95
12. P. B. Corkum: Phys. Rev. Lett. **71**, 1994 (1993)
13. G. G. Paulus, W. Nicklich, H. Xu, P. Lambropoulos, H. Walther: Phys. Rev. Lett. **72**, 2851 (1994)
14. B. Walker, B. Sheehy, K. C. Kulander, L. F. DiMauro: Phys. Rev. Lett. **77**, 5031 (1996)
15. W. Becker, J. K. McIver, M. Confer: Phys. Rev. A **40**, 6904 (1989)
16. A. Lohr, M. Kleber, R. Kopold, W. Becker: Phys. Rev. A **55**, R4003 (1997)
17. H. S. Antunes Neto, L. Davidovich: Phys. Rev. Lett. **23**, 2238 (1984)
18. M. Yu. Kuchiev: Pis'ma Zh. Eksp. Teor. Fiz. 45 (1987) [JETP Lett. **45**, 404 (1987)]
19. D. B. Milošević, F. Ehlotzky: Phys. Rev. A **58**, 3124 (1998)
20. S. P. Goreslavskii, S. V. Popruzhenko: Phys. Lett. A **249**, 477 (1998); Zh. Eksp. Teor. Fiz. **117**, 895 (2000) [JETP **90**, 778 (2000)]
21. F. H. M. Faisal, A. Becker, J. Muth-Böhm: Laser Phys. **9**, 115 (1999)
22. R. Kopold, W. Becker, M. Kleber: Opt. Commun. **179**, 39 (2000)
23. G. G. Paulus, W. Becker, W. Nicklich, H. Walther: J. Phys. B **27**, L703 (1994)
24. G. G. Paulus, W. Becker, H. Walther: Phys. Rev. A **52**, 4043 (1995)
25. D. B. Milošević, W. Becker, R. Kopold: Phys. Rev. A **61**, 063403 (2000)
26. R. Kopold, D. B. Milošević, W. Becker: Phys. Rev. Lett. **84**, 3831 (2000)
27. P. Dietrich, N. H. Burnett, M. Ivanov, P. B. Corkum: Phys. Rev. A **50**, R3585 (1994)
28. B. Gottlieb, A. Lohr, W. Becker, M. Kleber: Phys. Rev. A **54**, R1022 (1996)
29. B. Yang, K. J. Schafer, B. Walker, K. C. Kulander, P. Agostini, L. F. DiMauro: Phys. Rev. Lett. **71**, 3770 (1993)
30. M. Bashkansky, P. H. Bucksbaum, D. W. Schuhmacher: Phys. Rev. Lett. **59**, 274 (1987); ibid., **60**, 2458 (1988)
31. A. Jaroń, J. Z. Kamiński, F. Ehlotzky: Opt. Commun. **163**, 115 (1999)
32. M. J. Nandor, M. A. Walker, L. D. Van Woerkom, H. G. Muller: Phys. Rev. A **60**, R1771 (1999).

33. G. G. Paulus, F. Zacher, H. Walther, A. Lohr, W. Becker, M. Kleber: Phys. Rev. Lett. **80**, 484 (1998)
34. G. G. Paulus, F. Grasbon, A. Dreischuh, H. Walther, R. Kopold, W. Becker: Phys. Rev. Lett. **84**, 3791 (2000)
35. M. P. Hertlein, P. H. Bucksbaum, H. G. Muller: J. Phys. B **30**, L197 (1997)
36. P. Hansch, M. A. Walker, L. D. Van Woerkom, Phys. Rev. A **55**, R2535 (1997)
37. H. G. Muller, F. C. Kooiman, Phys. Rev. Lett. **81**, 1207 (1998)
38. R. Kopold, W. Becker: in *Multiphoton Processes*, ed. by L. F. DiMauro, R. R. Freeman, K. C. Kulander, AIP Conference Proceedings 525 (American Institute of Physics, New York, 2000) pp. 11-23
39. G. G. Paulus, F. Grasbon, H. Walther, R. Kopold, W. Becker: to be published
40. E. Cormier, P. Lambropoulos: J. Phys. B **30**, 77 (1997)

Micromaser Dynamics Beyond the Rotating-Wave Approximation

F. De Zela

Departamento de Ciencias, Sección Física,
Pontificia Universidad Católica del Perú, Ap.1761, Lima, Perú.

Abstract. We discuss the validity of the rotating-wave approximation (RWA) focussing in particular on the micromaser dynamics. We consider the full Hamiltonian - given by $H = \hbar \left(\omega_0 \sigma_z + \omega a^+ a + g(a + a^+)(\sigma^+ + \sigma^-) \right)$ - for the description of the atom-field interaction. We arrive at the conclusion that the so called counter-rotating terms, ($a\sigma^-$ and $a^+\sigma^+$) - which are neglected within the RWA - could give rise to physically observable effects for a parameter range not too far from current experiments. Early studies assumed that effects not described within the RWA could be experimentally observed by going from the optical to the microwave domain. In principle, the micromaser could be used for that purpose. We focus on atomic correlations, as the atoms exiting the micromaser cavity constitute the only physical entity that can be subjected to measurements. Within the RWA the only allowed transitions are the so-called energy-conserving transitions. We consider additional transitions which, despite being small, could give rise to measurable effects that we have investigated in the present work.

1 Introduction and Background

One of the most widely used approximations in quantum optics is the rotating-wave approximation (RWA). It is invoked in the context of the Jaynes-Cummings model (JCM) [1,2], which is a simplified though nontrivial model describing the interaction of a two-level atom with a single-mode electromagnetic field. Originally, the RWA was introduced in the context of magnetic resonance, where a spin-$\frac{1}{2}$ system is acted upon by two magnetic fields, one being uniform and the other being an oscillating field lying on a plane perpendicular to the first one. The uniform field causes a Larmor precession of the spin, whereas the oscillating field can be decomposed into a sum of two fields which rotate in opposite directions with respect to each other. One of these fields rotates in the same sense as the precessing spin and is the component of the oscillating field which is responsible for the spin-flip at resonance, i.e., when the rotation frequency approximately coincides with the Larmor frequency. The counter-rotating field exerts a time dependent torque on the magnetic moment associated to the spin. This torque reverses itself at a high frequency - approximately twice the Larmor frequency. Thus, the effective torque exerted by the counter-rotating field is assumed to be very small on average and is consequently neglected. In other words, from the two rotating fields out of which the oscillating field was composed, one neglects

M. Orszag, J.C. Retamal (Eds.): LNP 575, pp. 310–337, 2001.

the counter-rotating one: this is what was originally known as the RWA. On the other hand, every two-level system - or every Hamiltonian in a two-state Hilbert space - can be shown to be formally equivalent to a spin-$\frac{1}{2}$ system interacting with a magnetic field. The ground and the excited states are identified with the spin-down and spin-up states, respectively. One can formally invoke the RWA in this context as well, even though one is now dealing with fictitious spin and magnetic fields, so that a sound physical basis for the approximation might be lacking. In the case of a two-level atom interacting with a *quantized* single-mode electric field, it can be shown -as we will see below- that by invoking the RWA the original infinite-dimensional Hamiltonian of the system splits into a block-diagonal one, whereby each block pertains to a two-dimensional subspace. As a consequence, the problem becomes equivalent to that of an ensemble of spin-$\frac{1}{2}$ systems, each member of which interacts with a magnetic field on its own and evolves in time independently of the other members of the ensemble. In each two-dimensional subspace the Hamiltonian is exactly diagonalizable, so that one can extract all the information necessary for describing the dynamics of the total system. A word of warning should be said here. Even though any two-level system is formally equivalent to a spin-$\frac{1}{2}$ interacting with a magnetic field, we cannot infer from this fact that there is a complete equivalence with respect to any physical manifestation. As an example, consider the case in which one chooses a circularly polarized electric field applied to an atom obeying the electric-dipole selection rule $\Delta m = \pm 1$. It has been argued [36] that such a system would be a physical realization of the JCM, i.e., it would represent a case where the RWA becomes unnecessary. This would therefore represent the electric-dipole counterpart of the spin-$\frac{1}{2}$ system, to which a *rotating* magnetic field is applied. However, as it was subsequently proved [11], an electric-dipole transition driven with circularly polarized light turns out to be incompatible with a closed two-level system. This is so because there is an intrinsic difference between the magnetic- and the electric-dipole interactions: whereas in the first case the minimum number of levels involved is two, in the second case this number is four. If we want to reduce this four-level system to an effective two-level one, an appropriate polarization of the field must be chosen, in this case a linear one. This precludes our applying the circular polarization that would make the RWA unnecessary. If, on the contrary, a circular polarization has been chosen, then virtual stimulated transitions to a third level do play a role in the dynamics of the system when it is driven with a strong field. We would have thus no longer a closed two-level system.

Although the RWA has been often applied in quantum optics and its validity was only seldom questioned in the past, it has the peculiarity of not being like other approximations, which are based on sound mathematical or physical arguments. The neglect of the counter-rotating terms in the Hamiltonian turns out to be a very drastic recourse by which a nonintegrable Hamiltonian becomes a diagonalizable one. This fact gives to the approximation some of its attractiveness, the more considering that it in fact leads to predictions which has been experimentally confirmed under a variety of circumstances. However, there are cases in which the RWA ceases to be valid and properly designed experiments can show effects that cannot be explained within the RWA, like for example the Bloch-Siegert shift [5]. The possibility of a chaotic regime is also a feature that the RWA excludes in the case of a system of two-level atoms interacting with a non-quantized field [31]. There is therefore a qualitative shortcoming inherent to the RWA, a fact that prompted some authors to say that the "generally accepted use of the [RWA] Hamiltonian of the system has changed into a sort of symbol of faith" [30].

In recent times, there has been renewed interest in studying the limits of validity of the RWA, particularly within the framework of a two-level system interacting with a single-mode quantized field. The corresponding Hamiltonian is applicable to a variety of systems, among which the case of an atom inside a cavity and the case of ions interacting with the vibrational modes of a Paul trap [6] are two of the most salient ones. In the last case, the possibility of controlling the relative importance of the counter-rotating terms opens the way for experimental investigations of such effects which are not describable within the RWA [7]. In the case of cavity quantum-electrodynamics (QED) we are usually dealing with Rydberg atoms and the microwave domain. Here again, effects beyond the RWA might play a role, as some authors expected in the past. In the present article we deal with a special case of cavity-QED: the one-tom maser or micromaser. Here, very high-quality factors (Q) for the cavity have been achieved, which are on the order of 10^{10}. This allows one to study the coherent atom-field interaction for times much longer than the Rabi oscillation time, the time scale for the evolution between the two active atomic levels. The two-level atom coupled to a single-mode quantized radiation field via the electric-dipole interaction has been and still is the subject of a large amount of theoretical and experimental studies. Among these investigations, there is an increasing number of studies which do not assume the validity of the RWA. In the following, and for the purpose of giving some background for the present work, we give a brief account of some investigations, both experimental and theoretical.

Let us call ω_0 the transition frequency between the two atomic levels and ω the frequency of the cavity field. Within the RWA, resonance occurs when $\omega = \omega_0$. Beyond the RWA there is a shift of the resonance frequency, the already mentioned Bloch-Siegert shift, whose origin can be traced back to "virtual-photon" transitions stemming from the counter-rotating terms. Magnetic resonance experiments have been performed, giving results in good agreement with the predictions made within the semiclassical [3] as well as within the fully quantized theory [4]. We have already mentioned the appearance of chaos in the semiclassical case, i.e., in the case of a non-quantized field interacting with a system of two-level atoms. This is also the context where some work on superradiance [8] has been done. It was shown [9], for example, that the initial radiation rate produced by an ensemble of excited atoms could differ very much from the one predicted within the RWA, depending on the phase of the initial atomic dipole-moment. It is also interesting to remark in this context the following result: if we invoke the RWA on the Hamiltonian of the atom-field system and subsequently derive the master equation for the density matrix, then this equation is not the same as the one we obtain by applying the RWA to a master equation that has been derived from the full Hamiltonian [10]. The argument for neglecting rapidly oscillating terms can lead us therefore to different results, depending on when it is invoked. This is just one of several unsatisfactory features of the RWA.

The strong coupling necessary to bring into evidence effects beyond the RWA can be achieved not only by considering high atomic densities. It is also possible to experimentally produce a strong coupling by using so called "optical two-level atoms". Here, the two levels are represented by two optical modes which can be obtained by exploiting, e.g., the Sagnac effect [12]. The theoretically predicted intensity of one of the modes displays a "ripple structure" that has been experimentally confirmed. This structure becomes superposed to the Rabi oscillations predicted by the RWA and is caused by the counter-rotating terms. A similar high-frequency modulation appears - within a semiclassical treatment - in the occupation probability of a short lived state coupled to a long lived one through a dipole transition. Such a result follows from using the "oscillating-wave" approximation [13,14]. In contrast to the rotating-wave approximation, here one goes - through a series of unitary transformations - to an interaction picture, and then to a counter-rotating frame. The Hamiltonian splits into a time-independent and a time-dependent part. As a consequence of the procedure, the time-dependent part contains a substantive portion of the counter-rotating contributions. After neglecting the time-dependent part of the Hamiltonian - whose terms, besides being small, oscillate at frequencies like 2ω and 4ω - one can solve the resulting time-independent Schrödinger equation. By calculating the occupation probability of one of the levels, a high-frequency modulation of the line shape emerges. Another case of a ripple structure has been brought into light within a fully quantized formulation, by using a Hilbert-space basis of displaced harmonic-oscillator states [35]. The ripple structure appears here superimposed to otherwise smoothly varying energy eigenvalues, when these are plotted as a function of the atom-field coupling constant. The energy spectrum displays such a ripple, because of the presence of Laguerre polynomials in the coupling between coefficients entering the displaced-oscillator basis expansion. This brings us to mention a conjecture -supported by strong numerical evidence- raised some years ago [38] about the existence of exact solutions to the eigenvalue problem of the full Hamiltonian by known functions. These functions should be generalized spheroidal functions and appear by treating the problem in the framework of Bargmann's Hilbert space of analytical functions [49]. Such a framework was exploited by Schweber [48] in the years following its introduction by Bargmann. It seemed to be a promising approach to the eigenvalue problem: through the realization of the bosonic operators, as $a \to d/dz$ and $a^+ \to z$, in the space of entire functions, the eigenvalue problem reduces to solving a first-order differential equation. It is not surprising that orthogonal polynomials and recurrence relations leading to continued fractions appear in such a context. Recurrence relations leading to continued fractions can also be obtained by following a quite different but related approach. Swain developed in a series of papers a method that is generally applicable to a class of eigenvalue problems, as an alternative to perturbation expansions [50,51]. We use in this work the results obtained by Swain for some transition probabilities. In principle, results like those of Swain are exact, in the sense that one can achieve any prescribed accuracy by developing analytical expressions - which are in turn obtained in terms of continued fractions - as far as necessary. Such analytical expressions can be written down following simple rules. The approach resembles somewhat the Feynman graph method, but not so much as it is the case with the graph method developed by Ramirez and Orszag [54].

Alternative iterative procedures have been developed by Feranchuk et al [55]. They are based on the so-called operator method [56] and give rise to accurate numerical solutions of the eigenvalue problem. The accuracy of these and other methods can be tested by their application to those cases where exact solutions are available. These have been found for isolated values of the coupling constant [34,37]. Once in possession of a reliable numerical method, one can investigate several questions related to the model at hand, like for example the question of properly defining quantum chaos. Work on this point was done [32] shortly after the semiclassical case was studied by Milonni et al [31], leaving however several open questions. As a further example of investigations resting on numerical evaluations we can mention the work of Lo et al [47], who tested a conjecture raised by Bishop et al [46]. These last authors presented what they believed to be strong evidence for a second-order quantum phase transition in the JCM without RWA. According to their results, which were obtained by applying the coupled-cluster method [45], spontaneous symmetry breaking of the parity symmetry displayed by the JCM without RWA should arise under appropriate circumstances. In other words, the ground state of the system would not possess the parity symmetry that the Hamiltonian does possess. However, according to the exact numerical calculations performed by Lo et al [47], such a result should be an artifact of the coupled-cluster method, when applied to this case.

We have already mentioned that the Bloch-Siegert shift has been measured in magnetic resonance experiments that realize the magnetic counterpart of the electric-dipole case. For the quantum optical case there have been indications of the appearance of similar new phenomena [15], but they usually require the application of external, strong electromagnetic, i.e., classical fields. The pure quantum case without external assistance remains yet to be experimentally explored. On the other hand, among several theoretical investigations giving rise to predictions of new phenomena, we should mention the following ones: changes in atomic inversion [17,18,19], different phase [23] and squeezing properties of the field [22,24], raising of the laser threshold and broadening of the photon-number distribution [25]. The RWA has also been the subject of investigations dealing with cases in which the boson mode couples not to a dipole but to other boson modes; e.g., those of a reservoir [16]. In such a case, the master equation, which is of the Lindblad form within the RWA, can turn into one which presents an unphysical sector [26]. It is also worth mentioning that a "serious defect" of the RWA was pointed out for the Hamiltonian of an oscillator coupled to a reservoir described by boson operators [27]. Here, the corresponding spectrum has no lower bound and the reservoir acquires a non-passive character. There have also been extensions of the JCM to multiphoton interactions, which were studied without the RWA [28,29]. They show dramatic changes stemming from the counter-rotating terms.

Searching for the limits of applicability of the RWA in quantum optics, it is clear that the microwave domain offers itself as a natural arena where such limits could in principle be tested. This should be so because, as we said before, the RWA can be somewhat substantiated by going to an interaction picture where the counter-rotating terms are rapidly oscillating. Their effects are then likely to be averaged out, so as to remain unobservable. This is expected - and experimentally confirmed - to happen in the optical domain. Several authors, who investigated the effects of the counter-rotating terms, expressed the hope that a device like the one atom maser, or micromaser [42], would be capable of bringing such effects into evidence [20,35,39]. However, most studies dealing with the problem were developed within an idealized framework, where the electromagnetic field was assumed to be in a special state, e.g., in a coherent state. In other cases, where the micromaser dynamics has been invoked [43,44], it was described under simplifying assumptions - like a lossless cavity at zero temperature - making the situation somewhat idealized as well, and bringing the issue far from experimentally realizable schemes. In the present work we develop a calculational scheme where the micromaser dynamics is fully taken into account, without the assumption of the RWA. Such an approach has been hindered so far, perhaps due to difficulties in encompassing techniques used to deal with the atom-field interaction without the RWA, with the field dynamics of relaxation processes. When a closed form is available for the evolution operator describing the atom-field interaction, then it is possible to write down equations for the complete cycle between two successive atoms entering a lossy micromaser cavity. One obtains then straightforwardly the master equation for the micromaser. The problem becomes much more involved when one goes beyond the RWA, as there is no analytical solution available for the atom-field interaction. One has then to rely on approximations which can be either numerical, analytical, or a combination of both. The description of the full micromaser dynamics becomes, in such a case, a rather difficult problem that had been left open.

In the present work we use the approach developed by Swain [51,52,53]. It allows us to obtain a generalization of the standard micromaser theory in a form that can be used for numerical calculations similar to those of the conventional theory. This enables us to make predictions that could be tested by suitable designed experiments. Most of our results are not surprising and confirm that - at resonance - the RWA is perfectly justified within the parameter range usually employed in current experiments. We can however explore the domain where the RWA begins to break down. Moreover, we can test in a realistic environment whether the predictions made by other authors concerning collapse-revival phenomena in cavity QED without assuming the RWA, do occur in a micromaser setting. Such collapse and revival phenomena are among the most striking features in the atom-field interaction [40], being characteristic - at least the revival part of them - of the quantum nature behind the phenomenon. The micromaser has been used to bring such effects into evidence. Whether the micromaser can still be an appropriate tool for observing departures from standard predictions, that is the central question we address here. In a previous work [58] we undertook the study of the micromaser dynamics beyond the RWA, calculating linewidths and second-order field correlations. Differences with respect to the RWA appeared for parameter values within the range already accessible to current experiments. However, we limited ourselves to predictions that involved the photon field only. Such a quantity can only be sensed indirectly through the atoms leaving the cavity, which so serve to pump the maser field as well as to probe it. By measuring the state - either excited or not - in which the atoms leave the cavity, conclusions can be drawn about the photon field. We concentrate mainly on the atomic inversion and its behavior in connection with the atom-field interaction time, which is the time of flight of each atom traversing the cavity. Collapse and revival phenomena can thus be studied in detail within a full micromaser setting, i.e., with relaxation processes included. Besides that, we calculate the effect of the counter-rotating terms in the appearance of the so-called "trapping states". We obtain a remarkable result concerning departures from the RWA predictions for the standard deviation of the photon number, occuring at locations corresponding to trapping states. These departures appear for parameter values near those employed in current experiments.

2 The Micromaser

The micromaser [41] consists of a high-Q microwave cavity made of superconducting material - so as to approximate an ideal resonator - which is traversed by a thermal beam of atoms, whose velocity has been previously selected. In the usual setting all atoms enter the cavity in a highly excited, Rydberg state, to which they are brought by laser light tuned to the desired transition out of the ground level. Rydberg states have relatively long lifetimes and couple strongly to radiation resonant with transitions to neighboring levels. In the Garching experiments [42], to which we will refer often in this work, the upper atomic level is usually the 63 $P_{3/2}$ level of rubidium. The lower maser level is either the 61 $D_{5/2}$ or the 61 $D_{3/2}$ level, depending on the frequency to which the cavity has been tuned. The atomic beam is of such a low intensity, that at most one atom at the time can be inside the resonator. Mostly, the cavity is empty of atoms and the photon field that is build up, decays as a consequence of cavity losses. There is also some thermal noise, even though the cavity temperature can be kept at very low values, below 0.3 K. It has been shown that single atoms can maintain a continuous maser oscillation of the cavity with a mean number of photons between unity and several hundreds. This means that the continuous deposition of energy in the cavity, by single atoms, can bring the photon field from a thermal to a nonclassical statistical distribution. There is a threshold for this transition, essentially determined by the dimensionless pump-parameter $\Theta = g\tau\sqrt{N_{at}}$. Here, g means the atom-field coupling constant, τ the atomic flight or interaction time, and N_{at} the number of atoms entering the resonator during a time corresponding to the decay-time of the cavity. The first threshold is reached for $\Theta = 1$, for which the photon statistics becomes super-Poissonian. However, there are other first-order phase transitions besides the first one, for $\Theta = 2\pi$ and multiples thereof, this being a characteristic feature of the one-atom maser. Experimentally, one cannot test directly theoretical predictions concerning the photon field. This is so because photons leaking from the cavity would be incompatible with the very high quality factor Q and, besides that, detectors in the microwave domain are still not sensitive enough. Information about the field can thus be obtained only indirectly, by measurements on the exiting atoms through state-selective field ionization.

There are three independent time scales determining the dynamical behavior of the micromaser. We have, firstly, the average time interval t_p between two consecutive atoms, for which a Poissonian distribution can be assumed. Secondly, the atom-field interaction time τ. Third, the characteristic photon decay time $1/\gamma$, where γ means the decay rate due to cavity losses. For atoms being injected at a rate $R = 1/t_p$ (atoms per second), the dimensionless rate referred to above is given by $N_{at} = R/\gamma$. The experimental condition of having at most one atom a time inside the cavity translates itself into the condition $R\tau \ll 1$. As a further condition we have $\gamma\tau \ll 1$, meaning that during the atom-field interaction-time damping effects may be ignored. The atom-field evolution is ruled by the *Rabi Hamiltonian*

$$H = \frac{\hbar\omega_0}{2}\sigma_z + \hbar\omega a^+ a + \hbar g(a + a^+)(\sigma^- + \sigma^+), \tag{1}$$

in which ω_0 is the transition frequency connecting the two atomic levels, ω means the cavity frequency, a and a^+ are the annihilation and creation operators for the photon field respectively, and $\sigma^\pm$ are the raising and lowering Pauli operators. The dipole approximation has been assumed in writing such a Hamiltonian: the wavelength of the atomic transition is much larger than the dimensions of the atom, so that the electric field can be taken as position independent. The dipole coupling constant g is therefore a constant number, essentially fixed by the atomic transition and the cavity field.

The Rabi Hamiltonian reduces to the *Jaynes-Cummings* Hamiltonian in the rotating-wave approximation:

$$H = \frac{\hbar\omega_0}{2}\sigma_z + \hbar\omega a^+ a + \hbar g(a\sigma^+ + a^+\sigma^-) \tag{2}$$

One expects that being the system near resonance, i.e., $\omega \approx \omega_0$, and for values of the ratio g/ω sufficiently small, the RWA should be applicable.

Within the RWA, the time-evolution operator, $e^{-iHt/\hbar}$, can be calculated in closed form. Beyond the RWA we must rely on approximations based on different calculational schemes. We follow here the approach developed by Swain [52,53,51] and consider two arbitrary states, calling them the initial, $|i\rangle$, and the final one, $|f\rangle$. The probability that the initial state evolves under the action of the Hamiltonian H into the final state at time t, is given by

$$P_f^i(t) = |\langle f|i(t)\rangle|^2 = \left|\langle f|\, e^{-iHt/\hbar}\, |i\rangle\right|^2 \tag{3}$$

Swain brings this quantity into the form:

$$P_f^i(t) = \left|\frac{1}{2\pi i}\oint L_f^i(E) e^{-iEt/\hbar} dE\right|^2, \tag{4}$$

where $L_f^i(E)$ can be expressed in terms of the eigenbasis $\{|\alpha\rangle\}$ of H as follows:

$$L_f^i(E) = \sum_\alpha \frac{\langle i|\alpha\rangle\,\langle \alpha|f\rangle}{E - E_\alpha} \tag{5}$$

We will be interested in cases like $L_{b,n\pm1}^{a,n}, L_{b,n}^{a,n\pm1}$, etc. Here, $|a,n\rangle$ ($|b,n\rangle$) denotes a state in which the atom is in the upper a (lower b) state, with n photons in the field. Swain has derived a general procedure for calculating quantities like L_f^i, to any desired order, by continued fractions techniques. Integrals like those appearing in Eq.(4) can then be calculated by applying the theorem of residua.

The general approach ending up with the description of the micromaser dynamics is presently well known, so that we refer the reader to the literature for details [63]. We follow the standard procedure and consider an initially excited atom described by the density matrix ρ_{at} and a cavity-field given by ρ_c. Before the atom enters the cavity, the atom-field density matrix $\rho(0)$ is given by a direct product of the two noninteracting parts of the system:

$$\rho(0) = \rho_{at} \otimes \rho_c \tag{6}$$

During the time τ when one atom is inside the cavity, the photon-field evolves according to

$$\rho(\tau) = T(\tau)\rho(0) = Tr_{atoms}\left(e^{-iHt/\hbar}\rho(0)e^{iHt/\hbar}\right), \tag{7}$$

where we have dropped the index c in ρ_c, for brevity.

The transition operator $T(\tau)$ can be calculated in closed form *within the RWA*. In general, however, it can be decomposed as

$$T(\tau) = T^{(+)}(\tau) + T^{(-)}(\tau), \tag{8}$$

whereby $T^{(+)}$ and $T^{(-)}$ correspond to atoms outgoing in the upper and in the lower state, respectively.

During the time between two atoms, the cavity-field evolves according to the Liouvillian equation

$$\begin{aligned}\frac{d\rho}{dt} &= -\gamma L_C \rho = \\ &-\frac{\gamma}{2}(n_b+1)(a^+ a\rho + \rho a^+ a - 2a\rho a^+) \\ &-\frac{\gamma}{2}n_b(aa^+\rho + \rho aa^+ - 2a^+\rho a),\end{aligned} \tag{9}$$

with γ denoting the cavity decay constant and n_b the average number of thermal photons.

With the partial evolutions given by Eqs. (7) and (9) one obtains the master equation for the micromaser. The transition operator $T(\tau)$ cannot be calculated in closed form beyond the RWA. We can however calculate $T(\tau)$ to any desired order in the parameter g/ω by using continued fractions and cutting them at a prescribed order of approximation. Let us restrict ourselves to terms up to order $(g/\omega)^2$. To this order, the possible transitions out of a state $|a,n\rangle$, in which the atom is excited and there are n photons in the cavity, are the folowing ones:

$$\begin{array}{ccccc} |a,n-2\rangle & & & & |a,n\rangle \\ & \nwarrow & & \nearrow & \\ |b,n-1\rangle & \longleftarrow & |a,n\rangle & & \\ & \swarrow & & \searrow & \\ |a,n+2\rangle & & & & |b,n+1\rangle \end{array}$$

Expressing the density operator in the basis $\{|a,n\rangle, |b,n\rangle\}$, we have for the initial density operator the following expression:

$$\rho(0) = \sum_{\{i,j,n.m\}} \rho_{i,n;j,m} |i,n\rangle \langle j,m| \tag{10}$$

It then evolves to

$$\begin{aligned}&e^{-iHt/\hbar}\rho(0)e^{iHt/\hbar} \\ &= \sum \rho_{a,n;a,m} |k,n'\rangle \langle k,n'| e^{-iHt/\hbar} |a,n\rangle \langle a,m| e^{iHt/\hbar} |l,m'\rangle \langle l,m'|\end{aligned} \tag{11}$$

By taking the trace over the atomic degrees of freedom we obtain

$$\begin{aligned}\rho(\tau) &= Tr_{at}(e^{-iHt/\hbar}\rho(0)e^{iHt/\hbar}) \\ &= \sum \rho_{a,n;a,m}(|k,n'\rangle \langle k,m'|) \langle k,n'| e^{-iHt/\hbar} |a,n\rangle (\langle k,m'| e^{-iHt/\hbar} |a,m\rangle)^*\end{aligned}$$

In principle, we can calculate any matrix element $\rho_{n,m}(\tau) = \langle n| \rho(\tau) |m\rangle$. As we are interested in the populations, $p_n = \rho_{n,n}$, and specifically in steady-state properties of the micromaser field, we may correspondingly restrict the above calculation, by assuming that all coherences go to zero at steady-state [66], so that they can be dropped from the population equations. One obtains the following equation for the photon populations:

$$p_n(\tau) = (P_{a,n}^{a,n} p_n + P_{b,n}^{a,n-1} p_{n-1}) + \qquad (12)$$
$$P_{a,n}^{a,n+2} p_{n+2} + P_{b,n}^{a,n+1} p_{n+1} + P_{a,n}^{a,n-2} p_{n-2},$$

where we have written $p_n = p_n(0)$, etc., on the right-hand side, for brevity, and the probabilities P_y^x are given by Eq.(4).

Equation (12) for the populations differs from the standard one by the three last terms, as well as by the respective expressions for the transition probabilities. The three last terms correspond to additional ways of ending up with n photons in the cavity, after the interaction time τ. The term $P_{a,n}^{a,n+2} p_{n+2}$ for example, corresponds to a case in which, starting with $n+2$ photons in the cavity, we can end up with n photons, because there is now a probability $P_{a,n}^{a,n+2} \neq 0$ connecting the initial and final states, $|a, n+2\rangle$ and $|a, n\rangle$. Transitions like those corresponding to the terms $P_{a,n}^{a,n\pm 2}$and $P_{b,n}^{a,n+1}$ are commonly referred to as "energy non-conserving" ones. They are said to come from virtual-photon processes. We should stress that a transition like $|a, n+2\rangle \to |a, n\rangle$ is as real as the transition $|a, n\rangle \to |b, n+1\rangle$, even though the corresponding probabilities can differ from each other in several orders of magnitude. To say that one of such transitions is an "energy-conserving" one whereas the other is not, could be somewhat misleading: even within the RWA, neither $|a, n\rangle$ nor $|b, n+1\rangle$ are eigenstates of the Jaynes-Cummings Hamiltonian (JCH). Each one of these vectors is in fact a linear combination of two eigenvectors of the JCH. If the atom-field system has been prepared in, say, state $|a, n\rangle$, and we subsequently measure the energy of the system, then we could obtain any one, out of *two* possible (eigen)values. The same holds true for the state $|b, n+1\rangle$, so that we cannot properly say that the energy is conserved in such a transition. What is conserved in this case is the number of quanta. This follows from the fact that the operator $N_q = a^+ a + \sigma^+ \sigma^-$ whose eigenvalues are the number of quanta, is indeed a constant of the motion for the JCH. It is not a constant when we go beyond the RWA. For the Rabi Hamiltonian, what is conserved is instead the "parity" of the states. Indeed, it is the "parity" operator $P = \exp(i\pi N_q)$ which is now a constant of the motion. It enables us to split the Hilbert space into two unconnected subspaces. We may say therefore that the number of quanta is an approximately conserved quantity - an exact one if we invoke the RWA - the amount by which it is not conserved being measured by g/ω. Of course, the total energy of the system must be conserved, so that any gain or loss of energy, associated with the creation or annihilation of quanta, must be somehow compensated in the total energy balance. Within the micromaser scheme the source of energy in such cases must come from the kinetic atomic energy. This energy plays a substantial role in the dynamics of the system only for very cold atoms. In that case we would be within the framework of what has been called the one-atom "mazer" [67,68]. The atomic CM-motion has then to be included in a quantized form in the Hamiltonian of the system.

3 The Rotating and the Counter-rotating Wave Approximations

Before going further in the development of the micromaser theory beyond the RWA, let us analyze in more detail the structure of the Rabi Hamiltonian. Despite its apparent simplicity its diagonalization remains an open problem to date. In order to better analyze the role played by the counter-rotating terms, following Crisp [35] we rewrite the Hamiltonian as follows:

$$H = \frac{\hbar\omega_0}{2}\sigma_z + \hbar\omega a^+ a - \hbar(\alpha a\sigma^+ + \alpha^* a^+\sigma^- + \beta^* a\sigma^- + \beta a^+\sigma^+), \qquad (13)$$

with

$$\alpha = \frac{e}{m}\left(\frac{2\pi}{\hbar\omega V}\right)^{1/2}\left\langle a|e^{i\vec{k}.\vec{r}}\widehat{\varepsilon}\cdot\vec{p}|b\right\rangle,$$

$$\beta = \frac{e}{m}\left(\frac{2\pi}{\hbar\omega V}\right)^{1/2}\left\langle a|e^{-i\vec{k}.\vec{r}}\widehat{\varepsilon}\cdot\vec{p}|b\right\rangle$$

whereby $\vec{k}$ is the wave number vector corresponding to ω, V is the cavity volume, $\vec{p}$ the momentum operator, and e and m are the charge and mass of the electron, respectively. In the dipole approximation $\vec{k}.\vec{r} = 0$, so that $\alpha = \beta$; but we keep them as different labels to distinguish the counter- from the co-rotating terms. Setting $\beta = 0$ we have the RWA, when we like to invoke it.

Our Hilbert space has the basis $\{|a\rangle\otimes|n\rangle, |b\rangle\otimes|n\rangle\} \equiv \{|a,n\rangle, |b,n\rangle\}$. The matrix representing H in a basis whose vectors have been ordered as $\{|a,0\rangle, |b,0\rangle, |a,1\rangle, |b,1\rangle, |a,2\rangle, |b,2\rangle ...\}$, has a dispersed structure:

$$\hbar\begin{pmatrix} -\omega_0/2 & & & -\beta^* & & & \\ & \omega_0/2 & -\alpha & & & & \\ & -\alpha^* & \omega-\omega_0/2 & & & -\sqrt{2}\beta^* & \\ -\beta & & & \omega+\omega_0/2 & -\sqrt{2}\alpha & & \\ & & & -\sqrt{2}\alpha^* & 2\omega-\omega_0/2 & & \\ & & -\sqrt{2}\beta & & & \ddots & \end{pmatrix}$$

We see here that the dispersed structure stems from the counter-rotating terms. By putting $\beta = 0$, the above matrix turns into a block-diagonal one, which can be simply diagonalized. We only need to diagonalize a representative two-dimensional block

$$\hbar\begin{pmatrix} n\omega + \omega_0/2 & -\alpha\sqrt{n+1} \\ -\alpha^*\sqrt{n+1} & (n+1)\omega - \omega_0/2 \end{pmatrix}$$

giving as a result the well-known eigenvalues of the JCH, or energies of the "dressed-atom" [60]:

$$E_n^{\pm} = \hbar\omega(n+1/2) \pm \frac{\hbar}{2}\sqrt{(\omega-\omega_0)^2 + 4|\alpha|^2(n+1)}$$

The corresponding eigenvectors are

$$\left|\psi_n^+\right\rangle = \sin(\varphi_n)\left|a,n\right\rangle + \cos(\varphi_n)\left|b,n+1\right\rangle$$
$$\left|\psi_n^-\right\rangle = \cos(\varphi_n)\left|a,n\right\rangle - \sin(\varphi_n)\left|b,n+1\right\rangle$$

where

$$\cos(\varphi_n) \equiv \frac{\sqrt{(\omega-\omega_0)^2 + 4\left|\alpha\right|^2(n+1)}}{\left|(\omega-\omega_0)\right|}$$

Now, we observe that by decomposing the Hilbert space into two subspaces $\mathcal{E}=\mathcal{E}_a \oplus \mathcal{E}_b$, with $\mathcal{E}_a = \{\left|a,0\right\rangle, \left|b,1\right\rangle, \left|a,2\right\rangle, \left|b,3\right\rangle, \left|a,4\right\rangle, \left|b,5\right\rangle ...\}$ and $\mathcal{E}_b = \{\left|b,0\right\rangle, \left|a,1\right\rangle, \left|b,2\right\rangle, \left|a,3\right\rangle, \left|b,4\right\rangle, \left|a,5\right\rangle ...\}$, our Hamiltonian splits in two tridiagonal blocks:

$$\hbar \begin{pmatrix} \omega_0/2 & -\alpha & & & \\ -\alpha^* & \omega-\omega_0/2 & -\sqrt{2}\beta^* & & \\ & -\sqrt{2}\beta & 2\omega+\omega_0/2 & -\sqrt{3}\alpha & \\ & & -\sqrt{3}\alpha^* & 3\omega-\omega_0/2 & -\sqrt{4}\beta^* \\ & & & -\sqrt{4}\beta & 4\omega+\omega_0/2 \end{pmatrix}_{\mathcal{E}_a}$$

with a similar structure for $\mathcal{E}_b$. Here as well, by setting $\beta = 0$ we obtain a block-diagonal matrix. But we can now realize that also by setting $\alpha = 0$ a block-diagonal matrix arises. We call this the *counter-rotating-wave approximation* (CRWA). We stress that α and β are numerically the same. Thus, the RWA appears somewhat arbitrary in the present context.

The eigenvalue problem within the CRWA is easily solved as well, leading to the following eigenvalues:

$$E_n^{\pm} = \hbar\omega(n+1/2) \pm \frac{\hbar}{2}\sqrt{(\omega+\omega_0)^2 + 4\left|\beta\right|^2(n+1)},$$

whose eigenvectors are

$$\left|\psi_n^+\right\rangle = \sin(\phi_n)\left|b,n\right\rangle + \cos(\phi_n)\left|a,n+1\right\rangle$$
$$\left|\psi_n^-\right\rangle = \cos(\phi_n)\left|b,n\right\rangle - \sin(\phi_n)\left|a,n+1\right\rangle$$

with

$$\cos(\phi_n) \equiv \frac{\sqrt{(\omega+\omega_0)^2 + 4\left|\beta\right|^2(n+1)}}{\left|(\omega+\omega_0)\right|}$$

These expressions are similar to those of the RWA, the only difference being that $(\omega+\omega_0)$ replaces $(\omega-\omega_0)$ and that the couple of connected vectors are $\left|b,n\right\rangle$, $\left|a,n+1\right\rangle$, instead of $\left|a,n\right\rangle$, $\left|b,n+1\right\rangle$.

That the Hamiltonian splits in two unconnected subspaces, $\mathcal{E}_a$ and $\mathcal{E}_b$, is a consequence of the aforementioned fact that the parity operator, $P = \exp(i\pi N_q)$, with $N_q = a^+a + \sigma^+\sigma^-$, commutes with the Rabi Hamiltonian. Indeed, all vectors in $\mathcal{E}_{a(b)}$ have a definite parity: they are eigenvectors of P for the eigenvalue $-1(+1)$. On the other hand, within the RWA it is N_q which commutes with the corresponding Hamiltonian, the JCH. In this case both $|a,n\rangle$and $|b,n+1\rangle$ are eigenvectors of N_q for the same eigenvalue, $n+1$, so that the JCH connect only these two vectors among themselves and so splits into a set of two-dimensional blocks.

Coming back to the CRWA, let us take an initial atom-cavity state of the form $|\psi(0)\rangle = |a\rangle \otimes \sum_n p_n |n\rangle$, corresponding to an atom initially excited and the photon field in some (pure) quantum-mechanical state, which is defined through the occupation probabilities p_n. At time t the system will have evolved to the state

$$|\psi(t)\rangle = \sum_n a_n^+ e^{-iE_n^+ t/\hbar} |\psi_n^+\rangle + a_n^- e^{-iE_n^- t/\hbar} |\psi_n^-\rangle$$

with $a_n^+ = \langle \psi_n^+ | \psi(0)\rangle = p_{n+1}\cos(\phi_n)$, and $a_n^- = \langle \psi_n^- | \psi(0)\rangle = -p_{n+1}\sin(\phi_n)$.

The probability to find the system at time t with the atom being excited and with n photons in the cavity is given by

$$P(a,n;t) = |\langle a,n|\psi(t)\rangle|^2 = p_n^2 \left\{ \cos^2(\frac{R_{n-1}^+ t}{2}) + \cos^2(2\phi_{n-1})\sin^2(\frac{R_{n-1}^+ t}{2}) \right\}$$

where $R_n^+ = \sqrt{(\omega+\omega_0)^2 + 4|\beta|^2 (n+1)}$ is the *"Rabi frequency"* for the counter-rotating case.

Near resonance ($\omega \approx \omega_0$) and for a coupling constant $|\beta| << \omega_0$, we obtain $|\langle a,n|\psi(t)\rangle|^2 \approx p_n^2$. In such a case the field is almost unable to induce atomic transitions.

On the other hand, the collapse-revival phenomena predicted within the RWA appear also within the CRWA. Indeed, taking for the initial photon distribution the one corresponding to a coherent state, $p_n = \langle n\rangle^n \exp(-\langle n\rangle)/n!$, we obtain the collapse-revival behavior shown in Fig.(1). It is the same as in the RWA [63] but with much more rapid oscillations.

4 Diagonalization of the Rabi Hamiltonian by Continued Fractions

The tridiagonal matrices representing the Rabi Hamiltonian in the subspaces $\mathcal{E}_{a(b)}$ lead to eigenvalue problems of the form

$$\begin{vmatrix} Q_0-\lambda & Q_0^+ & & & \\ Q_1^- & Q_1-\lambda & Q_1^+ & & \\ & Q_2^- & Q_2-\lambda & Q_2^+ & \\ & & Q_3^- & Q_3-\lambda & Q_3^+ \\ & & & Q_4^- & Q_4-\lambda \end{vmatrix} = 0$$

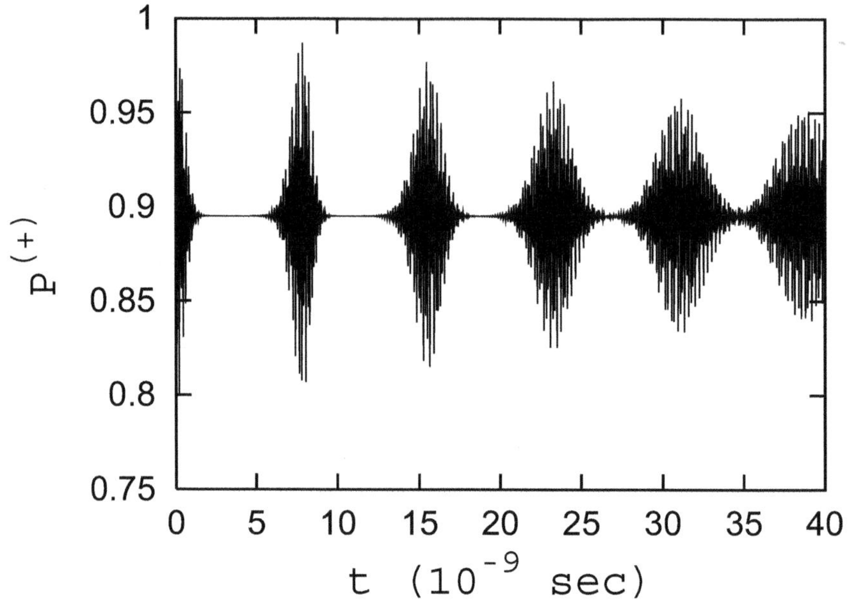

Fig. 1. Collapses and revivals of the upper atomic population in the CRWA. Parameter values are: $g = 4.4\ GHz$, $\omega = \omega_0 = 21\ GHz$, $\langle n \rangle = 6.25$

This secular equation can be solved either by applying techniques developed for the so-called "continuants" [64], or else by using iterative procedures leading to continued fractions [65]. In either case, one can write down an equation of the form:

$$Q_m - \lambda = C_1(Q_n^{\pm}, \lambda) + C_2(Q_n^{\pm}, \lambda),$$

where $C_{1,2}(Q_n^{\pm}, \lambda)$ represent two continued-fractions. We obtain for the eigenvalue λ the equation

$$n\omega + (-1)^n \frac{\omega_0}{2} - \lambda = \frac{(n+1)g^2}{(n+1)\omega + (-1)^{n+1}\frac{\omega_0}{2} - \lambda - \frac{(n+2)g^2}{(n+2)\omega + (-1)^{n+2}\frac{\omega_0}{2} - \lambda} \cdots} + \frac{ng^2}{(n-1)\omega + (-1)^{n-1}\frac{\omega_0}{2} - \lambda - \frac{(n-1)g^2}{(n-2)\omega + (-1)^{n-2}\frac{\omega_0}{2} - \lambda} \cdots}$$

where we have put $g = -\alpha = -\beta$, by exploiting the possibility of choosing the relative phase of the two atomic levels so as to make α - and consequently β - a real number.

Successive approximations lead to the following roots:

$$\lambda_n^{\pm} = n\omega + \frac{\omega + \omega_0}{2} + \frac{g^2}{2}\left(\frac{n}{\Omega \mp U} - \frac{n+2}{\Omega \pm U}\right) \mp V_{\pm},$$

where

$$U = \frac{1}{2}\sqrt{(\omega - \omega_0)^2 + 4g^2(n+1)},$$

$$V_\pm = [\frac{\omega - \omega_0}{2} + g^2(n+1) - \frac{g^2}{2}(\frac{n}{\Omega \mp U} + \frac{n+2}{\Omega \pm U})]^2$$

with $\Omega = \frac{1}{2}(\omega_0 + 3\omega)$.

This result was obtained some years ago by Swain [53], following quite different methods. For $g \ll \omega$, the expressions obtained for $\lambda_n^\pm$ reduce to those of the RWA (but with a different reference energy):

$$\lambda_n^{\pm(RWA)} = n\omega + \frac{\omega + \omega_0}{2} \mp U$$

5 Transition Probabilities

Either as an application of the techniques developed by Swain, or else by solving the complete eigenvalue problem, one can calculate transition probabilities from any initial state to any final state. In the following we will consider the resonant case, $\omega = \omega_0$. Swain [53] has obtained the transition probabilities relevant to our case, to second order in $g_\omega \equiv g/\omega$. That order of approximation is sufficient for our purposes. For completeness, we reproduce here his results:

$$P_{a,n}^{a,n}(\tau) = 1 - \frac{g_\omega^2}{2}n - \left(1 - \frac{g_\omega^2}{4}(3n+1)\sin^2(\omega\tau\theta_{n+1})\right) + \frac{g_\omega^2}{2}n\cos(\omega\tau\theta_{n+1})\cos(\omega\tau\theta_{n-1})\cos(2\omega\tau) \tag{14}$$

$$P_{b,n+1}^{a,n}(\tau) = \left(1 - \frac{3g_\omega^2}{4}(n+1)\right)\sin^2(\omega\tau\theta_{n+1}) \tag{15}$$

$$P_{a,n+2}^{a,n}(\tau) = \frac{g_\omega^2}{4}(n+2)\sin^2(\omega\tau\theta_{n+1}) \tag{16}$$

$$P_{a,n-2}^{a,n}(\tau) = \frac{g_\omega^2}{4}n\sin^2(\omega\tau\theta_{n-1}) \tag{17}$$

$$P_{b,n-1}^{a,n}(\tau) = \frac{g_\omega^2}{2}n\{1 - \frac{1}{2}\left[\sin^2(\omega\tau\theta_{n+1}) + \sin^2(\omega\tau\theta_{n-1})\right] - \cos(\omega\tau\theta_{n+1})\cos(\omega\tau\theta_{n-1})\cos(\omega\tau)\} \tag{18}$$

The generalized Rabi angle, θ_n, is given by

$$\theta_n = g_\omega\sqrt{n}(1 - \frac{1}{8}g_\omega^2 n) \tag{19}$$

and represents a small correction, of third order in g_ω to the RWA.

Having explicit expressions for the transition probabilities we can describe the micromaser dynamics by following the standard approach: We start defining the "micromaser mapping" as

$$\rho(t_{i+1}) = \exp\left(-\gamma L_C t_p\right) T(\tau)\rho(t_i), \tag{20}$$

where t_i represents the time at which the i-th atom is injected into the cavity, τ the interaction time, and $t_p = t_{i+1} - t_i - \tau \approx t_{i+1} - t_i$, the time between two successive atoms, for which it is assumed a Poissonian distribution. Introducing the corresponding Poissonian injection rate $R = 1/t_p$ and considering a so-called "coarse-grained" derivative [66] ($dt \ll \tau$) one obtains the continuous master equation

$$\frac{d\rho}{dt} = \{R[T(\tau) - 1] - \gamma L_C\}\rho \tag{21}$$

As we said before, the micromaser operational regime requires $R\tau \ll 1$ and $\gamma\tau \ll 1$. It is well known that the photon distribution can reach a steady-state. In order to find the steady-state photon distribution we need to solve Eq.(21) for the populations $p_n = \langle n|\,\rho\,|n\rangle$, setting $dp_n/dt = 0$. Following Elmfors [57], we seek to write the right-hand side of Eq.(21) in the form $J_{n+1} - J_n$, with J_n being some function of the populations. Now, the mapping

$$p(\tau) = T(\tau)p(0) \tag{22}$$

is given by Eq.(12), whereas the Liouvillian part of the evolution, which is ruled by the equation

$$\left(\frac{dp}{dt}\right)_{losses} = -\gamma L_C p, \tag{23}$$

has already the desired form. Indeed,

$$\left(\frac{dp}{dt}\right)_{losses} = \gamma\left[(n_b + 1)(n+1)p_{n+1} - n_b(n+1)p_n\right] - \gamma\left[(n_b + 1)np_n - n_b n p_{n-1}\right] \tag{24}$$

Our task reduces therefore to put the right-hand side of Eq.(22) in the form $p_n(0) + J_{n+1} - J_n$, so as to have $p_n(\tau) = \left[T(\tau)p(0)\right]_n = p_n(0) + J_{n+1} - J_n$, and hence

$$\left[(T(\tau) - 1)p(0)\right]_n = J_{n+1} - J_n \tag{25}$$

Even though the structure of $T(\tau)$ with the new terms is somewhat cumbersome, after some algebra the right-hand side of Eq.(22) can indeed be written in the desired form, $p_n(0) + J_{n+1} - J_n$, with

$$\begin{aligned} J_n = &\frac{g_\omega^2}{4}(n+1)\sin^2(\omega\tau\theta_{n+1})p_{n+1} \\ &-\frac{g_\omega^2}{2}n\cos(\omega\tau\theta_{n+1})\cos(\omega\tau\theta_{n-1})\cos(2\omega\tau)p_n \\ &+\frac{g_\omega^2}{2}n\left(1-\frac{1}{2}\sin^2(\omega\tau\theta_{n+1})\right)p_n \\ &-\left(1-\frac{g_\omega^2}{4}(2n-1)\right)\sin^2(\omega\tau\theta_n)p_{n-1} \\ &-\frac{g_\omega^2}{4}n\sin^2(\omega\tau\theta_{n-1})p_{n-2} \end{aligned} \tag{26}$$

Had we invoked the RWA from the start, the expression for J_n would have been the simple one:

$$J_n^{(RWA)} = -\sin^2(g\tau\sqrt{n})p_{n-1},$$

to which Eq.(26) reduces if we neglect contributions higher than first order in g_ω.

Finally, putting together the Hamiltonian and the Liouvillian contributions to the evolution equation for the populations, we obtain

$$\frac{dp_n}{dt} = \gamma\left(J'_{n+1} - J'_n\right), \tag{27}$$

with

$$J'_n = (n_b+1)np_n - n_b n p_{n-1} + N_{at}J_n \tag{28}$$

The equilibrium distribution $p_n^{(eq)}$ follows from setting $J'_{n+1} = J'_n$ and it can be found by solving the equation

$$J'_n = 0, \tag{29}$$

which holds true because the common value $J'_n = J$ must indeed vanish, on account of the fact that $p_n \to 0$ as $n \to 0$. Having now the equation which $p_n^{(eq)}$ has to satisfy, our next task is to solve it. We will have to rely on perturbation methods to fulfill our task.

6 The Steady-State Photon Distribution

Equation (29) for the equilibrium distribution $p_n^{(eq)}$ reads explicitly (writing p_n instead of $p_n^{(eq)}$, for brevity):

$$
\begin{aligned}
&(n_b+1)np_n - n_b np_{n-1} - N_{at}\sin^2(\omega\tau\theta_n)p_{n-1} \\
&+g_\omega^2 N_{at}\{\frac{(n+1)}{4}\sin^2(\omega\tau\theta_{n+1})p_{n+1} \\
&+\frac{n}{2}\left(1-\frac{1}{2}\sin^2(\omega\tau\theta_{n+1})-\cos(\omega\tau\theta_{n+1})\cos(\omega\tau\theta_{n-1})\cos(2\omega\tau)\right)p_n \\
&+\frac{2n-1}{4}\sin^2(\omega\tau\theta_n)p_{n-1} \\
&-\frac{n}{4}\sin^2(\omega\tau\theta_{n-1})p_{n-2}\} = 0,
\end{aligned} \tag{30}
$$

Eq.(30) has the form

$$
K_n(p_n) + g_\omega^2 K_n'(p_n) = 0 \tag{31}
$$

and can be solved by successive approximations, by writing

$$
p_n = p_n^{(0)} + g_\omega^2 p_n^{(1)} + \cdots \tag{32}
$$

and requiring that the resulting equation be satisfied for g_ω^2 small, but otherwise arbitrary:

$$
K_n\left(p_n^{(0)} + g_\omega^2 p_n^{(1)} + \cdots\right) + g_\omega^2 K_n'\left(p_n^{(0)} + g_\omega^2 p_n^{(1)} + \cdots\right) = 0 \tag{33}
$$

By following the usual steps in perturbation theory, we obtain the following set of equations to be solved iteratively:

$$
\begin{aligned}
K_n\left(p_n^{(0)}\right) &= 0 \\
K_n\left(p_n^{(1)}\right) + K_n'\left(p_n^{(0)}\right) &= 0 \\
K_n\left(p_n^{(2)}\right) + K_n'\left(p_n^{(1)}\right) &= 0 \\
&\vdots
\end{aligned} \tag{34}
$$

From the first one, $K_n\left(p_n^{(0)}\right) = 0$, we obtain a RWA-like result:

$$
p_n^{(0)} = \frac{n_b n + N_{at}\sin^2(\omega\tau\theta_n)}{(n_b+1)n} p_{n-1}^{(0)} \tag{35}
$$

This leads to

$$
p_n^{(0)} = p_0^{(0)} \prod_{m=1}^{n} \frac{n_b m + N_{at}\sin^2(\omega\tau\theta_m)}{(n_b+1)m}, \tag{36}
$$

whereby $p_0^{(0)}$ can be fixed through the normalization condition for the probabilities at zero-order $\left(\sum_{n=1}^{\infty} p_n^{(0)} = 1\right)$.

Inserting the $p_n^{(0)}$ into the second equation of the system (34), we obtain the recurrence relation

$$p_n^{(1)} = \frac{n_b n + N_{at} \sin^2(\omega\tau\theta_n)}{(n_b+1)n} p_{n-1}^{(1)} - \frac{K_n'(p^{(0)})}{(n_b+1)n} \equiv A_n p_{n-1}^{(1)} + f_n, \tag{37}$$

with

$$\begin{aligned} A_n &= \frac{n_b n + N_{at} \sin^2(\omega\tau\theta_n)}{(n_b+1)n} \\ f_n &= -\frac{K_n'(p^{(0)})}{(n_b+1)n} \end{aligned} \tag{38}$$

The first-order correction to $p_n^{(0)}$ is thus given by

$$p_n^{(1)} = \left(f_n + \sum_{j=1}^{n-1} f_j \prod_{m=j+1}^{n} A_m \right) + \left(\prod_{i=1}^{n} A_i \right) p_0^{(1)}, \tag{39}$$

where $p_0^{(1)}$ can be determined, as before, by requiring normalization at first-order $\left(\sum_{n=1}^{\infty} p_n^{(1)} = 1\right)$. It follows that

$$p_0^{(1)} = -\sum_{n=1}^{\infty} \left(f_n + \sum_{j=1}^{n-1} f_j \prod_{m=j+1}^{n} A_m \right) \left(1 + \sum_{n=1}^{\infty} \prod_{i=1}^{n} A_i \right)^{-1} \tag{40}$$

We stop with $p_n = p_n^{(0)} + g_\omega^2 p_n^{(1)}$, so as to be in accordance with the order up to which the transition probabilities have been calculated. Now that we have $p_n = p_n^{(eq)}$ to the desired order, it is possible to make predictions for measurable quantities, like the atomic inversion.

7 The Atomic Inversion

One important atomic observable in micromaser experiments is the probability $P^{(+)}$ of finding an outgoing atom in the upper state after the photon field has reached the steady- state. It is given by

$$P^{(+)} = \sum_{n=0}^{\infty} (T^{(+)} p^{eq})_n, \tag{41}$$

$P^{(-)}$ is defined analogously.

We can now introduce the "average inversion" $\langle s\rangle$ for states $s = +(-)$, with $+$ $(-)$ corresponding to the upper (lower) level, through

$$\langle s\rangle = P^{(+)} - P^{(-)} \tag{42}$$

Because $P^{(+)} + P^{(-)} = 1$, the inversion is determined by, e.g., $P^{(+)}$ alone, so that we can concentrate on investigating the properties of this quantity, which in our case is given by

$$\begin{aligned} P^{(+)} &= \sum_n P_{a,n}^{a,n} p_n + P_{a,n}^{a,n+2} p_{n+2} + P_{a,n}^{a,n-2} p_{n-2} \\ &= 1 - P^{(-)} = 1 - \sum_{n=0} P_{b,n}^{a,n+1} p_{n+1} + P_{b,n+1}^{a,n} p_n \end{aligned} \tag{43}$$

We see that $P^{(+)}$ differs from the RWA-result: it includes additional transition probabilities and, besides that, it depends on a somewhat different equilibrium distribution $p^{(eq)}$. Due to the complicated structure behind this last quantity we cannot say in advance, from which values of the parameters the RWA ceases to give accurate results. However, we can numerically compare our results with the ones obtained within the RWA. We present in the following our findings, which show that g/ω must be several orders of magnitude larger than the currently achievable value ($\sim 10^{-6}$) in order to see departures from the standard results concerning the atomic inversion. We stress that for each value of g/ω we have chosen the other parameter values in such a way as to satisfy the assumptions behind the micromaser standard operational scheme. In Fig. 3 we plot $P^{(+)}$ as a function of the pump parameter $\theta = g\tau\sqrt{N_{at}}$. The plots have been obtained by averaging the photon distribution over θ, i.e., over the interaction time. Such Gauss-averaged plotting procedure corresponds more closely to an actual experiment and brings into evidence the revival resonances. If we plot $P^{(+)}$ without averaging over θ, then the curves not always show clear evidence of revivals, as already observed by Elmfors et al [57]. We have numerically analyzed several cases [59]; for example, a case with $g/\omega = 0.1$, a value beyond present technological capabilities, but interesting for the purpose of comparison with other results, e.g., those of Seke [18], which were obtained under less realistic assumptions. The results obtained by assuming an ideal cavity and either a vacuum or a phase-diffused coherent photon-field, showed that the counter-rotating terms work against a complete collapse of the population inversion. The counter-rotating terms work also against revivals, whose appearance needs that the coherent atom-field interaction be able to compensate dephasing effects. However, in such ideal cases a value of $g/\omega = 0.1$ is not large enough so as to produce such noticeable differences as the ones observed in our results [59]. This demonstrates that the complete micromaser dynamics can enhance the incoherent effects due to the counter-rotating terms.

Fig. 3 shows a case where differences between the results with and without the RWA are quite large. The value of $g/\omega = 0.225$ taken here, is however very large too, and beyond the range of current experiments.

We see that for large interaction times the counter-rotating terms clearly inhibit the revivals, as already observed under more idealized conditions [18,19,20].

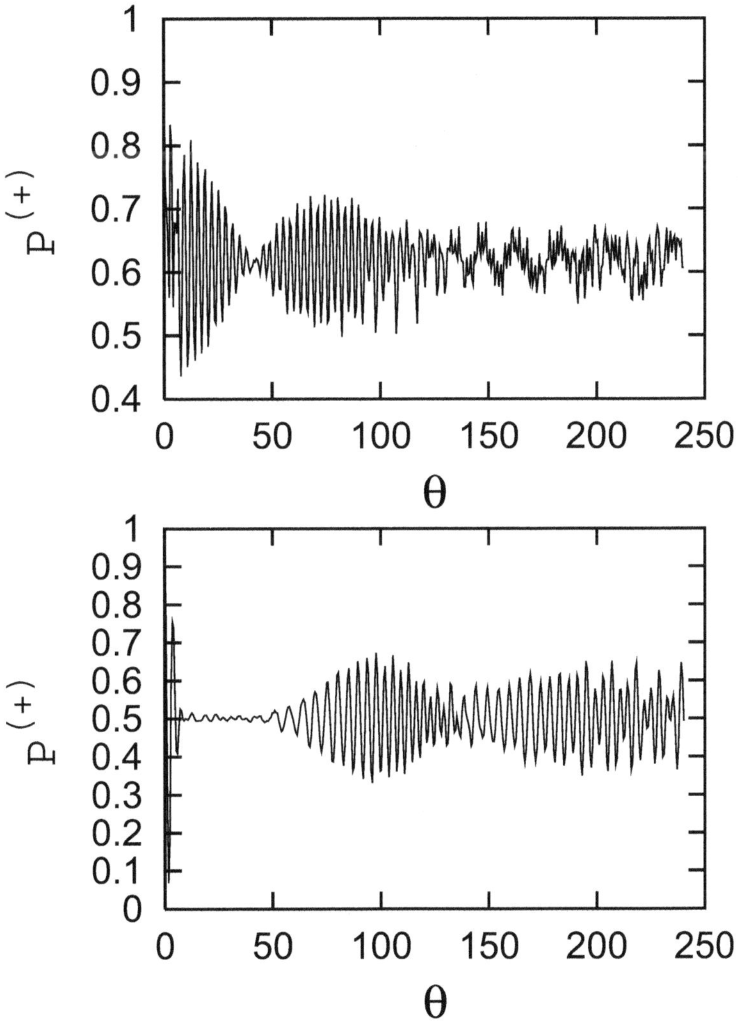

Fig. 2. Probability $P^{(+)}$ of finding the exiting atoms in the upper state, as a function of the Gauss-averaged pump parameter $\theta = g\tau\sqrt{N_{at}}$. The upper part corresponds to the case beyond the RWA, the lower part to the RWA. Parameter values in both cases are: $g = 4.5\ GHz$, $\omega = 20\ GHz$, $g/\omega = 0.225$, $\gamma = 10\ s^{-1}$, $N_{at} = 20$, $Q = 6.4 \times 10^8$, $n_b = 2$, $\sigma_\theta^2 = 6$

For shorter interaction times the counter-rotating terms are unable to destroy the coherent nature of the atom-field interaction, necessary for the revivals.

In Fig. 3 we have a case where $g/\omega \sim 10^{-3}$, a value not totally out of question within the microwave domain [21,20]. Although the differences between the RWA and the CRWA cannot be resolved from each other in this last case, by plotting the relative difference between them we see that it also displays collapses and revivals.

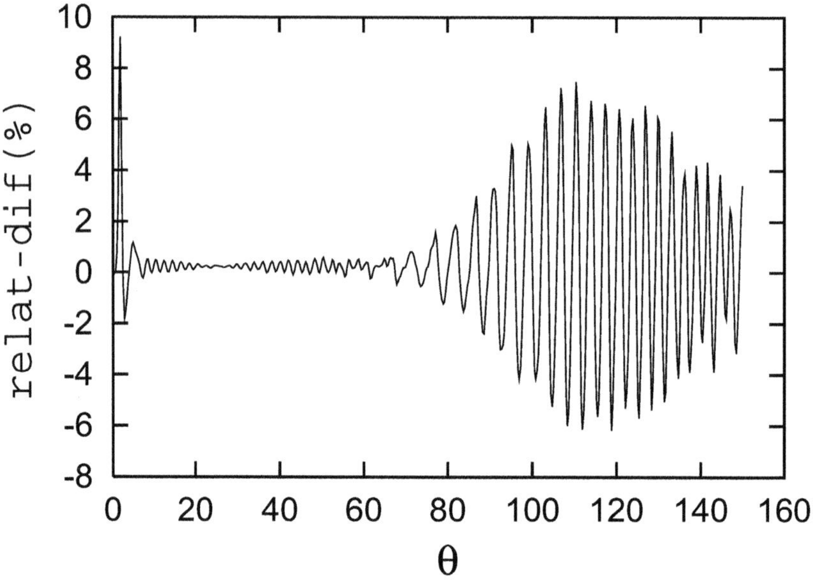

Fig. 3. For the case shown in this figure, there are not noticeable differences between the results obtained with (RWA) and without (WRWA) the RWA for $P^{(+)}$. However, plotting the relative difference, $(P^{(+)}_{WRWA} - P^{(+)}_{RWA})/P^{(+)}_{RWA}$, a collapse-revival structure in this quantity emerges. Parameter values are: $g = 50\ MHz$, $\omega = 20\ GHz$, $g/\omega = 0.025$, $\gamma = 10\ s^{-1}$, $N_{at} = 20$, $Q = 6.4 \times 10^8$, $n_b = 2$, $\sigma_\theta^2 = 4$

Our findings not only qualitatively confirm previous results obtained under idealized assumptions, but also show that the full micromaser dynamics might play a substantial role, by enhancing effects stemming from the counter-rotating terms. We stress, however, that the necessary values of g/ω have not been yet realized in actual experiments.

8 Trapping States

The so-called "trapping states", predicted within the conventional micromaser theory [61,62], constitute another promissory subject for testing the validity of the RWA. Such states are very sensitive to any departure from the conditions required for their production. Trapping states have been predicted for a micromaser working under very low temperatures, for which there is a very small number of thermal photons in the cavity. Putting $n_b = 0$ in Eqs.(36) and (37), the photon distribution reduces to

$$p_n^{(0)} = p_0^{(0)} \prod_{m=1}^{n} \frac{N_{at} \sin^2(\omega\tau\theta_m)}{m}$$

$$p_n^{(1)} = \frac{N_{at} \sin^2(\omega\tau\theta_n)}{n} p_{n-1}^{(1)} - \frac{K_n'(p^{(0)})}{n}$$

The trapping condition in the RWA is obtained by requiring that $p_n = 0$, for all $n > n_q$, so that the cavity cannot sample more than a prescribed number of photons. Thus, the following condition must be satisfied:

$$g\tau\sqrt{n_q + 1} = k\pi,$$

where n_q and k are integer numbers which serve to characterize each trapping state. In principle, by appropriately choosing the interaction time τ, it is possible to create different trapping states.

Beyond the RWA, the corresponding 0th-order condition for a trapping state reads:

$$\omega\tau\theta_{n_q+1} = k\pi$$

i.e.,

$$g\tau\sqrt{n_q + 1}\left(1 - \frac{g_\omega^2}{8}(n_q + 1)\right) = k\pi.$$

We see that by inserting this condition into the expression for $p^{(1)}$, we obtain

$$p_{n_q+1}^{(1)} = -\frac{K_{n_q+1}'(p^{(0)})}{n_q + 1} = -\frac{N_{at}(n_q + 1)}{4} \sin^2(\omega\tau\theta_{n_q}) p_{n_q-1}^{(0)}$$

It follows that populations beyond the Fock state $|n_q\rangle$ are not necessarily zero, as it would be the case within the RWA. Now they are given by

$$p_{n_q+1} = -g_\omega^2 \frac{N_{at}}{4}(n_q + 1) \sin^2(\omega\tau\theta_{n_q}) p_{n_q-1}^{(0)},$$

so that $p_{n_q+1} \neq 0$, in general. We would then have Fock states occupied beyond those which the standard trapping condition excludes. Please note that the factor g_ω^2 makes the violation of the RWA-trapping condition very small, in general. However, without performing numerical calculations it is hard to say in advance, exactly when the trapping condition begins to show some evidence of the contributions coming from the counter-rotating terms. A quantity that can be analyzed to this end is the mean photon number $< n >= \sum_n np_n$. Our results [59] show that we have to reach values of $g/\omega \sim 0.1$ in order to see noticeable departures from the RWA. However, for much smaller values of g/ω, even for some values on the order of $g/\omega \sim 10^{-6}$, we observe differences in the normalized deviation $\sigma = (< n^2 > - < n >^2)^{1/2} / < n >^{1/2}$. If we plot the relative difference for the standard deviation σ, i.e., $\Delta\sigma/\sigma = (\sigma^{(WRWA)} - \sigma^{(RWA)})/\sigma^{(RWA)}$, we observe the appearance of well-defined peaks.

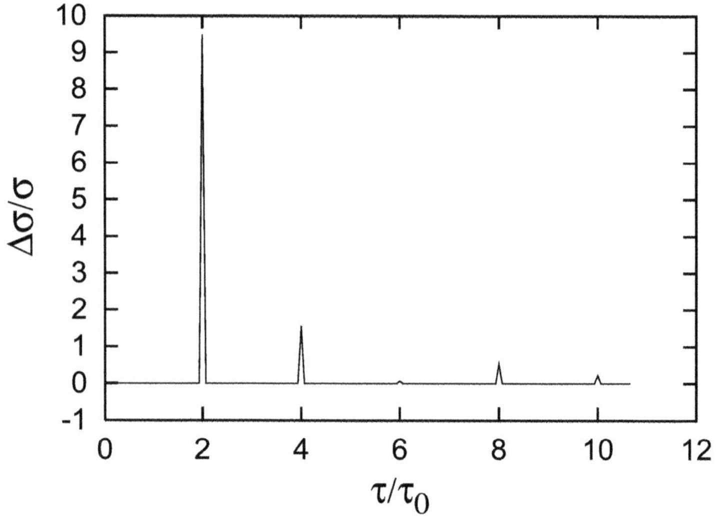

Fig. 4. By plotting the relative difference - in percentage - $\Delta\sigma/\sigma = (\sigma^{(WRWA)} - \sigma^{(RWA)})/\sigma^{(RWA)}$, as a function of a normalized interaction time τ/τ_0, we see that sharp peaks appear at times corresponding to trapping states: $\tau = 2\tau_0$, $4\tau_0$, $6\tau_0$, $8\tau_0$, ..., with $g\tau_0\sqrt{20} = \pi$. Parameter values are: $g = 0.4\,MHz$, $\omega = 80\,GHz$, $g/\omega = 5\times10^{-6}$, $\gamma = 1.25\ s^{-1}$, $N_{at} = 80$, $Q = 2. \times 10^{10}$, $n_b = 0$. The peaks correspond to trapping states with numbers $n_q = 19$, $k = 2, 4, 6, 8, 10$

In Fig. 4 we have plotted $100 \times \Delta\sigma/\sigma$ as a function of an interaction time τ being measured in units of a time τ_0 satisfying the trapping condition: $g\tau_0\sqrt{n_q + 1} = k\pi$. We have chosen $\tau_0 = \pi/g\sqrt{20}$, corresponding to a trapping state $k = 1$, $n_q = 19$. We observe peaks in $\Delta\sigma/\sigma$ occurring just for $\tau = 2\tau_0$, $4\tau_0$, $6\tau_0$, $8\tau_0$, ... For such times the standard trapping condition is correspondingly satisfied, with $n_q = 19$, $k = 2, 4, 6, 8, ...$We see therefore, at hand of the results presented in Fig.(5), that deviations with respect to the RWA predictions of approximately 10% can show up. This happens for parameter values not too far from those presently achieved in experiments: $g/\omega = 5.\ 10^{-6}$, $\gamma = 1.25\ s^{-1}$, $N_{at} = 80\ s^{-1}$, $Q = 2.\ 10^{10}$. Still larger deviations - of up to 30% - can be observed for $g/\omega \approx 10^{-5}$, a value that is not out of question for future experiments. Even though a high atomic detection accuracy is necessary to put in evidence the predicted effects, continuous improvements in current technology let us hope that such a task might be achieved in the near future.

9 Conclusions

We analyzed the validity of the RWA for highly excited, Rydberg atoms interacting with a microwave field, particularly in the context of the micromaser. To this end, we considered the atom-field dynamics of a two-level atom with dipole coupling, without invoking the RWA. The full micromaser dynamics was taken into account, extending the standard theory beyond the RWA. We have concentrated on those quantities which can be subjected to experimental testing. The relevant parameter - in terms of which successive approximations can be performed - is g/ω. By considering the micromaser dynamics, with expressions including g/ω up to second order, we start finding departures from the RWA in the following cases: for the atomic inversion there are differences with respect to the RWA when $g/\omega \approx 10^{-3}$. For larger values, typically on the order of 0.2, there are strong departures from the RWA in the collapse-revival structure of the atomic inversion. For $g/\omega \approx 0.1$ there are noticeable departures from the RWA-predictions in the structure of "trapping states". For a much smaller value of g/ω, i.e., $g/\omega \approx 10^{-6}$, we observe departures from the RWA-predictions in the normalized deviation $\sigma = (< n^2 > - < n >^2)^{1/2} / < n >^{1/2}$. This is a remarkable case in which well-defined peaks appear for parameters corresponding to trapping states and near the range employed in current experiments. Our results confirm therefore that the fragility of the trapping states can make them good candidates for testing the validity of the RWA.

On the other hand, we have shown that the parameter values necessary to bring into evidence the limits of the RWA, by measuring the atomic inversion, are still far from the range that is presently realizable. Even though our results do not yet support the conjectures raised by some authors, who questioned the validity of the RWA for Rydberg atoms and who saw the micromaser as a tool for bringing new effects into evidence, we should stress that we have restricted ourselves to the resonant case. The off resonance case is more involved, making an analytical extension of the conventional micromaser theory very difficult. A numerical study of the nonresonant case is a task one should accomplish before ruling out the possibility of employing the micromaser as an appropriate tool for testing the RWA.

References

1. E.T. Jaynes and F.W. Cummings: Proc. IEEE **51**, 89 (1963)
2. H. Paul: Ann. der Phys. **11**, 411 (1963)
3. E. Arimondo and G. Moruzzi: J. Phys. B: Atom. Molec. Phys. **6**, 2382 (1973)
4. C. Cohen-Tannoudji, J. Dupont-Roc and C. Fabre: J. Phys. B: Atom. Molec. Phys. **6**, L218 (1973)
5. F. Bloch and A. Siegert: Phys. Rev. **57**, 522 (1940)
6. J.I. Cirac, A.S. Parkins, R. Blatt, and P. Zoller: Phys. Rev. Lett. **70**, 556 (1993)
7. M. Feng, X. Zhu, X. Fang, M. Yang, and L. Shi: J.Phys.B: At. Mol. Opt. Phys. **32**, 701 (1999)

8. M. Gross and S. Haroche: Physics Reports **93**, 301 (1982)
9. G.S. Agarwal: Phys. Rev.A **4**, 1778 (1971)
10. G.S. Agarwal: 'Quantum Statistical Theories of Spontaneous Emission and their Relation to Other Approaches'. In: *Quantum Optics 70, Springer Tracts in Modern Physics* (Springer Verlag, 1974)
11. R.J.C. Spreeuw and J.P. Woerdman: Phys. Rev. A **44**, 4765 (1991)
12. R.J.C. Spreeuw and J.P. Woerdman: 'Optical Atoms' In: *Progress in Optics XXXI*, ed. by E. Wolf, (Elsevier Science, Amsterdam 1993), pp. 263-319
13. M.P. Silverman: *More than One Mystery, Explorations in Quantum Interference*, (Springer Verlag, 1995, New York, Berlin, Heidelberg)
14. M.P. Silverman and F.M. Pipkin: J.Phys.B: Atom. Molec. Phys. **5**, 1844 (1973)
15. W. Lange, H. Walther and G. S. Agarwal: Phys. Rev. A **50**, R3593 (1994)
16. S.S. Hassan, H.A. Batarfi, and R.K. Bullough: J. Opt. B: Quantum Semiclass. Opt. **2**, R35 (2000)
17. K. Zaheer and M-S. Zubairy: Phys. Rev. A **37**, 1628 (1988)
18. J. Seke: Quantum Opt. **3**, 127 (1991)
19. G.A. Finney and J. Gea-Banacloche: Phys. Rev. A **50**, 2040 (1994)
20. K. Zaheer and M-S. Zubairy: Optics Comm. **73**, 325 (1989).
21. M. Brune, J.M. Raimond, P. Goy, L. Davidovich, and S. Haroche: Phys. Rev. Lett. **59**, 1899 (1987)
22. J.S. Peng and G.X. Li: Phys. Rev. A **47**, 3167 (1993)
23. J.S. Peng and G.X. Li: Phys. Rev. A. **45**, 3289 (1992)
24. J. Seke: Quantum Opt. **7**, 161 (1995)
25. R. Vyas and S. Singh: Phys. Rev. A **33**, 375 (1986)
26. W.J. Munro and C.W. Gardiner: Phys. Rev. A **53**, 2633 (1996)
27. G.W. Ford and R.F. O'Connell: Physica A **243**, 377 (1977)
28. K.M. Ng, C.F. Lo and K.L. Liu: Physica A **275**, 463 (2000)
29. C.F. Lo, K.L. Liu and K.M. Ng: Physica A **265**, 557 (1999)
30. P.I. Belobrov, G.M. Zaslavskii and G.Kh. Tartakovskii: Sov. Phys. JETP **44**, 945 (1976)
31. P.W. Milonni, J-R. Ackerhalt, H.W. Galbraith: Phys. Rev. Lett. **50**, 966 (1983)
32. R. Graham and M. Höhnerbach: Phys. Lett. A **101**, 61 (1984)
33. R. Graham and M. Höhnerbach: Phys. Rev. Lett. **57**, 1378 (1986)
34. M. Kus and J. Lewenstein: J. Phys. A **19**, 305 (1986)
35. M.D. Crisp: Phys. Rev. A **46**, 4138 (1992)
36. M.D. Crisp: Phys. Rev. A. **43**, 2430 (1991)
37. H.G. Reik, H. Nusser and L.A.A. Ribeiro: J. Phys. A **15**, 3491 (1982)
38. H.G. Reik and M. Doucha: Phys. Rev. Lett. 57, 787 (1986).
39. H.G. Reik, P. Lais, M.E. Stuetzle, M. Doucha: J. Phys. A **20**, 6327 (1987)
40. J.H. Eberly, N.B Narozhny and J.J. Sanchez-Mondragon: Phys. Rev. Lett. **44**, 1323 (1980)
41. P. Filipowicz, J. Javanainen, P. Meystre: Phys. Rev. A **34**, 3077 (1986)
42. D. Meschede, H. Walther, G. Müller: Phys. Rev. Lett. **54**, 551 (1985)
43. Z. Liu, L. Zeng and P. Liu: Phys. Lett. A **217**, 219 (1996)
44. L. Zeng, Z. Liu, Y. Ling, S. Zhu: Phys. Lett. A **246**, 43 (1998)
45. R.F. Bishop: Theor. Chim. Acta. **80**, 95 (1991)
46. R.F. Bishop, N.J. Davidson, R.M. Quick and D.M. van der Walt: Phys. Rev. A **54**, R4657 (1996)
47. C.F. Lo, K.L. Liu, K.M. Ng and P.H. Yuen: Quantum Semiclass. Opt. **10**, L63 (1998)

48. S. Schweber: Ann. Phys., NY **41**, 205 (1967)
49. V. Bargmann: Commun. Pure. App. Math. **14**, 187 (1961)
50. S. Swain: Adv. At. Mol. Phys. **22**, 387 (1986)
51. S. Swain: Phys. Rev. A **49**, 2816 (1994)
52. S. Swain: J. Phys. A **6**, 192 (1973)
53. S. Swain: J. Phys. A **6**, 1919 (1973)
54. R. Ramirez and M. Orszag: J. Math. Phys. **22**, 1306 (1981)
55. I. D. Feranchuk, L.I. Komarov and A.P. Ulyanenkov: J. Phys. A **29**, 4035 (1996)
56. I. D. Feranchuk, L.I. Komarov, I.V. Nichipor and A.P. Ulyanenkov: Ann. Phys. **238**, 370 (1995)
57. P. Elmfors, B. Lautrup, and B-S. Skagerstam: atom-ph/9601004v2 (1996).
58. F. De Zela, E. Solano and A. Gago: Opt. Comm. **142**, 106 (1997)
59. F. De Zela: sub. to Phys. Rev. A
60. C. Cohen-Tannoudji, J. Dupont-Roc, G. Grynberg: *Atom-Photon Interactions* (Wiley, New York, 1992)
61. M. Weidinger, B.T.H. Varcoe, R. Heerlein, and H. Walther: Phys. Rev. Lett **82**, 3795 (1999)
62. P. Meystre, G. Rempe, and H. Walther: Optics Lett. **13**, 1078 (1988)
63. P. Meystre, M. Sargent III: *Elements of quantum Optics* (Springer-Verlag, Berlin, 1991)
64. T. Muir, W. H. Metzler: *A treatise on the theory of determinants* (Dover Pub., New York, 1938)
65. H. Risken: *The Fokker-Planck Equation: Method of Solution and Applications* (Springer-Verlag, Berlin, 1989)
66. M. Sargent, III, M.O. Scully, and W.E. Lamb Jr.: *Laser Physics* (Addison-Wesley, 1974)
67. M. Schröder, K. Vogel, W.P. Schleich, M.O. Scully, and H. Walther: Phys. Rev. A **56**, 4164 (1997)
68. J.C. Retamal, E. Solano, and N. Zagury: Optics Comm. **154**, 28 (1998)

What Is a Quantized Mode of a Leaky Cavity?

S.M. Dutra and G. Nienhuis

Huygens Laboratory, Leiden University
P.O. Box 9504, 2300 RA Leiden, The Netherlands

Abstract. Quantum optics conventionally represents the electromagnetic radiation in a leaky cavity by a normal-modes expansion, as if the cavity were closed and perfect, with the leakage being described by coupling these modes to a reservoir. This phenomenological approach is only a good approximation for stable resonators with very large quality factors. Here we show how the quantized field can be described by the actual natural cavity modes rather than these fictitious normal modes. Our approach leads to a radically different model for quantum dissipation, where reservoir and dissipative system operators no longer commute.

1 Introduction

Inside a cavity, the radiation field is constrained to assume only certain spatial patterns and to oscillate only at certain frequencies. These are the natural modes and frequencies of the cavity, respectively. A description of the radiation field in terms of such modes is often more convenient and offers more physical insight than a direct account of position dependence without a mode expansion. Semiclassical laser physics takes full advantage of this fact [1,2]. However, the situation is quite different in quantum optics. There, the idea of cavity mode is often linked to that of normal modes. Since Fermi [3] showed how easily the radiation field can be quantized by identifying each term in its normal modes expansion as a harmonic oscillator and then just quantizing each oscillator, the use of normal modes to describe the quantum electromagnetic field has been almost mandatory.

For perfect, closed cavities, the ideas of normal and cavity modes coincide. The problem only arises for leaky cavities, because they are open systems and so they do not have normal modes. This problem is often avoided by the adoption of a phenomenological Hamiltonian originally introduced by Sentizky [4] (based on earlier work by Callen and Welton [5] and Weber [6]), but which became known in the quantum optics community as the Gardiner-Collett Hamiltonian [7]. In this approach the cavity field is described in terms of normal modes as if the cavity were perfect and closed, with the leakage being accounted for by the ad hoc introduction of a coupling with a reservoir. The Gardiner-Collett Hamiltonian is an excellent approximation for cavities with high quality factors (Q), but it breaks down as the leakage increases. In this paper we present a formalism [8] where the quantized field is described in terms of natural cavity modes rather

M. Orszag, J.C. Retamal (Eds.): LNP 575, pp. 338–354, 2001.

than fictitious normal modes. Apart from the fundamental interest in knowing what lies beyond the Gardiner-Collett Hamiltonian, a good reason to develop this formalism is that some phenomena such as excess noise [9,10] cannot be described within a normal modes approximation of natural cavity modes. This formalism brings quantum optics closer to some useful intuitive ideas that were introduced in semiclassical laser physics. But it also gives rise to a surprise: a radically different model for quantum dissipation, where the dissipative system operators no longer commute with those of the reservoir.

In the next section, we discuss the general problem of treating an open system in quantum mechanics. This problem is connected to the even more fundamental problem of the classical limit of quantum mechanics through decoherence. We demonstrate the connection explicitly for the special case of the radiation field in a high-Q cavity, by using the correspondence principle to derive, from the classical dissipation dynamics, the same master equation that is obtained from the Gardiner-Collett Hamiltonian. To be able to generalize this Hamiltonian to arbitrary Q, the first question that must be answered is what is a mode of a leaky cavity. Section 3 deals with this question. Subsection 3 presents the answer: the concept of natural mode due to J. J. Thomson. We review the role of this concept in semiclassical laser theory and the confusing situation in quantum optics. To avoid unnecessary complications, we restrict our considerations to a simple one dimensional model of a leaky cavity presented in section 4. Then in section 5, we develop a Sturm-Liouville treatment for natural modes where Fox-Li modes are introduced in the same way as normal modes. The quantum theory with nonorthogonal Fox-Li modes is finally introduced in section 6. Section 7 summarizes our main results and discusses possible extensions of this theory.

2 Open Systems in Quantum Mechanics

In quantum mechanics, dissipation cannot be introduced in the same straightforward way as in Newtonian mechanics. If we were just to add a friction force proportional to the velocity, say, to Heisenberg's equations of motion for a given quantum system, the consequences would be disastrous: equal-time commutators would decay in time! The usual way out is simple: just avoid dealing with a dissipative system in the first place. Instead of describing the dissipative system alone, include its environment in the description as well. This way, the total system is closed and conservative, so that quantum mechanics can be applied without any problem. The environment then behaves as a large reservoir with an infinite number of degrees of freedom that damps the motion of the small dissipative system but also introduces just the right amount of noise to preserve the commutators.

This problem with friction is not peculiar to quantum mechanics alone. In fact, it was inherited from the variational formulation of classical mechanics upon which quantum mechanics is built. Unlike Newton's, the variational approach assumes that all forces are derivable from a scalar quantity, the "work function".

As friction forces have no work function, they are outside the realm of variational principles [11]. There have been attempts to include friction in the variational approach [12] and then eliminate the need for a reservoir in quantum mechanics [13], but so far the reservoir scheme is the most successful one. Some of the effective Lagrangians adopted in these alternative schemes do not even describe the dissipative system they are claimed to describe but a completely different system that is not dissipative [14]. More information about the introduction of dissipation in the simple case of a harmonic oscillator can be found in Dekker's review paper [15].

2.1 Quantum dissipation and the classical limit

Whereas in quantum mechanics every basis is equally valid, in classical reality certain basis seem to be preferred over others. For instance, classical light is basically a statistical mixture of coherent states not of squeezed states or of EPR states. According to Zurek's pointer-basis theory [16], this preference for a given basis is a consequence of decoherence [17]. In practice, decoherence is often caused by dissipation. So dissipation must be deeply connected to the classical limit of quantum mechanics. In fact, some authors have shown that certain ad hoc models for dissipation can lead to a preferred basis of highly nonclassical states such as Schrödinger cat states [18]. We will demonstrate this connection explicitly now for the specific case of a single mode cavity with a high quality factor. We will show that the knowledge that coherent states are the quasiclassical quantum states (the preferred basis), together with a classical description of the dissipation dynamics, is enough to determine the master equation governing the quantum dissipation dynamics. This section will follow closely the approach taken in [19].

Let us consider a one-dimensional cavity composed of two mirrors a distance l apart and with the same (amplitude) reflectivity $\sqrt{R}$. After a time $t = 2lN/c$ corresponding to N round trips in the cavity (c is the speed of light), the electric field of a classical electromagnetic wave at position x, $E(x)$, will decrease from its initial value at $t = 0$, due to the partial reflections at the mirrors, according to

$$E(x,t) = E(x,0)e^{-\frac{\kappa}{2}t}, \tag{1}$$

where $\kappa = -c\ln(R)/l$. As the radiation only leaks out at the mirrors, (1) is only valid for times that are integral multiples of the round-trip time. In this subsection, however, we will only consider high-Q cavities. Then the mirrors are sufficiently good reflectors so that electromagnetic radiation can complete many round trips before it loses a sizable fraction of its energy, i. e. the decay time is much longer than the time of flight of a light signal through the cavity. In that case, we can neglect the fact that the losses are localized at the mirrors and assume them to be homogeneously distributed in the cavity and occur continuously in time[1]. In this case, (1) is a good approximation for any time $t > 0$.

Another important approximation behind (1) is that thermal fluctuations can be neglected. When the mean energy of the field in the cavity becomes comparable to the mean energy of black-body radiation, (1) breaks down. This happens because the energy leaking out of the cavity will be balanced by energy being fed into the cavity by black-body radiation. At zero temperature, however, we would expect (1) to be valid all the way down to the quantum zero point energy of the radiation field.

Coherent states are the only quantum states that survive in the classical limit providing a bridge between the quantum and the classical regime. So if we assume that (1) can be carried over into the quantum regime, a coherent state $|\alpha\rangle$ at time t should decay to

$$|\alpha\rangle \to \left|e^{-\kappa t/2}\alpha\right\rangle . \tag{2}$$

Now, any density matrix $\rho(t)$ can be written as a linear combination of diagonal coherent state density matrices[2]

$$\rho = \int d^2\alpha \; P(\alpha) \; |\alpha\rangle\langle\alpha| , \tag{3}$$

where

$$P(\alpha) = \frac{1}{\pi^2} \int d^2\xi \; \mathrm{Tr}\left(\rho e^{\xi a^\dagger - \xi^* a}\right) e^{\alpha\xi^* - \alpha^*\xi} . \tag{4}$$

So the density matrix describing the single-mode cavity field at an arbitrary time t must be given by

$$\rho = \int d^2\alpha \; P(\alpha) \; |e^{-\kappa t/2}\alpha\rangle\langle e^{-\kappa t/2}\alpha| , \tag{5}$$

where $P(\alpha)$ is the same quasiprobability distribution that described the initial density matrix, i.e. before the field started to decay. For an infinitesimal time interval δt,

$$\rho(t+\delta t) = \int d^2\alpha \; P(\alpha) \; \left|e^{-\kappa\delta t/2}\alpha\right\rangle \left\langle \alpha e^{-\kappa\delta t/2}\right| . \tag{6}$$

---

[1] This implies coarse graining at a time scale of the order of the time of flight of a light signal through the cavity. Please, see [20].

[2] Off diagonal terms are not needed because the coherent state basis is over-complete. The diagonal representation was developed independently by Glauber and Sudarshan in the sixties [21]. Unfortunately, the Glauber-Sudarshan quasi-probability distribution P is often very singular; e.g., for a coherent state it is a Dirac delta function. In this work, however, P is always under an integral sign so that such possible singular behaviour poses no problem. The general validity of the diagonal coherent state expansion of every density matrix, in the context of distribution theory, was proved by Klauder, McKenna and Currie [22]. An excellent pedagogical exposition of all this can be found in [23].

A straightforward calculation reveals that

$$\left|e^{-\kappa\delta t/2}\alpha\right\rangle\left\langle e^{-\kappa\delta t/2}\alpha\right| = |\alpha\rangle\langle\alpha| - \frac{\kappa}{2}\left\{a^\dagger a, |\alpha\rangle\langle\alpha|\right\}\delta t + \kappa a|\alpha\rangle\langle\alpha|a^\dagger\delta t + O\left(\delta t^2\right). \tag{7}$$

Substituting (7) in (6), we obtain

$$\rho(t+\delta t) = \rho(t) - \frac{\kappa}{2}\left\{a^\dagger a, \rho(t)\right\}\delta t + \kappa a\rho(t)a^\dagger\delta t + O\left(\delta t^2\right), \tag{8}$$

where the curly brackets stand for the anti-commutator. The master equation is given by

$$\begin{aligned}\frac{d}{dt}\rho(t) &= \lim_{\delta t\to 0}\frac{\rho(t+\delta t)-\rho(t)}{\delta t}\\ &= -\frac{\kappa}{2}\left\{a^\dagger a, \rho(t)\right\} + \kappa a\rho(t)a^\dagger,\end{aligned} \tag{9}$$

which is exactly the well-known Lindblad term [24].

3 What Is a Mode of a Leaky Cavity?

This question was answered a few years after Maxwell predicted the existence of electromagnetic waves. It was then rediscovered around the beginning of the sixties, and it played a major role in the development of the laser. In quantum optics, however, all this progress has been largely ignored and many alternative "answers" have been proposed with very little consensus. Next, we briefly review the concept of mode of a leaky cavity and the situation in quantum optics.

3.1 The classical answer

The problem of what is a mode of a leaky cavity becomes even more dramatic for open cavities, where the border with the outside is ill-defined. So when Dicke [25], Schawlow and Townes [26], and Prokhorov [27] independently suggested the use of the Fabry-Perot interferometer as a cavity for the first laser, the key question was: can such an open device have any discrete cavity modes at all? As the Fabry-Perot is not enclosed by reflecting side walls, the general opinion at the time was that it could only have a continuum of modes. In fact, as its name says, before the advent of the laser, the Fabry-Perot had been traditionally used as an interferometer, not as a resonator. The essential point is that Fabry-Perot rings exist for any frequency, whereas a resonator is expected to have a large response only for a discrete spectrum of frequencies [28]. But Fox and Li [1] addressed this problem and showed that the diffraction losses, due to the finite surface area of the end mirrors, effectively turn the continuum into a discrete set of modes of unexpectedly high-Q (i.e. with low diffraction losses). It was then that the idea of mode (or quasimode) of a leaky cavity was first introduced in quantum electronics and quantum optics.

Fox and Li considered a propagating wave that was reflected back and forth by the two end mirrors of the Fabry-Perot. These mirrors were assumed to be perfect reflectors but of finite area. The propagated wave was calculated using the scalar formulation of Huygens' principle. Then they looked for field distributions whose profile was self-repeating (apart from a decay factor) in a complete round trip of the lossy cavity. This is their most natural and intuitive definition of mode. This concept, however, is not the same as that of normal modes adopted for perfect cavities [29,30,31]. The key difference is that the concept of normal modes makes no distinction between forward and backward propagating components whereas for Fox-Li modes such distinction is crucial. Normal modes are determined by solving Helmholtz equation and, for closed lossless cavities, they are *standing waves.* Fox-Li modes, on the other hand, are determined by finding the eigenfunctions of the Huygens propagation integral and they are *traveling waves.* Moreover, because the cavity is leaky, they turn out to be nonorthogonal [29,30,32].

Initially the Fox-Li result that open cavities have discrete high-Q modes, was used to justify an approximate treatment in terms of normal modes of a fictitious perfect cavity just as for closed cavities [33]. This treatment is still standard in quantum optics [34], but not any more in semiclassical theories. Because, in the eighties, Siegman showed [2] that formally simple yet exact (semiclassical) equations of motion for real laser cavities can be obtained by using the exact Fox-Li modes instead of the perfect modes approximation. The use of exact Fox-Li modes then turned out to be crucial to explain the origin of excess noise semiclassically [10].

The general concept of mode of a leaky cavity, however, dates from well before Fox and Li. It can be traced back to Thomson's [35] investigation of a simple idea for an electromagnetic oscillator. At the time, he was interested in doing what Hertz eventually did later: an experiment to test Maxwell's prediction of electromagnetic waves. For Thomson, modes are the natural (free) oscillations of the leaky cavity. His modes are *solutions of Helmholtz equation* for the actual boundary conditions on the cavity mirrors, but that contain only outgoing waves at infinity. They are standing waves in the cavity. These modes are the natural generalization for a leaky cavity of the normal modes of a perfect cavity. They can be interpreted in a similar way to decaying states in the Gamow-Condon-Gurney theory of α decay [36]. Three years after Fox and Li had published their work, Thompson's idea was taken up and refined by Vaĭnshteĭn [37]. Although Thomson-Vaĭnshteĭn modes are apparently very different from Fox-Li modes, they are just two facets of the same idea [38]. We show this explicitly in section 5 for the simple cavity model considered here. We also derive the inner product adopted in Fox-Li mode expansions [30], where forward and backward propagating components do not mix, as a direct generalization of the conventional inner

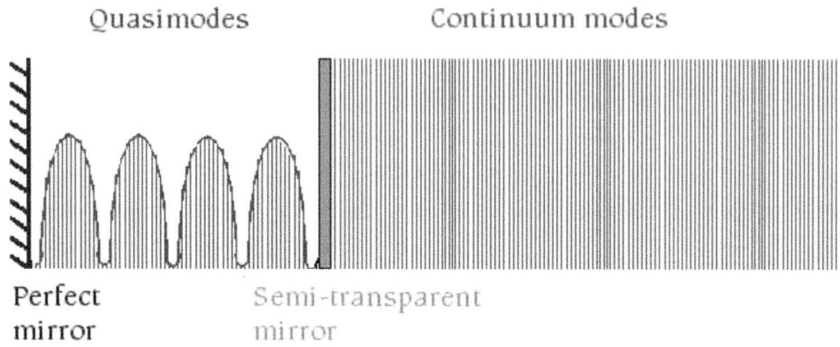

Fig. 1. Modes of the "universe" picture of cavity modes: Cavity modes are modulations in the continuum of modes of the "universe" due to the presence of the cavity walls.

product in normal mode expansions, where forward and backward propagating components do mix. These results show that *propagating wave* Fox-Li modes can be regarded as a direct generalization of *standing wave* normal modes for a leaky cavity. Fox-Li modes are then the natural candidates for quantization.

3.2 Quasimodes in the quantum theory

Even though quantum electrodynamics allows us to deal directly with the spatially dependent electric and magnetic fields, the annihilation and creation operators used in quantum optics require the introduction of normal modes. Leaky cavities, however, do not have normal modes: there lies the problem of quantum optics with these cavities. So quantum optics uses the normal modes of the closed system formed by the cavity and the outside, the modes of the "universe" [39]. To be able to talk about cavity modes this way, one must identify which subset of the modes of the "universe" constitutes a cavity mode (see Fig. 1). This complicates the problem tremendously and also makes it rather subjective. Most of the lack of consensus regarding cavity modes in quantum optics stems from all the alternative intuitive ideas of how to define leaky cavity modes in terms of normal (universe) modes.

As far as we know, the earliest such idea was proposed by Lang, Scully, and Lamb in a paper [39] where they also introduced the modes of the "universe" approach. In that paper and in a couple of more recent ones [40], a quasimode annihilation operator is defined as a linear combination of all modes-of-the-"universe" annihilation operators whose frequency lies in a certain interval centered at the cavity resonance corresponding to the quasimode. This interval is much larger than the width of the corresponding cavity resonance and much smaller than the separation in frequency between successive resonances but arbitrary otherwise. This way, annihilation and creation operators corresponding to different quasimodes always commute. The definition, however, only applies when the width of each cavity resonance is much smaller than their separation in frequency, which implies high-Q cavities.

Van der Plank and Suttorp [41] define a quasimode annihilation operator as Lang, Scully, and Lamb [39], except that they omit the requirement that the frequency interval must be much larger than the resonance width. This allows them to treat cavities of arbitrary Q, but their quasimode annihilation operators depend on position.

Bardroff and Stenholm [42] define quasimode annihilation operators as the linear combinations of the modes of the "universe" whose average value obey an equation of motion analogous to that followed by the electric field in class A lasers. Unlike the classical definition of mode of a lossy cavity, theirs depends on the existence of a gain medium and applies only in the specific case of a laser.

Grangier and Poizat [43] avoid the pitfalls of attempting to specify which linear combinations of the modes of the "universe" form a quasimode. Instead they assume that this problem was solved and that the modes of the "universe" can be split into two parts: cavity quasimodes and loss modes. Their main interest is developing a quantum theory of excess noise in lasers.

Knöll, Vogel, and Welsch [20] derive the propagators describing the time evolution of the electric field using the modes of the "universe" approach. From a coarse-graining approximation of these propagators, they define a quasimode annihilation operator. Within this approximation only times much larger than a cavity round-trip time but much smaller than the lifetime of a photon in a cavity mode are resolved. This is equivalent to the Lang-Scully-Lamb requirement on the interval of frequencies [39] and implies high-Q cavities.

In a series of papers on a nonperturbative approach to the decay of an atom in a cavity [44], Garraway and Knight have shown that the effect of the environment can be replaced by the notion of pseudomode. The pseudomode is determined by the poles in the complex plane of the frequency-dependent density-of-states function and has the properties of a finite-Q cavity mode. However, the relation between pseudomodes and the classical idea of mode as well as their connection with other definitions of quasimode has not been explored.

Dalton, Barnett, and Knight [45] define quasimodes as modes of the "universe" for idealized, rather than the actual, permittivity and permeability functions describing the cavity walls. The idea is that by choosing the idealized permittivity and permeability functions to *approach* those of perfect mirrors, the idealized modes of the "universe" will fall at least in two categories: the ones they call cavity quasimodes that practically vanish outside but are large inside, and what they call external quasimodes that almost vanish inside but are large outside. By adopting these idealized modes of the "universe" rather than the actual discrete Fox-Li modes of the cavity, they assure that their quasimodes remain orthogonal despite the fact that the cavity is leaky. But unlike the true modes of the "universe", the Hamiltonian for quasimodes is not diagonal: their quasimodes are coupled.

The authors mentioned so far have devised their own definitions of what a quasimode is, largely ignoring the classical solution to this problem (subsection 3.2). There are, nonetheless, three definitions of quasimode in the literature that build upon the progress achieved in the classical theory. Two of these, one due to Ujihara [46,47] and another due to Leung and collaborators [48], use the Thomson-Vaĭnshteĭn concept of mode[3]. The third definition, due to Lamprecht and Ritsch [50] uses the Fox-Li concept. Ujihara's quantum formalism, however, is constructed entirely with modes of the "universe". He uses the Thomson-Vaĭnshteĭn modes only for conciseness, to be able to refer to cavity modes without ambiguity. Leung and collaborators do construct a quantum formalism entirely based on Thomson-Vaĭnshteĭn modes. The connection between their work and the semiclassical theory developed by Fox-Li, Siegman and others is still unknown. Lamprecht and Ritsch, on the other hand, have built a quantum theory that almost carries the whole classical formalism through. However, they account for the damping of these modes by introducing an ad hoc Lindblad term [24] in their master equation. But there is no guarantee that the interaction between these Fox-Li modes and the reservoir is Markovian to allow this Lindblad description. In fact, there is good reason to expect the opposite[4].

4 A Simple Model of a Leaky Cavity

The key feature of a leaky cavity for the purposes of a quantum description is that it is an open system. This feature can be captured already in a simple one-dimensional model, where leakage is entirely due to a non-vanishing transmissivity. Our model is a modified version of a simple model of a one-dimensional cavity introduced by Ujihara [46] who analyzed it using a modes of the "universe" approach. In our model, the cavity is formed by a perfect mirror at $x = -L$ and a non-absorptive and non-dispersive dielectric extending all the the way from $x = 0$ to infinity and described by the permittivity

$$\epsilon(x) = \Theta(-x) + \Theta(x)n_d^2, \tag{10}$$

where n_d is the refractive index of the dielectric and $\Theta(x)$ is Heavyside's step function, which is unity for positive x and vanishes for negative x.

We consider only linearly polarized electromagnetic waves propagating in the x direction. The polarization of the electric field defines the y axis and that of the magnetic field, the z axis. To simplify the notation, we re-scale the fields

[3] Ujihara [46] refers to [49] where Vaĭnshteĭn's ideas are applied. Leung and collaborators [48] do not refer to either Vaĭnshteĭn or Thomson, but their definition of cavity modes as solutions of Helmholtz equation containing only outgoing waves from the cavity to the outside is clearly the Thomson-Vaĭnshteĭn concept.

[4] The Markovian approximation is good when the time scale of interest is much larger than the correlation time of the reservoir. In a one-dimensional cavity model, this correlation time is basically the round-trip time. Now excess noise only becomes non-negligible when the cavity modes start overlapping. As overlapping modes mean that the lifetime of a photon in the cavity is of the order of a round-trip time, the damping dynamics of such a photon is highly non-Markovian.

multiplying them by the square root of the transverse area in the yz plane. Then, Maxwell's equations take the following simple form

$$\frac{\partial}{\partial x}E(x,t) = -\frac{\partial}{\partial t}B(x,t), \tag{11a}$$
$$\frac{\partial}{\partial x}B(x,t) = -\frac{\epsilon(x)}{c^2}\frac{\partial}{\partial t}E(x,t). \tag{11b}$$

From (11b) and (11b), we obtain the following wave equation

$$\frac{\partial^2}{\partial x^2}E(x,t) - \frac{\epsilon(x)}{c^2}\frac{\partial^2}{\partial t^2}E(x,t) = 0. \tag{12}$$

5 Fox-Li Modes as Natural Modes

Hamel and Woerdman [51] applied the idea of Fox-Li mode to a one-dimensional model. The same notion of one-dimensional Fox-Li mode is also implicit in an earlier paper by Lugiato and Narducci [52]. In this section, we will show how one-dimensional Fox-Li modes can be derived in the same rigorous way as normal modes are derived: from a Sturm-Liouville treatment [32,53]. For normal modes, perfect cavity boundary conditions are applied. By using instead leaky cavity boundary conditions corresponding to natural modes, we show that a Sturm-Liouville treatment leads us naturally to Fox-Li modes and the Fox-Li inner-product.

Applying the standard method of separation of variables [32] to (12), we see that any solution of this equation can be written as a linear combination of the solutions $\chi(\omega,x)$ of the associated Helmholtz equation,

$$\frac{\partial^2}{\partial x^2}\chi(\omega,x) + \frac{\omega^2}{c^2}\epsilon(x)\chi(\omega,x) = 0, \tag{13}$$

with time-dependent coefficients $q(\omega,t)$ that obey a simple harmonic oscillator equation,

$$\ddot{q} + \omega^2 q = 0. \tag{14}$$

Usually, χ is defined in the whole space and has to satisfy the physical boundary conditions at $x=-L$, at $x=0$, and at infinity. The functions χ then form a basis that can represent any physical field configuration: the modes of the "universe". A different class of solutions χ of (13) that can also be used as a complete basis: the natural modes.

The natural modes for our one-dimensional cavity are solutions of (13) that satisfy the boundary conditions at $x=-L$ and $x=0$, but that contain only outgoing waves for $x>0$ (so they cannot satisfy the boundary condition at infinity). They are given by

$$g(c\kappa_n,x) = \begin{cases} e^{i\kappa_n x} + re^{-i\kappa_n x}, & \text{for } -L \le x < 0 \\ (1+r)e^{i\kappa_n n_d x}, & \text{for } x > 0 \end{cases} \tag{15}$$

where, unlike the modes of the "universe", the "frequency" $c\kappa_n$ is complex now and can no longer be interpreted as a physical frequency. As these modes are purely outgoing, they have to decay in time. Thus they are associated with the time dependence $\exp(-ic\kappa_n t)$ rather than with $\exp(-ick_n t)$ as an ordinary plane wave would.

As these natural cavity modes diverge at infinity, we will use them to represent the radiation field only in the cavity[5]. To cover the whole of space, we need to use external natural modes as well. These modes are solutions of (13) that satisfy the boundary condition at infinity and $x = 0$, but that contain only outgoing waves for $x < 0$ (so they cannot satisfy the boundary condition on the perfect mirror at $x = -L$, as that implies a reflected incoming wave). These solutions of (13) are given by

$$G(ck, x) = \begin{cases} (1-r)e^{-ikx}, & \text{for } x < 0 \\ e^{-ikn_d x} - re^{ikn_d x}, & \text{for } x > 0, \end{cases} \tag{16}$$

where k is real.

5.1 Sturm-Liouville with a Twist

To be able to use natural modes as a basis for the radiation field, we must have an inner product and show that such a basis is complete. Analogously to the case of normal modes [53], we obtain the appropriate inner product by first deriving an orthogonality relation. To find this relation, consider the following function that vanishes at the borders of our cavity (i.e., both at $x = -L$ and $x = 0$)

$$\begin{aligned} \zeta_{n,m}(x) = {} & g(c\kappa_n, x)\frac{1}{\kappa_m}\frac{\partial}{\partial x}\tilde{g}^*(c\kappa_m^*, x) \\ & - \tilde{g}^*(c\kappa_m^*, x)\frac{1}{\kappa_n}\frac{\partial}{\partial x}g(c\kappa_n, x), \end{aligned} \tag{17}$$

where

$$\tilde{g}(c\kappa_m^*, x) = \begin{cases} e^{i\kappa_m^* x} + \frac{e^{-i\kappa_m^* x}}{r}, & \text{for } -L \le x < 0 \\ \frac{1+r}{r}e^{-i\kappa_m^* n_d x}, & \text{for } x > 0 \end{cases} \tag{18}$$

is the adjoint of $g(c\kappa_m, x)$, i.e. the solution of (13) of "frequency" $c\kappa_m^*$ that satisfies the boundary conditions at the cavity and contains only incoming waves at infinity [32]. Differentiating (17) with respect to x, using (13), and

[5] Even though these mode functions diverge at infinity, it is still possible to expand the fields in the *whole of space* in terms of them, provided the initial excitation conditions are explicitly taken into account (see Beck and Nussenzveig [36]). However, such approach is not quite suitable for quantum electrodynamics.

then integrating the result over the cavity, we find

$$(\kappa_n - \kappa_m) \quad \int_{-L}^{0} dx \Big\{ n^2(x)\tilde{g}^*(c\kappa_m^*, x) g(c\kappa_n, x) + \frac{1}{\kappa_m}\frac{\partial}{\partial x}\tilde{g}^*(c\kappa_m^*, x)\frac{1}{\kappa_n}\frac{\partial}{\partial x} g(c\kappa_n, x)\Big\} = 0, \tag{19}$$

where $n(x) = \sqrt{\epsilon(x)}$ is the refractive index. Strictly speaking, (19) is a biorthogonality relation rather than an orthogonality relation, because it involves adjoint modes. Non-hermitean eigenproblems as this one yield biorthogonal instead of orthogonal eigenfunctions [32]. Another important point is that, unlike the orthogonality relation for normal modes [53], (19) involves spatial derivatives of the mode functions. This is what forces us to distinguish the direction of propagation when an inner product based on (19) is introduced. This is what leads to the Fox-Li traveling wave representation of these modes.

The mode functions $g(c\kappa_m, x)$ and $\tilde{g}(c\kappa_m^*, x)$ are related to the electric field. Their spatial derivatives $(-i/\kappa_n)\partial g(c\kappa_n, x)/\partial x$ and $(i/\kappa_m^*)\partial \tilde{g}(c\kappa_m^*, x)/\partial x$ are related to the product of the magnetic field by the speed of light. Now $E + cB$ is a purely forward propagating wave in the cavity, while $E - cB$ is a purely backward propagating wave, as can be seen from (11b) and (11b). This suggests that equation (19) should be rewritten in the following form which is completely analogous to the orthogonality relation for normal modes [53]

$$(\kappa_n - \kappa_m) \int_{-L}^{0} dx\, \tilde{\mathcal{G}}^\dagger(c\kappa_m^*, x) \cdot \mathcal{G}(c\kappa_n, x) n^2(x) = 0, \tag{20}$$

where

$$\mathcal{G}(c\kappa_n, x) = \frac{1}{\sqrt{8L}} \begin{bmatrix} g(c\kappa_n, x) - \frac{i}{n(x)\kappa_n}\frac{\partial}{\partial x} g(c\kappa_n, x) \\ g(c\kappa_n, x) + \frac{i}{n(x)\kappa_n}\frac{\partial}{\partial x} g(c\kappa_n, x) \end{bmatrix} \tag{21}$$

and

$$\tilde{\mathcal{G}}(c\kappa_m^*, x) = \frac{1}{\sqrt{8L}} \begin{bmatrix} \tilde{g}(c\kappa_m^*, x) - \frac{i}{n(x)\kappa_m^*}\frac{\partial}{\partial x} \tilde{g}(c\kappa_m^*, x) \\ \tilde{g}(c\kappa_m^*, x) + \frac{i}{n(x)\kappa_m^*}\frac{\partial}{\partial x} \tilde{g}(c\kappa_m^*, x) \end{bmatrix} \tag{22}$$

are the new mode and adjoint functions, respectively. Their upper components are forward propagating waves and the lower, backward propagating waves. The factor of $1/\sqrt{8L}$ was introduced so that

$$\int_{-L}^{0} dx\, \tilde{\mathcal{G}}^\dagger(c\kappa_m^*, x) \cdot \mathcal{G}(c\kappa_n, x) n^2(x) = \delta_{n,m}. \tag{23}$$

Thus the proper inner product under which the modes $\mathcal{G}$ are orthogonal to their adjoints $\tilde{\mathcal{G}}$ is

$$(\mathbf{F}_1, \mathbf{F}_2) = \int_{-L}^{0} dx\, \mathbf{F}_1^\dagger(x) \cdot \mathbf{F}_2(x) n^2(x), \tag{24}$$

where $\mathbf{F}_1$ and $\mathbf{F}_2$ are two "spinor fields". Associating E to g and cB to $-(i/\kappa_n)\partial g/\partial x$ in (21), we construct a "spinor field" $\mathbf{F}$ from given electric and magnetic fields as follows

$$\mathbf{F}(x) = \frac{1}{2}\begin{bmatrix} E(x) + \frac{c}{n(x)}B(x) \\ E(x) - \frac{c}{n(x)}B(x) \end{bmatrix}, \tag{25}$$

where the factor 1/2 is introduced just for later convenience.

These new mode functions are, in fact, the one-dimensional Fox-Li modes introduced by Hamel and Woerdman, which we have now put in the same Sturm-Liouville context as ordinary normal modes. The Fox-Li inner product, which seemed to appear out of the blue before, now follows naturally from Eq. (20).

To expand the field outside the cavity, we introduce external Fox-Li modes $\mathbf{G}(ck,x)$ and their adjoints $\tilde{\mathbf{G}}(ck,x)$ in an analogous way. They are defined just as (21) and (22) with the discrete normalization factor $1/\sqrt{8L}$ replaced by the continuum one $1/\sqrt{8\pi n_d}$, $g(c\kappa_n,x)$ replaced by $G(ck,x)$, $\tilde{g}(c\kappa_n^*,x)$ by $\tilde{G}(ck,x)$, and κ_n by k. The external Thomson adjoint modes $\tilde{G}(ck,x)$ are incoming solutions of Helmholtz equation. The external biorthogonality relation is a version of (23) for the continuum with the Kronecker delta replaced by a Dirac delta function.

It can be shown [8] that, even though none of the functions g, $\tilde{g}$, G, $\tilde{G}$ satisfies all the boundary conditions, completeness can be achieved, if we allow each side of the interface to have contributions from both outgoing modes and incoming adjoints.

6 Quantum Theory

As these modes form a complete set and have a well defined inner-product, they are just as good as normal modes. So we can use them to expand the quantized electromagnetic radiation field. These calculations are presented in detail in our recent paper [8]. Rather than repeating them here, we will only outline the key points. The expansion coefficients are operators in the quantum theory. The commutation relations obeyed by these new operators follow from the commutation relation[6] between $\hat{E}$ and $\hat{B}$

$$[\hat{E}(x), \hat{B}(x')] = i\frac{\hbar}{\epsilon_0\epsilon(x)}\frac{\partial}{\partial x'}\{\delta(x-x') - \delta(x+x'+2L)\} \tag{26}$$

and the inner product. The Hamiltonian describing the time evolution of these new operators can be obtained by substituting the mode expansions in the expression for the total energy

$$\hat{H} = \epsilon_0\int_{-L}^{\infty} dx\,\hat{\mathbf{F}}^\dagger(x)\cdot\hat{\mathbf{F}}(x)n^2(x), \tag{27}$$

[6] The last delta function in (26) is a consequence of the perfect mirror at the origin [54].

where $\hat{F}$ is given by (25) with E and B replaced by $\hat{E}$ and $\hat{B}$, respectively. This Hamiltonian assumes the following form

$$\hat{H} = \hat{H}_{\text{cav}} + \hat{H}_{\text{out}}, \tag{28}$$

where $\hat{H}_{\text{cav}}$ contains only cavity mode operators and their adjoints, whereas $\hat{H}_{\text{out}}$ contains only external mode operators and their adjoints.

Equation (28) appears to show that $\hat{H}_{\text{cav}}$ and $\hat{H}_{\text{out}}$ are uncoupled. They are not, however. The coupling is hidden on the noncommutability between cavity and external operators. As the outside plays the role of a reservoir where the energy stored in the cavity is damped, this gives rise to a very unusual model for quantum dissipation where dissipative-system and reservoir operators no longer commute[7].

7 Conclusions

Unifying the concept of normal modes used in quantum optics and that of Fox-Li modes from semiclassical laser physics, we have developed a quantum theory of dissipation in a leaky cavity of arbitrary Q. Our approach is based on the view that Thomson's idea of natural modes of oscillation [35] is the proper generalization of the concept of mode for an open system (see section 5). In this theory, the small system is the set of damped harmonic oscillators associated with the exact discrete cavity Fox-Li modes rather than the normal modes of a fictitious perfect cavity. The reservoir is the set of harmonic oscillators associated with the exact continuous external Fox-Li modes (a concept introduced here). Our Hamiltonian is obtained from first principles and has two distinct features that set it apart from conventional models used to describe dissipation in quantum mechanics. First, the modes associated with each oscillator are not orthogonal but rather biorthogonal to a set of adjoint modes. Second, the operators of the small system do not commute with the reservoir operators as in standard approaches.

The theory presented here can also be regarded as a first step towards a full quantum description of excess noise in lasers that is attracting growing interest [42,50,56]: to construct a quantum theory that builds on Siegman's semiclassical discovery [10] that excess noise is related to the nonorthogonality of true cavity modes, one must first deal with the problem, addressed here, of describing the dissipation of such modes quantum mechanically.

Finally, we would like to point out that a similar situation arises in the problem of tunneling of electrons through a finite barrier in the center of a one-dimensional potential well, where the Hamiltonians corresponding to the right and left sides of the barrier do not commute either [57]. Recently, Fransson and collaborators [58] re-examined this problem using a different method,

[7] For a derivation of the well-known Gardiner-Collett Hamiltonian as a high-Q approximation to the exact description, please see [55].

and showed that the separation into auxiliary subsystems unavoidably leads to eigenbases that are nonorthogonal to each other. However, as far as we know, a quantum theory of damping where this nonorthogonality occurs between system and reservoir has never been proposed before.

References

1. A. G. Fox and T. Li: Proc. IRE **48**, 1904 (1960) A. G. Fox and T. Li: Bell Sys. Tech. J. **40**, 453 (1961)
2. A. E. Siegman: Appl. Phys. Lett. **36**, 412 (1980)
3. E. Fermi: Rev. Mod. Phys. **4**, 87 (1932)
4. I. R. Senitzky: Phys. Rev. **115**, 227 (1959) I. R. Senitzky: Phys. Rev. **119**, 670 (1960)
5. H. B. Callen and T. A. Welton: Phys. Rev. **83**, 34 (1951)
6. J. Weber: Phys. Rev. **90**, 977 (1953)
7. M. J. Collett and C. W. Gardiner: Phys. Rev. A **30**, 1386 (1984) C. W. Gardiner and M. J. Collett: Phys. Rev. A **31**, 3761 (1985) C. W. Gardiner: *Quantum Noise* (Springer-Verlag, Berlin 1991)
8. S. M. Dutra and G. Nienhuis: Phys. Rev. A **62**, 063805 (2000)
9. K. Petermann: IEEE J. Quantum Electron. **15**, 566 (1979)
10. A. E. Siegman: Phys. Rev. A **39**, 1253 (1989) A. E. Siegman: Phys. Rev. A **39**, 1264 (1989)
11. H. Helmholtz: J. f. d. reine u. Math. **100**, 137 (1887) P. Havas: Sup. Nuovo Cimento **5**, 363 (1957) C. Lanczos: *The Variational Principles of Mechanics* (University of Toronto Press, Toronto 1952)
12. H. Bateman: Phys. Rev. **38**, 815 (1931) P. M. Morse and H. Feshbach: *Methods of Theoretical Physics* (McGraw-Hill, New York 1953) H. Feshbach and Y. Tikochinsky: Trans. N. Y. Acad. Sci. **38**, 44 (1977) F. Riewe: Phys. Rev. E **53**, 1890 (1996)
13. E. Kanai: Progr. Theoret. Phys. (Kyoto) **3**, 440 (1948) W. E. Brittin: Phys. Rev. **77**, 396 (1950) E. H. Kerner: Can. J. Phys. **36**, 371 (1958) W. K. H. Stevens: Proc. Phys. Soc. (London) **72**, 1027 (1958) V. W. Myers: Am. J. Phys. **27**, 507 (1959)
14. J. R. Ray: Am. J. Phys. **47**, 626 (1979)
15. H. Dekker: Phys. Rep. **80**, 1 (1981)
16. W. H. Zurek: Phys. Today **44**, 36 (1991) W. H. Zurek: Phys. Rev. D **24**, 1516 (1981) W. H. Zurek: Phys. Rev. D **26**, 1862 (1982) W. H. Zurek, S. Habib, and J. P. Paz: Phys. Rev. Lett. **70**, 1187 (1993)
17. W. Heisenberg: *Physics and Philosophy* (Harper & Row, New York 1962) pp. 177–178 R. Omnès: Rev. Mod. Phys. **64**, 339 (1992) R. Omnès: *The Interpretation of Quantum Mechanics* (Princeton University Press, Princeton 1994) E. Wigner: Am. J. Phys. **31**, 6 (1963) K. Gottfried: *Quantum Mechanics I* (W. A. Benjamin inc., New York 1966) pp. 165–232
18. B. M. Garraway and P. L. Knight: Phys. Rev. A **49**, 1266 (1994) B. M. Garraway and P. L. Knight: Phys. Rev. A **50**, 2548 (1994) L. Gilles, B. M. Garraway, and P. L. Knight: Phys. Rev. A **49**, 2785 (1994) C. Gerry and E. E. Hach III: Phys. Lett. A **174**, 185 (1993)
19. S. M. Dutra: Eur. J. Phys. **18**, 194 (1997)

20. L. Knöll, W. Vogel, and D. G. Welsch, Phys. Rev. A **43**, 543 (1991) L. Knöll and D.-G. Welsch: Prog. Quant. Electr. **16**, 135 (1992) W. Vogel and D.-G. Welsch: *Lectures on quantum optics* (Akademie Verlag, Berlin 1994) Chap. 8
21. R. J. Glauber: Phys. Rev. **131**, 2766 (1963) E. C. G. Sudarshan: Phys. Rev. Lett. **10**, 277 (1963)
22. J. R. Klauder, J. McKenna, and D. G. Currie: J. Math. Phys. **6**, 734 (1965)
23. H. M. Nussenzveig: *Introduction to Quantum Optics* (Gordon and Breach, London 1973), pp. 54–69
24. G. Lindblad: Commun. Math. Phys. **48**, 119 (1976)
25. R. H. Dicke, Molecular amplification and generation systems and methods, U.S. Patent 2 851 652, 9 September 1958, reprinted in *Lasers: A collection of reprints with commentary* ed. J. Weber (Gordon and Breach, New York, 1968) p. 645
26. A. L. Schawlow and C. H. Townes: Phys. Rev. **29**, 1940 (1958)
27. A. M. Prokhorov: Sov. Phys. JETP **7**, 1140 (1958)
28. S. R. Barone: J. Appl. Phys. **34**, 831 (1963)
29. A. E. Siegman: Opt. Commun. **31**, 369 (1979)
30. A. E. Siegman: *Lasers* (Oxford University Press, Oxford 1987)
31. M. A. Rippin and G. H. C. New: J. Mod. Opt. **43**, 993 (1996)
32. P. M. Morse and H. Feshbach: *Methods of Theoretical Physics* (McGraw-Hill, New York 1953)
33. W. E. Lamb Jr.: Phys. Rev. **134**, A1429 (1964)
34. P. Meystre and M. Sargent: *Elements of quantum optics* (Springer-Verlag, Berlin 1991)
35. J. J. Thomson: Proc. Lond. Math. Soc. **15**, 197 (1884) J. J. Thomson: *Notes on Recent Researches in Electricity and Magnetism: intended as a sequel to Professor Clerk-Maxwell's treatise on Electricity and Magnetism* (Clarendon Press, Oxford 1893) A. Sommerfeld: *Partial Differential Equations in Physics* (Academic Press, New York 1949), pp. 214–224 (this is Sommerfeld's pedagogic account of the original problem of a lossy electromagnetic resonator solved by Thomson)
36. G. Beck and H. M. Nussenzveig: Nouvo Cimento **16**, 416 (1960) H. M. Nussenzveig: *Causality and Dispersion Relations* (Academic Press, New York 1972) Chap. 4 R. Peierls: *More Surprises in Theoretical Physics* (Princeton University Press, Princeton 1991) pp. 76–82
37. L. A. Vaĭnshteĭn: J. Tech. Phys. **34**, 1541 (1964) L. A. Vaĭnshteĭn: High-Power Electronics **4**, 157 (1965) L. A. Weinstein: *Open Resonators and Open Waveguides* (The Golem Press, Boulder 1969)
38. L. Ronchi: 'Optical Resonators'. In: *Laser Handbook I.* ed. by F. T. Arecchi and E. O. Schulz-DuBois (North-Holland, Amsterdam 1972) pp. 151–190
39. R. Lang, M. O. Scully, and W. E. Lamb Jr.: Phys. Rev. A **7**, 1788 (1973)
40. J. Gea-Banacloche, N. Lu, L. M. Pedrotti, S. Prasad, M. O. Scully, and K. Wódkiewicz: Phys. Rev. A **41**, 369 (1990) J. Gea-Banacloche, N. Lu, L. M. Pedrotti, S. Prasad, M. O. Scully, and K. Wódkiewicz: Phys. Rev. A **41**, 381 (1990)
41. R. W. F. van der Plank and L. G. Suttorp: Phys. Rev. A **53**, 1791 (1996) R. W. F. van der Plank: Ph.D. thesis (University of Amsterdam, Amsterdam 1997)
42. P. J. Bardroff and S. Stenholm: Phys. Rev. A **60**, 2529 (1999) P. J. Bardroff and S. Stenholm: Phys. Rev. A **61**, 023806 (2000)
43. P. Grangier and J.-P. Poizat: Eur. Phys. J. D **1**, 97 (1998) P. Grangier and J.-P. Poizat: Eur. Phys. J. D **7**, 99 (1999)
44. B. M. Garraway and P. L. Knight: Phys. Rev. A **54**, 3592 (1996) B. M. Garraway: Phys. Rev. A **55**, 2290 (1997) B. M. Garraway: Phys. Rev. A **55**, 4636 (1997)

45. B. J. Dalton, S. M. Barnett, and P. L. Knight: J. Mod. Opt. **46**, 1315 (1999)
46. K. Ujihara: Phys. Rev. A **12**, 148 (1975)
47. K. Ujihara: Phys. Rev. A **20**, 1096 (1979) X.-P. Feng and K. Ujihara: IEEE J. Quantum Electron. **25**, 2332 (1989) X.-P. Feng and K. Ujihara: Phys. Rev. A **41**, 2668 (1990)
48. P. T. Leung, S. S. Tong, and K. Young: J. Phys. A **30**, 2139 (1997) P. T. Leung, S. S. Tong, and K. Young: J. Phys. A **30**, 2153 (1997) P. T. Leung, W. M. Suen, C. P. Sun, and K. Young: Phys. Rev. E **57**, 6101 (1998) E. S. C. Ching, P. T. Leung, A. M. V. D. Brink, W. M. Suen, S. S. Tong, and K. Young: Rev. Mod. Phys. **70**, 1545 (1998)
49. H. Ogura, Y. Yoshida, I. Iwamoto: J. Phys. Soc. Japan **22**, 1421 (1967)
50. C. Lamprecht and H. Ritsch: Phys. Rev. Lett. **82**, 3787 (1999)
51. W. A. Hamel and J. P. Woerdman: Phys. Rev. A **40**, 2785 (1989) W. A. Hamel: Ph.D. thesis (Leiden University, Leiden, The Netherlands 1991)
52. L. A. Lugiato and L. M. Narducci, Z. Phys. B **71**, 129 (1988)
53. D. J. Santos and R. Loudon: Phys. Rev. A **52**, 1538 (1995)
54. P. W. Milonni: Phys. Rev. A **25**, 1315 (1982) E. A. Power and T. Thirunamachandran: Phys. Rev. A **25**, 2473 (1982) S. M. Barnett, C. R. Gilson, B. Huttner, and N. Imoto: Phys. Rev. Lett. **77**, 1739 (1996)
55. S. M. Dutra and G. Nienhuis: J. Opt. B **2**, 584 (2000) S. M. Dutra and G. Nienhuis: Acta Phys. Slov. **50**, 275 (2000)
56. P. Goldberg, P. W. Milonni, and B. Sundaram: J. Mod. Opt. **38**, 1421 (1991) P. Goldberg, P. W. Milonni, and B. Sundaram: Phys. Rev. A **44**, 1969 (1991) P. Goldberg, P. W. Milonni, and B. Sundaram: Phys. Rev. A **44**, 4556 (1991) P. Grangier and J.-P. Poizat: Eur. Phys. J. D **1**, 97 (1998) P. Grangier and J.-P. Poizat: Eur. Phys. J. D **7**, 99 (1999)
57. R. E. Prange: Phys. Rev. **131**, 1083 (1963)
58. J. Fransson, O. Eriksson, B. Johansson, and I. Sandalov: Physica B **272**, 28 (1999)

The Quantum Jumps Approach for Infinitely Many States

D. Spehner[1] and J. Bellissard[2,3]

[1] Pontificia Universidad Católica de Chile, Facultad de física, Casilla 306, Santiago 22, Chile, dspehner@alfven.fis.puc.cl
[2] Université Paul Sabatier, Institut de Recherche sur les Systèmes Atomiques ou Moléculaires Complexes, F-31062 Toulouse cedex 4, France, Jean.Bellissard@irsamc.ups-tlse.fr
[3] Institut Universitaire de France

Abstract. A new class of models describing the dissipative dynamics of an open quantum system by means of random time evolutions of its wavefunction is considered. The random evolutions are linear and defined by Poisson processes. At the random Poissonian times, the wavefunction experiences discontinuous changes (quantum jumps). These changes are implemented by some non unitary operators satisfying a locality condition. The stochastic dynamics gives the Lindblad master equation back after averaging over the random times. If the Hilbert space of the system is infinite dimensional, the models involve an infinite number of independent Poisson processes and the total frequency of jumps is infinite. We show that the random evolutions are then given by some almost-surely defined unbounded random evolution operators obtained by a limit procedure. The relevance of the models in the field of electronic transport in Anderson insulators is briefly discussed.

1 Introduction

The dissipative dynamics of an open quantum system S can be described in two different ways. The first and most popular approach consists in coupling S with a reservoir R. The density matrix ρ_{tot} of the total system $S + R$ is assumed to follow a Liouville–von Neumann equation, i.e., one assumes that $S + R$ is a closed system. A state of S is specified by the reduced density matrix ρ, defined as the partial trace of ρ_{tot} over the reservoir's Hilbert space. ρ does not describe a single system but a statistical ensemble. By tracing out the degrees of freedom of R in the Liouville–von Neumann equation, one obtains an integro–differential equation for ρ (Nakajima–Zwanzig equation). Using a suitable Markov approximation to eliminate memory effects, this equation is then transformed into a simpler first-order linear differential equation, called the *master equation* [1]. The reduced dynamics does not conserve pure states. It has been shown by Lindblad [2] that the Markovian master equation is of the form:

$$\frac{\mathrm{d}\rho}{\mathrm{d}t} = \mathcal{L}\rho = -\mathrm{i}[H_S, \rho] + \frac{1}{2}\sum_{\ell} \left([L_\ell \, \rho, L_\ell^\dagger] + [L_\ell, \rho L_\ell^\dagger] \right) \; . \tag{1}$$

M. Orszag, J.C. Retamal (Eds.): LNP 575, pp. 355–376, 2001.

H_S is the Hamiltonian of S (including the energy shifts due to the coupling with the reservoir), and L_ℓ are some operators acting on the Hilbert space $\mathcal{H}_S$ of S, called the Lindblad operators in the sequel. An alternative approach for the same problem is based on stochastic evolutions of pure states. The state of S is specified by a random wavefunction (RW) in $\mathcal{H}_S$, evolving according to a linear or nonlinear *stochastic Schrödinger equation*. Different stochastic evolutions have been proposed in the last two decades in various fields of physics and mathematics, especially Quantum Optics [3,4], Quantum Measurement Theory [5,6,7,8,9,10,11,12,13,14] and Electronic Transport in Solids [15,16,17]. Consistency with the master equation approach requires that the pure state evolution gives the density matrix evolution back after averaging over the dynamical noise. Apart from being intuitively appealing, the RW models provide quite efficient tools for solving master equations numerically. Actually, one is led to integrate N coupled differential equations for the wavefunction, where N is the dimension of $\mathcal{H}_S$, for a large enough number of realizations of the dynamical noise. For large N, this is generally much more efficient than integrating the $N \times N$ coupled master equations for the density matrix. However, the RW models are more than simple mathematical or numerical tools: they describe the real evolution of the system S under continuous monitoring by means of measurements (photons counting, homodyne or heterodyne detections) [18]. The randomness of their dynamics is a consequence of our ignorance of the result of a measurement in quantum mechanics. At the end of the eighties, experiments on the fluorescence of single ions in magnetic traps have shown records of 'quantum jumps' between an excited atomic state and a lower state, occurring at random times [18]. These sudden jumps – which were already assumed to exist by Einstein in his paper on the A and B atomic coefficients [19] – correspond to the absorption or emission of a photon by the ion at the corresponding transition. Such direct observations have motivated the study of the RW models in Quantum Optics.

The aim of this paper is to study a particular quantum jumps model in the limit where infinitely many levels of the system are coupled by the dissipative dynamics. The model is built in such a way as to give the classical kinetic theory back, such as Boltzmann's equation, when quantum effects can be neglected. Unlike in the models studied in [3,4,6], the stochastic evolution for the wavefunction is linear, which makes the mathematical analysis easier. The price we pay for this convenience is the non conservation of the norm of the random wavefunction.

Many systems with non-degenerate spectra are correctly described by the master equation (1) with $\ell = (i,j)$, $i \neq j$, and:

$$\hat{L}_{i \to j} = \sqrt{\Gamma_{i \to j}}\, a_j^\dagger a_i \; . \tag{2}$$

$a_i^\dagger, a_i$ are the creation and annihilation operators in the one–particle energy eigenstate $|i\rangle$. $\Gamma_{i \to j}$ is the one–particle transition rate from $|i\rangle$ to $|j\rangle$. It can be calculated perturbatively by means of Fermi golden rule [20]. Examples of systems correctly described by this master equation are: (1) an atom coupled to the quantized electromagnetic field radiating in the vacuum [20]; (2) independent electrons in strongly disordered solids coupled to phonons at very low temperatures [21]. The inverse life-time of eigenstate $|i\rangle$ is $\Gamma_i = \sum_{j\neq i} \Gamma_{i\to j}$. If we ignore level shifts, the dynamics of S is completely characterized by its Hamiltonian H_S and by the set of the transition rates $\Gamma_{i\to j}$ for all pairs (i, j) of one–particle eigenstates. In the RW model studied in this paper, the rates $\Gamma_{i\to j}$ give the probability per unit time of occurrence of quantum jumps. For an infinite dimensional Hilbert space $\mathcal{H}_S$, it usually happens that the double sum $\sum_{i,j} \Gamma_{i\to j}$ is infinite, i.e., that an infinite number of jumps occur in any finite time interval. Our main result shows that, provided the discontinuous changes of the wavefunction at the jumps are sufficiently 'local', the stochastic dynamics in $\mathcal{H}_S$ is also well–defined in this case, and is given by an almost surely unbounded random evolution operator obtained by a limit procedure.

The physical situation which motivates our work is electronic transport in strongly disordered solids [22]. The electronic eigenfunctions are exponentially localized in such solids (Anderson localization [23]). The electrical conductivity thus vanishes at zero temperature. At non zero temperature $T > 0$, transport occurs via phonon–assisted hopping of electrons from one localized eigenstate into another. At small T, a phenomenological argument due to Mott [24] shows that the hopping conductivity σ is given by $\sigma = \sigma_0 \exp(-(T_0/T)^\gamma)$, where the exponent $\gamma = 1/(d+1)$ depends on the dimension d ($d = 1, 2, 3$) only and T_0 is a constant which depends on the localization length and the density of states at the Fermi energy. Efros and Shklovskii have shown that the presence of a wide Coulomb gap in the density of states modifies γ, which then equals $1/2$ for all dimensions [22]. Hopping transport occurs for instance in lightly doped compensated semiconductors at low temperature, in amorphous solids, in two–dimensional electron gases in zero or strong magnetic field, and in the quasicrystal i–AlPdRe [22,25]. The electrons in the disordered potential created by the ions, impurities or defects are coupled to low energy acoustic phonons. Since phonons do not carry current, the study of transport requires the knowledge of the electron dynamics only. The system S of all electrons is thus an open quantum system. If we ignore electron–electron interactions, it can be shown that, at low enough temperature, its dissipative dynamics is correctly described by the master equation (1) with the Lindblad operators (2) [21]. The transition rate $\Gamma_{i\to j}$ between two localized one–electron eigenstates $|i\rangle$ and $|j\rangle$ decreases exponentially with the distance $|i-j|$ between their localization centers, and depends strongly at low temperature on their energies E_i and E_j. The widely used relaxation time approximation, which amounts to replace all the $\Gamma_{i\to j}$'s by a single damping constant, is thus completely unjustified in hopping transport.

A fundamental question addressed by mathematical physicists in the theory of solids concerns the study of the spectrum and of the time evolution of the relevant electronic observables at the thermodynamic limit. Letting the volume of a strongly disordered solid tend to infinity, an infinite number of localized eigenstates $|i\rangle$ with energies close to the Fermi energy E_F comes into play. Thus, a well–defined RW dynamics in an infinite dimensional Hilbert space $\mathcal{H}_S$ must be specified. The quantum jump model studied in this paper is a candidate for such a RW dynamics in solids, giving the time evolution of the electrons under electron–phonon or another coupling.

2 The Model

2.1 The stochastic scheme

Let us consider a quantum system composed of identical non-interacting particles coupled to their environment. We denote by H the one–particle Hamiltonian, acting on the one–particle Hilbert space $\mathcal{H}$. $\hat{H}$ is the second quantized of H, acting on the Fock space $\mathcal{F} = \oplus_{n=0}^{\infty} P_{\pm}\mathcal{H}^{\otimes n}$, where $P_{\pm}$ is the particle (anti–)symmetrizer. We shall assume that $H = V + T$ is the sum of a possibly unbounded self–adjoint operator V, with dense domain $\mathcal{D}(H)$ and pure point spectrum, and of a bounded self–adjoint operator T. Let $\{|i\rangle; i \in \Lambda\}$ be the orthonormal basis of $\mathcal{H}$ formed by the eigenfunctions of V, where Λ is an infinite subset of $\mathbb{Z}^d$. If the system is a doped semiconductor, we can think of Λ as the impurity sites in the host crystal $\mathbb{Z}^3$; then H acts on the Hilbert space $\mathcal{H} = \ell^2(\Lambda)$ and $\{|i\rangle; i \in \Lambda\}$ is the canonical basis (see below). The annihilation and creation operators of a particle in state $|i\rangle$, $i \in \Lambda$, are denoted by a_i and $a_i^{\dagger}$, respectively. For each pair $(i, j) \in \Lambda^2$, $|i - j|$ is the Euclidean distance between $i \in \mathbb{Z}^d$ and $j \in \mathbb{Z}^d$. *Instantaneous jumps* take place at some random times

$$0 \le t_{i\to j}^1 \le \ldots \le t_{i\to j}^m \le \ldots \quad , \quad i, j \in \Lambda \, .$$

These jumps are labelled by pairs $(i, j) \in \Lambda^2$ of indices and by an integer m, which counts the number of jumps (i, j) that occurred since the initial time $t = 0$. The time delays between two consecutive jumps (i, j), $s_{i\to j}^m = t_{i\to j}^m - t_{i\to j}^{m-1}$, $m = 2, 3, \ldots$, and the first jump times $s_{i\to j}^0 = t_{i\to j}^1$, for all $i, j \in \Lambda$, are assumed to be mutually independent random variables distributed according to the exponential law $p(ds) = \Gamma_{i\to j} e^{-s\,\Gamma_{i\to j}} ds$, where $\Gamma_{i\to j} \ge 0$ depends on (i, j) but not on m. In other words, for any fixed (i, j), the jump times $t_{i\to j}^m$, $m \in \mathbb{N}^\star$ are given by a Poisson process with parameter $\Gamma_{i\to j}$. The *transition rates* $\Gamma_{i\to j}$ are considered here as phenomenological parameters. In concrete situations, they can be computed by using the Fermi golden rule. They contain all the quantitative physical information on the interaction of the system's particles with the environment (e.g., the coupling constant). If the environment is a thermal bath, they depend on its temperature.

Each jump modifies in a discontinuous way the wavefunction of the system in the Fock space $\mathcal{F}$. These discontinuous changes are implemented by some bounded operators $\hat{W}_{i\to j}$ (called *jump operators* in the sequel). More precisely, if the system's wavefunction is $|\psi\rangle \in \mathcal{F}$ just before a jump (i, j), it becomes $\hat{W}_{i\to j}|\psi\rangle$ just after it:

$$\text{jump } (i, j) : \qquad |\psi\rangle \to \hat{W}_{i\to j}|\psi\rangle \, . \tag{3}$$

The jump operators describe the qualitative effect of the interaction between S and its environment (e.g., the effect on S of the absorption or the emission of external particles such as photons, phonons,...). They do not depend on the damping rates or on the temperature of the bath.

Between two consecutive jumps, the system evolves according to the Schrödinger equation with Hamiltonian $\hat{H} + \hat{K}$, where $\hat{K}$ is a bounded operator describing both some damping effects and some energy renormalizations due to the coupling with the environment. We will restrict ourself in this work to systems with a norm–preserving average dynamics, namely, such that $\mathbb{E}\,\|\psi(t)\|^2 = 1$, where $\mathbb{E}$ is the average over all times $t^m_{i\to j}$. As we shall see below, in order that $\|\psi(t)\|^2$ be conserved in average, the damping operator $\hat{K}$ must be given, *up to a self–adjoint operator*, by:

$$\hat{K} = \hat{K}_1 \equiv \frac{1}{2\mathrm{i}} \sum_{i,j\in\Lambda} \Gamma_{i\to j}\big(\hat{W}^\dagger_{i\to j} + 1\big)\big(\hat{W}_{i\to j} - 1\big)\ . \tag{4}$$

This means that $\hat{K}$ is *not* self–adjoint.

The wavefunction at time t, $t_p \le t < t_{p+1}$, is thus formally given by:

$$\begin{aligned}|\psi(t)\rangle &= e^{-\mathrm{i}(t-t_p)(\hat{H}+\hat{K})}\hat{W}_{i_p\to j_p} e^{-\mathrm{i}(t_p-t_{p-1})(\hat{H}+\hat{K})}\dots \\ &\dots \hat{W}_{i_1\to j_1} e^{-\mathrm{i}t_1(\hat{H}+\hat{K})}|\psi\rangle\ ,\end{aligned} \tag{5}$$

where $|\psi\rangle \in \mathcal{F}$ is the wavefunction at time $t = 0$, $0 \le t_1 \le \ldots \le t_p \le \ldots$ are the times of occurrence of *any* jump, and (i_p, j_p) is the random pair of indices corresponding to the jump that takes place at time t_p. As it will be clear below, the formula (5) is only meaningful if:

$$\Gamma \equiv \sum_{i,j\in\Lambda} \Gamma_{i\to j} < \infty\ . \tag{6}$$

A way to define the random wavefunction when $\Gamma = \infty$ will be given in the next section.

From a mathematical point of view, it is convenient to represent each sequence of random times $(t^m_{i\to j})_{m\in\mathbb{N}^\star}$ by a counting process $(N_{i\to j}(t))_{t\ge 0}$ [26]. Here $t^m_{i\to j}$ are the left discontinuities of the counting function $N_{i\to j}(t)$. For any compact interval $I \subset \mathbb{R}_+$, $N_{i\to j}(I)$ is the (random) number of jumps (i,j) occurring at times $t \in I$, namely:

$$N_{i\to j}(I) = \sum_{m=1}^{\infty} \chi(t^m_{i\to j} \in I)\ , \tag{7}$$

where χ is the characteristic function ($\chi(\mathcal{P}) = 1$ if the property $\mathcal{P}$ is true, 0 otherwise). We set $N_{i\to j}(t) = N_{i\to j}([0,t])$. The stochastic scheme described above is thus specified by the infinite set of independent Poisson processes $(N_{i\to j}(t))_{t\geq 0}$ with parameters $\Gamma_{i\to j}$, for all $i,j\in\Lambda$. If (6) holds, the staircase function:

$$N(t)=\sum_{i,j\in\Lambda}N_{i\to j}(t)\;, \tag{8}$$

which counts the total number of jumps between 0 and t, define a Poisson process of parameter Γ. The left discontinuities of this function are the jump times t_p above. As follows from the independence of the Poisson processes $(N_{i\to j}(t))_{t\geq 0}$, the probability that the p–th jump is a jump (i,j) is:

$$\mathbb{P}\big((i_p,j_p)=(i,j)\big)=\frac{\Gamma_{i\to j}}{\Gamma}\;. \tag{9}$$

2.2 Examples

In order to be more concrete, let us give two examples corresponding to physical systems for which the above approach applies.

'Classical' model. Consider a system made of fermions initially in an eigenstate of the Hamiltonian $\hat{H}$, $H=V$. The eigenstates of $\hat{H}$ are of the form $|\underline{n}\rangle=\prod_i(a_i^\dagger)^{n_i}|0\rangle$, where $|0\rangle$ is the vacuum and $\underline{n}=(n_i)_{i\in\Lambda}\in\{0,1\}^\Lambda$. Let $\hat{n}_i=a_i^\dagger a_i$ be the number operator in state $|i\rangle$. Let $\Gamma_{i\to i}=0$ and:

$$\hat{W}_{i\to j}=1-\hat{n}_i+\hat{n}_i\hat{n}_j+a_j^\dagger a_i\;,\;i\neq j \tag{10a}$$

$$\hat{K}=0\;. \tag{10b}$$

The jump operator $\hat{W}_{i\to j}$ transforms an eigenstate $|\underline{n}\rangle$ into $a_j^\dagger a_i|\underline{n}\rangle$ if $n_i=1$ and $n_j=0$, and leaves $|\underline{n}\rangle$ unchanged otherwise. Let the system evolve from $t=0$ until time t according to the quantum jump scheme described above. If we restrict ourself to initial wavefunctions $|\psi(0)\rangle=|\underline{n}\rangle$, with $\underline{n}\in\{0,1\}^\Lambda$, the wavefunction $|\psi(t)\rangle$ at time $t\geq 0$ is in an eigenstate $|\underline{n}(t)\rangle$ at all time and its norm is conserved. Actually, $|\psi(t)\rangle$ remains unchanged up to a phase, as long as there is no jump. When a jump occurs, it may (or may not) jump from an eigenstate $|\underline{n}\rangle$ into another eigenstate $|\underline{n}'\rangle$. The RW at time t is thus completely specified by the sequence of random numbers $\underline{n}(t)\in\{0,1\}^\Lambda$. Define the average population $\rho_i(t)$ in the one–particle state $|i\rangle$ as the average of $n_i(t)$, $\rho_i(t)=\mathbb{E}\,n_i(t)$. In order to determine the equation satisfied by the ρ_i's, let us compute the amount of change $\mathrm{d}\rho_i=\mathrm{d}\rho_i^+-\mathrm{d}\rho_i^-$ of $\rho_i(t)$ between times t and $t+\mathrm{d}t$. The gain $\mathrm{d}\rho_i^+$ is equal to the sum over $j\in\Lambda$ of the probability $\Gamma_{j\to i}\,\mathrm{d}t$ that a jump (j,i) occurs between t and $t+\mathrm{d}t$, multiplied by the probability $\rho_j(t)(1-\rho_i(t))$ that this jump modifies the wavefunction. Similarly, the loss $\mathrm{d}\rho_i^-$ is equal to the sum over

$j \in \Lambda$ of the probability $\Gamma_{i\to j}\,\mathrm{d}t$ that a jump (i,j) occurs between t and $t+\mathrm{d}t$, multiplied by $\rho_i(t)(1-\rho_j(t))$. Hence, $\rho_i(t)$ satisfies the Boltzmann equation:

$$\frac{\mathrm{d}\rho_i}{\mathrm{d}t} = \sum_{j\in\Lambda, j\neq i} \Big(\Gamma_{j\to i}\,\rho_j(t)\big(1-\rho_i(t)\big) - \Gamma_{i\to j}\,\rho_i(t)\big(1-\rho_j(t)\big)\Big)\,. \tag{11}$$

This example shows that the RW model of sect.2.1 can be seen as a quantum generalization of a classical kinetic model. In the quantum case, the classical 'random collisions' between particles of the system and of the bath are replaced by 'random quantum jumps'. If we would have worked with single particle systems, the same Boltzmann equation (11) without the Pauli exclusion factors $(1-\rho_i(t))$ would have been obtained.

The Anderson model. Let us consider a crystal the atoms of which are located at the vertices of a Bravais lattice in dimension d. Using labelling of the lattice sites by integers, we can identify it with $\mathbb{Z}^d$. Some random sites are actually occupied by impurities instead of atoms of the original species. These sites form an infinite (random) set $L \subset \mathbb{Z}^d$. At low enough temperature, conducting electrons are almost all in the impurity band, i.e., they are in linear combinations of impurity orbitals. Neglecting the other electrons and assuming only one orbital par impurity, the one–electron Hilbert space $\mathcal{H}$ is identified with $\ell^2(L)$. The one–electron Hamiltonian can be chosen as the Anderson Hamiltonian:

$$H = \sum_{x\in L} \epsilon_x |x\rangle\langle x| + \sum_{x,y\in L} t_{xy} |x\rangle\langle y|\,.$$

$|x\rangle$, $x \in L$, are the canonical basis vectors, describing an electronic state in the impurity orbital at site x. ϵ_x are independent identically distributed random variables and t_{xy} are hopping terms. The randomness of the site energies ϵ_x, describing disorder in the solid, must be distinguished from the dynamical randomness above, which describes dissipation. As is well-known, for strong enough disorder, i.e., if $\langle \Delta\epsilon_x\rangle$ is large enough compared with t_{xy}, the eigenfunctions of H with energies close to the Fermi energy are exponentially localized (Anderson localization [23]). A first choice for the basis vectors $|i\rangle$ is $|i\rangle = |x\rangle$, for any $i \in \Lambda = L$. Then H is non-diagonal, i.e. $T \neq 0$. However, because the Lindblad operators (2) describing the electron-phonon interaction processes at very low temperature are expressed in terms of the eigenfunctions of H, another interesting choice is to take $|i\rangle$ equal to these eigenfunctions. Then $T = 0$ and $|i\rangle$ is an exponentially localized wavefunction differing from the canonical basis vectors $|x\rangle$. The set Λ can be considered as the set of the localization centers of $|i\rangle$, i.e., as the lattice points $i \in L$ where the amplitude of $|i\rangle$ is maximum.

We now describe how the electrons are kicked by phonons. We choose $\Gamma_{i\to i} = 0$ and

$$\hat{W}_{i\to j} = 1 + a_j^\dagger a_i \quad , \quad i \neq j\,. \tag{12}$$

The damping operator $\hat{K}$ is taken according to (4), i.e.

$$\hat{K} = -\frac{\mathrm{i}}{2} \sum_{i,j\in\Lambda} \Gamma_{i\to j}\, \hat{n}_i (1-\hat{n}_j) - \mathrm{i} \sum_{i\neq j\in\Lambda} \Gamma_{i\to j}\, a_j^\dagger a_i \tag{13}$$

(we have set $\hbar = 1$). $\hat{K}$ adds an imaginary part to the eigenenergies of $\hat{H}$ and introduce some new hopping terms $\Gamma_{i\to j}\, a_j^\dagger a_i$.

3 Case of Infinitely Many States

Suppose that the one–particle Hilbert space $\mathcal{H}$ is infinite dimensional, i.e., that the set Λ is infinite. As said in the introduction, if the double sum Γ in (6) diverges, the total number $N(I)$ of jumps in any finite interval $I \subset \mathbb{R}_+$ is infinite with probability one. One can even shows [27] that, if there exists $\beta > 0$ such that $\sum_j \Gamma_{i\to j} \geq \beta$ for infinitely many indices $i \in \Lambda$, then the sup over i of $\sum_j N_{i\to j}(I)$ is infinite with probability one. The random times t_p and the random indices (i_p, j_p) in (5) are therefore not well–defined. The idea for computing the random evolution in $\mathcal{F}$ when $\Gamma = \infty$ runs as follows: (1) restrict Λ to a finite 'box' $B \subset \Lambda$ and determine the wavefunction at time t using (5); (2) let the size of the box increase to reach the limit $B \uparrow \Lambda$.

The theorem stated in this section shows that the limit exists provided suitable conditions on the $\Gamma_{i\to j}$ and $\hat{W}_{i\to j}$ are made. We restrict our analysis to jump operators $\hat{W}_{i\to j}$ which are the second quantized of some one–particle operators $W_{i\to j}$:

Assumption 1
For any $i, j \in \Lambda$, $\hat{W}_{i\to j}$ is a single particle operator, i.e.,

$$\hat{W}_{i\to j} = \sum_{k,l\in\Lambda} \langle k|W_{i\to j}|l\rangle\, a_k^\dagger a_l\ .$$

Let $\{|\underline{n}\rangle, \underline{n} \in S\}$ be the occupation number basis of $\mathcal{F}$:

$$|\underline{n}\rangle = \prod_{i\in\Lambda} \frac{1}{\sqrt{n_i!}} \left(a_i^\dagger\right)^{n_i} |0\rangle \quad , \quad \underline{n} = (n_i)_{i\in\Lambda} \in S\ , \tag{14}$$

where $|0\rangle$ is the vacuum, $S = S_+ = \mathbb{N}^\Lambda$ if the the system's particles are bosons and $S = S_- = \{0,1\}^\Lambda$ if they are fermions. Assumption 1 implies that $\langle \underline{n}|\hat{W}_{i\to j}|\underline{n}'\rangle = 0$ if $|\underline{n}'\rangle$ is different from $c_{kl}^{(\pm)}(\underline{n})\, a_l^\dagger a_k |\underline{n}\rangle$ for some $k, l \in \Lambda$, where $c_{kl}^{(\pm)}(\underline{n})$ is a normalization constant:

$$c_{kl}^{(+)}(\underline{n}) = \left(n_k(1+n_l-\delta_{kl})\right)^{-1/2} \quad , \quad n_k \neq 0 \tag{15a}$$

$$c_{kl}^{(-)}(\underline{n}) = 1 \quad , \qquad n_k(1-n_l) \neq 0\ . \tag{15b}$$

The sign + refers to bosons and $-$ to fermions. δ_{kl} is the Kronecker delta ($\delta_{kl} = 1$ if $k = l$ and 0 otherwise). Similarly, let $c_{ijkl}(\underline{n}) = \|a_j^\dagger a_l^\dagger a_i a_k |\underline{n}\rangle\|^{-1}$. For any $n \in \mathbb{N}^\star$, we denote by $S^{(n)}$ the subset $\{\underline{n} \in S; \sum_{i\in\Lambda} n_i = n\}$ of S. The jump rates $\Gamma_{i\to j}$, the jump operators $\hat{W}_{i\to j}$, the damping operator $\hat{K}$ and the off–diagonal Hamiltonian $\hat{T}$ are assumed to fulfill the following requirements:

Assumption 2

There is $r_1 > 0$ such that:

$$\sup_{i\in\Lambda} \sum_{j\in\Lambda} \Gamma_{i\to j}\, e^{r_1|i-j|} < \infty \quad , \quad \sup_{i\in\Lambda} \sum_{j\in\Lambda} \Gamma_{j\to i}\, e^{r_1|i-j|} < \infty \, .$$

Assumption 3

For any $i, j, k, l \in \Lambda$, one has:

$$\left| \langle \underline{n}| (\hat{W}_{i\to j} - 1)\, a_l^\dagger a_k |\underline{n}\rangle \right| \leq \frac{1}{c_{kl}^{(\pm)}(\underline{n})} (f_{il} + f_{jl})(f_{ik} + f_{jk})$$

for all $\underline{n} \in S_\pm$, $n_k(1 \pm n_l) \neq 0$, where $f_{ij} = f_{ji} > 0$ is such that there is $r_2 > 0$,

$$\sup_{i\in\Lambda} \sum_{j\in\Lambda} f_{ij}\, e^{r_2|i-j|} < \infty \, .$$

Assumption 4

There is $r_3 > 0$ such that, for all $n \in \mathbb{N}^\star$,

$$\sup_{\underline{n}\in S^{(n)}} \Bigg\{ \sum_{k,l\in\Lambda} c_{kl}(\underline{n}) \left| \langle \underline{n}| \, (\hat{T} + \hat{K})\, a_l^\dagger a_k |\underline{n}\rangle \right| e^{r_3|k-l|}$$
$$+ \sum_{i,j,k,l\in\Lambda} c_{ijkl}(\underline{n}) \left| \langle \underline{n}|\hat{K}|a_j^\dagger a_l^\dagger a_i a_k |\underline{n}\rangle \right| e^{r_3 \min\{|i-j|+|k-l|,|i-l|+|k-j|\}} \Bigg\} < \infty \, .$$

Assumption 2 basically means that the rates $\Gamma_{i\to j}$ decrease exponentially with the distance $|i-j|$. Clearly, this does *not* imply that $\Gamma < \infty$ (for instance, $\sum_{ij} \exp(-r|i-j|) = \infty$). Assumption 3 means that a jump (i, j), up to small corrections, can remove a particle in state $|l\rangle$ and create one in a far state $|k\rangle$ only if l is 'exponentially close' to i and k is 'exponentially close' to j, or vice versa. For instance, the jump operators (12) satisfies this assumption with $f_{ij} = \delta_{ij}$. It will be shown below that assumptions 1 to 3 imply that the damping operator $\hat{K}_1$ in (4) satisfies assumption 4 with $r_3 = \min\{r_1/2, r_2\}$.

If $\hat{H}$ is an unbounded operator, with dense domain $\mathcal{D}(\hat{H})$, some care about domains must be taken. Let $\mathcal{B}(\mathcal{F})$ be the C^*–algebra of bounded operators on $\mathcal{F}$. Consider the subspace of $\mathcal{B}(\mathcal{F})$:

$$\{\hat{A} \in \mathcal{B}(\mathcal{F}); \hat{A}\,\mathcal{D}(\hat{H}) \subset \mathcal{D}(\hat{H}) \text{ and } [\hat{H}, \hat{A}] : \mathcal{D}(\hat{H}) \to \mathcal{F} \text{ is bounded } \} \, .$$

This subspace is denoted by $D(\mathcal{L}_{\hat{H}})$, as it is the ultraweakly dense domain of the Liouvillian $\mathcal{L}_{\hat{H}} : \hat{A} \in D(\mathcal{L}_{\hat{H}}) \mapsto \mathrm{i}[\hat{H}, \hat{A}] \in \mathcal{B}(\mathcal{F})$ [28].

Assumption 5
$\hat{W}_{i\to j}$, $\hat{T}$ and $\hat{K}$ belong to $D(\mathcal{L}_{\hat{H}})$ for any $i, j \in \Lambda$.

Let us denote $\mathcal{F}_n = P_\pm \mathcal{H}^{\otimes n}$ the n–particles subspace of $\mathcal{F}$. We say that a sequence $(|\psi_\nu\rangle)_{\nu\in\mathbb{N}}$ of random vectors in $\mathcal{F}_n$ converges *almost surely* to $|\psi\rangle$ if the set of outcomes of all the Poisson processes $(N_{i\to j}(t))_{t\geq 0}$ for which $|\psi_\nu\rangle \to |\psi\rangle$ has a probability one. Note that the limit $|\psi\rangle$ is also a random vector (it takes different values for different outcomes). Our main mathematical result is summarized in the theorem below.

Theorem: *Let $|\psi\rangle \in \mathcal{F}_n$, $n \in \mathbb{N}^\star$, and $B \subset \Lambda$ be a finite box of Λ. Consider the random variables:*

- $0 \leq t_1^B \leq \ldots \leq t_p^B \leq \ldots$, *the times of occurrence of* any *jump $(i, j) \in B^2$, i.e., the left discontinuities of the counting function*

$$N_B(t) = \sum_{i,j\in B} N_{i\to j}(t) \; ;$$

- (i_p^B, j_p^B), *the random pair of indices corresponding to the actual jump $(i, j) \in B^2$ that takes place at time t_p^B:*

$$t_p^B = t^m_{i_p^B \to j_p^B} \quad \textit{for some } m \in \mathbb{N}^\star .$$

Let $\hat{U}_B(t)$ be the random evolution operator defined by (5)*:*

$$\hat{U}_B(t) = e^{-\mathrm{i}(t-t_p^B)(\hat{H}+\hat{K})}\, \hat{W}_{i_p^B\to j_p^B}\, e^{-\mathrm{i}(t_p^B - t_{p-1}^B)(\hat{H}+\hat{K})} \ldots$$
$$\ldots \hat{W}_{i_1^B \to j_1^B}\, e^{-\mathrm{i}t_1^B(\hat{H}+\hat{K})} \quad \textit{for } 0 \leq t_p^B \leq t < t_{p+1}^B . \qquad (16)$$

Then, under assumptions 1 to 5, $\hat{U}_B(t)|\psi\rangle$ converges almost surely to a random vector $|\psi(t)\rangle = \hat{U}(t)|\psi\rangle \in \mathcal{F}_n$ as $B \uparrow \Lambda$. The convergence for a fixed outcome is uniform with respect to t on finite intervals of $\mathbb{R}_+$ (on a set of outcomes of probability one). Moreover, for any $t \geq 0$ and any bounded operator $\hat{A}$ on $\mathcal{F}_n$, the limit $B \uparrow \Lambda$ and the mean $\mathbb{E}$ over the times $t^m_{i\to j}$ can be inverted when calculating expectation values:

$$\mathbb{E}\,\langle\psi|\,\hat{U}(t)^\dagger \hat{A}\,\hat{U}(t)\,|\psi\rangle = \lim_{B\uparrow\Lambda} \mathbb{E}\,\langle\psi|\,\hat{U}_B(t)^\dagger \hat{A}\,\hat{U}_B(t)\,|\psi\rangle \; . \qquad (17)$$

Sketch of the proof. The theorem has been proved elsewhere [27] in the one-particle case, i.e., for RW dynamics in the one-particle Hilbert space $\mathcal{H}$. The proof in the n-particles case considered here is identical excepted that: (1) the basis vectors $|i\rangle \in \mathcal{H}$, $i \in \Lambda$, must be replaced by the basis vectors $|\underline{n}\rangle \in \mathcal{F}_n$, $\underline{n} \in S^{(n)}$; (2) the norms:

$$\|A\|_r = \max\Big\{\sup_{i\in\Lambda} \sum_{j\in\Lambda} |\langle i|A|j\rangle| e^{r|i-j|} \,,\, (A \leftrightarrow A^\dagger)\Big\} \quad , \quad r \geq 0 \,, \tag{18}$$

on sub-algebras of $\mathcal{B}(\mathcal{H})$ must be replaced by some appropriate norms $\|.\|_r^{(n)}$ on sub-algebras of $\mathcal{B}(\mathcal{F}_n)$. These norms may be constructed as follows. For any $\underline{n}, \underline{n}' \in S^{(n)}$, one can find $i_1, i_1, \ldots, i_n, i'_n \in \Lambda$ such that $|\underline{n}\rangle$ and $|\underline{n}'\rangle$ are respectively equal to $a^\dagger_{i_1} \ldots a^\dagger_{i_n}|0\rangle$ and $a^\dagger_{i'_1} \ldots a^\dagger_{i'_n}|0\rangle$ up to normalization constants. It is easy to see that:

$$d(\underline{n}, \underline{n}') = \min_{\sigma:\{i_1,\ldots,i_n\}\to\{i'_1,\ldots,i'_n\} \text{ one-to-one}} \sum_{\nu=1}^{n} |i_\nu - \sigma(i_\nu)| \,,$$

defines a distance on $S^{(n)}$. Clearly, $d(\underline{n}, \underline{n}') = |i-j|$ if $|\underline{n}'\rangle = c_{ij}(\underline{n})\, a^\dagger_j a_i |\underline{n}\rangle$. One sets for any $r \geq 0$ and $\hat{A} \in \mathcal{B}(\mathcal{F}_n)$:

$$\|\hat{A}\|_r^{(n)} = \max\Big\{\sup_{\underline{n}\in S^{(n)}} \sum_{\underline{n}'\in S^{(n)}} |\langle \underline{n}|\hat{A}|\underline{n}'\rangle| e^{r d(\underline{n},\underline{n}')} \,,\, (\hat{A} \leftrightarrow \hat{A}^\dagger)\Big\} \,. \tag{19}$$

If $\hat{A} = \sum_{k,l} \langle k|A|l\rangle a^\dagger_k a_l$ is a single–particle operator, then $\|\hat{A}\|_r^{(n)} \leq n^2 \|A\|_r$. Let F and G be the operators on $\mathcal{H}$ given by [27]:

$$F = \sum_{i,j\in\Lambda} f_{ij} |i\rangle\langle j| \quad , \quad G = \sum_{i,j\in\Lambda} \Gamma_{i\to j} (|i\rangle + |j\rangle)(\langle i| + \langle j|) \,.$$

Then assumption 2, the hypothesis on the f_{ij}'s in assumption 3 and assumption 4 can be rewritten as $\|G\|_{r_1} < \infty$, $\|F\|_{r_2} < \infty$ and $\|\hat{K} + \hat{T}\|_{r_3}^{(n)} < \infty$, respectively. □

The following estimates are obtained by using assumptions 1 to 3:

$$\Big\|\sum_{i,j\in\Lambda} \Gamma_{i\to j} (\hat{W}^\dagger_{i\to j} - 1)(\hat{W}_{i\to j} - 1)\Big\|_r^{(n)} \leq 8n^2 \|F\|_r^4 \|G\|_{2r}$$

$$\Big\|\sum_{i,j\in\Lambda} \Gamma_{i\to j} (\hat{W}_{i\to j} - 1)\Big\|_r^{(n)} \leq 2n \|F\|_r^2 \|G\|_r \,.$$

They show that assumptions 1 to 3 imply that $\|\hat{K}_1\|_{r_3}^{(n)} < \infty$ if $r_3 \leq r_1/2$ and $r_3 \leq r_2$, i.e, that the damping operator $\hat{K}_1$ in (4) satisfies assumption 4.

Let us come back to the Anderson model of sect.2.2. Since the one–electron eigenfunctions $|i\rangle$ are exponentially localized, the jump rates $\Gamma_{i\to j}$ decrease exponentially with the distance $|i-j|$ between their localization center [22]. Moreover,

$\hat{W}_{i\to j}$ and $\hat{K}$ in (12) and (13) and $\hat{T}=0$ satisfy assumptions 1, 3 and 4. Therefore, the theorem above shows that the stochastic dynamics in $\mathcal{F}$ is well–defined at the infinite volume limit $B \uparrow \Lambda$.

Remark: The theorem does not solve completely the question of the convergence at thermodynamic limit, since it is only valid for a finite number n of particles. The study of the limit $B \uparrow \Lambda$, $n \to \infty$, $n/|B| = const.$ is a difficult open problem.

4 Equivalence with the Master Equation

Let us show that the stochastic dynamics in $\mathcal{F}$ of sect.2 gives the Lindblad master equation (1) after averaging over the random times $t^m_{i\to j}$. Ignoring energy shifts, this master equation reads:

$$\frac{\mathrm{d}\rho}{\mathrm{d}t} = \mathcal{L}\rho = -\mathrm{i}[\hat{H},\rho] + \frac{1}{2}\sum_{i,j\in\Lambda}\Big([\hat{L}_{i\to j}\,\rho, \hat{L}^{\dagger}_{i\to j}] + [L_{i\to j}, \rho\hat{L}^{\dagger}_{i\to j}]\Big) \ . \tag{20}$$

We shall see that, for arbitrary Lindblad operators $\hat{L}_{i\to j}$, its solution $\rho(t) = e^{t\mathcal{L}}\rho$ for $\rho = |\psi\rangle\langle\psi|$ can be written as the mean value $\mathbb{E}\,|\psi(t)\rangle\langle\psi(t)|$, where $|\psi(t)\rangle$ is the random wavefunction (5).

4.1 Decomposition of the generator $\mathcal{L}$ into a jump and a damping parts

We assume in this section that $\Gamma < \infty$. We set:

$$\hat{L}_{i\to j} = \sqrt{\Gamma_{i\to j}}(\hat{W}_{i\to j} - \alpha_{i\to j}) \ , \tag{21}$$

where $\alpha_{i\to j}$, $i,j\in\Lambda$, are some complex numbers such that:

$$\lambda_\alpha = \sum_{i,j\in\Lambda}\Gamma_{i\to j}\big(1+\alpha^*_{i\to j}\big)\big(1-\alpha_{i\to j}\big) < \infty \ .$$

The arbitrariness of the parameters $\alpha_{i\to j}$ is linked to the fact that, for fixed Lindblad operators $\hat{L}_{i\to j}$, different choices for the jump operators $\hat{W}_{i\to j}$ are possible; it will be shown below that all these choices lead to the same Lindblad equation. Equation (20) yields:

$$\mathcal{L}\,\rho = -\mathrm{i}[\hat{H},\rho] + \sum_{i,j\in\Lambda}\Gamma_{i\to j}\Big(\hat{W}_{i\to j}\,\rho\,\hat{W}^{\dagger}_{i\to j} - \alpha_{i\to j}\,\rho\,\hat{W}^{\dagger}_{i\to j}$$
$$-\alpha^*_{i\to j}\hat{W}_{i\to j}\,\rho + |\alpha_{i\to j}|^2\rho - \frac{1}{2}\Big\{(\hat{W}^{\dagger}_{i\to j} - \alpha^*_{i\to j})(\hat{W}_{i\to j} - \alpha_{i\to j}), \rho\Big\}\Big) \ ,$$

with $\{A, B\} = AB + BA$. Let us set:

$$\hat{K}_\alpha = \frac{1}{2\mathrm{i}} \sum_{i,j\in\Lambda} \Gamma_{i\to j} \big(\hat{W}_{i\to j}^\dagger + \alpha_{i\to j}^*\big)\big(\hat{W}_{i\to j} - \alpha_{i\to j}\big) - \frac{\lambda_\alpha}{2\mathrm{i}} \tag{22}$$

and define the operators:

$$\mathcal{W}_{i\to j}\,\rho = \hat{W}_{i\to j}\,\rho\,\hat{W}_{i\to j}^\dagger \tag{23a}$$

$$\mathcal{L}_\alpha\,\rho = -\mathrm{i}[\hat{H}, \rho] - \mathrm{i}\hat{K}_\alpha\rho + \mathrm{i}\rho\hat{K}_\alpha^\dagger \tag{23b}$$

acting on density matrices. $\mathcal{W}_{i\to j}$ gives the discontinuous transformation of the density matrix during a jump (i, j) (compare with (3)). One can easily check by time differentiation that:

$$e^{t\mathcal{L}_\alpha}\rho = e^{-\mathrm{i}(\hat{H}+\hat{K}_\alpha)}\,\rho\,e^{\mathrm{i}(\hat{H}+\hat{K}_\alpha^\dagger)}\,. \tag{24}$$

$\hat{K}_\alpha$ coincides with $\hat{K}_1$ in (4) for $\alpha_{i\to j} = 1$. In the general case, $\hat{K}_\alpha$ and $\hat{K}_1$ differ by a self–adjoint operator. A simple computation gives:

$$\mathcal{L}\,\rho = \mathcal{L}_\alpha\,\rho + \sum_{i,j\in\Lambda} \Gamma_{i\to j}(\mathcal{W}_{i\to j} - 1)\,\rho\,. \tag{25}$$

It is assumed here that the $\hat{W}_{i\to j}$ are uniformly bounded in i, j, so that the sum converges in norm.

4.2 Average over quantum trajectories

Following Carmichael [4], the solution of (20) is expanded as an infinite Dyson series:

$$\rho(t) = e^{t\mathcal{L}}\rho = e^{t\mathcal{L}_\alpha}\rho + \sum_{p=1}^{\infty} \sum_{i_1,j_1,\ldots,i_p,j_p\in\Lambda} \Gamma_{i_1\to j_1}\ldots\Gamma_{i_p\to j_p} \int_0^t \mathrm{d}t_p \ldots \int_0^{t_2} \mathrm{d}t_1$$
$$e^{(t-t_p)\mathcal{L}_\alpha}(\mathcal{W}_{i_p\to j_p} - 1)\ldots e^{(t_2-t_1)\mathcal{L}_\alpha}(\mathcal{W}_{i_1\to j_1} - 1)\,e^{t_1\mathcal{L}_\alpha}\rho\,, \tag{26}$$

with $\rho = \rho(t = 0)$. The expression inside the sum over p is equal to:

$$\rho_p(t) = \sum_{q=0}^{p} \frac{(-\Gamma t)^q}{q!}\,\mathcal{J}_{p-q}(t)\,\rho\,, \tag{27}$$

with $\mathcal{J}_0(t) = e^{t\mathcal{L}_\alpha}$ and, for any $p \geq 1$,

$$\mathcal{J}_p(t) = \sum_{i_1,j_1,\ldots,i_p,j_p\in\Lambda} \Gamma_{i_1\to j_1}\ldots\Gamma_{i_p\to j_p} \int_0^t \mathrm{d}t_p \ldots \int_0^{t_2} \mathrm{d}t_1$$
$$e^{(t-t_p)\mathcal{L}_\alpha}\mathcal{W}_{i_p\to j_p}\ldots e^{(t_2-t_1)\mathcal{L}_\alpha}\mathcal{W}_{i_1\to j_1}e^{t_1\mathcal{L}_\alpha}\,. \tag{28}$$

In fact, (27) is clearly true for $p = 1$; let us assume that it holds for p and show that then it also holds for $p+1$. One has:

$$\rho_{p+1}(t) = \sum_{i,j \in \Lambda} \Gamma_{i \to j} \int_0^t \mathrm{d}t \, e^{(t-\tau)\mathcal{L}_\alpha} (\mathcal{W}_{i \to j} - 1) \rho_p(\tau) \, .$$

Moreover,

$$\frac{\mathrm{d}}{\mathrm{d}\tau} \Big(e^{-\tau \mathcal{L}_\alpha} \mathcal{J}_{p-q+1}(\tau) \Big) = \sum_{i,j \in \Lambda} \Gamma_{i \to j} \, e^{-\tau \mathcal{L}_\alpha} \mathcal{W}_{i \to j} \mathcal{J}_{p-q}(\tau) \, .$$

This identity and (27) yields:

$$\rho_{p+1}(t) = \sum_{q=0}^{p} e^{t\mathcal{L}_\alpha} \int_0^t \mathrm{d}\tau \, \frac{(-\Gamma\tau)^q}{q!} \Big\{ \frac{\mathrm{d}}{\mathrm{d}\tau} \Big(e^{-\tau \mathcal{L}_\alpha} \mathcal{J}_{p-q+1}(\tau) \Big) \rho$$
$$- \Gamma e^{-\tau \mathcal{L}_\alpha} \mathcal{J}_{p-q}(\tau) \, \rho \Big\} = \sum_{q=0}^{p+1} \frac{(-\Gamma t)^q}{q!} \mathcal{J}_{p-q+1}(t) \, \rho \, .$$

The last equality follows from an integration by parts. Hence (27) holds for $p+1$ and thus for all $p \in \mathbb{N}^\star$.

As a result,

$$\rho(t) = \sum_{p=0}^{\infty} \sum_{q=0}^{p} \frac{(-\Gamma t)^q}{q!} \mathcal{J}_{p-q}(t) \, \rho = \sum_{q=0}^{\infty} \sum_{p=q}^{\infty} \frac{(-\Gamma t)^q}{q!} \mathcal{J}_{p-q}(t) \, \rho$$
$$= e^{-\Gamma t} \sum_{r=0}^{\infty} \mathcal{J}_r(t) \, \rho \, . \tag{29}$$

We may therefore rewrite the Dyson series (26) as follows:

$$\rho(t) = e^{t\mathcal{L}_\alpha} \rho + \sum_{p=1}^{\infty} \sum_{i_1, j_1, \ldots, i_p, j_p \in \Lambda} \frac{\Gamma_{i_1 \to j_1}}{\Gamma} \cdots \frac{\Gamma_{i_p \to j_p}}{\Gamma} \int_{0 \le t_1 \le \ldots \le t_p \le t}$$
$$\mathrm{d}P_p(t_1, \ldots, t_p) \, e^{(t-t_p)\mathcal{L}_\alpha} \mathcal{W}_{i_p \to j_p} \cdots e^{(t_2 - t_1)\mathcal{L}_\alpha} \mathcal{W}_{i_1 \to j_1} e^{t_1 \mathcal{L}_\alpha} \rho \, , \tag{30}$$

with $\mathrm{d}P_p(t_1, \ldots, t_p) = \Gamma^p \, e^{-\Gamma t} dt_1 \ldots dt_p$. We now come back to the Poisson process (8) of parameter Γ. Actually, $dP_p(t_1, \ldots, t_p)$ is precisely the joint probability that p first jump times immediately following $t = 0$ occur respectively in the time intervals $[t_1, t_1 + \mathrm{d}t_1]$, ..., $[t_p, t_p + \mathrm{d}t_p]$, and that no other jump occurs between 0 and t. Moreover, by (9), the quantity before the integrals in (30) is the probability that these p jumps are jumps (i_1, j_1), ..., (i_p, j_p). We may therefore interpret (30) as an average over all the independent Poisson processes $(N_{i \to j}(t))_{t \ge 0}$:

$$\rho(t) = \mathbb{E} \, e^{(t-t_p)\mathcal{L}_\alpha} \mathcal{W}_{i_p \to j_p} \cdots e^{(t_2 - t_1)\mathcal{L}_\alpha} \mathcal{W}_{i_1 \to j_1} e^{t_1 \mathcal{L}_\alpha} \rho \, . \tag{31}$$

The remarkable fact in this formula is that the random operator acting on the density matrix inside the mean conserve pure states. More precisely, it follows from (23a) and (24) that, if $\rho = |\psi\rangle\langle\psi|$, then:

$$\rho(t) = \mathbb{E}\,|\psi(t)\rangle\langle\psi(t)|\,, \tag{32}$$

with $|\psi(t)\rangle$ given by (5).

4.3 Comments

The calculation above shows that the average of the random pure state $|\psi(t)\rangle\langle\psi(t)|$ given by (5) defines a density matrix $\rho(t)$ satisfying the Lindblad equation (20). Using the terminology of [4], the map $t \mapsto |\psi(t)\rangle$ for a fixed outcome of the Poisson processes is called a *quantum trajectory*. The average over all quantum trajectories thus gives the density matrix evolution back. The formula (21) connects the Lindblad operators $\hat{L}_{i\to j}$ with the jump rates and operators. For given $\hat{L}_{i\to j}$, one has many different random evolutions, all giving the same master equation, which correspond to different choices for $\hat{W}_{i\to j}$ and $\alpha_{i\to j}$, in accordance with (21). These different random evolutions are characterized by different damping operators $\hat{K} = \hat{K}_\alpha$, defined by (22). However, there exists only one choice for the complex numbers $\alpha_{i\to j}$ such that assumption 3 holds. Actually, changing $\alpha_{i\to j}$ is the same as adding a constant to $\hat{W}_{i\to j}$; but if assumption 3 is true for $\hat{W}_{i\to j}$, then it is not true for $\hat{W}_{i\to j} + c$ if the constant c differs from zero. As a result, $\hat{K}_\alpha$ and the random evolution may not be defined when $\Gamma = \infty$ except for particular values of the $\alpha_{i\to j}$'s.

For instance, if $\hat{L}_{i\to j}$ is given by (2), one must take $\alpha_{i\to j} = 1$ for all but a finite number of pairs (i,j). Taking $\alpha_{i\to j} = 1$ for any $i,j \in \Lambda$, one ends up with the model discussed at the end of sect.2.2. The corresponding Lindblad equation describes the electronic dynamics at very low temperature due to phonon absorption/emission processes; it does not take into account the elastic electron–phonon scattering. It should be noted that the equation for the diagonal and for the off-diagonal elements of $\rho(t)$ in the eigenbasis $\{|i\rangle\}$ are decoupled. This is a special property of this Lindblad generator, related to an adiabatic approximation [21]. The diagonal elements $\rho_{ii}(t)$ satisfy the Boltzmann equation (11).

5 Stochastic Hamiltonians

Let us assume that the jump operators can be expressed as exponentials, i.e. that there exists some bounded operators $\hat{V}_{i\to j} \in D(\mathcal{L}_{\hat{H}})$ such that:

$$\hat{W}_{i\to j} = e^{-\mathrm{i}\hat{V}_{i\to j}}\ ,\ i,j \in \Lambda\,. \tag{33}$$

This is the case for instance in the model for electrons in strongly disordered solids of sect.2.2. Actually, $(a_i)^2 = 0$ for fermions, hence the operators (12) satisfy (33) with $\hat{V}_{i\to j} = \mathrm{i}\, a_j^\dagger a_i$. It is shown in this section that the random dynamics of the preceding sections can be found by solving formally the stochastic Schrödinger equation:

$$\mathrm{i}\,\frac{\mathrm{d}|\psi\rangle}{\mathrm{d}t} = \Big(\hat{H} + \hat{K} + \sum_{i,j\in\Lambda} \hat{V}_{i\to j} \sum_{m=1}^{\infty} \delta\big(t - t^m_{i\to j}\big)\Big)\, |\psi(t)\rangle\ . \tag{34}$$

The last term in the time–dependent stochastic Hamiltonian is a random kicked Hamiltonian (noise term). The second term, which is time independent and non random, can be interpreted as a 'damping term', by analogy with the Langevin equation for Brownian motion.

Let us restrict as before Λ to a finite box $B \subset \Lambda$ and compute the solution of the corresponding stochastic Schrödinger equation,

$$\mathrm{i}\,\frac{\mathrm{d}|\psi_B\rangle}{\mathrm{d}t} = \Big(\hat{H} + \hat{K}_B + \sum_{p=1}^{\infty} \hat{V}_{i_p^B\to j_p^B}\, \delta\big(t - t_p^B\big)\Big)\, |\psi(t)\rangle\ . \tag{35}$$

As usual (see e.g. [29]), this solution is found in two steps: (1) replace the Dirac distribution δ by a smooth function δ_ε of compact support $\operatorname{supp}\delta_\varepsilon \subset [-\varepsilon, \varepsilon]$ and integral unity:

$$\int_{\mathbb{R}} \mathrm{d}t\, \delta_\varepsilon(t) = 1 \quad , \quad \varepsilon > 0\ ;$$

(2) find the limit as $\varepsilon \to 0+$ of the corresponding solution $|\psi_{B,\varepsilon}(t)\rangle$.

Let us substitute δ_ε to δ in (35). The solution at time t is expressed in terms of its value at time $\tau < t$ with the help of a Dyson series [30]. For $t_{p-1}^B < \tau \le t_p^B \le t < t_{p+1}^B$ and $\varepsilon \le (t_{p+1}^B - t)/2$, $\varepsilon \le (\tau - t_{p-1}^B)/2$, this yields:

$$\begin{aligned}|\psi_{B,\varepsilon}(t)\rangle = \sum_{q=0}^{\infty} \frac{(-\mathrm{i})^q}{q!}\, \mathcal{T} \int_{[\tau,t]^q} \prod_{r=1}^{q} \big[\mathrm{d}\tau_r\, \delta_\varepsilon(\tau_r - t_p^B)\big]\, e^{-\mathrm{i}(\hat{H}+\hat{K})(t-\tau_q)}\, \hat{V}_{i_p^B\to j_p^B} \cdots \\ \cdots e^{-\mathrm{i}(\hat{H}+\hat{K})(\tau_2-\tau_1)}\, \hat{V}_{i_p^B\to j_p^B}\, e^{-\mathrm{i}(\hat{H}+\hat{K})(\tau_1-\tau)}\, |\psi_{B,\varepsilon}(\tau)\rangle\ ,\end{aligned} \tag{36}$$

where $\mathcal{T}$ is the time ordering operator. Letting $\varepsilon \to 0+$, the integrals tend to:

$$e^{-\mathrm{i}(\hat{H}+\hat{K})(t-t_p^B)}\, \big(\hat{V}_{i_p^B\to j_p^B}\big)^q\, e^{-\mathrm{i}(\hat{H}+\hat{K})(t_p^B-\tau)} |\psi_{B,\varepsilon}(\tau)\rangle\ .$$

It can be easily shown [27] that the series (36) converges strongly uniformly with respect to ε. As a result,

$$\begin{aligned}|\psi_{B,\varepsilon}(t)\rangle &\to \sum_{q=0}^{\infty} \frac{(-\mathrm{i})^q}{q!} e^{-\mathrm{i}(\hat{H}+\hat{K})(t-t_p^B)}\, \big(\hat{V}_{i_p^B\to j_p^B}\big)^q\, e^{-\mathrm{i}(\hat{H}+\hat{K})(t_p^B-\tau)} |\psi_{B,\varepsilon}(\tau)\rangle \\ &= e^{-\mathrm{i}(\hat{H}+\hat{K})(t-t_p^B)} \hat{W}_{i_p^B\to j_p^B}\, e^{-\mathrm{i}(\hat{H}+\hat{K})(t_p^B-\tau)} |\psi_{B,\varepsilon}(\tau)\rangle\ .\end{aligned} \tag{37}$$

By repeating this operation until $\tau = 0$, we obtain the solution of (35):

$$|\psi_B(t)\rangle \equiv \lim_{\varepsilon \to 0+} |\psi_{B,\varepsilon}(t)\rangle = \hat{U}_B(t)|\psi\rangle \ , \tag{38}$$

where $\hat{U}_B(t)$ is the random evolution operator (16). The (formal) solution of the Schrödinger equation (34) is obtained by letting the size of the box B tend to infinity:

$$|\psi(t)\rangle = \lim_{B\uparrow\Lambda} |\psi_B(t)\rangle = \hat{U}(t)|\psi\rangle \ . \tag{39}$$

The limit exists with probability one by the theorem above. Note the importance of the order of the limits: we have first taken $\varepsilon \to 0+$ and then $B \uparrow \Lambda$.

Remark: Equation (35) may be rewritten as an Ito stochastic differential equation in the following way:

$$\mathrm{i}\,\mathrm{d}|\psi_B\rangle = \Big((\hat{H} + \hat{K}_B)\mathrm{d}t + \mathrm{i} \sum_{i,j\in B} (\hat{W}_{i\to j} - 1)\,\mathrm{d}N_{i\to j}(t) \Big) |\psi_B(t)\rangle \ . \tag{40}$$

To see that this equation has the same solutions as (35), one can compute the values of the discontinuities of $|\tilde{\psi}_B(t)\rangle = \exp(\mathrm{i}(\hat{H} + \hat{K}_B))\hat{U}_B(t)|\psi\rangle$ at the jump times t_p^B, and use the fact that $|\tilde{\psi}_B(t)\rangle$ is constant between jumps. It should be noticed that, although it might be tempting to replace $\mathrm{d}N_{i\to j}(t)$ by $\sum_m \delta(t - t^m_{i\to j})\mathrm{d}t$, the operator multiplying the stochastic differential in (40) is $\mathrm{i}(\hat{W}_{i\to j} - 1)$, whereas $\hat{V}_{i\to j}$ multiplies the Dirac distributions in (35).

6 Comparison with Other Stochastic Schemes

Other stochastic dynamical schemes with Poisson processes have been introduced by Dalibard, Castin and Mølmer [3], Carmichael [4], Ghirardi, Riminiand Weber [6] and Barchielli and Belavkin [8,9]. Different schemes using Wiener processes have been studied by Gisin and Percival [5], Ghirardi, Pearle and Rimini [7], van Kampen [10] and by other authors [8,11,12,13,14]. We outline in this section the main similarities and differences of these models with the model presented above. The reader can find more information and other relevant references in the reviews [4,18,31].

6.1 Quantum jump schemes

To our knowledge, the first quantum jump scheme is due to Ghirardi *et al.* [6], who introduced the following phenomenological model in connection with the problem of the linear superpositions of macroscopically distinguishable states (Schrödinger cat states). The authors consider some jump operators

$$L_{x,\nu} = (\sqrt{\pi}\alpha)^{-1/2} \exp\Big(-\frac{(x - X_\nu)^2}{2\alpha^2} \Big)$$

which implement 'spontaneous collapses' in the position space around some point $x \in \mathbb{R}^3$. X_ν is the position operator of the ν–th particle of a composite system of n particles. The collapse (x, ν) localizes the ν–th particle around x with an accuracy $\alpha > 0$. The operators $L_{x,\nu}$ are self–adjoint and satisfy:

$$\int_{\mathbb{R}^d} dx\, L_{x,\nu}^2 = 1 \quad , \quad \nu = 1, \ldots, n \; . \tag{41}$$

In the wavefunction formulation of the model [32], the collapses (jumps) modify discontinuously the wavefunction of the system according to the non-linear transformation:

$$\text{collapse } (x, \nu) : |\psi\rangle \to \big\| L_{x,\nu} |\psi\rangle \big\|^{-1} L_{x,\nu} |\psi\rangle \; .$$

The probability that a collapse (x, ν) (resp. that *any* collapse) occurs between times t and $t + \mathrm{d}t$ is equal to $\mathrm{d}p_\nu(x) = \lambda \| L_{x,\nu} |\psi(t)\rangle \|^2 \mathrm{d}t$ (resp. to $\mathrm{d}p = \sum_\nu \int dx\, \mathrm{d}p_\nu(x) = \lambda n \| \psi(t) \|^2 \mathrm{d}t$), where λ is a characteristic frequency. Between jumps, the composite system evolves according to Schrödinger equation (with no damping operator; this fact is related to assumption (41)). As shown in [6], for macroscopic systems ($n \gg 1$), the stochastic collapses kill very rapidly the coherences between states localized a distance greater than α. Provided one chooses λ small enough, they have little effect on the dynamics of microscopic systems ($n \simeq 1$) at times accessible in a laboratory experiment.

Dalibard *et al.* [3] and Carmichael [4] have studied a similar but different model in order to describe photon–atom interactions in Quantum Optics. The original motivation was to reproduce experimental data on the fluorescence of single atoms [18]. Quantum jumps occur as a result of a continuous measurement of photon emission from the atom. The discontinuous change of the wavefunction occurring at a jump (i, j) is given, as in the collapse model above, by a non-linear transformation:

$$\text{jump } (i, j) : |\psi\rangle \to \big\| \hat{L}_{i \to j} |\psi\rangle \big\|^{-1} \hat{L}_{i \to j} |\psi\rangle \; . \tag{42}$$

The Lindblad operators $\hat{L}_{i \to j}$ are arbitrary and do not need to satisfy a relation similar to (41), as in the case considered in this work. A jump (i, j) occurs if a photon of energy equal to the Bohr frequency $\omega_{ij} = \langle i|V|i\rangle - \langle j|V|j\rangle$ is detected. Between jumps, the atom evolves in the following way. Successive measurements on the fluorescence of the atom are performed at short time intervals δt, with the result of no photon detected. These measurements increase our knowledge on the state of the system; it can be shown that the wavefunction evolves between two measurements according to Schrödinger equation but with a non self–adjoint Hamiltonian $\hat{H} + \hat{K}'$. Perturbation theory gives [3]:

$$\hat{K}' = \frac{1}{2\mathrm{i}} \sum_{i,j} \hat{L}_{i \to j}^\dagger \hat{L}_{i \to j} \tag{43}$$

(note the difference with the damping operator $\hat{K}_\alpha$ defined by (22) and (21)). Immediately after a zero–photon measurement, the wavefunction is normalized,

$|\psi\rangle \to \|\psi\|^{-1}|\psi\rangle$. If, on the contrary, a photon is detected, the wavefunction is transformed as in (42). The probability of detection of a photon of frequency ω_{ij} is $\delta p_{i\to j} = \|\hat{L}_{i\to j}|\psi(t)\rangle\|^2\,\delta t$. It depends upon the wavefunction $|\psi(t)\rangle$ before the jump, and thus upon t. As a consequence, the time delays $s_{i\to j}$ between consecutive jumps (i,j) are not given by simple exponential laws. The quantum jumps scheme of Dalibard *et al.* and Carmichael is therefore more involved than the one given by a set of independent Poisson processes (this conclusion also holds for the collapse model of ref. [6]). Despite this mathematical complexity, its dynamics is very simple to implement numerically [3]. On time scales greater than δt, the stochastic dynamics is norm–preserving. The same model has been derived by a completely different and more abstract method using quantum stochastic calculus by Barchielli and Belavkin [8].

The main difference between the models of ref. [3,4,6,8] with the model presented in sect.2 is that the stochastic dynamics is linear and not norm–preserving in the latter, and vice versa in the formers. A linear non norm–preserving model based on Poisson processes has also been discussed by Belavkin [9]. This author consider a stochastic differential equation of the form (40) with a damping operator $\hat{K}_B$ given by (22) for $\alpha_{i\to j} = 0$. Correspondingly, the jump operators are related to the Lindblad operators by $\hat{L}_{i\to j} = \sqrt{\Gamma_{i\to j}}\,\hat{W}_{i\to j}$, which means that the discontinuous changes of the wavefunction are given by (42) without normalization. This model thus coincides with the stochastic scheme of sect.2 in the particular case $\alpha_{i\to j} = 0$. For the Lindblad operators (2), this gives the jump operators $\hat{W}_{i\to j} = a_j^\dagger a_i$ which do not satisfy assumption 3 above. Thus, if $\Gamma = \infty$, the limit $B \uparrow \Lambda$ is not defined in this model.

6.2 Quantum diffusion schemes

As in the schemes of ref. [3,4,6,8], the stochastic scheme investigated by Gisin *et al.* [5], Ghirardi *et al.* [7] and Barchielli *et al.* [8] has a norm–preserving and non-linear stochastic quantum evolution. However, this evolution is given by a stochastic Schrödinger equation with Wiener processes (white noise). Another model based on Wiener processes but with linear stochastic dynamics has been studied in [7,10]. Its wavefunction satisfies the Ito stochastic Schrödinger equation:

$$\mathrm{i}\,\mathrm{d}|\psi\rangle = \Big((\hat{H} + \hat{K}')\,\mathrm{d}t + \mathrm{i}\sum_{i,j} \hat{L}_{i\to j}\,\mathrm{d}\xi_{i\to j}(t)\Big)|\psi(t)\rangle \;, \tag{44}$$

where $\hat{K}'$ is given by (43) and $(\xi_{i\to j}(t))_{t\geq 0}$ are independent complex Wiener processes. The Ito differentials $\mathrm{d}\xi_{i\to j}$ satisfy:

$$\mathrm{d}\xi_{i\to j}\,\mathrm{d}\bar{\xi}_{k\to m} = \delta_{ik}\delta_{jl}\,\mathrm{d}t \quad , \quad \mathrm{d}\xi_{i\to j}\,\mathrm{d}\xi_{k\to m} = 0 \;.$$

The link between this linear non norm–preserving model and the norm–preserving non linear one has been emphasized in [7].

As shown in [3,4,5,7,10], provided that the appropriate damping operator $\hat{K}$ is added to the Hamiltonian $\hat{H}$, all the above models lead to the same trace–preserving average dynamics, given by the Lindblad equation (1). We have seen above that the same holds true for our model. More general stochastic dynamical models, which lead to non-Markovian master equations and use correlated noise, have been introduced recently by several authors [11,12,13,14]. In ref. [11,12], two nice derivations of these models by means of a path integral and a coherent states methods have been proposed.

6.3 Comparison with the model of Sect. 2

The main advantage of the stochastic scheme presented in this work compared with the non linear quantum jumps scheme of e.g. Dalibard *et al.* is its simplicity. Because of the use of Poisson processes and of the linearity of the dynamics, the solution of the stochastic Schrödinger equation is known exactly: it is given by formula (5). The mathematical analysis is also more easy, as the operator theory framework can be used to study the stochastic evolution in the Hilbert space. In the most interesting case $\alpha_{i\to j} = 1$, one should probably look for a physical interpretation of the stochastic dynamics in connection with scattering theory (see [17,33]). From the point of view of the average dynamics (i.e., for statistical ensembles of systems), our model is equivalent to the model of Dalibard *et al.* and to the other models discussed in the preceding subsections. Actually, as seen in sect.4, the density matrix $\rho(t) = \mathbb{E}\,|\psi(t)\rangle\langle\psi(t)|$ follows the same Lindblad equation as in these models.

For general (non unitary) jump operators $\hat{W}_{i\to j}$, the norm of the wavefunction for a fixed outcome is not constant (and not continuous) in time (although, as said before, the square norm is conserved *on average*). Indeed, if one insists in describing an open system by a wavefunction, its norm may not be necessarily conserved by the dynamics, since, unlike in a closed system, the dynamics is not unitary. This is related to the fact that the interaction with the environment may provide us with some information on the system.

From the numerical side, the stochastic dynamics in our model could be of interest if the exponentials $e^{-is(\hat{H}+\hat{K})}$ were known on a broad interval of times s and if its eigenvalues have a modulus smaller than one. This happens, for example, if $\hat{H}+\hat{K}$ can be diagonalized analytically and has eigenvalues with negative imaginary parts. Then, the computation of the wavefunction at time t requires a multiplication of $2p$ matrices, where p is the number of jumps between $t=0$ and t (formula (5)), whereas the non linear quantum jumps and the Wiener schemes involve a time integration between 0 and t.

7 Conclusion

We have studied in this work a model describing dissipation in a system of quantum particles by means of a random evolution in time. This model can be seen as a quantum generalization of a classical kinetic model, the classical collisions being replaced by quantum jumps. The input parameters of the models are: (1) a set of transition rates $\Gamma_{i\to j} \geq 0$, for all pairs (i, j) of one–particle states; (2) some operators $\hat{W}_{i\to j}$ acting on the Fock space $\mathcal{F}$ of the system, which describe the discontinuous changes of the wavefunction at a jump. The linear random time evolution $t \mapsto |\psi(t)\rangle \in \mathcal{F}$ is specified by a set of independent Poisson processes, a different Poisson process, with parameter $\Gamma_{i\to j}$, being associated to each pair (i, j). The average density matrix $\rho(t) = \mathbb{E}\, |\psi(t)\rangle\langle\psi(t)|$ obeys the general Lindblad equation. Our main result is that, if the transition rates decay exponentially with the distance $|i - j|$ and the jump operators satisfy the locality condition 1 and 3 of sect.3, then the stochastic evolution of the system is well–defined as some limit if the double sum $\sum_{i,j} \Gamma_{i\to j}$ diverges. This result is only valid for systems with a finite number of particles. The limit of an infinite number of particles with a finite density (thermodynamic limit) requires a more abstract algebraic approach [28]. For aperiodic solids like strongly disordered solids, one should define a stochastic dynamics on the C^*-algebra of the electronic observables in second quantization, which is the crossed product of a continuous field of C^*-algebras by a groupoid [21].

The use of Poisson processes is natural from a physical point of view, especially if the dissipation mechanism under study is due to absorption and emission of external particles by the system (phonons, photons,...). It is also convenient because of its mathematical simplicity. Unlike in the model defined by Dalibard *et al.* [3] and Carmichael [4], the random time evolution in our model is linear, but not norm–preserving. The linearity simplifies greatly the mathematical analysis. The random evolution operators can be computed directly from formula (5). The model thus provides an example of quantum jump scheme for which one can handle rigorously the case where infinitely many levels are coupled together by the environment. It can be applied to study electronic transport in disordered or aperiodic solids. A simple example was given in sect.2.2. However, a theory of linear response similar to that elaborated in [15,16,17] is still lacking within our stochastic wavefunctions framework. Investigation in this direction will be the object of a future publication [34].

Acknowledgements: We are grateful to Rolando Rebolledo, Miguel Orszag, Pierre Gaspard, Hermann Schulz–Baldes, Italo Guarneri and Boris Shklovskii for valuable discussions and remarks. D.S. acknowledge support by a Fondecyt postdoctoral grant 3000035.

References

1. F. Haake: *Statistical Treatment of Open Systems by Generalized Master Equations*, Springer tracts in modern physics **66** (Springer–Verlag, 1973)
2. G. Lindblad: Comm. Math. Phys. **48**, 119 (1976)

3. J. Dalibard, Y. Castin, K. Mølmer: Phys. Rev. Let. **68**, 580 (1992); K. Mølmer, Y. Castin, J. Dalibard: J. Opt. Soc. Am. B **10**, 524 (1993)
4. H. Carmichael: *An open system approach to Quantum Optics*, Lecture Notes in Physics **m18** (Springer–Verlag, 1991)
5. N. Gisin, I.C. Percival: J. Phys A: Math. Gen. **25**, 5677 (1992) and references therein
6. G.C. Ghirardi, A. Rimini, T. Weber: Phys. Rev D **34**, 470 (1986)
7. G.C. Ghirardi, P. Pearle, A. Rimini: Phys. Rev. A **42**, 78 (1990)
8. A. Barchielli, V.P. Belavkin: J. Phys. A: Math. Gen. **24**, 1495 (1991)
9. V.P. Belavkin: J. Phys. A.: Math. Gen. **22**, L1109 (1989)
10. N.G. van Kampen: *Stochastic Processes in Physics and Chemistry*, 2nd ed. (North–Holland, 1992)
11. W.T. Strunz: Phys. Lett A **224**, 25 (1996)
12. L. Diósi, W.T. Strunz: Phys. Lett. A **235**, 569 (1997)
13. W.T. Strunz, L. Diósi, N. Gisin: Phys. Rev. Lett. **82**, 1801 (1999)
14. M.O. Caceres, A.K. Chattah: Physica A **234**, 322 (1996)
15. J. Bellissard, A. van Elst, H. Schulz–Baldes: J. Math. Phys. **35**, 5373 (1994).
16. H. Schulz–Baldes, J. Bellissard: Rev. Math. Phys. **10**
17. H. Schulz–Baldes, J. Bellissard: J. Stat. Phys. **91**, 991 (1998)
18. M.B. Plenio, P.L. Knight: Rev. Mod. Phys. **70**, 101 (1998)
19. A. Einstein: Phys. Zeits **18**, 121 (1917)
20. C. Cohen–Tannoudji, J. Dupond–Roc, G. Grynberg: *Atom–photon interactions: basic processes and applications* (Wiley, New York 1992)
21. D. Spehner: Contributions à la théorie du transport électronique dissipatif dans les solides apériodiques. PhD Thesis, Université Paul Sabatier, Toulouse, France (2000)
22. B.I. Shklovskii, A.L. Efros: *Electronic properties of doped semiconductors* (Springer–Verlag, 1984)
23. P.W. Anderson: Phys. Rev. **109**, 1492 (1958)
24. N.F. Mott: J. Non–Crystal. Solids **1**, 1 (1968)
25. J. Delahaye, J.P. Brison, C. Berger: Phys. Rev. Lett. **98**, 4204 (1998)
26. I.N. Kovalenko, N.Y. Kuznetsov, V.M. Shurenkov: *Models of random processes: a handbook for mathematician and engineers* (CRC press, 1996)
27. D. Spehner, J. Bellissard: 'A Kinetic Model of Quantum Jumps'. Submitted to J. Stat. Phys. (2000)
28. O. Bratteli, D.W. Robinson: *Operator Algebras and Quantum Statistical Mechanics*, vol.1 and 2 (Spinger–Verlag, 1987)
29. D.R. Grempel, R.E. Prange, S. Fishman: Phys. Rev. A **29**, 1639 (1984)
30. M. Reed, B. Simon: *Methods of Modern Mathematical Physics*, vol.2 (Academic press, 1975)
31. D. Giulini, E. Joos, C. Kiefer, J. Kupsch, I.–O. Stamatescu, H.D. Zeh: *Decoherence and the Appearance of the Classical World in Quantum Theory*, chap. 8 (Springer–Verlag, 1996)
32. J.S. Bell: *Speakable and Unspeakable in Quantum Mechanics* (Cambridge University Press, 1987)
33. H. Schulz–Baldes: Etude du transport électronique dans les solides apériodiques par les méthodes de la géométrie non commutative. PhD Thesis, Université Paul Sabatier, Toulouse, France (1997)
34. J. Bellissard, R. Rebolledo, D. Spehner, W. von Waldenfels. In preparation

Part V

Short Contributions

Coherent Population Trapping and Resonance Fluorescence in a Closed Four Level System

M.L. Ladrón de Guevara

Universidad de Chile, Facultad de Ciencias, Departamento de Física,
Casilla 653, Santiago, Chile

Abstract. We consider a four–level atom, where an excited closely–lying pair is driven coherently from a lower state. The pair decays spontaneously to the latter directly, as well as via an auxiliary level, so that quantum interference exists along two different couples of spontaneous emission pathways. We show that this system exhibits coherent population trapping under conditions more general to those found in the driven three-level V-type configuration. The so generated coherent superpositions originate fluorescence spectra where just two of the sidebands are present. In particular situations, the spectrum exhibits only one intense and ultranarrow line, as well as quenching of resonance fluorescence.

1 Introduction

Part of the research in quantum optics in the last two decades has been focused in the study of different schemes of quantum interference in atom–field systems, what has lead to novel effects such as lasing without inversion, enhancement of a refraction index without absoption or electromagnetically induced transparency [1]. One of the known mechanisms of quantum interference is the one taking place between different spontaneous emission pathways of two closely-lying levels decaying to a common state. This may originate, in driven three–level systems, quenching of resonance fluorescence [2] a strong narrowing of the central peak [3], as well as transparency, narrow resonances, and gain without inversion of a weak probe [4]. Supression and narrowing of one or more lines in spontaneous emission are predicted in multilevel systems, either when the driven transitions do not coincide with the interferring ones or when there is present more than one driving field [5]. The cancellation partial or total of the spontaneous decay in the above examples is a manifestation of coherent population trapping, and was experimentally demostrated by Xia and coworkers [6], which results are analyzed in [7]. In this work we consider a driven V-type atom where an auxiliary level exists apart from the ground state and the excited closely–lying pair, so that quantum interference occurs between two different couples of spontaneous emission pathways, both leading the atom (direct or indirectly) to the same final state. This atomic system differs from others already studied in that the auxiliary level and the lowest of the laser–coupled levels are conected by an incoherent process. We focus our attention in the study of coherent population trapping and of the resonance fluorescence emitted when the former takes place.

M. Orszag, J.C. Retamal (Eds.): LNP 575, pp. 379–383, 2001.

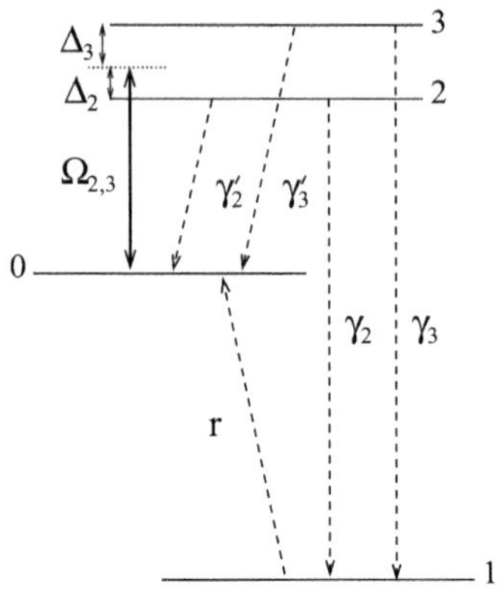

Fig. 1. Atomic level scheme

2 Model

Fig. 1 shows a diagram of the studied system. The closely spaced levels $|2\rangle$ and $|3\rangle$ are coupled to $|0\rangle$ by a laser, being Ω_2 and Ω_3 the respective Rabi frequencies. The transitions between $|2\rangle$ and $|3\rangle$ are dipole forbidden. The constants γ_k' and γ_k are the decay rates from the k-th state of the pair ($k = 2,3$) to states $|0\rangle$ and $|1\rangle$, respectively, and r the rate of the incoherent pumping from $|1\rangle$ to $|0\rangle$. The frequencies of the different levels, say ω_l ($l = 0,1,2,3$), are such as $\omega_3 - \omega_2 \ll \omega_3 - \omega_0 \ll \omega_3 - \omega_1$, the latter inequality needed to neglect the laser–induced transitions between the excited pair and the auxiliary level. The driven atom is governed by a Liouville equation of the form

$$\dot{\varrho} = i[H, \varrho] + \mathcal{L}\varrho, \tag{1}$$

where $H = \sum_{k=2,3} \Delta_k \sigma_{0k}^\dagger \sigma_{0k} + i\Omega_k(\sigma_{0k}^\dagger - \sigma_{0k})$ is the atom–laser Hamiltonian in a rotating frame, with $\Delta_k = \omega_k - \omega_0 - \omega_L$, ω_L being the laser frequency. $\mathcal{L}\varrho$ describes the spontaneous emission processes, where terms proportional to

$$\sum_{k,l=2,3\ (k \neq l)} p_j (2\sigma_{jk}\varrho\sigma_{jl}^\dagger - \varrho\sigma_{jl}^\dagger\sigma_{jk} - \sigma_{jl}^\dagger\sigma_{jk}\varrho), \qquad (j = 0,1), \tag{2}$$

where $p_j \equiv \mathbf{d}_{j2} \cdot \mathbf{d}_{j3} / |\mathbf{d}_{j2}||\mathbf{d}_{j3}|$ (with $\mathbf{d}_{jk}$ ($k = 2,3$) the dipole matrix element for the transition $|j\rangle \to |k\rangle$) accounts for the magnitude of the quantum interference; this will be maximal when $\mathbf{d}_{j2}$ and $\mathbf{d}_{j3}$ are parallel or antiparallel and zero when they are perpendicular.

3 Results

To study coherent population trapping in steady state we neglect the spontaneous emission along the driven transitions. Solving (1) for $\dot{\varrho} = 0$ and forcing the resulting state to be pure we found trapping conditions which for $\Delta_3 = -\Delta_2 \equiv \Delta$ and $\gamma_2 = \gamma_3 \equiv \gamma$ can be written as

$$\Omega_3 = p_1 \Omega_2 \,, \tag{3a}$$

$$\Omega_3 = -p_1 \Omega_2 \pm \sqrt{2}\Delta \,, \tag{3b}$$

where $p_1 = \pm 1$. Both conditions (3a) and (3b) produce a cancellation of the spontaneous decay to the state $|1\rangle$, independently of the initial conditions. In what follows we assume that $p_1 \geq 0$, being analogous the $p_1 \leq 0$ case. For $\Omega_2 = \Omega_3$ ($p_1 = 1$) the atom evolves to the state

$$|\psi_0\rangle = \frac{1}{\mu}(\Delta|0\rangle + i\Omega_2|2\rangle - i\Omega_3|3\rangle), \quad \mu \equiv (2\Omega_2^2 + \Delta^2)^{1/2} \,, \tag{4}$$

which corresponds to a dressed state with zero energy. Conditions (3b), in turn, lead the atom to

$$|\phi_\pm\rangle = \frac{1}{2}(\sqrt{2}|0\rangle \pm i|2\rangle \mp i|3\rangle) \,, \tag{5}$$

which are dressed states with energies $w_\pm = \pm\sqrt{2}\Omega_2 - \Delta$. Such steady states are reached whatever is p_0, but even though none of the trapping conditions depends on r, the inequality $\gamma_2', \gamma_3' \ll r, \gamma\Delta, \Omega_2$ and Ω_3 must be obeyed. When γ_2' and γ_3' are arbitrary, the trapping condition (3b) is no longer valid, while $\Omega_2 = p_1\Omega_3$ ($p_1 = \pm 1$) remains valid only when: i) $p_0 = p_1$, condition under which a quenching of the spontaneous decay processes from the driven transitions is produced, and ii) $p_0 \neq 1$ and $\mu \gg \gamma, \gamma_2', \gamma_3'$, where the fluorescence is not cancelled.

The incoherent part of the resonance fluorescence spectrum is given by

$$S(\omega) \propto \mathrm{Re} \int_0^\infty d\tau \langle \Delta\mathbf{P}^{(-)}(0) \cdot \Delta\mathbf{P}^{(+)}(\tau)\rangle_{ss} e^{i\omega\tau}, \tag{6}$$

where $\mathbf{P}^{(+)}(\tau) = \mathbf{d}_{02}\sigma_{02}(\tau) + \mathbf{d}_{03}\sigma_{03}(\tau)$, $\mathbf{P}^{(-)} = (\mathbf{P}^{(+)})^\dagger$ and $\Delta\mathbf{P}^{(+)}(\tau) = \mathbf{P}^{(+)}(\tau) - \langle\mathbf{P}^{(+)}\rangle_{ss}$. The correlation function $\langle\Delta\mathbf{P}^{(-)}(0) \cdot \Delta\mathbf{P}^{(+)}(\tau)\rangle_{ss}$ is evaluated numerically via the quantum regression theorem [8]. We assume the laser polarized along the $\hat{\mathbf{x}}$ direction and $\mathbf{d}_{k0} = d_{k0}(\cos\theta_k\hat{\mathbf{x}} + \sin\theta_k\hat{\mathbf{y}})$ ($k = 2, 3$) so that $p_0 = \cos(\theta_3 - \theta_2)$. Figs. 2(a)-(c) show some typical examples of incoherent spectra for cases in which the system evolves, or very much approaches, to some of the states (4) and (5). In the three plots $\Omega_3 = -\Omega_2 + \sqrt{2}\Delta$ and $p_1 = 1 \neq p_0$ and all the curves are normalized to unity. In (a) $\Omega_2 = \Omega_3$ (that is, (3a) is also satisfied), two identical lines are onserved at $\omega = -\mu$ and μ. The central peak and outer sidebands (present in the $p_1 = 0$ case) have been cancelled. We have chosen negligible γ_2' and γ_3', but similar curves are found whenever the steady state of the atom is (5), even though such rates are finite (the widths of the lines ($\sim \gamma/2$) and their heights are observed not to depend on γ_2' and γ_3', but only on Ω_2, Δ and γ). When Δ is made smaller, as in Fig. 2(b), one of the lines becomes comparatively higher, and both widths are modified from their values

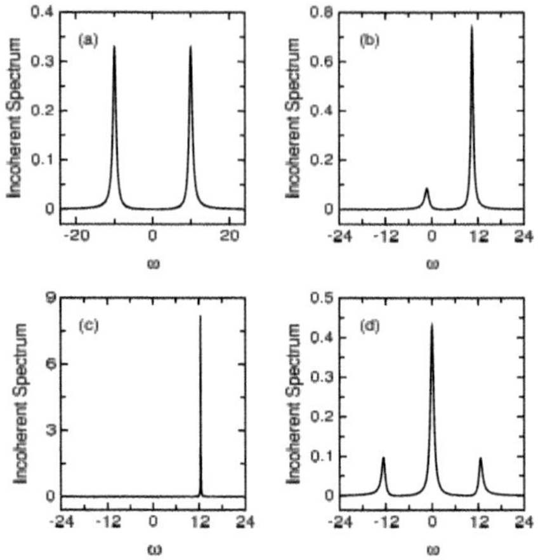

Fig. 2. Incoherent spectrum for fixed Rabi frequency Ω_2 and different values of Δ ($\Omega_3 = -\Omega_2 + \sqrt{2}\Delta$), with $p_1 = 1$, $\gamma_1 = 6$, $\Omega_2 = 5$, $\gamma_2' = 10^{-4}$, $\theta_2 = -\pi/4$ and (**a**) $\Delta = 7$, (**b**) $\Delta = 4$ and (**c**) $\Delta = 1$. In (**d**) $\Delta = 1$ but $p_1 = 0$

when $\Omega_2 = \Omega_3$ (in the example $\sim 1/2$). The higher line has become narrower (its width $\Delta\omega \sim 0.37$) and the smaller wider ($\Delta\omega \sim 0.65$). In Fig. 2(c) just one of the lines is visible, which linewidth is about 0.08 times the width of such line when $p_1 = 0$, case shown in Fig. 2(d). It is interesting to notice that the intensity of the ultranarrow line in (c) is of the same order than those of the lines in (d), in fact, $I_{inc}(p_1 = 1)/I_{inc}(p_1 = 0) \simeq 0.54$. Approximate expressions of the above spectra (namely, when $p_1 = 1$) can be obtained by making $\gamma_2 \equiv \gamma_3 = 0$ and evaluating (6) in the dressed states approach. We find that for $\Omega_2 = \Omega_3$ ($p_1 = 1$) the incoherent spectrum is composed by two Lorentzians of equal heights and widths ($= \gamma/2$) and is proportional to $|\mathbf{d}_{20} - \mathbf{d}_{30}|^2$, what is in total agreement with our numerical results. For $\Omega_3 = -\Omega_2 + \sqrt{2}\Delta$ ($p_1 = 1$) this can be written as

$$S(\omega) \propto \left[\mathcal{N}_+^2 \frac{\kappa_{0+}}{(\omega - \omega_+)^2 + \kappa_{0+}^2} + \mathcal{N}_-^2 \frac{\kappa_{0-}}{(\omega - \omega_-)^2 + \kappa_{0-}^2}\right] |\mathbf{d}_{20} - \mathbf{d}_{30}|^2, \quad (7)$$

where $\omega_\pm = (3\varepsilon_0 \mp \delta)/2$, $\varepsilon_0 = \sqrt{2}\Omega_2 - \Delta$, $\delta = [9\Delta^2 - 2\sqrt{2}\Omega_2\Delta + 2\Omega_2^2]^{1/2}$,

$$\kappa_{0\pm} = \frac{\gamma}{2}\mathcal{N}_\pm^2 \left(\frac{\Omega_2}{\varepsilon_\pm + \Delta} + \frac{-\Omega_2 + \sqrt{2}\Delta}{\varepsilon_\pm - \Delta}\right)^2, \quad (8)$$

with $\mathcal{N}_\pm^{-2} = 1 + \Omega_2^2[1/(\varepsilon_\pm + \Delta)^2 + 1/(\varepsilon_\pm - \Delta)^2]$ and $\varepsilon_\pm = (-\varepsilon_0 \pm \delta)/2$. Fig. 3(a) shows that for a given value of Δ the linewidth of one of the lines, namely κ_0-, decays to zero as Ω_2 is increased, at the same time the other approaches to 1. Such a behaviour is accompanied by a fall of the ratio I_+/I_- between the intensities of the lines indicating that the narrower one (labeled by "-") becomes strongly dominant over the other.

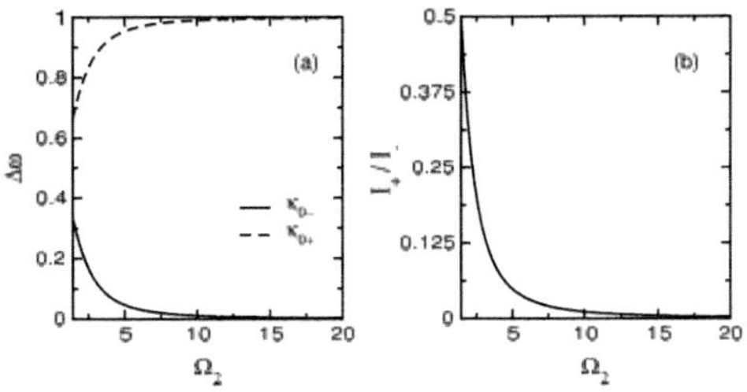

Fig. 3. (**a**) Linewidths κ_O- (*solid line*) and κ_O+ (*dashed line*) vs the Rabi frequency Ω_2, for $\Delta = 1$ ($\Omega_3 = -\Omega_2 + \sqrt{2}\Delta$). (**b**) Cuotient between the intensities of each of the Lorentzians composing the incoherent spectrum, I_+/I_- vs Ω_2, for $\theta_2 = -\pi/8$, $\theta_3 = 3\pi/4$ and $\Delta = 1$

In summary, we have seen that the existence of an alternative way through which the excited doublet decays to the ground state of the driven transitions opens new possibilities of coherent population trapping as well as new features in the resonance fluorescence spectrum, different to those discussed in previous works [3,5]. We show for example the incoherent spectrum can be reduced to just one, ultranarrow and intense line, which is out of resonance with the driving field and appears when the quantum interference is maximal and coherent trapping is produced.

Acknowledgments

This work was supported by Fondecyt grant 3990040.

References

1. S. E. Harris; Phys. Rev. Lett. **62**, 1033 (1989) O. Kocharovskaya: Phys. Rep. **219**, 175 (1992) M. O. Scully: Phys. Rev. Lett. **67**, 1855 (1991) M. O. Scully and Shi-Yao Zhu: Opt. Commun. **87**, 134 (1992) S. E. Harris: Physics Today, **50**, 36 (1997)
2. D. A. Cardimona, M. G. Raymer and C. R. Stroud Jr.: J. Phys. B: At. Mol. Phys. **15**, 55 (1982)
3. Peng Zhou and S. Swain: Phys. Rev. Lett. **77**, 3995 (1996)
4. Peng Zhou and S. Swain: Phys. Rev. Lett. **78**, 832 (1997)
5. Shi-Yao Zhu and M. 0. Scully: Phys. Rev. Lett. **77**, 1032 (1996) Fu-li Li and Shi-Yao Zhu: Phys. Rev. A **59**, 2330 (1999)
6. Hui-Rong Xia, Cen-Yun and Shi-Yao Zhu: Phys. Rev. Lett. **77**, 1032 (1996)
7. J. Wang, H. M. Wiseman and Z. Ficek: Phys. Rev. A **62**, 013818 (2000)
8. M. Lax: Phys. Rev. **129**, 2342 (1963)

Dynamics of Bose–Einstein Condensation for Negative Scattering Length

V.S. Filho[1], A. Gammal[1], L. Tomio[1], and T. Frederico[2]

[1] Instituto de Física Teórica - Universidade Estadual Paulista, Rua Pamplona, 145, 01405-900, São Paulo, SP, Brazil. vsf@ift.unesp.br
[2] Departamento de Física, Instituto Tecnológico da Aeronáutica, Centro Técnico Aeroespacial, 12228-900, São José dos Campos, SP, Brazil.

Abstract. We investigate the dynamics of the trapped Gross-Pitaevskii equation, with negative two-body scattering length and nonconservative terms related to atomic feeding and nonlinear three-body recombination. Our aim is to explore the dynamics of collapses, that occurs in Bose-Einstein condensates of ultra-cold atoms, using a wide range of parameters for the nonconservative terms. A chaotic regime is found for certain class of parameters, using a criterion proposed by Deissler *et al.* to characterize spatio-temporal chaos in nonlinear Schrödinger equations.

The Bose-Einstein Condensation (BEC) of trapped dilute gas of ^{7}Li atoms was observed experimentally by Hulet *et al.* [1], confirming a numerical simulation of Ruprecht *et al.* [2] for systems with negative two-body scattering length ($a < 0$). With the assumption of spherical symmetry, the condensate of such systems can be described by the following dimensionless generalization of the Gross-Pitaevskii equation:

$$\mathrm{i}\frac{\partial \Phi}{d\tau} = \left[-\frac{d^2}{dx^2} + \frac{1}{4}x^2 - \frac{|\Phi|^2}{x^2} - 2\mathrm{i}\xi\frac{|\Phi|^4}{x^4} + \mathrm{i}\frac{\gamma}{2}\right]\Phi, \tag{1}$$

where x and τ are, respectively, related to the space and time variables $\boldsymbol{r}$ and t by $x \equiv \sqrt{2m\omega/\hbar}|\boldsymbol{r}|$ and $\tau \equiv \omega t$, with m being the mass of a single atom and ω the frequency of a spherically symmetric harmonic trap. The wave-function $\Phi \equiv \Phi(x,\tau)$, normalized to the reduced number of atoms $n(\tau) \equiv 2N(t)|a|\sqrt{2m\omega/\hbar}$, is related to the physical wave-function $\Psi(\boldsymbol{r},t)$ by $\Phi(x,\tau) = \sqrt{8\pi N(t)|a|}|\boldsymbol{r}|\Psi(\boldsymbol{r},t)$, where $N(t)$ is the number of atoms at a given instant t. The nonconservative parameters, γ and ξ, respectively, refer to atomic feeding from the thermal cloud and three-body recombinations.

The eq. (1) corresponds to a mean-field approximation to the quantum $N-$body problem of a dilute gas at low temperature [3], where the interatomic distances are much larger than $|a|$, which is also much smaller than the corresponding thermal de Broglie wavelength. Under such conditions, the exact shape of the two-atom interaction becomes unimportant (see also [4]). This equation was numerically solved in the static case to obtain the energy eigenvalues, using a method described in ref. [5], that allows easy access to the unstable branch of the solutions, where the critical number of atoms N_c for stability is determined,

M. Orszag, J.C. Retamal (Eds.): LNP 575, pp. 384–388, 2001.

in agreement with ref. [2] (N_c is given by n_c =1.625). In a complete dynamical analysis (with the nonconservative terms), the eq. (1) was solved using the Crank-Nicolson method in ref. [4], with the same initial condition as given by Kagan et al. [6] ($N(0) = 0.75N_c$).

As we have observed that the eq. (1), with the parameters given in ref. [6], turns out to be chaotic in the long time interval after a series of initial collapses, with strong indication of higher radial excited states being populated [7] (that apparently are not supported by experimental verifications), we start an investigation of the regime of parameters that lead the equation to such a behavior. In our general study, we have also considered a parameter related to lost of atoms due to two-body dipolar relaxation. In this communication we present a resume of our results considering only the parameters presented in eq. (1). This study is timely, for recent reported experiments have shown the possibility of altering the two-body interaction by means of light or induced magnetic field [8], that can strongly affect the value of ξ.

The chaotic behavior in spatio-temporal nonlinear Schrödinger equations can be characterized by following a criterion suggested by Deissler *et al.* [9], given by a function ζ that is related to the largest Lyapunov exponent:

$$\lambda \equiv \ln \zeta(\tau) = \ln \left(\int_0^L |\delta\Phi(x,\tau)|^2 dx \right)^{1/2}, \tag{2}$$

where L is the length of the system and $\delta\Phi(x,\tau)$ is an infinitesimal spatial separation between two nearby states. The spatio-temporal chaos is characterized by an average positive slope of λ.

In our dynamical study, we obtain the time evolution of the number of atoms $N(t)$, in terms of the corresponding static critical limit N_c. We have also obtained the root-mean-square radius, $X \equiv \sqrt{\langle x^2 \rangle}$, for each set of parameter ξ and γ that we have considered, in a wide spectrum of values for both, from 10^{-1} to 10^{-5} (including the values used by [6]). We evolved N and X up to $\omega t = 1000$, with enough accurate and precise numerical discretizations in time and space. Presently, our results are shown for $\omega t \leq 500$.

We start our numerical analysis by examining two cases in Figs. 1a and 1b, where we have a feeding rate smaller than the dissipation rate. In Fig. 1a we have $\gamma = 0$ with ξ between 10^{-5} and 10^{-1}; and, in Fig. 1b, $\gamma = 0.1$ with ξ between 0.1 and 0.8. As expected, the smaller ξ, the larger is the time for eliminating the atoms from the system. As we increase γ, comparing Figs.1a and 1b, we observe that the decaying rate is larger when we use ξ of the same order. Figure 1b also shows that the time for N to decay to a stable value is practically independent on the value of the dissipation rate ξ (for $\xi \geq \gamma$).

In Fig. 2 we consider a fixed and small parameter $\gamma = 8 \times 10^{-4}$, in order to study the corresponding magnitude of the dissipation that allows the collapse. The number of atoms is shown in Fig. 2a and the corresponding mean-square radius X^2 in Fig. 2b. We note in this figure that, for very small values, γ and

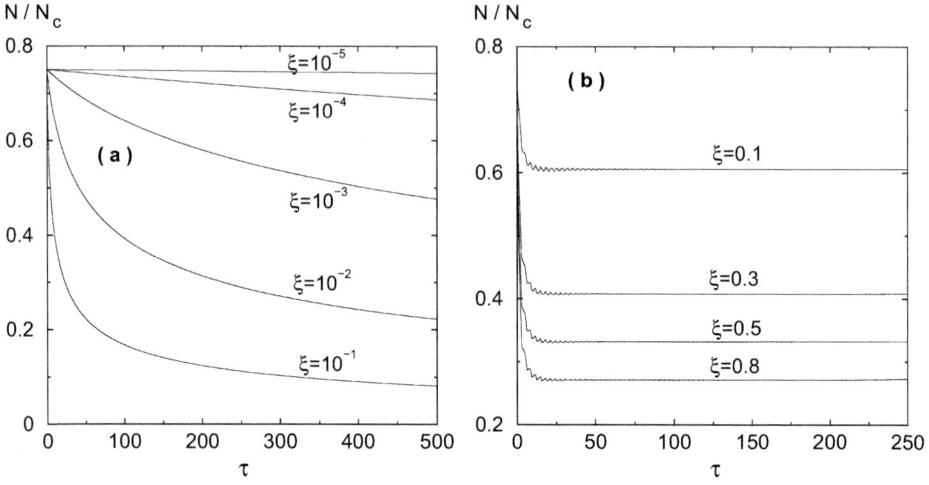

Fig. 1. $N(t)/N_c$ versus $\tau \equiv \omega t$ for $\gamma = 0$ (a) and $\gamma = 0.1$(b), for sets of $\xi \geq \gamma$.

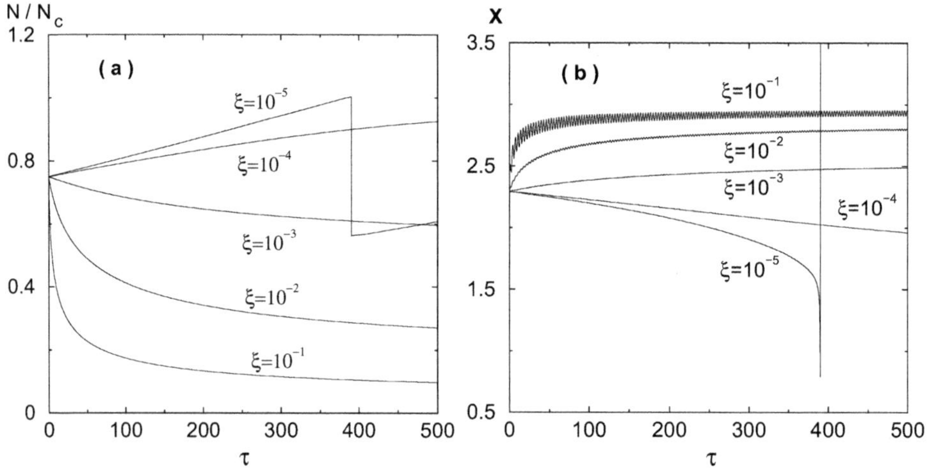

Fig. 2. (a) N/N_c and (b) X^2 versus τ for $\gamma = 8 \times 10^{-4}$ and a set of values of ξ.

ξ are consistent with their direct association with addition (γ) and subtraction (ξ) of atoms in the condensate. However, as one can infer from Fig. 1b, in the regime of larger values (as γ =0.1), we need ξ enough smaller than γ for N to start increasing; the system collapses when the critical limit N_c is reached, a result expected on theoretical grounds. This is illustrated in Fig. 2 for $\xi = 10^{-5}$, when $\tau \cong 390$. Corresponding to Fig. 2a, the mean-square radius X^2 is shown in Fig. 2b, for the same values of ξ. At the "collapsing" point, N decreases from N_c to $\cong 0.58 N_c$. The radius X shrinks in a very short time (with the finite ξ preventing the full collapse $X \to 0$), followed by a jump to large values (of the order of 100 times the initial X^2), with intense and large amplitude oscillations (not shown in this figure).

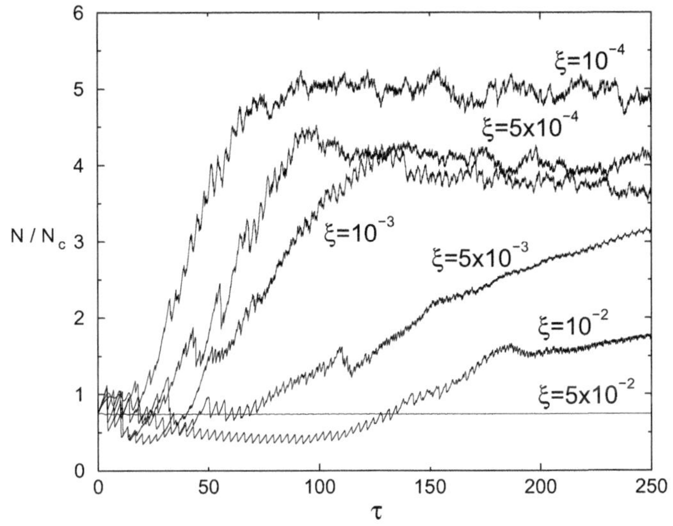

Fig. 3. N/N_c versus τ for $\gamma = 0.1$ and a set of values of ξ.

The collapsing scenario is quite different in Fig. 3, where γ is the same as in Fig. 1b, but with dissipation rate much smaller than the feeding rate. As in Fig. 2a, when the ratio between feeding and dissipation favors an increasing in the number of atoms, the system collapses; but, in this case, this happens more frequently due to the ratio between feeding and dissipation. We also note that exists an oscillator behavior (related to the frequency of the trap) that, after each collapse, starts to approximately double the frequency that is found in the previous cycle. And one can observe that, after each cycle and just before the collapse, the maximum number of atoms reaches values larger than N_c and larger than the corresponding maximum number in the previous cycle, with a clear indication that excited states are being populated. This also differs from what is found in Fig. 2, where the critical limit is approached much slowly (compare, e.g., Fig. 2a with the cases $\xi \leq 0.001$ of Fig. 3).

After the results we have obtained, particularly the ones shown in Fig. 3, with the fractal kind of doubling in the frequencies after the collapses, we start a study to characterize a possible chaotic behavior. As in the present case we are analyzing spatio-temporal NLSE, we found appropriate to use the criterion given in eq. (2), where the chaos is characterized by an average positive slope of λ (see, also, ref. [7]).

In Fig. 4 we select the significative results we have obtained for the ζ function, with particular attention to the results of ξ represented in Fig. 3, where γ =0.1. For the smaller values of γ that we have represented in Figs. 1 and 2, the results have not indicated any chaotic behavior, which is characterized by an exponential grow of ζ. We have calculated the extend Lyapunov exponent λ, given by eq. (2), in all cases analyzed. As shown in Fig. (4), for $\gamma = 0.1$ there are cases with slow

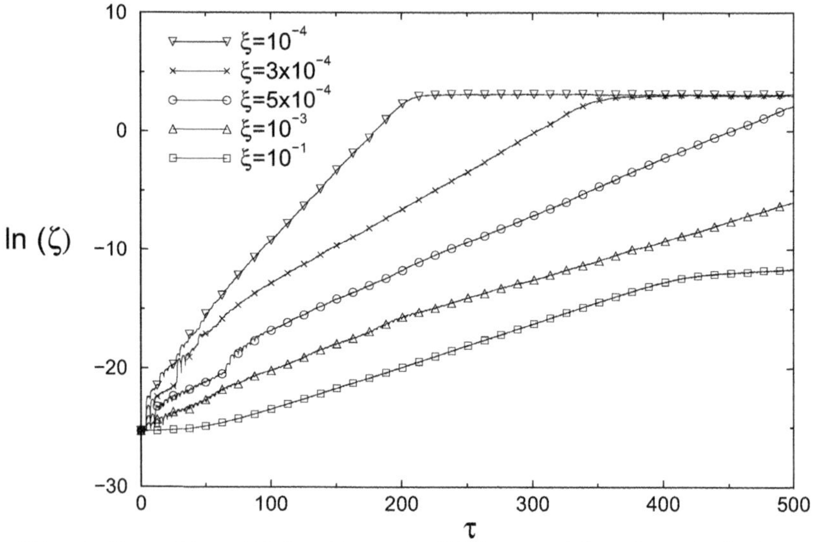

Fig. 4. Logarithm of ζ function as in (2) for $\gamma = 0.1$ and a set of values of ξ.

growth of λ and cases with exponential grow. The chaos is characterized by the latter. When $\xi \cong 10^{-1}$, after enough large interval of time, a clear disorder starts to appear in the system (a sign of the onset of chaos), which is confirmed by the (small) positive slope of λ. The chaotic behavior for that high γ can be easily recognized when $10^{-5} < \xi \leq 10^{-3}$.

In summary, our main result is a characterization of chaotic behavior, for certain class of nonconservative parameters, in the solutions obtained for the generalized Gross-Pitaevskii equation, that is usually considered for trapped BEC atoms with negative scattering length.

References

1. C.C. Bradley, C.A. Sackett, J.J. Tollet and R.G. Hulet, Phys. Rev. Lett. **75**, 1687 (1995). C.A. Sackett, J.M. Gerton, M. Welling and R.G. Hulet, Phys. Rev. Lett. **82**, 876 (1999).
2. M. Edwards and K. Burnett, Phys. Rev. A **51**, 1382 (1995); P.A. Ruprecht, M.J. Holland, K. Burnett, and M. Edwards, Phys. Rev. A **51**, 4704 (1995).
3. K. Huang, *Statistical Mechanics*, 2nd. edition (John Wiley and Sons, New York, 1987).
4. A. Gammal, T. Frederico, L. Tomio, and Ph. Chomaz, J. Phys. B **33**, 4053 (2000); Phys. Rev. A **61**, 051602(R) (2000).
5. A. Gammal, T. Frederico, and L. Tomio, Phys. Rev. E **60**, 2421 (1999).
6. Y. Kagan, A.E. Muryshev, and G.V. Shlyapnikov, Phys. Rev. Lett. **81**, 933 (1998).
7. V.S. Filho, A. Gammal, L. Tomio and T. Frederico, Phys. Rev. A **62**, 033605 (2000).
8. S.L. Cornish *et al.*, Phys. Rev. Lett. **85**, 1795 (2000).
9. R.J. Deissler and K. Kaneko, Phys. Lett. A **119**, 397 (1987); R.J. Deissler, Journal of Stat. Phys. **54**, 1459 (1989); H.R. Brand and R.J. Deissler, Phys. Rev. E **58**, R4064 (1998).

Quantum Gates with a Selective Interaction

E. Solano[1,2], M. França Santos[1], and P. Milman[3]

[1] Instituto de Física, Universidade Federal do Rio de Janeiro, Caixa Postal 68528, 21945-970 Rio de Janeiro, RJ, Brazil
[2] Sección Física, Departamento de Ciencias, Pontificia Universidad Católica del Perú, Apartado 1761, Lima, Peru
[3] Laboratoire Kastler Brossel, Département de Physique de l'Ecole Normale Supérieure, 24 rue Lhomond, F-75231 Paris Cedex 05, France

Abstract. We present a proposal for implementing universal quantum gates by means of a selective interaction in trapped ions and Cavity Quantum Electrodynamics (CQED). Selectivity arises from the possibility of tuning to resonance transitions inside a chosen Hilbert subspace, while leaving other transitions dispersive. In particular, a quantum phase (QPG) gate might be realized, in trapped ions and CQED, making use of a single pulse of the associated selective interaction.

In the last decades a set of logical and mathematical problems found their solution through classical computation. These problems can be classified in different subsets according to their calculation time (number of steps) as a function of their size (N). It is a relative easy task to solve computational problems whose solution is found in polinomial time in their dependence on N and a very hard task when this dependence is polynomial. In the last two decades, a new kind of computers were proposed, the so-called quantum computers [1], where the usual bits are replaced by qubits (quantum bits) and the logic gates replaced by quantum operations (quantum gates). They proved their utility when hard problems, with calculation time exponentially dependent on N like the large numbers factorization problem, were shown to have a fast solution, with calculation time polynomially dependent on N, using quantum computers. The fundamental building blocks of quantum algorithms are the universal quantum gates. In fact, it was shown that any quantum computational problem could be implemented with a sequence of operations consisting of a universal two-qubit gate and local operations [2]. Therefore, the search for physical systems where these operations could be implemented has increased significantly. Recent works [3] suggest practical realizations of two equivalent universal quantum gates: a quantum controlled not gate (CNOT) and a quantum phase gate (QPG) [4]. These gates differ from each other only by local operations (single qubit rotations). In a given bipartite system, consisting of the so called control and target qubits, a CNOT operation changes the target qubit state only if the control qubit state is in a specific state. For instance, the initial general pure state $|\Psi\rangle = \alpha|\downarrow,\downarrow\rangle + \beta|\uparrow,\downarrow\rangle + \gamma|\downarrow,\uparrow\rangle + \delta|\uparrow,\uparrow\rangle$ is taken into state $|\Psi'\rangle = \alpha|\downarrow,\downarrow\rangle + \beta|\uparrow,\uparrow\rangle + \gamma|\downarrow,\uparrow\rangle + \delta|\uparrow,\downarrow\rangle$, where the first label in the kets refers to the control qubit and the second one refers to

M. Orszag, J.C. Retamal (Eds.): LNP 575, pp. 389–393, 2001.

the target qubit. A QPG acting on the same initial state produces the state $|\Psi''\rangle = -\alpha|\downarrow,\downarrow\rangle + \beta|\uparrow,\downarrow\rangle + \gamma|\downarrow,\uparrow\rangle + \delta|\uparrow,\uparrow\rangle$.

In this work, we suggest simple alternative ways to implement a QPG in two common experimental situations, namely, trapped ions and high-Q cavity modes. Our method is based on the use of a selective interaction that perform a quantum phase gate with a single pulse, which has the advantage of not changing experimental parameters during the computational process.

The first experimental system we will consider is an array of N ions of mass m in a Paul trap. The degrees of freedom involved in our proposal will be two of their electronic levels and their collective motion. We consider they interact with each other through a coulombian force, so that stretch and center of mass (CM) vibrational modes can be conveniently introduced. The manipulation of both electronic states and vibrational collective modes can be done by means of laser beams tuned to appropriate frequencies.

Our task is then to produce an interaction that performs a QPG in this system. This interaction must act effectively on two chosen ions (the control qubit ion j and the target qubit k). To achieve this, we address ion j with a Raman laser pair described by the electric field $\boldsymbol{E}_I = \boldsymbol{E}_{oI} e^{i(qz-\omega_I t)}$ and ion k with another Raman laser pair described by the electric field $\boldsymbol{E}_{II} = \boldsymbol{E}_{oII} e^{i(qz-\omega_{II} t)}$. The Raman systems have different effective frequencies, ω_I and ω_{II}, that are quasiresonant to an electronic transition associated with the angular frequency ω_o. The Hamiltonian corresponding to this situation is $\hat{H} = \hat{H}_o + \hat{H}_{\text{int}}$, with

$$\begin{aligned}\hat{H}_o = {} & \hbar\omega_o\left(\hat{S}_{+j}\hat{S}_{-j} + \hat{S}_{+k}\hat{S}_{-k}\right) \\ & + \hbar\nu\hat{a}^\dagger\hat{a} + \sum_n \hbar\nu_n\hat{b}_n^\dagger\hat{b}_n \end{aligned} \tag{1}$$

and

$$\begin{aligned}\hat{H}_{\text{int}} = {} & \hbar\Omega\{(\hat{S}_{+j} + \hat{S}_{-j})(e^{i(q\hat{z}_j-\omega_I t)} + e^{-i(q\hat{z}_j-\omega_I t)}) \\ & + (\hat{S}_{+k} + \hat{S}_{-k})(e^{i(q\hat{z}_k-\omega_{II} t)} + e^{-i(q\hat{z}_k-\omega_{II} t)})\}. \end{aligned} \tag{2}$$

Here, $\widehat{S}_{+i} = |\uparrow_i\rangle\langle\downarrow_i|$, $\widehat{S}_{-i} = \widehat{S}_{+i}^\dagger$ and the state $|\uparrow_i\rangle$ corresponds to the i-th ion in the excited state. $\hat{a}^\dagger$ is the creation operator associated with the harmonic oscillation, with frequency ν, of the CM mode. $\hat{b}_n^\dagger$ is the creation operator associated with the harmonic oscillation, with frequency ν_n, of the other collective modes. All frequencies ν_n are bigger than ν. $\hat{z}_j$ and $\hat{z}_k$ are the operators corresponding to the positions of the ions j and k, respectively, and can be rewritten in terms of the collective coordinates. $\hat{H}_o$ is the free Hamiltonian that corresponds to the internal energy of the two ions plus the energy of the CM mode and of the other collective modes. $\hat{H}_{int}$ is the interaction Hamiltonian describing the position dependent dipolar interaction of ions j and k with the two Raman beams.

We choose $\omega_I = \omega_o + \nu - \delta$ and $\omega_I = \omega_o + \delta$, where δ is the detuning of each Raman beam with respect to the mentioned resonant vibronic transitions, such that $\omega_I + \omega_{II} = 2\omega_o + \nu$. Similarly to what was done in [5,6], the Hamiltonian in Eq. (2) can be expanded in terms of creation and anihilation operators of the normal modes and rewritten in the interaction picture. Then, following the standard procedure described in [7], we can make the rotating wave approximation (RWA) and discard the terms that oscillate with higher frequencies in the dispersive limit $\Omega \ll \delta \ll \nu$. In this way, the following effective time independent Hamiltonian

$$\begin{aligned} H_{\text{eff}} &= \hbar\Omega_o \hat{G}_o^2 \{ i\eta \hat{S}_{+j}\hat{S}_{+k} e^{2i\phi}[\hat{a}^\dagger \hat{F}_o - \hat{F}_o \hat{a}^\dagger]\hat{F}_1 \\ &+ \frac{1}{2}\hat{S}_{+j}\hat{S}_{-j}[\eta^2 \hat{a}^\dagger \hat{F}_1^2 \hat{a}] - \frac{1}{2}\hat{S}_{-j}\hat{S}_{+j}[\eta^2 \hat{F}_1 \hat{a}\hat{a}^\dagger \hat{F}_1] \\ &- \frac{1}{2}\hat{S}_{+k}\hat{S}_{-k}[\hat{F}_o^2] + \frac{1}{2}\hat{S}_{-k}\hat{S}_{+k}[\hat{F}_o^2] + \text{H.c.}\} \end{aligned} \tag{3}$$

can be derived. $\Omega_o = \frac{\Omega^2}{\delta}$ and $\eta = q\sqrt{\hbar/2Nm\nu}$ is the Lamb-Dicke parameter associated to the CM motion. The functions

$$\hat{F}_k = \sum_n f(n)|n\rangle\langle n|, \tag{4}$$

with

$$f(n) = e^{-\eta^2/2} \frac{n!}{(n+k)!} L_n^k(\eta^2), \tag{5}$$

where $L_n^k(\eta^2)$ are the generalized Laguerre polynomials, can always be written in terms of the CM mode number operator $\hat{n}$. $\hat{G}_o$ is a similar non-linear function, involving the number operators related to all other normal modes, that will only contributes to the effective Rabi frequency with a constant numerical factor of the order of 1. We will suppose that all collective modes are cooled down to the ground state and that only the CM mode will be excited. The first term of H_{eff}, in Eq. (3), and its Hermitian conjugate describe the common excitation of both electronic states and the CM mode, similar to a non linear anti-Jaynes-Cummings interaction.

The other terms are motional dependent self energy terms. Notice that, in this situation, the subspace $\{|\downarrow\uparrow\rangle, |\uparrow\downarrow\rangle\}$ remains untouched, which is a natural consequence of the ionic individual addressing claimed in the present scheme [6]. Selectivity arises from the $\hat{n}$ dependence of the self energy corrections in Eq. (3). The stark shift of the electronic states of the ions depends explicitly on the number of vibrational quanta of the CM mode through $\widehat{F_0^2}$. We can adjust the laser beams frequencies to compensate this shift for one particular subspace transition, tuning it to resonance. This will yield a selective interaction that only performs resonant transitions inside the closed subspace $\{|\downarrow\downarrow\rangle|n\rangle, |\uparrow\uparrow\rangle|n+1\rangle\}$. For the sake of simplicity, we only label the CM mode excitacions, since the other vibronic states are cooled down to the vacuum state, in which case they are not affected by H_{eff} in Eq. (3). After adjusting the laser beams frequencies, to select subspace $\{|\downarrow\downarrow\rangle|0\rangle, |\uparrow\uparrow\rangle|1\rangle\}$, and the time of interaction to be $t_{int} = \frac{2\pi}{\Omega_{eff}}$ (where

$\Omega_{eff} = i\eta\Omega_o g_o^2(0) f_1(0)[f_o(0) - f_o(1)]$), to perform a π pulse in this subspace, evolution operator $\hat{U}$ takes state $|\Phi\rangle = [\alpha|\downarrow\downarrow\rangle + \beta|\downarrow\uparrow\rangle + \gamma|\uparrow\downarrow\rangle + \delta|\uparrow\uparrow\rangle]|0\rangle$ into state $|\Phi''\rangle = [-\alpha|\downarrow\downarrow\rangle + \beta|\downarrow\uparrow\rangle + \gamma|\uparrow\downarrow\rangle + \delta|\uparrow\uparrow\rangle]|0\rangle$. This process realizes a QPG for the electronic states of ions j and k with a single pulse.

The fact that we are dealing here with a dispersive interaction enables us to implement a universal quantum logic gate with a single collective Rabi flip, permitting the realization of the gate with fixed experimental parameters. This fact may be considered as an advantage when comparing our proposal to other ones requiring, in general, more than three different operations [3,8,9]. However, the nonresonant character of the proposed interaction produces a slower process. On the other hand, we are less exposed to errors arising from the sequential switching of lasers. Recently, four ions in a linear array were cooled down to their collective ground state in the NIST group [10] and individual ionic addressing has been reached in the Innsbruck group [11].

With this recent results in hand, we believe that our proposal will be soon available to present technology.

In the domain of CQED, we could also make use of a dispersive selective interaction for implementing a QPG. We will suppose that the two qubits are encoded in two non degenerated quantized modes of the electromagnetic field inside a high Q cavity, having each mode either one or zero photons. In this case, atoms will serve only as catalizers of the logical operation. We will show that, again, a QPG can be implemented with only a single interaction pulse of a three level atom crossing the cavity.

The experimental setup considered is easily identified as the one corresponding to a non degenerated two-photon micromaser [12]. The cavity is crossed by a three level atom that interacts with the cavity field during a time interval much smaller than the atomic and field decay time. Here, an effective Hamiltonian corresponding to a two-photon transition can be derived in the limit where transitions from the lower and upper level, $|g\rangle$ and $|e\rangle$, to an intermediate level $|i\rangle$ are not resonant, the detuning being such that $\Delta = \omega_I - (E_e - E_i)/\hbar \gg \frac{\Omega_{ei}^2 + \Omega_{ig}^2}{\Delta}$. E_e and E_i are the energies of the levels $|e\rangle$ and $|i\rangle$, respectively. Ω_{ei} (Ω_{ig}) is the Rabi frequency corresponding to the transition $|e\rangle \to |i\rangle$ ($|i\rangle \to |g\rangle$). These calculations and the details of the non degenerated two-photon micromaser are found in [12]. Explicitly, the effective two-photon interaction Hamiltonian reads

$$\hat{H}_{\text{int}} = \hbar\Omega \left(|e\rangle\langle g|\hat{a}_1\hat{a}_2 + \text{H.c.}\right), \tag{6}$$

where $\Omega = \frac{\Omega_{ei}\Omega_{ig}}{\Delta}$ is the effective Rabi frequency, Ω_{ei} and Ω_{ig} taken as real. Supposing that each cavity mode has either one or zero photons and that the decoupled atom is in the ground state, the most general initial pure state for the combined atom-field system is

$$|\psi\rangle = [\alpha|0,0\rangle + \beta|1,0\rangle + \gamma|0,1\rangle + \delta|1,1\rangle]\,|g\rangle. \tag{7}$$

States associated with the field subspace $\{|0,1\rangle, |1,0\rangle\}$ and the vacuum field state $|0,0\rangle$ will not evolve under the action of $\hat{H}_{\text{int}}$. Only the state $|1,1\rangle|g\rangle$ will suffer Rabi oscillations, $|1,1\rangle|g\rangle \rightarrow \cos(\Omega t)|1,1\rangle|g\rangle + \sin(\Omega t)|0,0\rangle|e\rangle$. By selecting the atom velocity, in such a way that the interaction time corresponds to a π pulse, the final state

$$|\psi''\rangle = [\alpha|0,0\rangle + \beta|1,0\rangle + \gamma|0,1\rangle - \delta|1,1\rangle]\,|g\rangle \tag{8}$$

is produced when the atom leaves the cavity. This corresponds to the action of a QPG on the two-mode field state.

To summarize, we have shown that a selective interaction can be useful for implementing quantum logic schemes in trapped ions and in the domain of CQED. Specifically, we made a proposal for implementing a QPG with a single pulse of a selective interaction in the trapped ions system and identify the two-photon micromaser action as a natural analogy in CQED. In both cases, we find that our scheme reduces significantly the number of steps required for the gate operation, which may be attractive when considering scalable quantum logical processes.

We acknowledge the support of Conselho Nacional de Desenvolvimento Científico e Tecnológico (CNPq).

References

1. C. H. Bennett and D. P. DiVincenzo, Nature (London) **404**, 247 (2000).
2. D. P. DiVincenzo, Phys. Rev. A **51**, 1015 (1995).
3. J.J. Cirac and P. Zoller, Phys. Rev. Lett. **74**, 4091 (1995); J. I.Cirac and P. Zoller, Nature (London) **404**, 579 (2000).
4. For reviews, see D.P. Di Vicenzo, Science **270**, 255 (1995); A. Ekert and R. Josza, Rev. Mod. Phys. **68**, 733 (1996); J. Preskill, Physics Today **52**, 24 (1999).
5. E. Solano, R. L. de Matos Filho, and N. Zagury, Phys. Rev. A **59**, R2539 (1999); **61**, 029903(E) (2000).
6. E. Solano, P. Milman, R. L. de Matos Filho, and N. Zagury, Phys. Rev. A **62**, R021401 (2000).
7. W. Vogel and R. L. de Matos Filho, Phys. Rev. A **52**, 4214 (1995).
8. D. Jonathan, M. Plenio, and P. Knight, Phys. Rev. A **62**, 042307 (2000).
9. A. Sørensen and K. Mølmer, Phys. Rev. Lett., **82**, 1971 (1999).
10. C. A. Sackett, D. Kielpinski, B. E. King, C. Langer, V. Meyer, C. J. Myatt, M. Rowe, Q. A. Turchette, W. M. Itano, D. J. Wineland and C. Monroe, Nature **404**, 256 (2000).
11. H. C. Nägerl, D. Leibfried, H. Rohde, G. Thalhammer, J. Eschner, F. Schmidt-Kaler, and R. Blatt, Phys. Rev. A **60**, 145 (1999).
12. P. A. Maia Neto, L. Davidovich, and J. M. Raimond, Phys. Rev. A **43**, 5073 (1991).

Measuring Entanglement Through the Wigner Function

M. França Santos and L. Davidovich

Instituto de Física, Universidade Federal do Rio de Janeiro, Caixa Postal 68528, 21945-970 Rio de Janeiro, RJ, Brasil

Abstract. We show how to measure, through four points of a two-mode Wigner function, the quantity of entanglement of a pure two-qubit state involving two modes of the electromagnetic field, with only the zero- and one-photon Fock states populated.

1 Introduction

The recent development of quantum information theory and its applications can be directly related to the enhanced understanding of quantum entanglement, the property that lies in the core of the theory. From information sharing, manipulation and communication such as in quantum cryptography [1] to teleportation schemes [2], quantum entanglement is the main responsible for many new and interesting results in theoretical and experimental physics in the last years. Therefore, measuring and quantifying entanglement has become an important task for many physicists, yet to be fully completed. Much work has been done in the direction of finding a good quantity to completely characterize entanglement, and some progress has been made in discovering the properties any definition should obey [3,4,5,6]. However, there is not yet a full definition for the quantity of entanglement of an arbitrarily large quantum system or even for bipartite mixed states (except when the two subsystems are two-level qubits - see Ref. [4,5]). For the simplest case of a pure state of two two-level systems (A, B), a measurement of the von Neumann entropy of the reduced density operator of one subsystem ($\mathrm{Tr}\{\rho_A \log \rho_A\}$) is enough to quantify the entanglement of the whole system [3]. In this case we can define the quantity of entanglement as

$$E\{\rho_{AB}\} = -2\mathrm{Tr}\left\{\rho_A \log \rho_A\right\}. \tag{1}$$

Since phase-space distributions have the same amount of information as the density matrix, it should be possible, in principle, to quantify and characterize entanglement in phase space.

In this work, we consider two entangled modes of the electromagnetic field, such that each mode is occupied by at most one photon. Under this condition, the two-mode electromagnetic field is reduced to a bipartite system containing a pair of two-level subsystems. This situation is easily realized in current experiments in cavity quantum electrodynamics [7].

M. Orszag, J.C. Retamal (Eds.): LNP 575, pp. 394–398, 2001.

For a similar situation, K. Banaszek and K. Wódkiewicz have found Bell-type inequalities for linear combinations of four points of the two-mode Wigner function [8], for the pure maximally entangled Fock state $|1\rangle|0\rangle - |0\rangle|1\rangle/\sqrt{2}$ [9].

This work shows that, for pure states, four points of the two-mode Wigner function actually contain all the entanglement information of the system. Furthermore, this information is easily recovered by using an extension of a procedure previously proposed to measure the Wigner function of a cavity field [10].

2 Entanglement in the Two-Mode Wigner Function

We will adress generic pure states of the form

$$|\Psi_{AB}\rangle = a|0_A\rangle|0_B\rangle + b|0_A\rangle|1_B\rangle + c|1_A\rangle|0_B\rangle + d|1_A\rangle|1_B\rangle, \tag{2}$$

where a, b, c, d are arbitrary complex numbers restrained by the normalization condition $\langle\Psi_{AB}|\Psi_{AB}\rangle = 1$, and we have restricted ourselves to the state space generated by the possibility of having only the Fock states, $n = 0, 1$ in each of the two modes, A and B. Our method can be easily extended to superpositions of any two Fock states.

For state (2), the reduced density matrix of mode A is given by

$$\rho_A = \mathrm{Tr}_B\{\rho_{AB}\} = \begin{pmatrix} |a|^2 + |b|^2 & ac^* + bd^* \\ a^*c + b^*d & |c|^2 + |d|^2 \end{pmatrix}. \tag{3}$$

The von Neumann entropy corresponding to this density operator is easily expressed in terms of its eigenvalues ($\lambda_\pm$):

$$E\{\rho_{AB}\} = -2\left(\lambda_+ \log\lambda_+ + \lambda_- \log\lambda_-\right). \tag{4}$$

On the other hand, it follow from (3) that these eigenvalues are the roots of the equation $\lambda^2 - \lambda + n1 + n2 = 0$, where

$$n_1 = |a|^2|c|^2 + |b|^2|d|^2 - ac^*b^*d - a^*cbd^*, \tag{5}$$

$$n_2 = \left(|a|^2 - |b|^2\right)\left(|d|^2 - |c|^2\right). \tag{6}$$

We show now that n_1 and n_2, and therefore the eigenvalues of ρ_A, can be obtained from four points of the two-mode Wigner function, which can be written, for the pure state $|\Psi_{AB}\rangle$, as [11]

$$\langle\Psi_{AB}|\,\widehat{\Pi}(\alpha,\beta)\,|\Psi_{AB}\rangle, \tag{7}$$

where $\widehat{\Pi}(\alpha,\beta)$ is the product of the displaced parity operators for each mode (A and B),

$$\widehat{\Pi}(\alpha,\beta) = D(\alpha)(-1)^{\hat{n}_A}D^\dagger(\alpha)D(\beta)(-1)^{\hat{n}_B}D^\dagger(\beta).$$

The Wigner function $W(\alpha,0)$, on the plane $\beta=0$, is easy to calculate for state (2) and results in:

$$\frac{W(\alpha,0)}{e^{-2|\alpha|^2}} = |a|^2 - |b|^2 + (4|\alpha|^2-1)(|c|^2-|d|^2) \qquad (8)$$
$$+2\alpha^* ac^* + 2\alpha a^* c - 2\alpha^* bd^* - 2\alpha b^* d.$$

Eq. (8) may be separated in two parts. The first one embraces the phase-independent terms, given by some combination of the populations of state (2), and clearly related to the quantity n_2. These populations are obtained by adding up the values of the Wigner function $W(\alpha,0)$ at two points, symmetrically localized with respect to the origin of phase space:

$$\frac{W(\alpha,0)+W(-\alpha,0)}{2e^{-2|\alpha|^2}} = |a|^2 - |b|^2 + (4|\alpha|^2-1)(|c|^2-|d|^2). \qquad (9)$$

As we are only interested in the product of the populations differences, $|d|^2-|c|^2$ and $|a|^2-|b|^2$, two of these equations, (9), are enough to determine n_2, as long as they are linearly independent. The only free parameter, in each equation, is the magnitude of the chosen phase-space point, α. Therefore, to assure linear independence, we must choose two pairs of symmetric points lying in different circumferences on the plane $\beta=0$. These are the first properties of the points we need to measure to obtain the quantity of entanglement. We will show next that, in order to obtain n_1 and fully determine the entanglement of state (2), we only need to establish phase relations between these four points.

The quantity n_1 is a function of the real part of ac^*bd^* which is a function of both real and imaginary parts of ac^* and bd^*. On the other hand, the phase-dependent part of (8) is proportional to the real part of ac^* and bd^* if α is a pure real number, or to the imaginary one if α is a pure imaginary number. Therefore, the set of points $\{|\alpha|, |\alpha|e^{i\pi}, |\alpha'|e^{i\frac{\pi}{2}}, |\alpha'|e^{i\frac{3\pi}{2}}\}$ provides the information necessary and sufficient to find n_1, which is then given by

$$n_1 = \frac{\left[\frac{W(|\alpha|,0)-W(|\alpha|e^{i\pi},0)}{2|\alpha|e^{-2|\alpha|^2}}\right]^2 + \left[\frac{W(|\alpha'|e^{i\frac{\pi}{2}},0)-W(|\alpha'|e^{i\frac{3\pi}{2}},0)}{2|\alpha'|e^{-2|\alpha'|^2}}\right]^2}{4}. \qquad (10)$$

Of course, this set of points may include those used to determine n_2. More generally, any set of points of the two-mode Wigner function containing the vertices of a losangle located in the plane $\beta=0$, and centered at $\alpha=0$, allows one to calculate the eigenvalues of the reduced density matrix for mode A for the state ρ_{AB}. The only restrictions are that the diagonals of this losangle must be different ($|\alpha'| \neq |\alpha|$), i.e. it must not be a square. Once the eigenvalues of ρ_A are obtained, through these four points, we can fully determine the quantity of entanglement for the proposed system. Obviously the same conditions apply to mode B.

From (9) and (10) it is clear that sums and differences of the values of the Wigner function evaluated at these points completely determine the quantity of entanglement of the state.

An interesting application of our result is the quantum diagnosis of cavity fields. It is now possible to build entanglement between two cavity modes in the same cavity using cavity QED schemes [12]. For example, if we pass through an empty cavity a two-level atom ($\{|g\rangle, |e\rangle\}$), initially in the excited state $|e\rangle$, that will interact with only two of its modes (modes A and B), and we let the atom interact resonantly with these two modes, in such a way that it undergoes a $\pi/2$ Rabi rotation in the first mode and a π Rabi rotation in the second mode (the interaction time with each mode can be controlled by Stark shifting the atomic levels, through external applied electric fields), at the end of these interactions the state of the system should be $|g\rangle \left(|1_A\rangle |0_B\rangle + |0_A\rangle |1_B\rangle\right)/\sqrt{2}$. Extensions of this technique would allow the production of generic states like (2). Our scheme would allow a diagnosis of the produced state, through the direct measurement of the corresponding Wigner function at four points of the phase space. While in this case it is not easy to measure directly the density matrix elements, the two-mode Wigner function can be directly accessed, by using a simple extension of the method proposed in [10]. The same technique could be applied to analyze the entangled motion of two ions in a trap.

3 Conclusions

We have shown that phase-space distributions can be useful tools to characterize entanglement. In particular, for a pure two-qubit state, four points are enough to determine the quantity of entanglement of the state. This result generalizes previous work on the use of phase-space representations to establish Bell-type inequalities, in the sense that not only are four points sufficient to write down these inequalities, but they can actually completely characterize the entanglement of the state.

We acknowledge the partial support of Conselho Nacional de Desenvolvimento Científico e Tecnológico (CNPq), Fundação de Amparo à Pesquisa do Estado do Rio de Janeiro (FAPERJ), Fundação Universitária José Bonifácio (FUJB), and Programa de Apoio a Núcleos de Excelência (PRONEX).

References

1. See, for instance, C.H. Bennett, G. Brassard, and A. Ekert, Sci. Am. **267**, No. 4, 50 (1992).
2. C. H. Bennett, G. Brassard, C. Crepeau, R. Jozsa, A. Peres, and W. Wooters, Phys. Rev. Lett. **70**, 1895 (1993).
3. C.H. Bennett, H.J. Bernstein, S. Popescu, and B. Schumacher, Phys. Rev. A **53**, 2046 (1996).

4. A. Peres, Phys. Rev. Lett. **77**, 1413 (1996).
5. M. Horodecki, P. Horodecki, and R. Horodecki, Phys. Lett. A **223**, 1 (1996).
6. V. Vedral, M.B. Plenio, M.A Rippin, and P.L. Knight, Phys. Rev. Lett. 78, 2275 (1997).
7. Q. A. Turchette, C. J. Hood, W. Lange, H. Mabuchi, and H. J. Kimble, Phys. Rev. Lett. **75**, 4710 (1995); A. Rauschenbeutel, G. Nogues, S. Osnaghi, P. Bertet, M. Brune, J.M. Raimond, and S. Haroche, Phys. Rev. Lett. **83**, 5166 (1999).
8. E. Wigner, Phys. Rev. **40**, 749 (1932); for a review, see M. Hillery, R.F. O'Connel, M.O. Scully, and E.P. Wigner, Phys. Rep. **106**, 121 (1984).
9. K. Banaszek and K. Wódkiewicz, Phys. Rev. Lett. **82**, 2009 (1999).
10. L.G. Lutterbach and L. Davidovich, Phys. Rev. Lett. **78**, 2547 (1997).
11. This differs by a factor four from the natural extension to two modes of the expression proposed by K.E. Cahill and R.J. Glauber, Phys. Rev. **177**, 1857 (1969); *ibid* **177**, 1882 (1969).
12. S. Haroche, private communication.

Reflection of a Slow Atom by a Cavity

A. Delgado[1], L. Roa[2], and C. Saavedra[2]

[1] Departamento de Física, Universidad de Concepción, Casilla 160-C, Concepción
[2] Abteilung für Quantenphysik, Universität Ulm, D-89069 Ulm, Germany

Abstract. We study the center of mass movement of a cold two-level atom crossing through an electromagnetic cavity. This cavity contains both a quantized field mode and a set of A fixed two-level atoms resonant with the field. We find that, even in the absence of fixed atoms, the atomic coherence allows us to switch the interaction from an exclusive barrier potential to an attractive potential. In such a way that the initial atomic phase behaves as a quantum switch. The set of fixed atoms, initially prepared in semiclassical eigenstate, gives origin to an effective growing of the coupling constant of the moving atom and the field, allowing an important increase of the threshold value of kinetic energy of moving atom where a total reflection can be experimentally observed.

1 Introduction

Recently has been considered the injection of a beam of ultra-cold two level atoms into an electromagnetic cavity with a high quality factor, which is called a micromazer [1]. The main feature of the atom-field interaction is that when the atom reach a resonant cavity the strength of the interaction quickly grows. For the atom, this growing of the coupling manifests itself as a position dependent force, which affects the center of mass motion [2]. If the atom is slow enough, the cavity will reflect it with a probability of almost one half. This statement is equivalent to saying that the kinetic energy of the moving atom is lower than the interaction energy, so in the interaction time the atom can be object of sufficient momentum transfer to be reflected. However, this condition imposes a strong constraint on the atomic velocities where the effect arises: a typical velocity of $v_o = 5$ mm/s is required for rubidium atoms [2]. This restriction could experimentally imply considering special microgravity and also extremely high cavity quality factors, which are far from the present state of art of high-Q cavities. The above mentioned limit is obtained from a set of approximations, for instance, by neglecting the field decay during the flight time of the atom inside the cavity. This effect strongly reduces the probability of reflection for present experimental setups by a factor of 10 [3].

The main goal of this communication is to show that it is possible to drastically modify the probability of atomic reflection (transmission) via initial atomic coherence of the moving atom. In addition, we propose a mechanism to increase the threshold value for the kinetic energy required to obtain a high reflection probability. In order to achieve this second goal, in addition to the moving atom we consider a set of fixed two-level atoms.

M. Orszag, J.C. Retamal (Eds.): LNP 575, pp. 399–405, 2001.

2 The Model

The Hamiltonian of a set of A two-level atoms interacting with a quantized mode of the electromagnetic field is the so called Dicke model [5]. We consider a slow atom crossing the cavity containing the Dicke system. For sake of simplicity we assume that the fixed atoms are also resonant with the cavity field mode. The full Hamiltonian, in the standard dipolar and rotating wave approximation, is given by:

$$H = \frac{p^2}{2m} + \hbar\left(\omega_a\sigma_z + \omega_c\hat{S}_z\right) + \hbar g(x)(\sigma_- a^\dagger + \sigma_+ a)$$

$$+\hbar g_o(\hat{S}_- a^\dagger + \hat{S}_+ a) + \hbar\omega_f a^\dagger a,$$

where $a^\dagger$ and a are the usual field operators; σ_z, σ_- and σ_+ are atomic operators of the moving atom. $\hat{S}_z$, $\hat{S}_-$, $\hat{S}_+$ are collective atomic operators of the fixed atoms, which are defined as: $\hat{S}_{z,-,+} = \sum_{i=1}^{A}\hat{S}^{(i)}_{z,-,+}\otimes I^{(j\neq i)}$. For simplicity, we assume all the frequencies to be equals, i.e., $\omega_a = \omega_c = \omega_f = \omega$. A closely related scheme has recently been proposed for atomic Schrödinger-cat states generation [5].

We assume that the cavity is initially prepared in an intense field, and the tail of the distribution is far from the vacuum. For instance, this can be achieved with an intense coherent field. Thus, to simplify the Hamiltonian, we introduce the following transformations Ref. [4]:

$$\hat{Q}_c = \exp\left[i\hat{\phi}\,(\hat{S}_z + A/2)\right],\ \hat{Q}_a = \exp\left[i\hat{\phi}\,(\sigma_z + 1/2)\right],$$

where $\exp(\pm i\hat{\phi})$ are the lower and raise photon field operators respectively [8]. For field intense enough they are well behaved. This transformation directly enables us to diagonalize the Hamiltonian in the field space variables, and it has allowed analytically studing many properties [6] and the effects of field dissipation [7].

The operators $\hat{S}_z + A/2$ and $\sigma_z + 1/2$ have an integer spectrum when acting on symmetric atomic states, for instance, $(\hat{S}_z + A/2)|k\rangle_c = k|k\rangle_c$, and the operator $\hat{Q}_c$ can be expressed as a direct sum of powers of the phase operators, $\hat{Q}_c = \sum_{k=0}^{A}|k\rangle_c e^{i\hat{\phi}k}\,{}_c\langle k|$. In general the operators $\hat{Q}'$s are well defined when acting on field states which satisfy $n > A$. We consider that this condition holds, i.e., $\bar{n} > (A+1)$, where $\bar{n}$ is the initial average number of photons.

By applying the transformations to H we get:

$$\underline{H} \equiv \hat{Q}_a^\dagger\hat{Q}_c^\dagger H\hat{Q}_c\hat{Q}_a$$

$$= \frac{p^2}{2m} + \hbar\hat{\nu}\left(g_o\hat{S}_+ + g(x)\sigma_+\right) + \hbar\left(g_o\hat{S}_- + g(x)\sigma_-\right)\hat{\nu} \quad (1)$$

where

$$\hat{\nu} = \sqrt{\hat{n} - (A+1)/2 - \hat{S}_z - \sigma_z + 1}.$$

In the limit of intense (high) fields we can expand the transformed Hamiltonian (1) as a series of powers of the parameter $(\hat{n} - A/2)^{-1} \ll 1$. We have chosen this parameter because it is unsure that the first order corrections to eigenvalues are zero, so we can write:

$$\underline{H} = \frac{p^2}{2m} + 2\sqrt{\hat{n} - \frac{A}{2}}\left[g_o \hat{S}_x + g(x)\sigma_x\right] + \Theta[\frac{A+1}{\sqrt{\hat{n}}}].$$

In the potential barrier case we consider two limiting of the field structure inside the cavity for a moving atom, which has a sine spatial dependence, $g(x) = g_o \sin(\pi x/d)$, where d is the length of the cavity. As the limiting cases we consider stepwise and triangular approximations, which correspond respectively to the maximum and minimum interaction energy restricted to the condition to have the same values of the maximum coupling constant g_o, so that both appear as limiting cases of the actual spatial dependence of the coupling constant.

In the triangular potential case we restrinct ourselves to the case of $A = 0$, i.e. no static atoms inside the cavity. Of course, these ideas can be directly generalized to the case $A \neq 0$. Thus, firstly we assume the following dependence:

$$g(x) = \begin{cases} g_o, & 0 \leq x \leq d \\ 0, & x < 0\text{yd} < \text{x} \end{cases}$$

and place it into the transformed Hamiltonian

$$\underline{H} \simeq \frac{p^2}{2m} + 2g_o\sqrt{\hat{n} - A/2}\left(\hat{S}_x + \sigma_x\right).$$

In order to solve this Hamiltonian it is convenient to use the total angular momentum basis J of the system, with $J_x = \hat{S}_x + \sigma_x$. This corresponds to an addition of a spin $A/2$ and another spin $1/2$. To simplify the procedure we assume that the quantization axis is the $\hat{x}$ axis, i.e.,

$$\sqrt{\hat{n} - A/2}J_x|n; j, j_x\rangle = \sqrt{n - A/2}j_x|n\rangle \otimes |j, j_x\rangle,$$

where $j = (A+1)/2, (A-1)/2$ and $j_x = -j, ..., j$. The system will evolve inside the subspace with $j = (A+1)/2$ when the atoms are prepared as eigenstates of the operators $\hat{S}_x$ and σ_x with eigenvalues $A/2$ and $1/2$, respectively, i.e., $j_x = (A+1)/2$. This case corresponds to a potential barrier. Of course, this is also achieved in the attractive potential case, i.e., when the atoms are prepared in eigenstates of operator $\hat{S}_x$ and σ_x with eigenvalues $-A/2$ and $-1/2$ respectively. Other initial atomic conditions will give origin to mixed dynamics of the total angular momentum subspaces. Thus, the wave function of each moving atom satisfies the following equation:

$$\left(\frac{\partial^2}{\partial x^2} - \kappa^2\right)\Psi(x) = 0,$$

where we defined,

$$\kappa^2 = \chi_{_A}^2 - k^2$$

$$\chi_{_A}^2 = 2m(A+1)g_o\sqrt{n - A/2}/\hbar.$$

As a first consequence, we observe that the effect of fixed atoms is an effective increasing of the average photon number inside the cavity due to the initial atomic coherence, which can be also seen as an increase of the coupling constant, i.e., $\sqrt{\bar{n}} \to \sqrt{A^2\bar{n}/2}$ or $g_o \to Ag_o/2$. This allows increase the kinetic energy of moving atom and so get reflection. Furthermore, this scaling is independent of the shape of the potential. As usual, the solution of above equation must satisfy the condition of free propagation of the incident, reflected and transmitted wave functions:

$$\Psi(x) = \begin{cases} \exp(ikx) + \rho_o(k)\exp(-ikx), & x < 0 \\ \tau_o(k)\exp(ik(x-L)), & x > L \end{cases}$$

where the label o means that we have assumed a stepwise coupling constant inside the cavity. The coefficients are a solution of a linear system of equations determined by boundary conditions and the solution. The reflection coefficient is given by:

$$\rho_o(k) = \frac{e^{-2\sqrt{1-\alpha_A^2}\chi_{_A} d} - 1}{\Lambda_+^2 e^{-2\sqrt{1-\alpha_A^2}\chi_{_A} d} - \Lambda_-^2},$$

where,

$$\alpha_A = k/\chi_{_A}, \qquad \Lambda_\pm = i\alpha_A \pm \sqrt{1-\alpha_A^2}.$$

If the atom has a well defined wave number, k_o, it will observe a potential barrier when $j_x > 0$ and an attractive potential when $j_x < 0$. In the potential barrier case the condition $k_o^2 < \chi_{_A}^2$ should hold. In the case of $j_x = -A/2$ the corresponding reflection coefficient is given by:

$$\rho_o(k) = \frac{i(2\alpha_A^2 - 1)\sin(\sqrt{1-\alpha_A^2}\chi_{_A} d)}{i\sin(\Omega\chi_{_A} d) + 2g\alpha_A\Omega\cos(\Omega\chi_{_A} d)},$$

where

$$\Omega = \sqrt{1-\alpha_A^2}.$$

In order to address the question of the influence of the specific spatial profile of the coupling constant, we consider a triangular spatial dependence for the coupling constant. We only concentrate on the cases where the kinetic energy is far enough from the borders of the potential with $\sigma_x = 1/2$. We consider the case of non-static atoms inside the cavity, $A = 0$. The spatial dependence of the electromagnetic field can be described as:

$$f(x) = \begin{cases} \frac{2x}{d} & 0 \le x \le d/2 \\ 2(1 - x/d) & d/2 \le x \le d. \end{cases}$$

In this case, the equation of the wave function of each moving atom is:

$$\left(\frac{\partial^2}{\partial x^2} + k^2 - \chi_0^2 f(x)\right) \Phi(x) = 0.$$

We concentrate our attention on the situation of reflections in the cavity. The condition for this phenomenon is:

$$k^2 - \chi_0^2 \max\{g(x)\}_{x\in[0,l]} < 0,$$

and from the potential we obtain $k^2 < \chi_0^2$. This equation can be put into the differential form for Airy functions. The solution is:

$$\Phi(y) = \begin{cases} \sqrt{y}\left[AI_{\frac{1}{3}}[\frac{2y^{2/3}}{3}] + BK_{\frac{1}{3}}[\frac{2y^{2/3}}{3}]\right], & y > 0 \\ \sqrt{-y}\left[\mathrm{A}\mathrm{J}_{\frac{1}{3}}[\frac{2(-\mathrm{y})^{2/3}}{3}] + \mathrm{B}\mathrm{Y}_{\frac{1}{3}}[\frac{2(-\mathrm{y})^{2/3}}{3}]\right], & y < 0 \end{cases}$$

where J_μ, I_μ, K_μ and Y_μ are respectively the Bessel and modified Bessel functions.

As in the stepwise potential, the solutions outside the cavity correspond to the free - particle wave functions. The usual boundary conditions at points of discontinuity must be supplemented by four additional conditions. This comes from the fact that in both intervals $[0, d/2]$ and $[d/2, d]$ the transformation changes the sign because $\alpha_0^2 \leq 1$, so that in every interval there exists a solution for $y > 0$ and also for $y < 0$. In order to write the boundary conditions we need the derivatives of the above solution. They are

$$\frac{\partial\Phi(y)}{\partial x} = \begin{cases} \left[AI_{\frac{-2}{3}}[\frac{2y^{2/3}}{3}] - BK_{\frac{-2}{3}}[\frac{2y^{2/3}}{3}]\right] \frac{y\partial y}{\partial x} & y > 0 \\ \left[AJ_{\frac{-2}{3}}[\frac{-2iy^{2/3}}{3}] + BY_{\frac{-2}{3}}[\frac{-2iy^{2/3}}{3}]\right] \frac{y\partial y}{\partial x} & y < 0, \end{cases}$$

After a straightforward calculation, we arrive at the following expressions for the reflection coefficient:

$$\rho = -\frac{4e^{-2\eta} - 3e^{2\eta} - i\left(2 + \sqrt{3}e^{2\eta}\right)}{4e^{-2\eta} - 3e^{2\eta} - i\left(2 - \sqrt{3}e^{2\eta}\right)} e^{2i\tau},$$

where

$$\tau = \frac{\alpha_0^3 \chi_0 d}{3} + \frac{\pi}{12}, \qquad \eta = \frac{(1-\alpha_0^2)^{3/2}}{3}\chi_0 d.$$

To derive the above expressions for the reflection coefficient, we have explicitly used the asymptotic expressions of Bessel functions. This is because we have assumed that the kinetic energy of the moving atom is far away from the borders of the potential, i.e., $0 < \alpha_0 \leq 1$, and $\chi_0 d \gg 1$. Here this explains that we basically arrive at the same result when we consider both a step and a triangular potential. The stepwise and triangular potentials are bounding the real shape of the field inside the cavity, which is well described by a sine potential (well). We have considered the limiting cases of sine spatial dependence of the coupling constant with minimum and maximum effective interaction area.

3 Summary

We would like to indicate some direct results. Firstly, we observe that the presence of fixed atoms allows increasing the coupling constant between the moving atoms and the field, that is, $g_0 \to g_0(A+1)$.

Secondly, note that the choice of the initial conditions as eigenstates of S_x and σ_x allows a symmetry breaking on the barrier potential and the attractive potential behavior of this interaction with a simple choice of total spin along the x axis j_x. This symmetry breaking is due to the presence of the initial atomic coherence, which still survives in the absence of fixed atoms. If $l_x = (A+1)/2$, then

$$|\rho_A|^2 = 1 - \Theta\left((A+1)/\sqrt{\bar{n}}\right),$$

where we have replaced the photon number dependence n by the initial average photon number $\bar{n}$. We can justify this replacement by means of wave function factorization for the chosen initial condition of the internal atomic degrees of freedom and the field. The factorization is expressed as:

$$|\Psi(t)\rangle \simeq |\Phi_p(t)\rangle \otimes |A_p(t)\rangle,$$

$$|\Phi_{\frac{A+1}{2}}(t)\rangle = \exp\left[-igt\frac{A+1}{2}\sqrt{\bar{n}-\frac{A}{2}}\right] \otimes |\zeta\rangle,$$

$$|\Lambda_{\frac{A+1}{2}}(t)\rangle = \exp\left[-i\xi\left(L_z + \frac{A+1}{2}\right)\right] \otimes |\frac{A+1}{2}, \frac{A+1}{2}\rangle,$$

where $\xi = g(A+1)/\sqrt{2\bar{n}-A}$ and ζ is the initial coherent state parameter. The factorization holds for times that are short compared to $t_0 \simeq \bar{n}/g$ with an accuracy in coefficients of order $O\left((A+1)/\sqrt{\bar{n}}\right)$.

Because of the initial atomic coherence some modification for reflection (transmission) properties can appear because of spontaneous emission. However, these effects can be overcame by employing three-level Λ-type atom, where the lower levels are coupled by Ramman transitions. However, by considering this type of atoms we must rederive the results because of the presence of dynamical stark shift, i.e., in order to suppress spontaneous emission, a large detuning limit must be considered.

Finally we mention that the effect of field dissipation does not strongly modify these results. This has been discussed in Ref. [6], where in presence of field dissipation an approximate density matrix factorization still survives instead of wave function factorization.

Acknowledgment: This work has been partially supported by Fondecyt 1990838.

References

1. M. O. Scully, G. M. Meyer, H. Walther, Phys. Rev. Lett. **76**, 4144 (1996).
2. B.G. Englert, J. Schwinger, A.O. Barut, M.O. Scully, Europhys Letters **14**, 25 (1991).
3. Battocletti, M. and Englert, B.G., Journal de Physique **24**, 1939 (1994). A. B. Klimov and S. M. Chumakov, Phys. Lett. A **202**, 145 (1995).
4. C. Garry and R. Grobe, Phys. Rev. A **56**, 2390 (1997).
5. J. C. Retamal, C. Saavedra, A. B. Klimov and S. M. Chumakov, Phys. Rev. A **55**, 2413 (1997).
6. H. Castro, S. M. Chumakov and J. Sanchez-Mondragón, Phys. Rev. A **53**, 4420 (1996).
7. R. Loudon, The quantum theory of light (Clarendon Press, Oxford, UK, 1973), pp. 141 145.
8. S. Haroche, Il Nuovo Cimento **110** B, 545 (1995).

Lecture Notes in Physics

For information about Vols. 1–536
please contact your bookseller or Springer-Verlag

Vol. 537: S. Cotsakis, G. W. Gibbons (Eds.), Mathematical and Quantum Aspects of Relativity and Cosmology. Proceedings, 1998. XII, 251 pages. 1999.

Vol. 538: Ph. Blanchard, D. Giulini, E. Joos, C. Kiefer, I.-O. Stamatescu (Eds.), Decoherence: Theoretical, Experimental, and Conceptual Problems. Proceedings, 1998. XII, 345 pages. 2000.

Vol. 539: A. Borowiec, W. Cegła, B. Jancewicz, W. Karwowski (Eds.), Theoretical Physics. Fin de Siècle. Proceedings, 1998. XX, 319 pages. 2000.

Vol. 540: B. G. Schmidt (Ed.), Einstein's Field Equations and Their Physical Implications. Selected Essays. 1999. XIII, 429 pages. 2000.

Vol. 541: J. Kowalski-Glikman (Ed.), Towards Quantum Gravity. Proceedings, 1999. XII, 376 pages. 2000.

Vol. 542: P. L. Christiansen, M. P. Sørensen, A. C. Scott (Eds.), Nonlinear Science at the Dawn of the 21st Century. Proceedings, 1998. XXVI, 458 pages. 2000.

Vol. 543: H. Gausterer, H. Grosse, L. Pittner (Eds.), Geometry and Quantum Physics. Proceedings, 1999. VIII, 408 pages. 2000.

Vol. 544: T. Brandes (Ed.), Low-Dimensional Systems. Interactions and Transport Properties. Proceedings, 1999. VIII, 219 pages. 2000

Vol. 545: J. Klamut, B. W. Veal, B. M. Dabrowski, P. W. Klamut, M. Kazimierski (Eds.), New Developments in High-Temperature Superconductivity. Proceedings, 1998. VIII, 275 pages. 2000.

Vol. 546: G. Grindhammer, B. A. Kniehl, G. Kramer (Eds.), New Trends in HERA Physics 1999. Proceedings, 1999. XIV, 460 pages. 2000.

Vol. 547: D. Reguera, G. Platero, L. L. Bonilla, J. M. Rubí(Eds.), Statistical and Dynamical Aspects of Mesoscopic Systems. Proceedings, 1999. XII, 357 pages. 2000.

Vol. 548: D. Lemke, M. Stickel, K. Wilke (Eds.), ISO Surveys of a Dusty Universe. Proceedings, 1999. XIV, 432 pages. 2000.

Vol. 549: C. Egbers, G. Pfister (Eds.), Physics of Rotating Fluids. Selected Topics, 1999. XVIII, 437 pages. 2000.

Vol. 550: M. Planat (Ed.), Noise, Oscillators and Algebraic Randomness. Proceedings, 1999. VIII, 417 pages. 2000.

Vol. 551: B. Brogliato (Ed.), Impacts in Mechanical Systems. Analysis and Modelling. Lectures, 1999. IX, 273 pages. 2000.

Vol. 552: Z. Chen, R. E. Ewing, Z.-C. Shi (Eds.), Numerical Treatment of Multiphase Flows in Porous Media. Proceedings, 1999. XXI, 445 pages. 2000.

Vol. 553: J.-P. Rozelot, L. Klein, J.-C. Vial Eds.), Transport of Energy Conversion in the Heliosphere. Proceedings, 1998. IX, 214 pages. 2000.

Vol. 554: K. R. Mecke, D. Stoyan (Eds.), Statistical Physics and Spatial Statistics. The Art of Analyzing and Modeling Spatial Structures and Pattern Formation. Proceedings, 1999. XII, 415 pages. 2000.

Vol. 555: A. Maurel, P. Petitjeans (Eds.), Vortex Structure and Dynamics. Proceedings, 1999. XII, 319 pages. 2000.

Vol. 556: D. Page, J. G. Hirsch (Eds.), From the Sun to the Great Attractor. X, 330 pages. 2000.

Vol. 557: J. A. Freund, T. Pöschel (Eds.), Stochastic Processes in Physics, Chemistry, and Biology. X, 330 pages. 2000.

Vol. 558: P. Breitenlohner, D. Maison (Eds.), Quantum Field Theory. Proceedings, 1998. VIII, 323 pages. 2000

Vol. 559: H.-P. Breuer, F. Petruccione (Eds.), Relativistic Quantum Measurement and Decoherence. Proceedings, 1999. X, 140 pages. 2000.

Vol. 560: S. Abe, Y. Okamoto (Eds.), Nonextensive Statistical Mechanics and Its Applications. IX, 272 pages. 2001.

Vol. 561: H. J. Carmichael, R. J. Glauber, M. O. Scully (Eds.), Directions in Quantum Optics. XVII, 369 pages. 2001.

Vol. 562: C. Lämmerzahl, C. W. F. Everitt, F. W. Hehl (Eds.), Gyros, Clocks, Interferometers...: Testing Relativistic Gravity in Space. XVII,507 pages. 2001.

Vol. 563: F. C. Lázaro, M. J. Arévalo (Eds.), Binary Stars. Selected Topics on Observations and Physical Processes. 1999.IX, 327 pages. 2001.

Vol. 564: T. Pöschel, S. Luding (Eds.), Granular Gases. VIII, 457 pages. 2001.

Vol. 565: E. Beaurepaire, F. Scheurer, G. Krill, J.-P. Kappler (Eds.), Magnetism and Synchrotron Radiation. XIV, 396 pages. 2001.

Vol. 566: J. L. Lumley (Ed.), Fluid Mechanics and the Environment: Dynamical Approaches. VIII, 412 pages. 2001.

Vol. 567: D. Reguera, L. L. Bonilla, J. M. Rubí (Eds.), Coherent Structures in Complex Systems. IX, 465 pages. 2001.

Vol. 568: P. A. Vermeer, S. Diebels, W. Ehlers, H. J. Herrmann, S. Luding, E. Ramm (Eds.), Continuous and Discontinuous Modelling of Cohesive-Frictional Materials. XIV, 307 pages. 2001.

Vol. 569: M. Ziese, M. J. Thornton (Eds.), Spin Electronics. XVII, 493 pages. 2001.

Vol. 570: S. G. Karshenboim, F. S. Pavone, F. Bassani, M. Inguscio, T. W. Hänsch (Eds.), The Hydrogen Atom: Precision Physics of Simple Atomic Systems. XXIII, 293 pages. 2001.

Vol. 572: H. Latal, W. Schweiger (Eds.), Methods of Quantization. XI, 224 pages. 2001.

Vol. 574: J. Bricmont, D. Dürr, M. C. Galavotti, G. Ghirardi, F. Petruccione, N. Zanghi (Eds.), Chance in Physics. XIII, 276 pages, 2001.

Vol. 575: M. Orszag, J. C. Retamal (Eds.), Modern Challenges in Quantum Optics. XXIII, 405 pages. 2001.

Monographs

For information about Vols. 1–24
please contact your bookseller or Springer-Verlag

Vol. m 25: A. V. Bogdanov, G. V. Dubrovskiy, M. P. Krutikov, D. V. Kulginov, V. M. Strelchenya, Interaction of Gases with Surfaces. XIV, 132 pages. 1995.

Vol. m 26: M. Dineykhan, G. V. Efimov, G. Ganbold, S. N. Nedelko, Oscillator Representation in Quantum Physics. IX, 279 pages. 1995.

Vol. m 27: J. T. Ottesen, Infinite Dimensional Groups and Algebras in Quantum Physics. IX, 218 pages. 1995.

Vol. m 28: O. Piguet, S. P. Sorella, Algebraic Renormalization. IX, 134 pages. 1995.

Vol. m 29: C. Bendjaballah, Introduction to Photon Communication. VII, 193 pages. 1995.

Vol. m 30: A. J. Greer, W. J. Kossler, Low Magnetic Fields in Anisotropic Superconductors. VII, 161 pages. 1995.

Vol. m 31 (Corr. Second Printing): P. Busch, M. Grabowski, P.J. Lahti, Operational Quantum Physics. XII, 230 pages. 1997.

Vol. m 32: L. de Broglie, Diverses questions de mécanique et de thermodynamique classiques et relativistes. XII, 198 pages. 1995.

Vol. m 33: R. Alkofer, H. Reinhardt, Chiral Quark Dynamics. VIII, 115 pages. 1995.

Vol. m 34: R. Jost, Das Märchen vom Elfenbeinernen Turm. VIII, 286 pages. 1995.

Vol. m 35: E. Elizalde, Ten Physical Applications of Spectral Zeta Functions. XIV, 224 pages. 1995.

Vol. m 36: G. Dunne, Self-Dual Chern-Simons Theories. X, 217 pages. 1995.

Vol. m 37: S. Childress, A.D. Gilbert, Stretch, Twist, Fold: The Fast Dynamo. XI, 406 pages. 1995.

Vol. m 38: J. González, M. A. Martín-Delgado, G. Sierra, A. H. Vozmediano, Quantum Electron Liquids and High-Tc Superconductivity. X, 299 pages. 1995.

Vol. m 39: L. Pittner, Algebraic Foundations of Non-Com-mutative Differential Geometry and Quantum Groups. XII, 469 pages. 1996.

Vol. m 40: H.-J. Borchers, Translation Group and Particle Representations in Quantum Field Theory. VII, 131 pages. 1996.

Vol. m 41: B. K. Chakrabarti, A. Dutta, P. Sen, Quantum Ising Phases and Transitions in Transverse Ising Models. X, 204 pages. 1996.

Vol. m 42: P. Bouwknegt, J. McCarthy, K. Pilch, The W3 Algebra. Modules, Semi-infinite Cohomology and BV Algebras. XI, 204 pages. 1996.

Vol. m 43: M. Schottenloher, A Mathematical Introduction to Conformal Field Theory. VIII, 142 pages. 1997.

Vol. m 44: A. Bach, Indistinguishable Classical Particles. VIII, 157 pages. 1997.

Vol. m 45: M. Ferrari, V. T. Granik, A. Imam, J. C. Nadeau (Eds.), Advances in Doublet Mechanics. XVI, 214 pages. 1997.

Vol. m 46: M. Camenzind, Les noyaux actifs de galaxies. XVIII, 218 pages. 1997.

Vol. m 47: L. M. Zubov, Nonlinear Theory of Dislocations and Disclinations in Elastic Body. VI, 205 pages. 1997.

Vol. m 48: P. Kopietz, Bosonization of Interacting Fermions in Arbitrary Dimensions. XII, 259 pages. 1997.

Vol. m 49: M. Zak, J. B. Zbilut, R. E. Meyers, From Instability to Intelligence. Complexity and Predictability in Nonlinear Dynamics. XIV, 552 pages. 1997.

Vol. m 50: J. Ambjørn, M. Carfora, A. Marzuoli, The Geometry of Dynamical Triangulations. VI, 197 pages. 1997.

Vol. m 51: G. Landi, An Introduction to Noncommutative Spaces and Their Geometries. XI, 200 pages. 1997.

Vol. m 52: M. Hénon, Generating Families in the Restricted Three-Body Problem. XI, 278 pages. 1997.

Vol. m 53: M. Gad-el-Hak, A. Pollard, J.-P. Bonnet (Eds.), Flow Control. Fundamentals and Practices. XII, 527 pages. 1998.

Vol. m 54: Y. Suzuki, K. Varga, Stochastic Variational Approach to Quantum-Mechanical Few-Body Problems. XIV, 324 pages. 1998.

Vol. m 55: F. Busse, S. C. Müller, Evolution of Spontaneous Structures in Dissipative Continuous Systems. X, 559 pages. 1998.

Vol. m 56: R. Haussmann, Self-consistent Quantum Field Theory and Bosonization for Strongly Correlated Electron Systems. VIII, 173 pages. 1999.

Vol. m 57: G. Cicogna, G. Gaeta, Symmetry and Perturbation Theory in Nonlinear Dynamics. XI, 208 pages. 1999.

Vol. m 58: J. Daillant, A. Gibaud (Eds.), X-Ray and Neutron Reflectivity: Principles and Applications. XVIII, 331 pages. 1999.

Vol. m 59: M. Kriele, Spacetime. Foundations of General Relativity and Differential Geometry. XV, 432 pages. 1999.

Vol. m 60: J. T. Londergan, J. P. Carini, D. P. Murdock, Binding and Scattering in Two-Dimensional Systems. Applications to Quantum Wires, Waveguides and Photonic Crystals. X, 222 pages. 1999.

Vol. m 61: V. Perlick, Ray Optics, Fermat's Principle, and Applications to General Relativity. X, 220 pages. 2000.

Vol. m 62: J. Berger, J. Rubinstein, Connectivity and Superconductivity. XI, 246 pages. 2000.

Vol. m 63: R. J. Szabo, Ray Optics, Equivariant Cohomology and Localization of Path Integrals. XII, 315 pages. 2000.

Vol. m 64: I. G. Avramidi, Heat Kernel and Quantum Gravity. X, 143 pages. 2000.

Vol. m 65: M. Hénon, Generating Families in the Restricted Three-Body Problem. Quantitative Study of Bifurcations. XII, 301 pages. 2001.

Vol. m 66: F. Calogero, Classical Many-Body Problems Amenable to Exact Treatments. XIX, 749 pages. 2001.

GPSR Compliance

The European Union's (EU) General Product Safety Regulation (GPSR) is a set of rules that requires consumer products to be safe and our obligations to ensure this.

If you have any concerns about our products, you can contact us on ProductSafety@springernature.com

In case Publisher is established outside the EU, the EU authorized representative is:

Springer Nature Customer Service Center GmbH
Europaplatz 3
69115 Heidelberg, Germany

Batch number: 10682207

Printed by Printforce, the Netherlands